AF323315

# Thermodynamics and Equations of State for Matter

## From Ideal Gas to Quark-Gluon Plasma

# Thermodynamics and Equations of State for Matter

## From Ideal Gas to Quark-Gluon Plasma

**Vladimir Fortov**
*Russian Academy of Sciences, Russia*

NEW JERSEY · LONDON · SINGAPORE · BEIJING · SHANGHAI · HONG KONG · TAIPEI · CHENNAI · TOKYO

*Published by*

World Scientific Publishing Co. Pte. Ltd.

5 Toh Tuck Link, Singapore 596224

*USA office:* 27 Warren Street, Suite 401-402, Hackensack, NJ 07601

*UK office:* 57 Shelton Street, Covent Garden, London WC2H 9HE

**British Library Cataloguing-in-Publication Data**
A catalogue record for this book is available from the British Library.

**THERMODYNAMICS AND EQUATIONS OF STATE FOR MATTER**
**From Ideal Gas to Quark-Gluon Plasma**



ISBN 978-981-4749-19-0

Desk Editor: Song Yu

Typeset by Stallion Press
Email: enquiries@stallionpress.com

Printed in Singapore

# Preface

This monograph is based on the lectures which the author has delivered for a number of recent years to the students of the Moscow Institute of Physics and Technology, as well as plenary, review, and invited reports to scientific conferences and symposia. It reflects the experience of many years' work in the areas of equations of state, the physics of extreme states, physical mechanics, and substance structure.

Equations of state express in a compact mathematical form the representation of the specific data of many scientific disciplines using rigorous mathematical language. Beginning from the XVIIIth century, the number of papers on equations of state is huge and permanently growing, simultaneously with the arrival of new scientific data. However, this progress has not been smooth. It accelerated dramatically in the middle of the XXth century, and was closely related to the entry of our civilization into the atomic and space era. The optimal operation of rocket engines and reliable heat shielding of space vehicles relies on the study and description of hot, chemically reactive plasmas of complex and constantly varying composition. In nuclear charges, the high-energy densities generated by high-power shock waves are employed for initiating nuclear chain reactions in compressed nuclear fuel; in thermonuclear charges, and in microtargets for controlled nuclear fusion the high-energy states are the main instrument for compressing and heating the thermonuclear fuel and for initiating fusion reactions in it. The equations of state of the substance under these extreme conditions are the requisite element for the design and operation analysis of these devices.

Although the limiting pressures of laboratory plasmas are so far different by 20–30 orders of magnitude from the maximal astrophysical values, this gap is rapidly shortening; the physical processes in laboratories and the cosmos quite frequently exhibit an amazing diversity and, at the same time, deep analogies, thereby testifying, at the very least, to the uniformity of the physical principles of substance behavior throughout an extremely broad pressure (approximately 42 orders of magnitude) and temperature (up to $10^{13}$ K) range.

It is significant that the range of technical problems, along with the basic ones, whose solution necessitates precise equations of state is permanently broadening. These substance states determine the operation of pulsed fusion reactors with inertial confinement of hot plasma, high-power magnetic explosion and magnetohydrodynamic generators, power installations and rocket engines with gas-phase nuclear reactors, plasmachemical and microwave reactors, plasmatrons and high-power sources of optical, and X-ray radiation. In the energy installations of the future, strongly compressed and heated plasmas will be employed as the working medium, like water vapor in modern thermal power stations.

Extreme states emerge when a substance is subjected to intense shock, detonation, and electroexplosion waves, concentrated laser radiation, and electron and ion beams, as well as in powerful chemical and nuclear explosions, in the pulsed vaporization of liners in pinches and magnetic cumulation generators, in the hypersonic motion of bodies in dense planetary atmospheres, in high-velocity impacts, and in many other situations characterized by ultimately high pressures and temperatures. The physics of near-electrode, contact, and electroexplosion processes in a vacuum breakdown is intimately related to the high-energy plasma which defines the operation of high-power pulsed accelerators, microwave radiation generators, and plasma switches.

High-energy densities determine the behavior of a substance in a vast domain of the phase diagram, which occupies the range from a solid and a fluid to a neutral gas, covers the phase boundaries of melting and boiling, as well as the metal–dielectric transition domain. The solution of these problems of modern high-pressure high-temperature physics calls for the construction of reliable wide-range equations of state for structural materials in all four aggregate states.

Nowadays general scientific interest, along with a pragmatic one, has rekindled in equations of state at high energies, because the overwhelming part of visible matter in the Universe is in precisely this exotic state. For, according to modern estimates, about 95% of the mass (neglecting dark matter) is the plasma of ordinary and neutron stars, pulsars, black holes, and the giant planets of the Solar system, as well as the recently discovered hundreds of exoplanets — the planets beyond the Solar system.

Prior to becoming a star, the substance of the Universe undergoes diverse physical transformations: from quarks and elementary particles to complex molecules and again to atoms and particles, from relativistic energies to the absolute zero and again to the state of high-energy dense plasma, from tremendous densities to deep vacuum and again to the ultrahigh densities of atomic nuclei and quarks. That is why our basic knowledge about the structure, evolution, and history of the Universe is directly dependent on the understanding and correct description of the behavior of matter in all its transformations up to ultrahigh energy densities, which forms not only specific physical models, but also the vision of contemporary natural science.

Today we can clearly see that studying the equation of state of substances under extreme conditions is one of the hottest and vigorously developing basic scientific disciplines at the junction of plasma physics, nonlinear optics, condensed state physics, nuclear, atomic, and molecular physics, relativistic and magnetic hydrodynamics with a vast diversity of compression- and heating-induced physical effects and a permanently broadening set of objects and states, where interparticle interactions quite often play the decisive role. Despite the extraordinary diversity of objects, experimental and physical situations, they are all united by the decisive role of high energy densities in their physical behavior and by the necessity to describe them thermodynamically.

These circumstances are a permanent stable incentive for intensive theoretical and experimental investigations, which have recently yielded a wealth of new and, which is most important, reliable data about the thermodynamic properties of substances under extreme conditions. These specific data are contained in the vast stream of original publications, a part of which is not readily available to Russian specialists, especially to young people.

The author has endeavored to systematize, generalize, and set forth from a unified viewpoint the theoretical and experimental material relating to this new realm of science and to show, following Titus Lucretius Carus, "Thus from the mixture of the elements there emerge infinite multitudes of creatures, which are Strange and highly diversified in appearance" (*Titus Lucretius Carus. O Prirode Veshchei* (Of the Nature of Things) Ser.: Great Thinkers. Moscow: Mir, 2006).

In this book, the author intended to discuss the widest possible range of problems related to precisely the equations of state of substances. That is why many interesting astrophysical, laser, and nuclear physics problems, as well as technical applications, are only briefly outlined and the reader is referred to original works and monographs. In doing this the author did not set himself the task of including everything known about the equations of state to date. The emphasis was placed on those issues which appeared to be most interesting to the author and which he happened to directly work on. Realizing that the material touched upon in this book would steadily broaden, become defined more precisely, and inevitably become obsolete owing to an extremely rapid development of the science of equations of state, I would be thankful to the readers who let me know their critical remarks and suggestions.

Hopefully this book will be beneficial to a wide readership of scientists, post graduates, and students in natural sciences by offering them access to original papers and being a helpful guide through the exciting problems of modern equations of state.

In the preparation of the manuscript I took advantage of the assistance, stimulating discussions, and advice of V.K. Gryaznov, I.L. Iosilevskii, M.B. Kozintsova, P.R. Levashov, E.G. Maksimov, A.Yu. Potekhin, A.N. Starostin, V.G. Sultanov, K.V. Khishchenko, G.V. Shpatakovskaya, D.G. Yakovlev, and V.A. Yakovleva, to whom I express my sincere gratitude.

Abramtsevo.
April 2015.

# Contents


*Preface*                                                                    v

**Introduction**                                                             1

**1.  Phase States of Matter, their Classification**                         7

    1.1   Electron component . . . . . . . . . . . . . . . . . . . .   7
    1.2   Nuclear component . . . . . . . . . . . . . . . . . . . .  19
    1.3   Nuclear transformations . . . . . . . . . . . . . . . . .  22
    1.4   Nuclear transformations and new forms of matter . . . .  26
    1.5   Experiment and astronomical observations . . . . . . . .  32

**2.  Equations of State of Gases and Liquids**                             49

    2.1   Virial expansion . . . . . . . . . . . . . . . . . . . . .  49
    2.2   Short-range order. Integral equations . . . . . . . . . .  52
    2.3   Dust plasma for fluid theory . . . . . . . . . . . . . . .  68

**3.  Quantum-Mechanical Models of a Solid**                                85

    3.1   Hartree and Hartee–Fock approximations . . . . . . . . .  85
    3.2   Asphericity of cells . . . . . . . . . . . . . . . . . . .  92
    3.3   Pseudopotential models . . . . . . . . . . . . . . . . . .  98

**4.  Plasma Thermodynamics**                                              101

    4.1   Historical remarks . . . . . . . . . . . . . . . . . . . . 101
    4.2   Hierarchy of models . . . . . . . . . . . . . . . . . . . 104
    4.3   Chemical model. Dimensionless parameters . . . . . . . . 107
    4.4   Physical model. Thermodynamic relations . . . . . . . . . 115

| | | |
|---|---|---|
| 4.5 | Thermodynamics of nonideal solar plasma | 123 |
| 4.6 | Bound atom model | 159 |
| 4.7 | Thermodynamic calculations | 164 |
| 4.8 | Thermodynamics of shock compressed plasma of megabar pressures. Nonideality and degeneracy | 175 |
| 4.9 | Gas thermodynamics | 185 |
| 4.10 | Hydrogen thermodynamics | 190 |
| 4.11 | Metal plasma thermodynamics | 195 |

**5. Monte Carlo and Molecular Dynamics Methods**    **211**

| | | |
|---|---|---|
| 5.1 | Monte Carlo method. Pseudopotentials | 211 |
| 5.2 | Quantum MCM. Path integrals | 214 |
| 5.3 | H and H + He plasmas | 217 |
| 5.4 | Shock adiabat of deuterium | 228 |
| 5.5 | Electron–hole plasma of semiconductors | 231 |
| 5.6 | Crystallization of holes | 237 |
| 5.7 | Electron–hole plasma of germanium | 243 |
| 5.8 | Quantum MD | 245 |
| 5.9 | QGP simulations | 248 |

**6. Statistical Substance Model**    **259**

| | | |
|---|---|---|
| 6.1 | On the quantum-mechanical many-electron structure calculations | 259 |
| 6.2 | Statistical model | 267 |
| 6.3 | Oscillation effects | 273 |
| 6.4 | Quantum and exchange corrections | 282 |
| 6.5 | Application of TFM to an isolated atom | 284 |
| 6.6 | Thomas–Fermi equation of state | 288 |
| 6.7 | Limits of TFM applicability. Experiment | 293 |

**7. Density Functional Method**    **301**

| | | |
|---|---|---|
| 7.1 | Density functional method | 301 |
| 7.2 | Atomic and molecular structures | 307 |
| 7.3 | Condensed media | 311 |

**8. Phase Transitions**    **317**

| | | |
|---|---|---|
| 8.1 | Melting | 317 |
| 8.2 | Polymorphic and electronic transformations | 328 |
| 8.3 | High-temperature boiling | 335 |

8.4 Plasma phase transitions . . . . . . . . . . . . . . . 336

8.5 Noncongruent phase transitions . . . . . . . . . . . . . 358

**9. Semi-Empirical Equations of State** **373**

9.1 Quasi-harmonic approximation . . . . . . . . . . . . . 374

9.2 Anharmonicity. Melting of the lattice . . . . . . . . . . 383

9.3 Thermal excitation of electrons . . . . . . . . . . . . . 387

9.4 Evaporation. Wide-range EOS . . . . . . . . . . . . . 390

9.5 The Fermi–Zeldovich problem. Construction of the thermodynamically complete EOS using the results of dynamic measurements . . . . . . . . . . . . . . . . 394

**10. Relativistic Plasma. Wide–Range Description** **399**

10.1 Description of electrons . . . . . . . . . . . . . . . 400

10.2 Ions . . . . . . . . . . . . . . . . . . . . . . . 402

10.3 Fully ionized plasma . . . . . . . . . . . . . . . . 404

10.4 Ion liquid. OCP . . . . . . . . . . . . . . . . . . 407

10.5 Coulomb crystal . . . . . . . . . . . . . . . . . . 410

**11. Nuclear Transformations Under Strong Compression** **417**

11.1 Extreme states of neutron stars . . . . . . . . . . . . 418

11.2 Compression. Nuclear structures . . . . . . . . . . . . 430

11.3 Thomas–Fermi model . . . . . . . . . . . . . . . . 433

11.4 Meson, pion, and kaon condensations . . . . . . . . . . 438

11.5 Nucleons and hyperons under supercompression . . . . . . 442

**12. Quark–Gluon Plasma and Strange Matter** **457**

12.1 Quark deconfinement and quark–gluon plasma . . . . . . 457

12.2 Observed manifestations of the QGP . . . . . . . . . . 460

12.3 Analogies between the QGP and EMP . . . . . . . . . . 470

12.4 EOS of the QGP . . . . . . . . . . . . . . . . . . 474

12.5 Neutron crystallization and strange matter . . . . . . . . 478

**13. Semi-Empiric Nuclear Models** **483**

13.1 "Cold" components . . . . . . . . . . . . . . . . . 484

13.2 Thermal excitations . . . . . . . . . . . . . . . . . 487

13.3 Hydrodynamics of nuclear collisions . . . . . . . . . . 492

*Bibliography* 497

# Introduction

The equation of state (EOS) which is a functional dependence between the parameters characterizing the state of matter is its basic quantitative characteristic permitting the application of the general formal apparatus of thermodynamics and continuous media dynamics (mathematical physics) to physical objects and processes of extremely diverse nature from thermal machines and biological structures to ultra-extreme conditions of Big Bang and relativistic hadron collisions.

The conservation laws reflect the most fundamental properties of space–time symmetry (Noether theorem); the laws of conservation (of mass, momentum, and energy) in the most general mathematical form reflect the whole ensemble of processes in the surrounding nature, whereas the EOS introduces in this general formalism also the quantitative specificity of a given particular state of the substance (gas, liquid, solid, electromagnetic, or quark–gluon plasma, nuclear matter radiation), that is, the whole diversity of processes and phenomena in nature. Therefore, the interest in the EOS of matter has always been strong enough from the viewpoint of not only the numerous pragmatic (technical and energetic) applications, but also of the comprehension and description of the processes and phenomena involving extremely high energy concentrations, i.e., the Big Bang stages, as well as processes proceeding in astrophysical objects (neutron stars, black holes) at final stages of their evolution when ultrahigh nuclear-matter pressures and temperatures occur under the action of gravitational compression and thermonuclear energy release.

The description of the state of matter under extreme conditions has always been the forefront of research into the EOS since it is precisely the progression up the scale of temperatures and pressures of working substances that is responsible for the increase in heat engine efficiency

Table 0.1   Energy transformation. $\vec{\pi} \sim \sqrt{T} \cdot p$ Umov-Pointing vector.

| Generators | $\vec{\pi}$ (au) | $T$ (K) | $p$ (bar) | Working substance |
|---|---|---|---|---|
| Rankine cycle – VGI | $10^3$ | 800 | 250 | $H_2O$ |
| Rankine cycle + MHD | $10^4$ | 3 000 | 200 | gas, coal |
| Explosive MHD generators | $10^5$ | 30 000 | 1 000 | Ar, Xe |
| Magnetic thermonuclear reactor (TOKAMAK) | $10^6$ | $10^8$ | 100 | DT |
| Explosive magnetic generators | $10^8$ | $10^4$ | $10^6$ | metal |
| Thermonuclear installation $Z$ pinch magnetic | $10^{10}$ | $10^8$ | $10^8$ | DT |
| Inertial thermonuclear fusion | $10^{14}$ | $10^8$ | $10^{10}$ | DT |

characterized by $\mathrm{div}\vec{W}$, where $\vec{W}$ is the Umov-Pointing energy flux density (Table 0.1). Hence, the EOS of working substances of heat engines is the scientific basis of high-temperature thermal physics and the ideological basis of modern energetics. The scope of research on this subject is huge and so is the number of publications, in internet one can find tens of thousands of references to papers, monographs, arrays of thermodynamical data on thermodynamic properties of individual substances and thousands of compounds, especially in the range of extremely high pressures and temperatures.

In our review we will not touch upon these vast quantitative studies, but will restrict ourselves to only general approaches of thermodynamic description of nowadays extreme states of matter in the ultimately broad range of the phase diagram.

The striking advances in contemporary astronomy have led to a great amount of qualitatively new and, what is most important, reliable information [345] obtained through ground-based observations and in automatic space stations. This changed qualitatively our notions of the structure and evolution of a wide class of astrophysical objects of the type of neutron, strange (quark–gluon) stars, black holes where relativistic electromagnetic and quark–gluon plasma of matter is compressed to ultrahigh pressures by gravitational forces and is heated by thermonuclear energy release to temperatures of hundreds of billions of degrees [294, 295].

The description of the structure of the dynamics and evolution of these astrophysical objects and the Universe as a whole calls for construction of new fairly sophisticated EOS of superextreme state of matter, and on the other hand the astrophysical data are a gratuitous "experimental" source of physical information for creation of EOS models under conditions far from the resources of laboratory experiments. According to the figurative expression of academician Ya. B. Zeldovich, "Space is the laboratory for the poor". True, acquisition of extreme "cosmic" parameters in laboratory or quasi-laboratory experiments would require absolutely unreal economic efforts [345].

Of particular importance is simulation of EOS of the early Universe after the Big Bang, where the most extreme, "Planck" conditions are realized to describe the early stages.

Owing to the appearance of modern high-power sources of local energy concentration (lasers, electron and ion lows, shock and superintense electromagnetic waves, etc.) the states of matter with the earlier inaccessible extremely high pressures and temperatures corresponding to the new regions of phase diagram became a subject of laboratory research. This substantially increased in number the range of specialists interested in the physics of high energy concentrations [294, 295, 727].

At the same time the progress in numerical methods based on new computer petaflopic and exaflopic performance has led to the development of efficient difference systems of computations of nonstationary gas-dynamic phenomena, which considerably increased requirements on an adequate and detailed description of thermodynamic properties of matter as the accuracy of gas-dynamic calculations is determined not only by errors in solution of conservation equations, but in the first place in EOS of the medium, which stand for "closing" equations for a complete system of equations of motion.

Computation of thermodynamic characteristics of different substances becomes necessary in the solution of almost any problem of contemporary high energy-density physics [294, 295, 727], namely, realization of the idea of controlled thermonuclear fusion, design of powerful magnetohydrodynamic and magnetocumulative generators [294, 295], computation of the dynamics of strong shock waves, meteorite protection of spaceships, determination of the construction and evolution of stars and planets. All this, along with problems of the promising high-pressure technology (synthesis of diamond phases of graphite and boron nitride, explosion, electron-beam and laser welding, metal working, etc.), is a strong stimulus for intense experimental and theoretical research aimed at both advance to new phase diagram regions and obtaining detailed data on the already familiar range of parameters.

The scales of the states of matter realized by nature or humans are capable of striking the most daring imagination. At the bottom of the Mariana Trench (11 km deep) the water pressure reaches 1.2 kbar, at the Earth center it is 3.6 Mbar, $T \approx 0.5\,\mathrm{eV}$, the density $\rho \approx 10\text{--}20\,\mathrm{g/cm^3}$; at the Jupiter center $p \approx 40\text{--}60\,\text{Mbar}$, $\rho \approx 30\,\text{g/cm}^3$, $T \approx 2 \cdot 10^4\,\text{K}$; at the center of the Sun $p \approx 240\,\text{Gbar}$, $T \approx 1.6 \cdot 10^3\,\text{eV}$, $\rho \approx 150\,\text{g/cm}^3$; in cooling stars, i.e., white dwarfs, $p \approx 10^{10}\text{--}10^{16}\,\text{Mbar}$, $\rho \approx 10^6 - 10^9\,\text{g/cm}^3$, and $T \approx 10^3\,\text{eV}$. In targets of controlled thermonuclear fusion with inertial plasma confinement $p \approx 200\,\text{Mbar}$, $\rho \approx 150\text{--}200\,\text{g/cm}^3$, and $T \approx 10^8\,\text{eV}$. Neutron stars which are elements of pulsars, black holes, gamma bursts, and magnetars have evidently record-high parameters: $p \approx 10^{19}\,\text{Mbar}$, $\rho \approx 10^{11}\text{--}10^{14}\,\text{g/cm}^3$, and $T \approx 10^4\,\text{eV}$ for the mantle and the core: $p \approx 10^{22}\,\text{Mbar}$, $\rho \approx 10^{14}\,\text{g/cm}^3$, and $T \approx 10^4\,\text{eV}$ with a giant magnetic-field induction from $10^{11}$ to $10^{16}\,\text{Gs}$. The study and description of these exotic states makes up the subject of the impetuously developing branch of science — high energy-density physics [294, 295, 727]. The lower boundary of "high energy density" is normally understood as energy densities comparable with condensed-substance binding energy $10^4\text{--}10^5\,\text{J/cm}^3$, which corresponds to the binding energy of valence electrons equal to several electron-volt and pressure of about $100\,\text{kbar}$ to $1\,\text{Mbar}$. These pressures exceed greatly the limits of mechanical strength of materials and make necessary the allowance for their compressibility and, therefore, the hydrodynamic motion upon pulsed energy release.

As has already been noted, the EOS of matter is a "closing" external equation relative to the conservation laws and is based on either the results of experiments or the methods of statistical physics [598]. The main difficulty of a successive theoretical calculation of the EOS of matter by the methods of statistical physics lies in the necessity of a correct allowance for a complex-structure interparticle interaction in the quantum-mechanical many-body problem for any coupling constant values and any type of statistics (in spite of the impressive advances in modern parameter-free Monte Carlo methods of molecular dynamics).

For this reason, when performing calculations one has to consider simplified models whose applicability is limited and is established in each particular case either on the basis of intrinsic characteristics of the model or by comparison with more accurate solutions or experimental results. The latter way is obviously more constructive since a lot of examples are known (van der Waals theory, integral equations, nonideal plasma, etc.) where the actual range of applicability of the model exceeds greatly the limits determined by smallness of the corresponding alphabetic dimensionless numerical parameters and criteria.

At the next stage of development of thermodynamic models the available experimental data are applied to select the basic numerical parameters in functional dependences based on the exact solutions or simplified models.

The semi-empirical models thus obtained are used to describe and formulate adequate models of zeroth approximation in particularly difficult situations (liquids, solids, dense electromagnetic and quark–gluon plasma, nuclear and neutron matter) which do not admit separation of small parameters of the perturbation theory. The success of construction of semi-empirical models is verified by both the quality of description of a possibly broad range of diverse experimental data and by the feasibility of extrapolational computations. Clearly, for semi-empirical EOS models experiment is not only a necessary supplement, but, in fact, the basis of their existence.

Finally, a considerable part of the diagram of the states of matter important for practical applications is now inaccessible for theoretical and experimental methods [294, 295, 727]. In this situation there is ground for the most diversified hypotheses typically formulated in the language of heuristic models which are extrapolated into a strongly nonideal region of results of simplified physical notions obtained for the case of small nonidealities. Being exceptionally qualitative such models often predict the occurrence of new phases and states with exotic properties ("plasma" crystals and liquids, plasma phase transitions, and phase transitions due to quark deconfinement, electrically charged phases in neutron stars, noncongruent phase transitions, etc.), which necessitates an experimental verification of these intriguing possibilities.

The present monograph is aimed at comparative analysis of different thermodynamic models and discussion of their applicability limits in a wide range of parameters. Bearing in mind the extensive thermodynamical literature we shall only consider the main ideological premises of the theoretical methods and very briefly the experiment. The reader interested in the details of calculations and the exhaustive bibliography will be referred to specialized reviews and monographs [38, 57, 59, 63, 152, 174, 221, 287, 291, 292, 296, 345, 346, 434, 435, 439, 547, 589, 625, 648, 719, 727, 817, 997, 1073] devoted to a thorough study of the properties of the medium in particular domains of the phase diagram and intended for physicists involved in the corresponding area of knowledge. In view of the huge number of papers and books on models of the EOS, references in the text are given mostly to the works of recent years which offer also the history of the problems. The main emphasis in the proposed monograph is laid on the description of states which are now of greatest interest for the physics of high energy concentrations which are now either already gained or can in the foreseeable future be realized in controlled conditions on Earth, or are realized in astrophysical objects at different stages of their evolution. Hence, along with the states of pragmatic earthly interest we also consider superextreme astrophysical and nuclear-physical applications, where the thermodynamics of a substance is considerably affected by relativism, high-power gravitational and magnetic fields, thermal radiation, transformation of nuclear particles, and neutronization of nucleons, quark deconfinement, etc. Directing our attention to extreme physical conditions, we shall not touch upon the strength effects in solids ($p = 10$–$100$ kbar) [294,295,509] and the problems of thermodynamics of complex chemical compounds. Beyond the framework of our consideration will also be the details of experimental methods (see Refs. [294, 295] and the references therein).

# Chapter 1

# Phase States of Matter, Their Classification

As a substance advances along the scale of pressures and temperatures, its composition, structure, properties, etc. undergo radical changes from the ideal state of noninteracting neutral particles described by the classical statistical Boltzmann function to the exotic forms of baryon and quark–gluon matter some of whose features now began to show up in record-parameter experiments on relativistic ion collisions [294, 295] and manifest themselves in the interpretation of astronomical observations of so-called "compact" ("neutron" and "strange") stars [414].

The pressures and temperatures realized in nature are presented schematically in Fig. 1.1.

In the second section of this chapter we shall touch upon ultrahigh compressions when the nuclear properties of matter come out to the foreground.

## 1.1 Electron component

To begin, we shall consider the states of matter under "moderate" compressions when its properties are determined by electrons resident in atomic and ionic shells and by free electrons. We shall discuss compressed atomic and molecular systems and electron-ion "electromagnetic" plasma [287,294,295]. All these interesting and essentially nonlinear phenomena appear in both astrophysical and laboratory plasma, and in spite of giant differences in spatial scales, they have much in common and are a subject of "laboratory astrophysics" [221,871].

In Chapter 2 we consider the range of low ($T \ll J$, $J$ is ionization potential) temperatures and low ($\rho \gg \rho_c$ is density at the critical point)

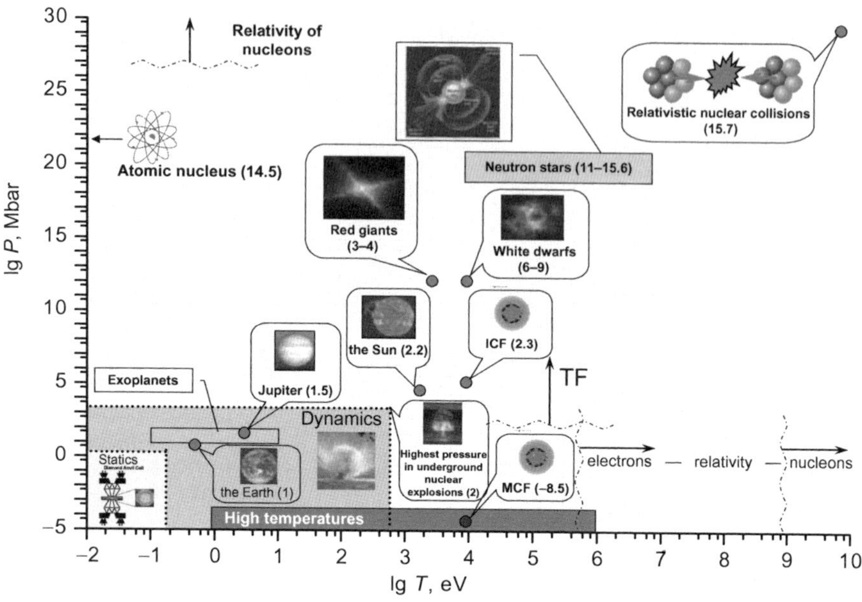


Fig. 1.1   Extreme states [545] in nature and laboratory. The numbers in brackets indicate the density logarithm in g/cm$^3$. The region "statics" corresponds to the static and "dynamics" to the dynamic methods of obtaining high pressures, the "high temperatures" domain corresponds to experiments at high temperatures.

densities, where the interparticle interaction is small and can be taken into account within perturbation theory. With increasing density the interaction strengthening in the system requires the use of purely empirical descriptions and in supercritical conditions — the models of liquids (Sec. 2.3) based on either integral equations or the parameter-free methods of molecular dynamics or Monte Carlo method.

The liquid state region (Sec. 2.3) has traditionally been thought of as the most difficult and poorly investigated since the strong interaction and disorder make theoretical predictions of the properties of real liquids extremely ambiguous [656]. It however appeared possible to formulate here realistic zero-approximation models based on the use of exceedingly simplified hard- and soft-sphere potentials permitting the use of the numerical methods of molecular dynamics and the Monte Carlo method. Thus, the main qualitative features of experimentally observed structural factors were reproduced and the nature of melting and crystallization was analyzed (Sec. 7.3). The equations of state (EOS) of real liquids are found from machine calculations applying the variational method of perturbation theory, the parameters of interparticle pair potential being found from comparison of the calculations with experiment.

The absence of reliable experimental data at high pressures and temperatures very largely obstructs the development of realistic models of liquids valid up to the boiling point converting liquid into vapor or (metals) directly into the plasma phase.

As the temperature rises, dissociation and ionization of first external $(T \sim Z^{4/3}\mathrm{Ry})$ and then internal $(T \sim Z^2\mathrm{Ry})$ electron shells of atoms proceed successively, and we find ourselves in the plasma region (Sec. 3.3) which is the most widespread state of matter in nature [287,294,295]. After all, plasma state is inherent in about 95 to 98% of the baryon matter in nature with the exception of dark matter and dark energy.

The thermodynamic properties of plasma are described by the ionization equilibrium equations involving the experimental and computational data on the spectra of atoms and ions and the computational values of ionization potentials. The applicability limits of such an approach for low densities are determined by violation of the condition of local thermodynamic equilibrium [287].

In the region of higher compressions, the strong interparticle interaction affects not only free electrons but also those bound in atoms and ions thus causing deformation of the discrete energy spectrum of plasma; along with thermal, pressure-induced or "cold" ionization of matter becomes dominant. The condition of this effect is the closeness of atomic and ionic sizes to the average interparticle distance, $a_0 \sim n^{1/3}$. This phenomenon is described by the "restricted" atom models that take into consideration the final actual ion size and the influence of plasma surrounding upon its discrete spectrum [287].

The phase diagram of a substance corresponding to high energy densities is presented in Fig. 1.2 [287, 293, 319, 727] where the conditions in astrophysical objects, in technical and laboratory experimental setups are given. One can see that being the most wide-spread state of matter, nature plasma occupies almost the whole region of phase diagram.

The greatest difficulty in the physical description of such a medium lies in the region of nonideal plasma where the interparticle Coulomb interaction energy $e^2 n^{1/3}$ is comparable with or exceeds the kinetic energy $E_k$ of particle motion. In this region, $\Gamma = e^2 n^{1/3}/E_k > 1$, the plasma nonideality effects cannot be described by perturbation theory [287,545], and application of machine parameter-free Monte Carlo or molecular dynamics

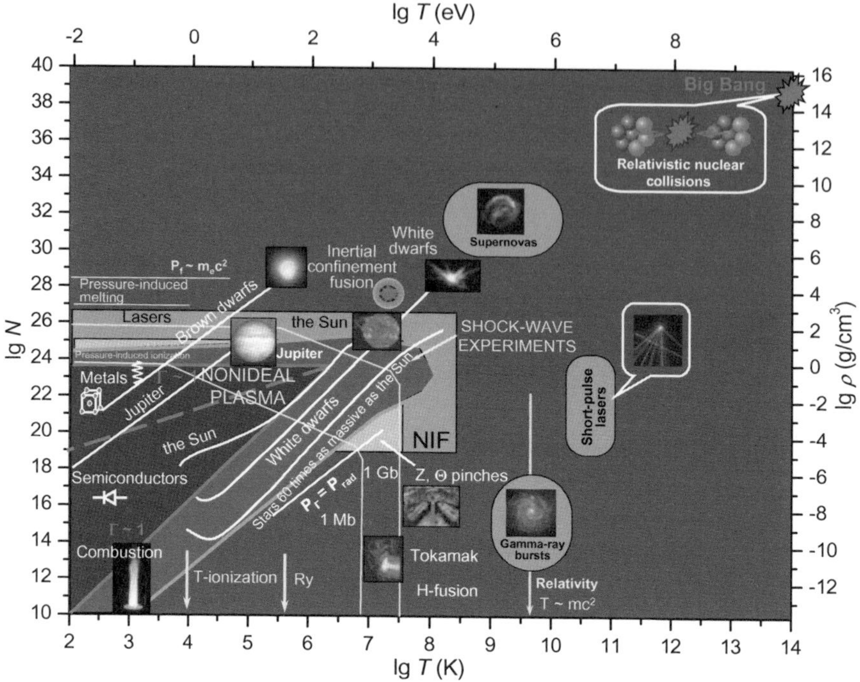


Fig. 1.2    Phase diagram of the states of matter [293, 727].

methods [287,545] runs into the difficulties of the choice of adequate pseudo-potentials and correct allowance for quantum effects.

The electron relativism effects in the equation of state and in the transport properties of plasma, when $m_e c^2 \sim kT$, correspond to $T \approx 0.5\,\mathrm{MeV} \approx 6 \cdot 10^6\,\mathrm{K}$. Above this temperature the substance becomes unstable under spontaneous electron–positron pair production.

The quantum effects are determined by the degeneracy parameter $n\lambda^3$ ($\lambda \sim \sqrt{\hbar^2/2mkT}$ is the thermal de Broglie wavelength). For degenerate plasma $n\lambda^3 \gg 1$ the scale of kinetic energy is the Fermi energy $E_\mathrm{F} \approx \hbar^2 n^{2/3}/2m$ which increases with increasing plasma density thus making it more and more ideal in the course of compression, i.e., as $n \longrightarrow \infty$; $\Gamma = me^2/(\hbar^2 n^{1/3}) \to 0$. The relativism condition corresponding to the condition $m_e c^2 \sim E_\mathrm{F} \approx 0.5\,\mathrm{MeV}$ yields the density $\rho \approx 10^6\,\mathrm{g/cm}^3$.

Similar asymptotics is also observed in another limiting case $T \to 0$ of classical ($n\lambda^3 \ll 1$) plasma, where $E_k \sim kT$ and the plasma becomes increasingly ideal ($\Gamma \sim e^2 n^{1/3}/(kT)$) when heated. One can see that the

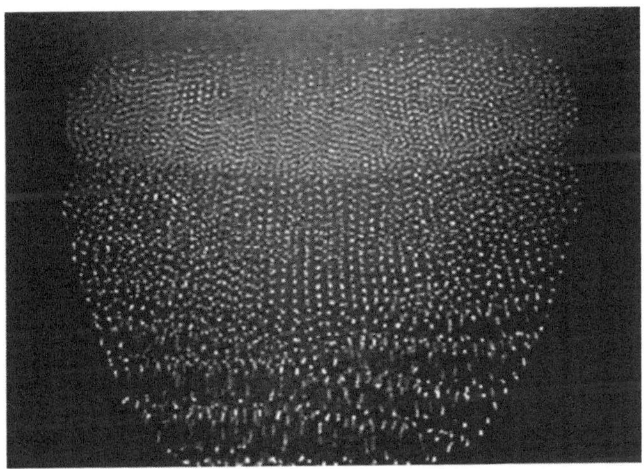

Fig. 1.3   Plasma dust crystal and plasma liquid.

periphery of the phase diagram of a substance (Fig. 1.2) is occupied by ideal ($\Gamma \ll 1$) Boltzmann ($n\lambda^3 \ll 1$) or degenerate ($n\lambda^3 \gg 1$) plasma for the description of which adequate physical models [102, 264, 270, 287, 291, 293, 302, 545] have been developed.

Electron plasma in metals and semiconductors corresponds to the degenerate case with the interaction energy $E_{\text{int}} \sim e^2/r_0$, $r_0 \sim \hbar/k_{\text{F}}$, $E_k \sim k_{\text{F}}^2/m$; $\Gamma \sim e^2/\hbar v_{\text{F}} \approx 1 - 5$, where $v_{\text{F}} \sim 10^{-2} - 10^{-3}\,\text{m/s}$, the index $F$ refers to the parameters on the Fermi boundary.

Of particular interest are plasma phase transitions in strongly non-ideal Coulomb systems, namely, crystallization of dust plasma (Fig. 1.3) [302, 304] and ions in electrostatic traps [223] and cyclotrons [888, 893] in electrolytes [792, 870], in colloidal systems [724], in two-dimensional electron systems on the liquid helium surface [902], exciton condensation in semi-conductors [264], and so forth. We shall particularly emphasize the recently discovered phase transition in thermal deuterium plasma [102, 264, 270, 301] compressed quasi-adiabatically to megabar pressures by a series of rever-berating shock waves.

For quark–gluon plasma (QGP) $E_{\text{int}} \sim g^2/r_0$, $r_0 \sim 1/T$, $E_k \sim T$; $\Gamma \sim g^2 \approx 1 - 50$. For ultracold plasma in traps $\Gamma \sim (n/10^9)^{1/3}/T_k$. The greatest difficulty for the theory is a vast region of nonideal plasma $\Gamma \geq 1$ occupied by numerous technical applications (plasma of metals and semiconductors, impulse power engineering, explosions, arches, electric discharges, etc.), where the theory predicts qualitatively new physical effects (metallization "cold" ionization, dielectrization, plasma phase transitions, etc. [287,292,293]) whose investigation needs considerable experimental and theoretical efforts.

The search for such kind of qualitatively new effects in a nonideal region of parameters is a strong and always alive stimulus of research into matter at high energy densities [287,293,727].

Of particular interest under extreme energetic impact are nonstationary hydrodynamic phenomena such as instabilities of shock waves and laminar flows [249], going over to turbulent regime [249], turbulent mixing, the dynamics of strings and solitons [221,249].

Another characteristic feature of plasma with high energy concentrations is a collective character of its behavior and strong nonlinearity of its reaction to external energetic effects such as shock and electromagnetic waves, solitons, fluxes of laser radiation, and fast particles. Electromagnetic wave propagation in plasma excites a number of parameter instabilities (Raman, Thompson, and Brillouin radiation scatterings), self-focusing and filamentation of radiation, the development of relativistic-nature instabilities, generation of fast particles and jets, and upon higher intensities — vacuum "boiling" with electron–positron pair production [50, 59, 144, 182, 292, 569, 589, 625, 642, 647, 719, 727, 817, 871].

The structure and thermodynamic description of substance get much simplified for extremely high pressures and densities when the electron shells of atoms turn out to be crushed, and the electron density in atoms experiences quasi-homogeneous distribution described by a quasi-classical approximation to the self-consistent field method (Thomas–Fermi model), Sec. 6.

As the energy density heightens, the substance acquires an increasingly universal structure [545, 547]. The distinctions between neighboring elements of the Periodic System are smoothed and the substance properties become increasingly smooth functions of its composition. The energy density heightening induces an apparent "universalization" or simplification of the substance properties. The pressure and temperature rise destroys the molecular complexes generating atomic states which then lose outer shell electrons responsible for the chemical selfhood of the substance because of thermal and/or pressure-induced ionization. The electron shells of atoms and ions get restructured demonstrating more and more regular level occupation, and the crystal lattice passes over into a closely packed volume-centered cubic structure after a series of polymorphic transitions (this typically happens for $p < 0.5\,\mathrm{Mbar}$) [545,547].

These "simplifications" of substance properties occur for energy densities comparable with the characteristic energies of the indicated "universalization" processes. When the characteristic energy density becomes of the order of valence shell energy and equals $e^2/a_0^4 \approx 3 \cdot 10^{14}\,\mathrm{erg/cm^3}$ ($a_0 = \hbar/(me^2) = 5.2 \cdot 10^{-9}\,\mathrm{cm}$ is Bohr radius), this determines the order of magnitudes of the lower boundary of substance "universalization", $T \approx 10\,\mathrm{eV}$, $p_a \sim e^2/a_0^4 \approx 300\,\mathrm{Mbar}$. Here, $p_a$ is the so-called "atomic" pressure which is the quantity following from dimensionless estimations. With respect to this quantity the pressure in an atomic system is considered to be "high" or "low". An exact quantitative setting of "universalization" boundaries is a crucial problem of experimental high energy-density physics, the more so as the theory [547] also predicts more diversified behavior of matter in the ultramegabar pressure range (shell effects [547], electron and plasma phase transitions [27, 40, 102, 223, 264, 270, 291, 302, 304, 435, 637, 1081], etc.).

The upper boundary of the domain of extreme states is determined by the level of present-day knowledge in high-energy physics [345, 754, 997] and observed astrophysical data and is perhaps limited to nothing but our imagination.

As the pressure and temperature lowers, the behavior of matter is determined by a specific shell structure and the character of occupation of the electron energy bands as well as by the crystal lattice symmetry. The classification of possible states is fairly cumbersome and the description of different situations calls for the application of modern methods of condensed state quantum theory (Secs. 3.3, 4.11 and 6.7).

At sufficiently high pressures ($p > 0.5\,\mathrm{Mbar}$), after a series of polymorphic transitions in solids, structures with close packing are realized, which makes it possible to apply the Wigner–Seitz model of spherical cells in which the solution of the wave equation with Bloch boundary conditions is sought [545, 547].

Further density or temperature heightening causes a relative decrease in the Coulomb interaction, which permits the use of formulas of respectively the ideal Fermi or Boltzmann electron gas. The lower applicability limit of the Thomas–Fermi model is determined by approximate allowance for quantum and exchange corrections and by the necessity of considering the shell structure of elements.

For a thorough analysis of specific features of the Fermi surface and the description of nondense structures one should take into account nonsphericity of the problem and use more refined quantum-mechanical models of adjoint plane waves and linear MT orbitals, the Green function method, etc.

The calculations of band structures in these approximations are inevitably cumbersome and have been performed for only a few elements in a restricted region of the phase diagram [545, 547].

The main factor determining the thermodynamic properties of nontransition metals is the character of the behavior of valence electrons which depends weakly on the interion potential [545, 547]. This permits the introduction of pseudo-potential of electron–ion interaction with an extremely simplified description of the ion core, the choice of its numerical parameters being based on the experimental data on the shape of the Fermi surface. The absence of such kind of data for high pressures and temperatures restricts the use of the pseudo-potential model in a sufficiently wide parameter range.

A revolutionary advance in the quantum-mechanical description of complex many-electron structures was a result of elaboration by Kohn and Sham of the density functional method, which provisioned a pithy computation of the broad spectrum of elements and compounds, including those in compressed state.

The most extensive and reliable experimental data on the EOS of "electromagnetic" substance (considered in this section) at high pressures have now been obtained by dynamic methods based on exploiting strong shock waves for generation and thermodynamic diagnostics of occurring states. These experimental results underlie the construction of semi-empirical EOS (Sec. 8.5) based on simplified notions of the spectrum of thermal lattice vibrations and the character of interparticle interaction in a substance and containing some free parameters for a quantitative description of experiments. In spite of pure empiricism, such models describe to high precision a considerable part of the phase diagram, including high-temperature melting and evaporation and exhibit regular asymptotics in the region of Thomas–Fermi gas and ideal plasma.

The performed analysis shows that a rigorous theoretical description of substance properties is possible in the range of extreme pressures and temperatures occupying the periphery of the phase diagram, whereas its inner part, which is most interesting and important for practical applications, is accessible for at best model theories. This fact is an effectual stimulus for the presently performed intense theoretical and experimental investigations of the physical properties of substance under conditions of strong interparticle interaction.

A more detailed classification of extreme states is presented in paper [545], and we shall follow it in our further exposition.

While a substance is in the ordinary electron–nuclear form, many of its properties are mainly determined by electrons, i.e., the lightest structural units of substance. In view of a substantial (3 − 4 orders of magnitude) difference between the electron and ion masses their contribution to the physical properties can be considered [545] separately, which makes up the basis of the "adiabatic" approximation.

It has already been noted that upon crossing the boundary of the universal state of the substance (curves *1* in Figs. 1.4 and 1.5) the external electrons of its atoms appear to be completely collectivized. Any substance that remains solid under such transition possesses metallic properties. In the language of band theory substance "metallization" is due to energy band broadening as a result of which the Fermi level of electrons finally gets inside the allowed band. For a substance in the supercritical gaseous–liquid state crossing curve *1* leads to the plasma state. It should be noticed that the degree of ionization on curve *1* itself is already rather high [598]. Analogously, metallization of semiconductors possessing a narrow forbidden band

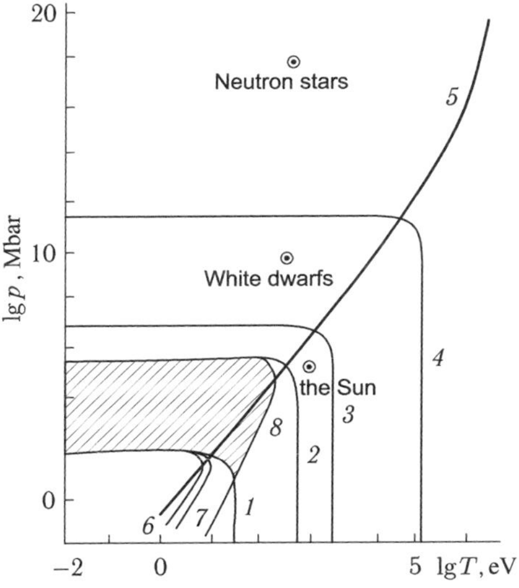


Fig. 1.4 The states of the electron component of a substance. *1* Boundary of the region of universality (collectivization of outer shell electrons), *2* of collectivization of bulk electrons ($Z = 10$), *3* of collectivization of internal electrons ($Z = 10$), *4* boundary of relativistic region, *5* boundary of degeneration region, *6* boundary of quasi-classicality region, *7* range of applicability of self-consistent field approximation, *8* boundary of the electron gas ideality (homogeneity) region. Dashed is the range of Thomas–Fermi model applicability. From Ref. [545].

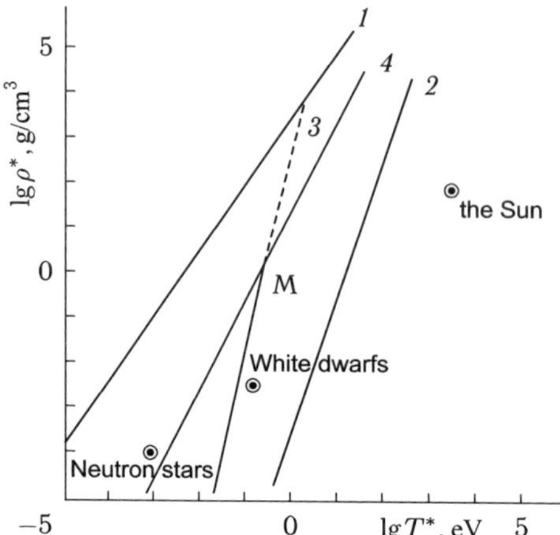


Fig. 1.5   States of the nuclear component of the substance. *1* boundary of the nucleus degeneracy region, *2* boundary of the ideality region, *3* melting curve, *4* boundary of the region in which the lattice can be considered classical. From paper [545].

begins far below this curve. On the contrary, such substances as lithium and sodium first pass over to a dielectric state and only under megabar pressures are they finally transformed into metal [322–324, 338, 1027]. At not very high temperatures plasma will be strongly nonideal and close in the character of ion motion to an ordinary liquid (see Sec. 3.3).

The other electrons of the atoms with not a low $Z$ value will remain near curve *1* in the bound state. Their greater part has binding energy of the order of $Z^{4/3}e^2/a_0$ and is localized in a volume of the order of $Z^{-1}a_0^3$ (these estimates appear from the Thomas–Fermi model). Accordingly, the boundary where the greater part of atomic electrons get unbound with the nucleus is determined by the temperature of the order of $Z^{4/3} \cdot 10\,\mathrm{eV}$ and pressure of about $Z^{10/3} \cdot 10^2\,\mathrm{Mbar}$ (curve *2*, Fig. 1.4).

Electrons that are most strongly bound to the nucleus and are the closest to it have binding energy of the order of $Z^2e^2/a_0$ and the volume of their localization is $Z^{-3}a_0^3$. These electrons become collective at a temperature of the order of $Z^2 \cdot 10\,\mathrm{eV}$ and pressure of the order of $Z^5 \cdot 10^2\,\mathrm{Mbar}$ (curve *3* in Fig. 1.4). As a result the substance becomes either a completely ionized electron-nuclear plasma or an ideal metal with lattice of "bare" nuclei.

At relatively low pressures and temperatures, $T < m_e c^2$, the electron motion may be thought of as nonrelativistic. The effects of the Relativity Theory become notable when the energy increment due to compression or heating reaches about $mc^2$ (the corresponding characteristic length is nearly $\sim \hbar/mc$). This corresponds to a pressure of about $10^{11}$ Mbar and temperature of approximately $10^5$ eV (curve *4* in Fig. 1.4).

The dynamic description of the electron subsystem is conveniently accomplished in the language of characteristic lengths. Apart from the Bohr radius $a_0$ four such lengths exist. This is first of all the average interelectron spacing $d \sim n^{-1/3}$ ($n$ is the mean electron concentration) and the average distance $D \sim Z^{1/3}d$ between nuclei. Next, this is the de Broglie electron wavelength $\lambda \sim \hbar/p$ ($p$ is its average momentum); for low and high temperatures we respectively have $\lambda \sim n^{-1/3}$ and $\lambda \sim \hbar/(mT)^{1/2}$. Finally, this is the characteristic length $l$ of inhomogeneity along which the electron spatial distribution changes noticeably. This quantity coincides with the electron screening radius and is expressed in terms of the already introduced lengths $l \sim \left(a_0d^3\right)^{1/2}/\lambda$; in a cold substance $l \sim a_0^{1/2}n^{-1/6}$, and in a hot substance $l \sim (T/e^2n)^{1/2}$.

For different ratios of the characteristic lengths we arrive at different states of the electron subsystems. The inequality $\lambda \ll d$ (or $n\lambda \ll 1$) corresponds to the classical (Boltzmann) electron gas; for $\lambda \sim d$ (or $n\lambda \gtrsim 1$) the degeneracy state holds (Fig. 1.2). The curve bounding the region $\lambda \sim d$ is given in Fig. 1.4 (curve *5*).

The condition $\lambda \ll d$ means that the electron wavelength varies little along it. This is known to correspond to the condition of quasi-classical electron motion. In the opposite limiting case the electron behavior is of essentially wave character. The relation $\lambda \sim l$ is demonstrated by curve *6*, Fig. 1.4.

Under condition $\lambda \ll (a_0d)^{1/2}$ or, which is the same, $e^2n^{1/3} \ll p^2/m$ the Coulomb interaction energy of a pair of electrons is low compared to their kinetic energy. This however does not yet lead to smallness of interaction effects in general. The point is that a stronger electronuclear interaction exists. Moreover, owing to the long-range action of Coulomb forces an electron can interact with a large number of its neighbors. These interactions are exhaustively described in the language of mean quantities, i.e., in the framework of a self-consistent Hartree field. Therefore, the discussed inequality means in fact smallness of the effects that cannot be described within the Hartree approximation. They are called correlation effects and include, along with the exchange effects related to the Pauli principle, also the correlation effects proper that describe the deviation of a true interaction from the mean one (for more details see Ref. [543]). The dependence $\lambda \sim (a_0d)^{1/2}$ is illustrated in Fig. 1.4 (curve *7*).

For still higher pressures and temperatures the just-now mentioned effects of electronuclear and collective electron–electron interaction become inessential. The electron gas can then be described by the ideal-gas formulas, and its distribution in space becomes practically homogeneous. This happens when the length $l$ of inhomogeneity exceeds the largest parameter of length dimension; it appears to be the mean distance $D$ between the nuclei. The curve $l \sim D$ for $Z = 10$ is presented in Fig. 1.4 (curve *8*). For arbitrary $Z$ values it can be obtained from curve *7* assuming that on the axes are the quantities $Z^{-10/3}p$ and $Z^{-4/3}T$. Within the homogeneity region a considerable part of electrons can still be in the bound state (curve *8* lies inside curve *2* in Fig. 1.4).

In the region where the electron gas is ideal, the equation of state of the substance has the well-known form [598]: $p \sim \frac{\hbar^2}{m}n^{5/3}$ (nonrelativism, degeneracy region), $p \sim \hbar c n^{4/3}$ (ultrarelativism, degeneracy region), $p \sim nT$ (Boltzmann region). At lower temperatures and pressures the EOS changes under the effects of Coulomb interaction, the contribution of nuclei, etc. Detailed quantum-mechanical computations or semi-empirical models are needed here.

Figure 1.4 shows that curves *6* and *7* lie outside the region under consideration. Hence, in the description of the electron subsystem one can neglect the correlation and quantum effects (reflecting inaccuracy of quasi-classical approximation) and employ the Thomas–Fermi model. This refers to the region dashed in the figure, while in the remaining part of the extreme state region of interest the electrons can generally be thought of as an ideal gas. The details referring to the issues touched upon in this section can be found in papers [539, 543], and the results of numerical calculations in paper [498].

Figure 1.4 also shows several typical objects depicted in Fig. 1.1. We can see that the electrons in the central part of the Sun form a nonrelativistic ideal gas which can be taken as classical, although not very far from degeneration. Atoms of hydrogen which are the dominant components of the solar substance are completely ionized, while the atoms of heavier elements can retain some fraction of electrons [387]. In the central part of white dwarfs the electron gas is ideal and degenerate. Applicable in this case is the well-studied model of one-component electron gas against the neutralizing background of compensating charge [287]. The relativistic effects play a remarkable role. Substance in these conditions consists of electrons and "bare" nuclei. The same refers to the substance in the core of neutron stars [414] where the electron gas can be considered ultrarelativistic.

## 1.2 Nuclear component

The nuclear subsystem determines the aggregate state of a substance and the run of nuclear processes. Deferring for a time nuclear transformations at ultrahigh pressures, we shall consider the contribution of a nuclear component assuming the nuclei themselves to be invariable and determining the chemical composition of the substance.

At high temperatures, when the thermal energy is high compared to the Coulomb energy, the substance is close to plasma in its properties to an ideal gas (Fig. 1.4).

The aggregate state determines specific ways of describing the thermodynamics of heavy components of the substance and the position of phase transition boundaries, while nuclear reactions determine stability of the elemental composition.

With lowering temperature or heightening pressure (the latter causes shortening of spacing between nuclei) the role of Coulomb interaction between nuclei increases. For this reason the nuclear subsystem ordering, i.e., substance transition into the crystal state turns out to be energetically advantageous. On the other side of the transition line into the crystal state we shall deal with fluidal plasma [140]. In particular, at zero temperature (and not very low and not very high pressures) any substance will be in a solid state [540, 684].

We will only be concerned with a substance containing one type of nuclei, and for simplicity will neglect the change of the ion effective charge in the region inside curve *3* in Fig. 1.4. To make the consideration unified for all the elements, we shall proceed from the natural "nuclear" units of density and temperature $-AM/a_0^3$ and $Z^2e^2/a_0$, where $a_0 = \hbar^2/AMZ^2e^2$ is the Bohr radius of the nucleus, and introduce the "reduced" quantities

$$\rho^* = \rho Z^{-6} A^{-4}, \quad T^* = T Z^{-4} A^{-1},$$

in the language of which the description will have a universal character. However, for pressure which is mainly determined by the electron subsystem, such self-similarity will not already exist. Hence, we shall have to deal with the "density-temperature" diagram.

In the plasma state the nuclear subsystem is described by the following characteristic lengths: the average internuclear distance $D \sim N^{-1/3}$ ($N = n/Z$ is the mean nuclear concentration), the de Broglie nucleus wavelength $\Lambda \sim \hbar/p$ ($p$ is the mean nucleus momentum) and the inhomogeneity length $L \sim (A_0 D^3)^{1/2}/\Lambda$ coincident with the Debye plasma screening radius. The same as in the case of electrons, the line separating the Boltzmann region from the nuclear degeneracy region (see curve *1* in Fig. 1.5) is defined by the condition $D \sim \Lambda$.

The role of the Coulomb interaction effects in plasma is determined by the relative values of the lengths $D$ and $L$. The curve $D \sim L$ is given in Fig. 1.5 (curve *2*). Within the range of high temperatures and pressures, light nuclei (hydrogen, helium) "burn out" as a result of thermonuclear fusion reactions and for heavier nuclei we get into the relativistic region relative to nucleons.

The line of phase transition into the crystal state (which is our concern) must lie to the left of curve *2*, i.e., in the region where the Coulomb effects play a significant role. If both the plasma and the crystal appearing from it are considered to be classical objects, then the equation for the melting curve will not contain the Planck constant. Only one combination of characteristic length exists which, being dimensionless, does not contain $\hbar$. This is the ratio $D/L$. Obviously for this region the equation for the melting curve must have the form $D/L = $ constant. The value of this constant for strongly compressed substance is close to 10. This corresponds to the Coulomb energy $Z^2 e^2 N^{1/3}$ which is two orders of magnitude higher than the thermal energy $T$.[1] It should be emphasized that what has been said above exactly corresponds to the well-known Lindemann criterion: on the melting curve the ratio $\delta_m$ of the mean amplitude of nuclear oscillations to the mean internuclear distance is a constant quantity (see Refs. [949, 1091]). In our case $\delta_m \approx 1/4$ [1006].

The phase transition line is depicted in Fig. 1.5 (curve *3*). Although it lies to the right of curve *1* which determines the boundary of the classical plasma region, the original assumptions concerning applicability of classical statistics are valid not along the entire curve *3*. The point is that the character of nuclear motion in a crystal differs essentially from that in plasma. The criterion of Boltzmann statistics applicability will also be different: the amplitude of zero nuclear oscillations must be small compared to the total amplitude including thermal vibrations. This criterion comes down to the obvious condition $\hbar \omega_D \ll T$, where $\omega_D$ is the Debye frequency. In the considered conditions $\omega_D \approx 0.45 \omega_0$, where $\omega_0 = (4\pi N Z^2 e^2/AM)^{1/2}$ is the plasma frequency of nuclei [159]. The curve $\hbar \omega_D \sim T$ is presented in

---

[1] The density variation under phase transition proceeding at a constant pressure is quite negligible in the considered conditions (not above 1/100 of percent; see Ref. [1005]). This permits the use of density as a single-valued variable.

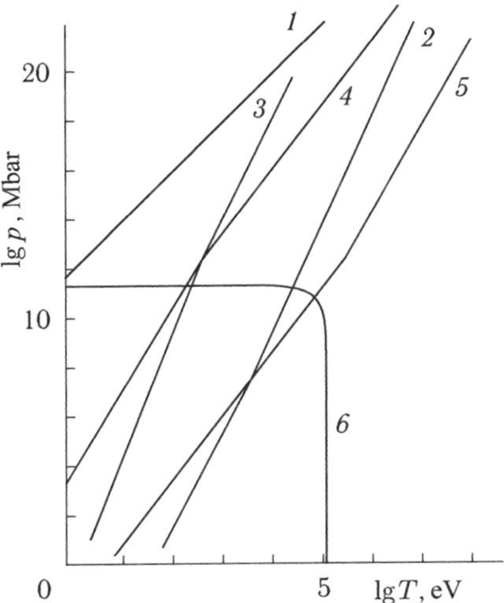



Fig. 1.6   The states of the electron component of a substance in the coordinates of $p$, $T$ for carbon. *5* boundary of electron degeneracy region, *6* boundary of electron relativism region. From Ref. [545].

Fig. 1.5 (curve *4*). The intersection point $M$ of curves *3* and *4* restricts the reliable portion of the melting curve shown by a thick line. We also present the "pressure-temperature" diagram for $Z = 6$, $A = 12$ (Fig. 1.6) similar to the diagram of Fig. 1.5.

For a roughly qualitative description of the melting curve run in the quantum region the Lindemann criterion and the known formulas for quantum oscillator [540] can be employed. Figure 1.7 demonstrates an ordinary phase diagram with a triple point.

The quantum effects lead to a knee in the melting curve towards high pressures and low temperatures, which in turn causes the appearance of limiting values of temperature, density, and pressure above which the crystal state is impossible. We shall now dwell on the "cold melting" effect which must in principle take place also at zero temperature. It is precisely this effect that explains the existence of liquid helium at low temperatures and atmospheric pressure. Cold melting is due to zero oscillation of nuclei: for sufficiently strong compressions the energy of such oscillations $\hbar\omega_{\mathrm{D}} \sim N^{1/2}$ exceeds the Coulomb binding energy of the lattice, proportional to $N^{1/3}$. A reliable description of this phenomenon is hampered by an exceedingly sharp dependence of the melting density $\rho_m$ on the Lindemann constant $\rho_m = \delta_m^{12}$, the more so as the constant itself appears to be changeable

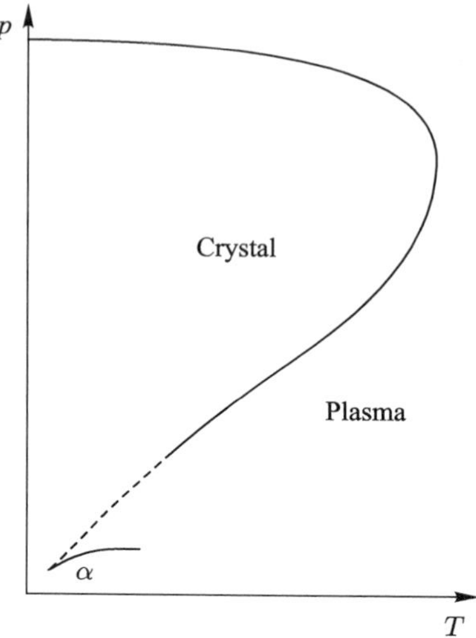



Fig. 1.7 Approximate form of the phase diagram, $\alpha$ is the triple point. From Ref. [545].

to a great extent [225]. The estimates available in the literature give $\rho_m$ values in the range of $10^3 - 10^8\,\text{g/cm}^3$ [3, 159, 202, 748, 1003]. Excluding the lightest nuclei, the $\rho_m$ value perhaps falls onto the nuclear densities $\rho_m \approx 2.8 \cdot 10^{14}\,\text{g/cm}^3$. Hence, the very reality of the considered phenomenon is still open to question.

We shall now dwell on some astrophysical applications. Figures 1.1 and 1.5 show the points corresponding to the conditions in the center of the Sun (hydrogen), in the center of a typical white dwarf (carbon), and in the neutron star — pulsar crust (iron). In Fig. 1.5 one can see that the solar matter is in the plasma state, the white dwarf matter is in a state close to crystallization [540], and the pulsar crust matter is in the solid state [86].

To make our presentation more comprehensive we note that the anomalous form of the melting curves with temperature maximum is also observed in experiments for a number of substances (alkali metals, etc.) and has a different character. This effect will be considered in detail in Sec. 7.3.

## 1.3 Nuclear transformations

When the temperature or pressure become sufficiently high, considerable exothermic nuclear transformations occur in substances [545]. The importance of these processes, not to mention thermonuclear fusion in terrestrial conditions, is determined by the fact that they are the main source of stellar (and also solar) energy, represent an essential factor of stellar body evolution and, finally, form the chemical composition of the substance. Typical examples of exothermic reactions may be the transformations $4p \to \mathrm{He}^4$ and $3\mathrm{He}^4 \to \mathrm{C}^{12}$; the former proceeds by the scheme of carbon or hydrogen cycle.

Although the considered reactions are accompanied by energy release, their rates should be notable in order that the external conditions be sufficiently extreme. This is needed for the Coulomb barrier obstructing the approach of nuclei-reagents to be overcome. As the substance is heated, the barrier penetrability grows as a result of the relative reagent energy heightening and under compression this happens as a result of distortion (narrowing) of the barrier itself. These two regimes of nuclear reactions, called respectively thermo- and pyknonuclear, have different kinetic mechanisms. The rate of thermonuclear reaction in plasma, along with the factor of penetration through the barrier, is determined by the number of partners encountered by a given nucleus per unit time when it moves inside the substance. The rate of pyknonuclear reaction in condensed matter is determined by the number of approaches of nuclei-reagents oscillating near neighboring equilibrium positions per unit time, i.e., by the frequency of such oscillations. Following Ref. [545], we shall consider the classification of regimes of nuclear reactions at different temperatures and pressures.

For simplicity, we shall now restrict our consideration to a substance consisting of one kind of nuclei which enter into a reaction with one another. At relatively high temperatures and low densities, when plasma differs little from an ideal gas, the factor of penetration through the barrier nearly coincides with the corresponding expression for a pair of isolated nuclei [330]:

$$f \sim \exp(-\tau^{1/3}), \quad \tau = \frac{27\pi^2}{4} \frac{Z^2 e^2}{A_0 T} \sim \frac{1}{T}.$$

In the assumption that $\tau \gg 1$ (otherwise the substance will burn out too fast) the effective energy of reacting nuclei (Gamow peak energy) lies in the tail of Maxwellian distribution $T_{\mathrm{eff}} \sim \tau T \gg T$. Since the mean internuclear distance is small as compared to the Debye radius, the influence of the rest of the substance comes down to the appearance in the exponent of a small correction which is due to the effect of Coulomb interaction screening by particles that do not participate in the reaction. This is the so-called thermonuclear regime with weak screening. The corresponding region lies to the right of curve *1* in Fig. 1.8 coincident with curve *2* in Fig. 1.5 and meeting the condition $D \sim L$.

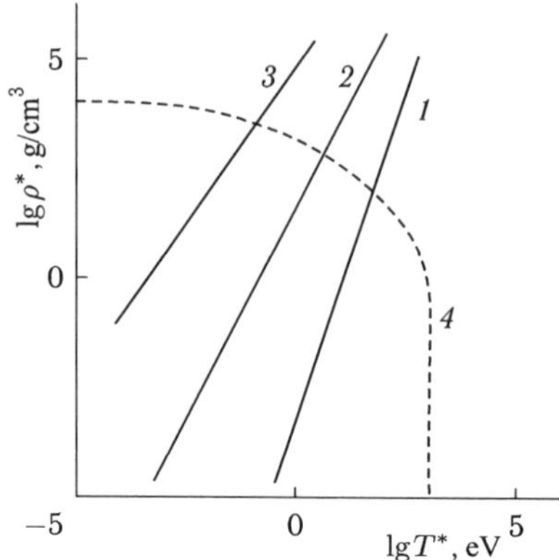

Fig. 1.8   Regimes of exothermic nuclear reactions. *1* boundary between thermonuclear regimes with strong and weak screening, *2* boundary between thermonuclear regime with strong screening and pyknonuclear regime with hot nuclei, *3* the boundary of pyknonuclear regime proper, *4* approximate form of the "threshold" for the hydrogen reaction. From Ref. [545].

In this connection we should mention paper [273] predicting a weaker (power-law compared with exponential) fall of the distribution function for large pulsed reagents noticeably increasing the thermonuclear reaction rate at low temperatures.

After crossing this boundary, the Coulomb interaction effects in plasma become significant, and we get into the region of fluidal plasma, and after crossing curve *3* of Fig. 1.5 into the region of solid body. However, the difference between these states is inessential from the point of view of interest because important for nuclear reactions is only the short-range order. Transition of a substance into a condensed state affects strongly the form of the internuclear interaction potential, but the nuclei themselves are still far beyond the boundary of the considered region and can in a sense be thought of as free. The point is that as has already been said the effective energy $T_{\text{eff}}$ of colliding nuclei exceeds significantly the actual temperature of the system. Hence, the Coulomb effects will actually become material from the viewpoint of reaction kinetics when $D$ becomes of the order of the effective Debye radius obtained from the ordinary equation by substitution of $T_{\text{eff}}$ for $T$. This determines the left boundary of the considered region (see curve *2* in Fig. 1.8); this curve coincides with curve *4* of Fig. 1.5. The regime under discussion is referred to as thermonuclear regime with strong screening. The penetration factor for this regime is changed greatly by the screening effects; its expression is presented in paper [875].

At still lower temperatures the nuclei-reagents may pass for "frozen". The nuclear oscillations about the equilibrium position then begin playing a decisive role. While $T_{\text{eff}}$ is large compared to $\hbar\omega_{\text{D}}$, these oscillations are mainly of thermal nature. This is the pyknonuclear regime with hot nuclei. The left boundary of the considered region is defined by the condition $\hbar\omega_{\text{D}} \sim T_{\text{eff}}$.

Finally, to the left of this curve we fall into the region of pyknonuclear regime proper [875, 1003, 1041, 1076] where the reaction is due to zero oscillations of nuclei. Under sufficiently strong compressions the reaction can well go at zero temperature thanks to the Coulomb barrier narrowing to a width of the order of $D \sim N^{-1/3}$. The corresponding penetration factor has the form

$$f \sim \exp(-\chi), \quad \chi \approx \frac{2.8}{(A_0 N^{1/3})^{1/2}} \sim \rho^{*-1/6},$$

where we assume $\chi \gg 1$ as above.

Although these reactions have not got a threshold in the exact sense of the word, a substantial reaction yield only occurs at high enough temperatures or pressures. For the hydrogen reaction it is $T \approx 10^2 - 10^3\,\text{eV}$ or $\rho \approx 10^4 - 10^5 \text{g/cm}^3$, for the helium reaction the temperature rises to $10^4\,\text{eV}$. Under a joint action of heating and compression the corresponding boundary could be found from the cumbersome expressions given in Ref. [875] for the penetration factor under all the enlisted regimes. We shall not go beyond a purely qualitative curve for the hydrogen reaction (curve *4* of Fig. 1.8).

As an example we shall say that nuclear reactions in the interior of the Sun and similar stars go in the regime of thermonuclear reaction with weak screening; this is seen directly from the diagram of Fig. 1.5. Concluding this section we shall briefly touch upon the reactions $d + d \to T - p$ (or $\to \text{He}^3 + n$) and $d + T \to \text{He}^4 - n$ important for "terrestrial" applications. These reactions, as distinct from the above-mentioned hydrogen reaction, are due to the strong interaction and for this reason their purely nuclear cross-section is by 20 orders of magnitude larger than the cross-section of the hydrogen reaction due to weak interaction. However, the "threshold" temperature and "threshold" density for these reactions are comparatively close (in the logarithmic sense) to the values of Fig. 1.8. This is because of the sharp dependence of the penetration factor on $T^*$ and $\rho^*$. Note that the muon catalysis of these reactions proposed earlier by Ya. B. Zeldovic and A.D. Sakharov might be regarded as a peculiar way of substance "contraction"; since the mesoatom radius is approximately 200 times smaller than the ordinary atom radius, we deal with an increase in the effective density by almost seven orders of magnitude.

## 1.4 Nuclear transformations and new forms of matter

The final result of the processes analyzed in the preceding section consisted only in nucleon regrouping leading to transformation of the nuclei to other nuclei without a change in the structural composition of the substance in what concerns the elementary particles and the appearance of its new forms. Meanwhile, at fairly high temperatures and pressures such transformations are inevitable and are of importance under superextreme conditions. We digress from the occurrence due to the hydrogen reaction of neutrinos and positrons that either leave the substance or annihilate. Therefore, they anyway fail to produce a new form of matter.

We shall begin with the simplest process of the appearance at high temperatures of equilibrium thermal radiation as a separate component of matter that makes an appreciable contribution to its EOS. Using for the radiation pressure the known formula $p \sim T^4/\hbar^3 c^3$ (see curve $1$ in Fig. 1.9), one can readily see that near this curve the photon component of matter makes a large and even decisive contribution to pressure.

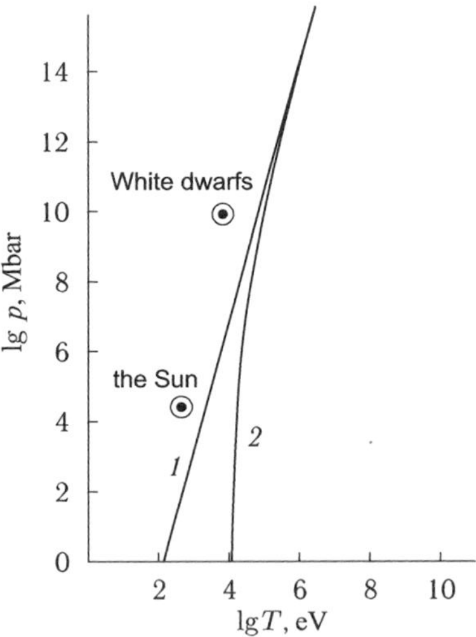


Fig. 1.9 Photon and positron components of matter. $1$ photon component pressure, $2$ positron component pressure. From Ref. [545].

Experiments with underground nuclear explosions [63] make it possible to closely approach this "radiation-dominating" regime.

A fairly similar picture might be obtained for the neutrino component of matter. However, neutrinos leave the fire "ball" volume upon relativistic collisions of ions or celestial bodies and normally are not therefore in local thermodynamic equilibrium necessary for the application of thermodynamic formalism and, hence, EOS and could play the role of a peculiar matter component in only the early Universe (see Ref. [1079]).

We now proceed to the electron–positron pair production and to the birth of the positron component of matter. This process is of endothermic nature (the corresponding threshold is — $2mc^2 \approx 1$ MeV). For $T \gg mc^2$ the expression for positron pressure is almost the same as that for radiation, and for $T < 2mc^2$ the exponential threshold factor $\exp(-2mc^2/T)$ appears. The boundary of the region where the positron pressure is substantial is given by curve *2* in Fig. 1.9. To the production of muon, baryon, and other pairs there correspond higher thresholds.

The above-mentioned processes are characterized by the fact that for them to proceed the system should be thermally excited. Accordingly, at low temperatures these processes exhibit very low intensity. They also include thermal dissociation of nuclei, in particular, nuclei of the most stable isotope $Fe^{56}_{26}$. This reaction comes down to the transformations $Fe \rightarrow 13He^4 + 4n$ and $He^4 \rightarrow 2p + 2n$ and induces the emergence of the neutron component of matter. Figure 1.10 presents the diagram borrowed from paper [462] (see also Ref. [1079]). Curve *1* corresponds to a half of iron dissociation, curve *2* is the same for helium.

A very important type of transformation proceeding also in a cold substance is its neutronization, that is, the electron capture by the nucleus

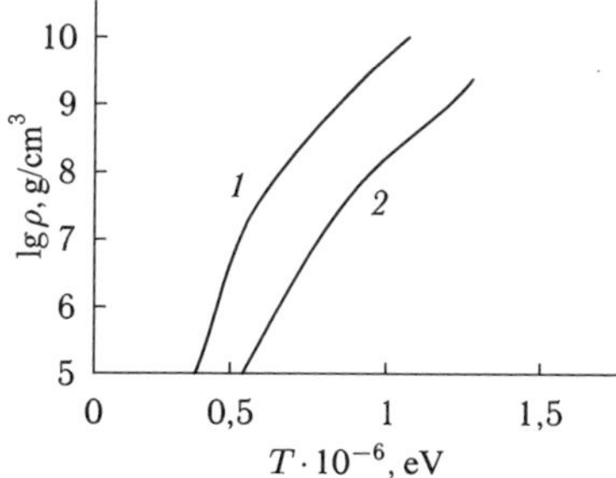

Fig. 1.10   "Iron-helium-nucleons" thermal dissociation. *1* iron half-dissociation, *2* helium half-dissociation. From Ref. [545].

with transformation of intranuclear proton to a neutron [414]. This reaction follows the scheme $(A, Z) - e^- \to (A, Z - 1)$. Since the original nucleus is considered to be stable and the occurring excessive-neutron nucleus has a higher energy, the neutronization process is endothermic and has the threshold $\Delta M c^2 = (M_{Z-1}^A - M_Z^A)c^2$. The energy needed to overcome the threshold is drawn in cold matter[2] from a gravitational source: the gravitational forces compress the star, heighten the Fermi energy of electrons and "speed them up" to the required energy. At high temperatures neutronization is due to the increase in the thermal energy of electrons and relates to the ordinary chemical equilibrium mechanism. Sufficiently extensive literature is devoted to matter neutronization (see Refs. [43,44,154,184,414,598,602,1038,1079]). Many related problems however call for a quantitative solution because of imperfection of the available methods of description of multinucleon nuclear systems and especially those containing still heavier particles. Therefore, below we shall restrict ourselves to several remarks and a few quantitative estimates. With some exceptions, the latter are of tentative character.

The neutronization threshold is defined by the formula $\rho \sim \frac{Mc^2}{\hbar^3}$ $[(\Delta M)^2 - m^2]^{3/2}$. The corresponding pressure for $\Delta M \gg m$ has the form $p \sim \frac{c^5(\Delta M)^4}{\hbar^3}$. As an example we note that the threshold values of density and pressure make up for the transition $C_6^{12} \to B_5^{12}$ respectively, $4 \cdot 10^{10}\,\mathrm{g/cm^3}$ and $6 \cdot 10^{16}\,\mathrm{Mbar}$ and for the transition $Si_{14}^{28} \to Al_{13}^{28}$ respectively, $2 \cdot 10^9\,\mathrm{g/cm^3}$ and $8 \cdot 10^{11}\,\mathrm{Mbar}$.

At pressures above the threshold nuclei will appear "overloaded" with neutrons. Up to density of nearly $10^{11}\,\mathrm{g/cm^3}$ and pressure of the order of $10^{18}\,\mathrm{Mbar}$ these nuclei preserve their mass number, and it is only their charge that decreases because instability of such nuclei outside the substance would show up as their $\beta$-decay. In a substance, however, such a decay is impossible because of the high boundary of Fermi electrons. For high densities and pressures, disintegration of nuclei unloading their excessive neutrons becomes advantageous, and a separate neutron component of the substance appears.

For still higher densities (presumably of nearly $5 \cdot 10^{13}\,\mathrm{g/cm^3}$) the nuclei ultimately break up and the substance is transformed into a mixture of neutrons, protons, and electrons, the charged particle concentration being

---

[2]The term "cold" is rather conditional and refers to low temperatures compared to $\Delta M c^2$.

approximately two orders of magnitude lower than the neutron concentration [43,184]. A further increase in density is accompanied by the emergence in the substance of new elementary particles which are unstable under ordinary conditions. In the first place, these are muons (their occurrence threshold is nearly $10^{14}$ g/cm$^3$) whose decay is obstructed by the high boundary of Fermi electrons, then come hyperons, resonances, etc. (the thresholds lie in the range of $10^{14} - 10^{15}$ g/cm$^3$) [43,154,602].

Ultimately high energy densities are nowadays reached upon head-on collisions of heavy ions (Fig. 1.11) accelerated in synchrotrons up to sublight velocities. These experiments are aimed at the search for new particles, at experimental study of fundamental problems of high-energy physics under hadron collisions accompanied by the formation of superdense nuclear matter, i.e., QGP (Fig. 1.12). The unique experiments have been carried

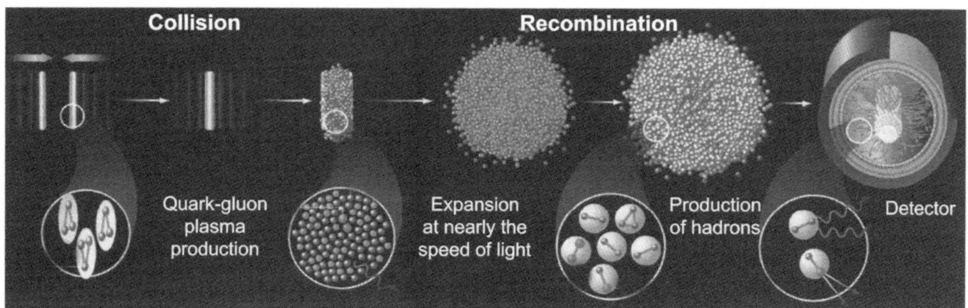


Fig. 1.11   Dynamics of collisions of relativistic heavy nuclei on accelerators [843].

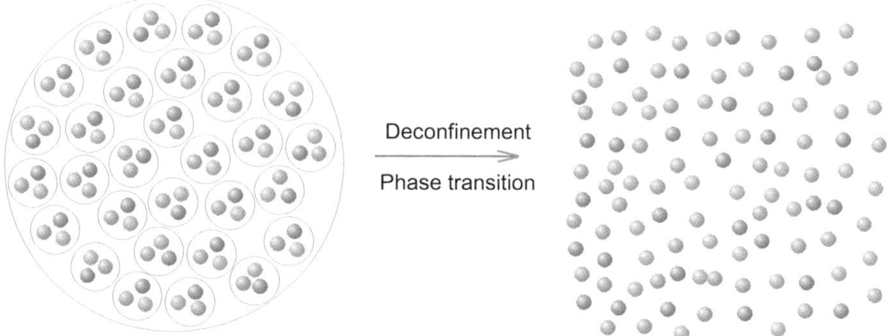


Fig. 1.12   Formation of QGP. Under ordinary conditions (on the left) quarks (balls) are bound in hadrons. At temperatures $T > T_c$ quark deconfinement takes place and they stop being bound in hadrons and form QGP.

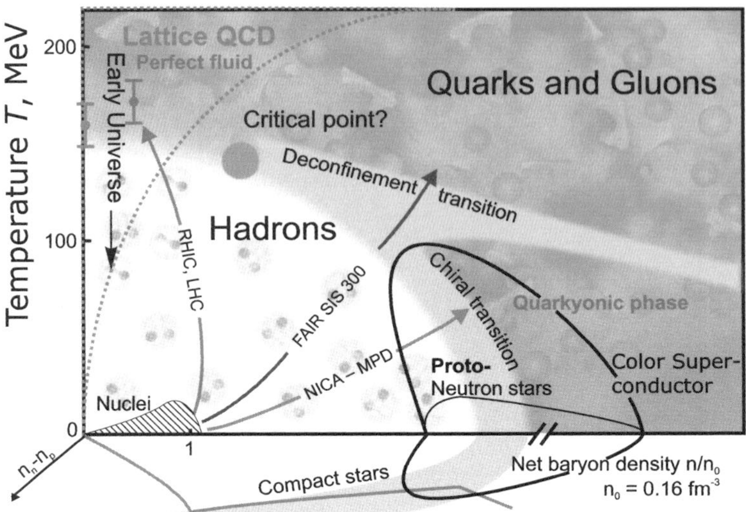


Fig. 1.13   Phase diagram of ultradense matter [84, 409].

out using the accelerators at CERN, Brookhaven, and Darmstadt on the generation of an ultraextreme baryon substance in a superdense and heated state with the density of the order of $10^{15}\,\mathrm{g/cm^3}$, pressure of $10^{30}\,\mathrm{bar}$, and temperature of approximately $200\,\mathrm{MeV}$ (Fig. 1.13) in individual collisions of heavy nuclei Cu–Cu and Au–Au. According to the present-day notions this particular state was inherent in the matter of the Universe within already the first microseconds after the Big Bang (Figs. 1.14 and 1.15) and is also inherent in the matter of astrophysical objects such as $\gamma$-bursts, neutron stars, and black holes.

QGP consists of quarks, antiquarks, and gluons [350, 419, 746, 867]. Such plasma is sometimes called the "oldest" form of matter since it existed within already the first microseconds after the Big Bang, and hadrons appeared from it as it expanded and cooled down. QGP possesses the maximum density, approximately $9-10\rho_0$ ($\rho_0 = 2.8 \cdot 10^{14}\,\mathrm{g/cm^3}$ is nuclear density) and can emerge in the center of neutron stars, black holes or, under collapse of ordinary stars (see Sec. 12.1).

Another "experimental" manifestation of QGP are the compact astrophysical objects under discussion, i.e., "strange" stars [294, 295] consisting of high-density QGP which is the next, deeper stage of neutron stars [294, 295, 414].

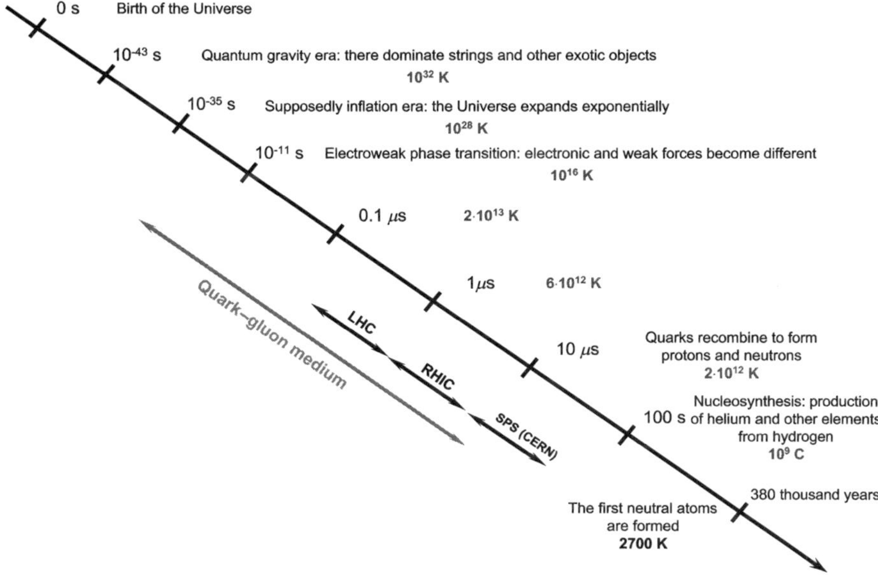


Fig. 1.14   Cosmic time scale. Universe expansion after the Big Bang [843].

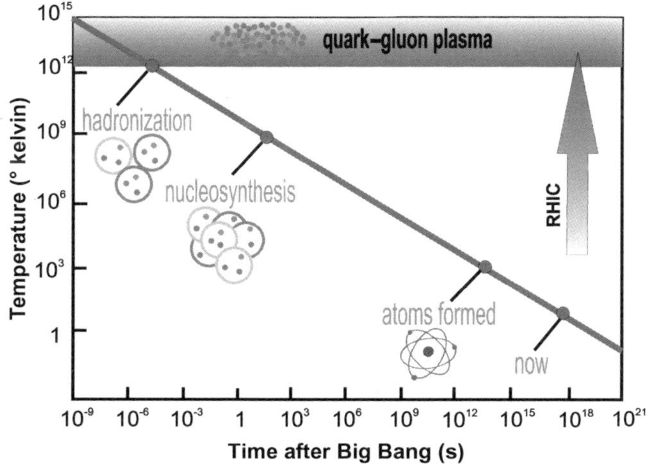


Fig. 1.15   Matter transformation after the Big Bang. The arrow gives the parameters of relativistic heavy-ion collider (RHIC) accelerator.

## 1.5  Experiment and astronomical observations

We shall present the scheme of matter transformation for high energy densities (Fig. 1.16), which in a sense illustrates the thesis concerning matter simplification during motion towards extremely high pressures and temperatures (Fig. 1.1).

Although rapidly progressing, our experimental resources are still capable of only partly breaking into the area of extreme states. Accessible for them turn out to be the left bottom and the right top portions of the diagram of states of matter (Fig. 1.2), whereas its larger, intermediate part is occupied by different astrophysical objects. These stellar objects are not only the points of application of EOS for the description of their structure and observed characteristics, but in some cases are a source of information for the construction and verification of compressed hot matter models. Examples may be the papers on geoseismology [387] and the studies of neutron and strange stars [414]. The strength of a substance sharply restricts the application of static methods of research of high energy densities since

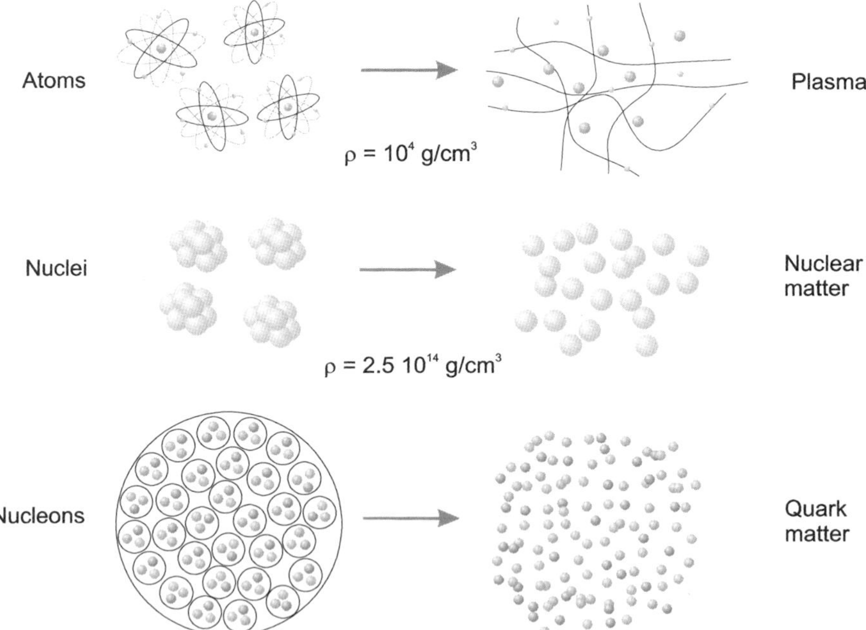


Fig. 1.16   Matter transformation under high energy densities.

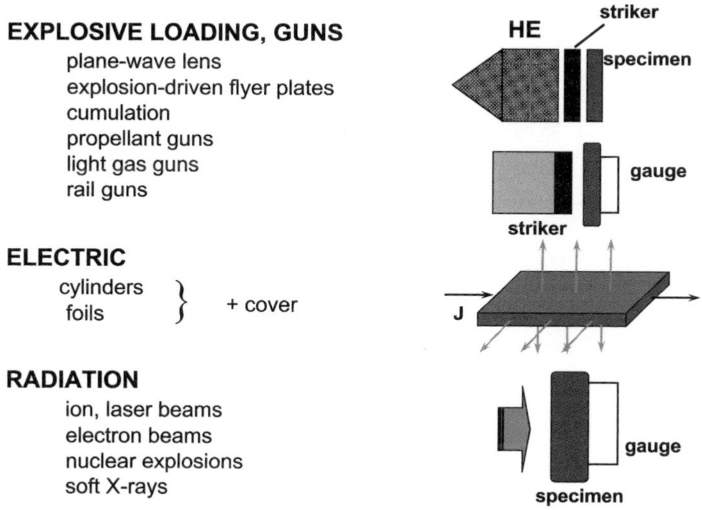

Fig. 1.17 Schemes of generation of high-power shock waves for investigation of macroscopic quantities of electromagnetic plasma.

the overwhelming majority of constructional materials are unable to withstand the pressures of interest [294,295]. An exception is the diamond which is the record-holder in solidity ($\sigma_n \approx 500\,\text{kbar}$), which permits attaining pressures of 3–5 Mbar in diamond anvils in static experiments [434,435,637].

The palm of supremacy is now with the dynamic methods [27,40,287, 293–295,319,734,1081] based on pulsed cumulation of high energy densities in a substance. Here, two different directions of research exist. This is an experimental study of the physical properties of electromagnetic plasma of megabar pressure range upon collision of macroscopic strikers at velocities of kilometers, tens of kilometers a second (Fig. 1.17) and the study of the properties of compressed nuclear matter upon head-on collision of nuclei moving at relativistic velocities (Figs. 1.11 and 1.18). We shall begin with considering the first of these approaches, namely, the examination of macroscopic compressed substance volumes [294,295]. The lifetime of such high-energy states is determined by the time of inertial plasma expansion with a characteristic scale of $10^{-10} - 10^{-6}\,\text{s}$, which needs application of refined fast means of diagnostics. The physical conditions corresponding to the lower boundary of the states of interest are presented in Table 1.1 [152,287,727]. One can see that obtaining high energy densities in plasma puts serious claims on the means of generation, making necessary the effective space and time power compression.

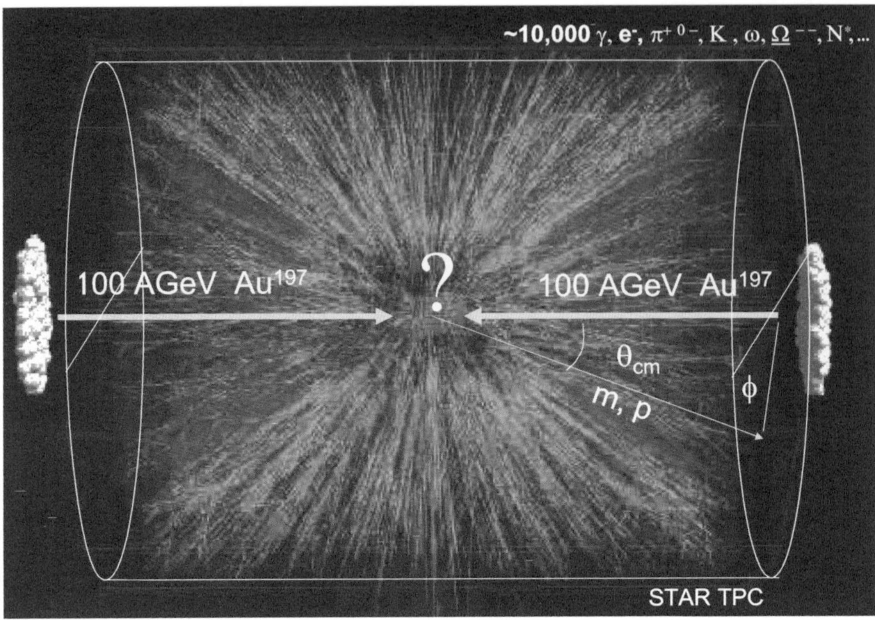


Fig. 1.18    Collision of relativistic hadrons [601].

High-power shock waves [63, 287, 296] induced by nonlinear hydrody-namic phenomena during the substance motion are a convenient tool for this purpose. In a viscous compression shock called the shock front the kinetic energy of the incident flux is converted into the thermal energy of a com-pressed and irreversibly heated medium. This way of generation (Hugoniot adiabat $H$ in Fig. 1.19 has no restrictions on the obtained pressure values) is limited to short lifetimes of shock-compressed substance. The diagnostic methods for such states must therefore have a high time resolution. It is precisely such time resolution that is normally used in electro-contact and optical recording of time intervals of motion of shock discontinuities and contact surfaces, pyrometric, spectroscopic, X-ray diffraction, and adsorp-tion interferometric laser measurements, as well as registration of low- and high-frequency Hall electrical conduction and fixation of piezoelectric and magnetoelectric phenomena [292, 293].

The characteristic feature of shock wave technique consists of the fact that they permit obtaining high pressure and temperature in compressed media, whereas the lowered density range (including the boiling curve and the neighborhood of the critical point) turns out to be inaccessible for

Table 1.1 Physical conditions corresponding to high energy densities $10^4 - 10^5$ J/cm$^3$ [727].

| Physical conditions | Values of physical parameters |
| --- | --- |
| Energy density, $W$ | $\approx 10^4 - 10^5$ J/cm$^3$ |
| Pressure, $p$ | $\approx 0.1 - 1$ Mbar |
| Condensed explosive substances | $W \approx 10^4$ J/cm$^3$ |
|     Pressure | $\approx 400$ kbar, |
|     Temperature | $\approx 4000$ K, |
|     Density | $\approx 2.7$ g/cm$^3$, |
|     Velocity of detonation | $\approx 9 \cdot 10^5$ cm/s |
| Impact of an aluminum plate against aluminum, | |
|     Velocity | $5 - 13.2 \cdot 10^5$ cm/s |
| Impact of a molybdenum plate against | |
|     molybdenum, Velocity | $3 - 7.5 \cdot 10^5$ cm/s |
| Electromagnetic radiation | |
|     laser, intensity, $q$ $(W \sim q)$ | $2.6 \cdot 10^{15} - 3 \cdot 10^{15}$ W/cm$^2$ |
|     Black body temperature, $T$ $(p \sim T^4)$ | $2 \cdot 10^2 - 4 \cdot 10^2$ eV |
| Electric field strength, $E$ $(W \sim E^2)$ | $0.5 \cdot 10^9 - 1.5 \cdot 10^9$ V/cm |
| Magnetic field induction, $B$ $(W \sim B^2)$ | $1.6 \cdot 10^2 - 5 \cdot 10^2$ T |
| Plasma density at the temperature, $T = 1$ keV | |
|     $(p = nkT)$ | $6 \cdot 10^{19} - 6 \cdot 10^{20}$ cm$^{-3}$ |
| Laser radiation intensity $q$, | |
|     for $\lambda = 1$ $\mu$m, $W \sim q^{2/3}$ | $0.86 \cdot 10^{12} - 4 \cdot 10^{12}$ W/cm$^2$ |
|     Black body temperature, $T$ $(p \sim T^{3.5})$ | $66 - 75$ eV |

them. The intermediate plasma states between a solid and a gas can be examined by the method of isoentropic expansion based on plasma generation upon adiabatic expansion of condensed matter (curve $S$ in Fig. 1.19) preliminarily compressed and irreversibly heated in front of a high-power shock wave.

We can see that the dynamic methods in their different combinations suggest realization and study of a wide spectrum of extreme states of matter with diverse and strong interparticle interaction. Then the conditions with high energy concentrations can actually be realized and fairly comprehensive diagnostics of these exotic states appears to be possible because shock and adiabatic waves are, apart from being the tool of generation, are also a specific instrument of diagnostics of extreme states of matter [27,306,1081].

The dynamic methods of diagnostics are applied for the study of both the macroscopic objects of electromagnetic plasma matter and the compressed nuclear matter, i.e., QGP [293–295]. These methods are based on the use of relation between the thermodynamic properties of the

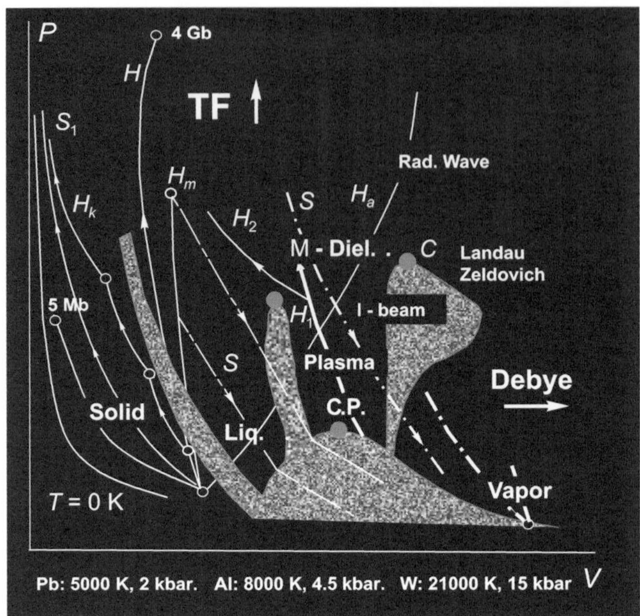


Fig. 1.19   Thermodynamic trajectories of the dynamic methods of matter research [287]. At the bottom are the parameters of the critical points (C.P.) of a number of metals.

investigated shock-compressed medium and the experimentally observed hydrodynamic phenomena caused by cumulation of high energy density in matter [27,1081]. In the general form this relation is expressed by a system of nonstationary gas-dynamic nonlinear (three-dimensional in spatial coordinates) differential equations whose complete solution is difficult even for the most mighty modern supercomputers. For this reason in the dynamical studies of macroscopic substance (EMF) researchers endeavor to use self-similar solutions of the type of a stationary shock wave and a centered rarefaction Riemann wave [27,1081] that express conservation laws in simple algebraic or integral forms. To apply such simplified relations, one should provide experimental conditions of self-similarity of the corresponding flow regimes.

With propagation of a stationary shock-wave discontinuity through a substance, the laws of conservation of mass, momentum, and energy [1081] hold on its front, which relate the kinematic parameters (the shock wave velocity $D$ and the mass velocity $u$ of the substance behind the shock front) to the thermodynamic quantities (specific internal energy $E$, pressure $p$, and specific volume $V$:

$$\frac{V}{V_0} = \frac{(D-u)}{D}; \quad p = \frac{p_0 + Du}{V_0},$$

$$E - E_0 = \frac{1}{2}(p + p_0)(V_0 - V),$$

$$(1.1)$$

where the index 0 marks the parameters of the substance at rest before the shock-wave front.

Using these equations one can find the hydro- and thermodynamic characteristics of shock-compressed matter by recording any two of the five parameters $E$, $p$, $V$, $D$, $u$, characterizing the shock-wave discontinuity. Most easily and exactly measured by the basis methods is the shock wave velocity $D$. The choice of the second recorded parameter depends on particular experimental conditions. Usually this is the mass velocity $u$ of shock-wave discontinuity [27, 1081] or the shock-compressed plasma density $\rho = V^{-1}$ [306].

Analysis of the errors of relation (1.1) shows that in the case of strongly compressed ("gas") media it is reasonable to choose the shock-compressed matter density as the second parameter. The methods of such measurements based on fixation of "soft" X-ray radiation absorption by cesium, argon, and air plasma have now been developed [287]. For lower matter compressibility (condensed media) admissible accuracies are provided by way of recording the mass velocity of motion $u$. These methods were employed to examine the states of degenerate metal plasma and dense Boltzmann plasma of argon and xenon [27, 287, 296, 1081].

In experiments on fixation of isoentropic shock-compressed matter expansion curves, the states in a centered unloading wave are described by Riemann integrals [1081]:

$$V = V_H + \int_p^{p_H} \left(\frac{du}{dp}\right)^2 dp; \quad E = E_H - \int_p^{p_H} \left(\frac{du}{dp}\right)^2 dp,$$

$$(1.2)$$

which are calculated along the measured isentrope, $p_s = p_s(u)$.

The coloristic EOS $E = E(p, V)$ in the region of $p - V$-diagram overlapped by Hugoniot or/and Poisson adiabats can be obtained by recording shock waves and extension waves under different initial conditions and intensities. In the experiments accomplished by now on the dynamic effect on plasma, the intensity of shock waves was varied by varying the excitation source power, i.e., varying the pressure of driver gas, types of explosives, missiles and target densities. Furthermore, different ways were used to vary the parameters of initial states, namely, the variation of initial temperatures and pressures (plasma of inert gases, cesium, and liquid) and the application of fine-dispersed targets intended to increase the irreversibility effects [27, 1081].

Thus, on the basis of general conservation laws, the dynamic methods of diagnostics make it possible to reduce the problem of finding the coloristic EOS $E = E(p, V)$ to measurement of kinematic parameters of motion of shock waves and contact surfaces, i.e., to recording the distances and times, which can be performed with high accuracy. The internal energy $E$ is however not a thermodynamic potential relative to the variables $p$, $V$, and closed thermodynamics of the system can only be constructed with an additional dependence of temperature on the parameters of state $T(p, V)$. In optically transparent and isotropic media (gases, ion crystals) the temperature can be measured together with other parameters of shock compression. Condensed media and in the first place metals are as a rule nontransparent, and so the light radiation of shock-compressed medium is inaccessible for direct recording.

A thermodynamically complete EOS can be constructed directly from the results of dynamic measurements without introducing *a priori* considerations of the properties and character of investigated substance [306] but using the Fermi procedure, proceeding from the first law of thermodynamics and the experimentally known dependence $E = E(p, V)$ (for more details see Refs. [287, 306]). This method of constructing the EOS of matter applied in dynamic experiments is free of any restricting assumptions concerning the properties, character, and phase composition of investigated substance since it is based on the very general conservation laws (1.1) and the first law of thermodynamics. Stationarity and one-dimensionality of shock-compressed medium flow should then be provided in experiment for the conservation laws to be used in the simple algebraic (1.1) or integral (1.2) forms.

It is important that a shock wave not only compresses but also heats the substance to high temperatures, which is especially significant for obtaining plasma, i.e., an ionized state of matter. In experimental study of strongly nonideal plasma a number of dynamic methods are now used (Fig. 1.19).

The shock compression of matter in a solid or liquid initial state permits obtaining behind the shock front the states (Hugoniot adiabats $H$ in Fig. 1.19) of nonideal degenerate (Fermi statistics) and classical (Boltzmann statistics) plasma compressed to maximum pressures $p \approx 4$ Gbar and temperatures $T \approx 10^7$ K [63, 1019] for which the specific density of internal plasma energy is comparable with the nuclear energy density and the temperatures are close to the conditions when the energy and pressure of equilibrium radiation begin playing a substantial role in summary thermodynamics of such exotic states.

To lower the irreversible heating effects it is reasonable to compress substances by a sequence of incident and reflected shock waves [319, 323, 725, 967, 1037]. As a result, such multi-stage compression ($H_k$) becomes close to a "softer" isentropic one ($S_1$) thus providing much higher ($10 - 50$ times) degrees of compression and much (about 10 times) lower temperatures as compared to a single shock-wave compression. Repeated shock compression was successfully applied for experimental examination of plasma ionization by pressure [319, 967, 1037] and substance dielectrization [323] under megabar pressures. Quasi-adiabatic compression was also realized under explosive highly symmetric cylindric compression of hydrogen and inert gases [319]. In experiments on "soft" adiabatic plasma compression, explosive squeezing of samples by a megagauss magnetic field was applied [431, 774].

In the other limiting case, when high-temperature plasma should be obtained, one can employ shock-wave compression of lowered-density (compared to solid-state) targets, namely, porous metals (the $H_m$ curve [27, 287, 389, 1081]) or aerogels [403] (the $H_a$ curve in Fig. 1.19). This allows a sharp increase of shock compression irreversibility effects, thus exercising a rise of compressed state entropy and temperature.

The shock compression of noble gases and saturated vapors of alkali metals by incident $H_1$ and reflected $H_2$ (Fig. 1.19) shock waves enables plasma research in the region with developed thermal ionization [149, 286, 287, 306, 318, 319, 393, 591, 633].

The adiabatic expansion of matter (curves $S$ in Fig. 1.19) preliminarily compressed by a shock wave to megabar pressures makes it possible to explore an interesting region of plasma parameters lying between a solid body and a gas, including the metal–dielectric transition region and the high-temperature portion of the metal boiling curve with their critical point [287, 296, 308, 611].

In view of considerably high metallic bond energy, the parameters of the critical points of metals are exceedingly high (4.5 kbar, 8000 K for Al and 15 kbar, 21000 K for W) and inaccessible for technology of static experiments. For this reason, the characteristics of critical points have up to now been measured for only three of all the metals that make up 80% of the elements of Periodic system [287]. On the other hand, because of high critical temperatures of metals comparable with their ionization potential they, when in the near-critical state, evaporate directly into the ionized state and not into gas as is normally the case with the other chemical elements.

This circumstance can lead to exotic "plasma" phase transitions predicted in the case of metallization by Landau and Zeldovich [1078] and other theoreticians for strongly compressed Coulomb systems (see Refs. [230, 267, 745, 881, 1039] and the references therein). An experimental search for these exotic plasma phase transitions is now one of the most interesting problems of the dynamic physics of high energy densities. The recently revealed sharp (five orders of magnitude) increase in electrical conductivity [318, 319, 377, 734, 967, 1037] and adiabatic compressibility [701] of nonideal hydrogen and deuterium plasma confirmed by quantum-mechanical Monte Carlo computations [267] evidently testifies to an experimental realization of such plasma phase transition.

Without considering here other methods of generation and study of the physical properties of matter under extreme conditions (see monographs [287, 294, 295]) we shall pass over to the discussion of experimental investigation of superextreme ($p \approx 10^{30}$ bar, $T \approx 200\,\text{MeV}$, $\rho \approx 10^{15}\,\text{g/cm}^3$) states of nuclear matter generated upon collision of nuclei sped up in accelerators to relativistic velocities [294, 295].

These experiments are aimed at a search for new particles (Fig. 1.20), at experimental study into the fundamental problems of high-energy physics upon hadron collisions accompanied by the formation of superdense nuclear matter, i.e., QGP.

And while particle production can be fixed in these experiments by direct or indirect detection, the collection of data on equilibrium (or quasi-equilibrium) properties of compressed nuclear matter encounters great difficulties as one deals with exceedingly small space and time scales that makes impossible the application of dynamic methods based on the conservation laws in self-similar (1.1)–(1.2) type flows. In RHIC experiments, the longitudinal Lorentzian reduction (Figs. 1.11 and 1.18) of the size of colliding nuclei is of order 100. The characteristic volume of U + U collision region amounts to about $3000\,\text{fm}^3$, containing nearly 10,000 quarks and gluons, and the characteristic collision time is $\tau_0 \approx 0.2{-}2\ \text{fm}/c \approx (5{-}50) \cdot 10^{-25}\,\text{s}$. For this reason part of high-energy processes evidently proceed in the expanding matter already after the nuclear bunches have flown through each other.

During collision and in the course of nuclear matter expansion and cooling, the emerging quarks and gluons become thermalized (the time

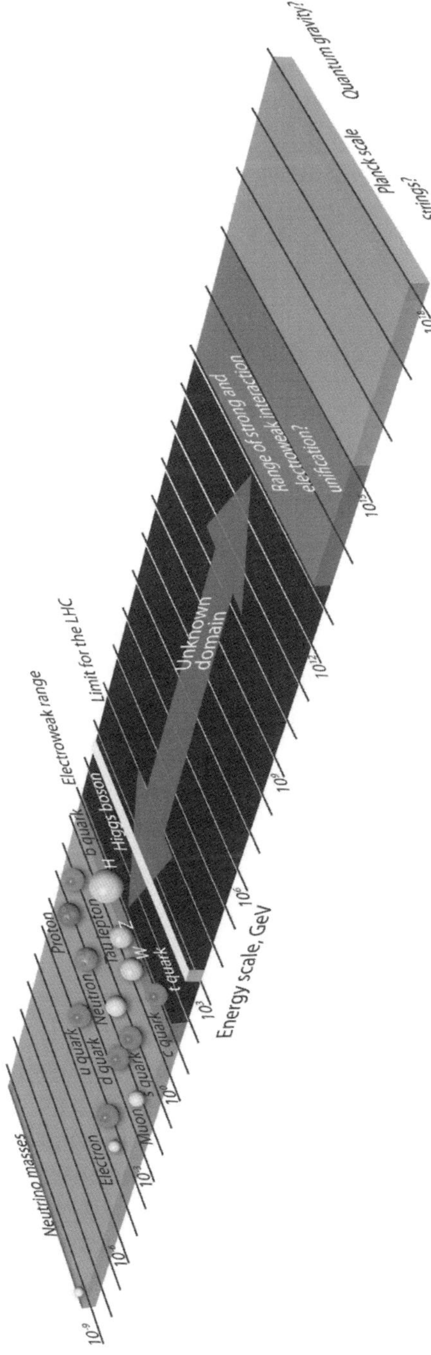


Fig. 1.20  Problem of particle hierarchy [818].

$\tau_{\mathrm{eq}} \leq 1\,\mathrm{fm}/c \approx 3 \cdot 10^{-24}$ s) and can get into local thermodynamic equilibrium within the plasma lifetime $\tau_0 \approx (2)\,R/c \approx 10\,\mathrm{fm}/c$. In this case the medium will get into hydrodynamic motion whose fixation bears experimental information about the properties of hadron or quark–gluon matter and also about the boundaries of mutual transition which, according to quantum electrodynamics, must proceed at energy density of the order of $\mathrm{GeV}/\mathrm{fm}^3$.

In these exotic conditions of the absence of direct means of diagnostics of compressed baryon matter properties, the same as for macroscopic electromagnetic plasma, dynamic methods are also used when the properties of the medium are judged by its hydrodynamic manifestations, i.e., by the dynamics of expansion of relativistic collision products, by the dissipative properties of the hydrodynamic flow, etc. [294, 295]. In the framework of this formalism it is assumed that colliding nuclei can be described as a continuous medium with corresponding EOS, whose motion obeys relativistic hydrodynamic equations. Such kind of representation of nuclear collisions is given in Fig. 1.21. The picture illustrates the elegant Stocker's idea [955] concerning generation of Mach shock waves whose properties show the characteristics of compressed nuclear matter.

Another striking example displaying manifestations of ultraextreme states is neutron stars [414].

The size of neutron stars is exceedingly small, and therefore it is fairly difficult to measure their parameters. One can measure here their mass by their motion in the composition of stellar structures [414]. Moreover, the size of neutron stars is determined by the shift of their emission lines. This information available for observation together with the radiation characteristics in different wavelengths is just the basis underlying rather sophisticated

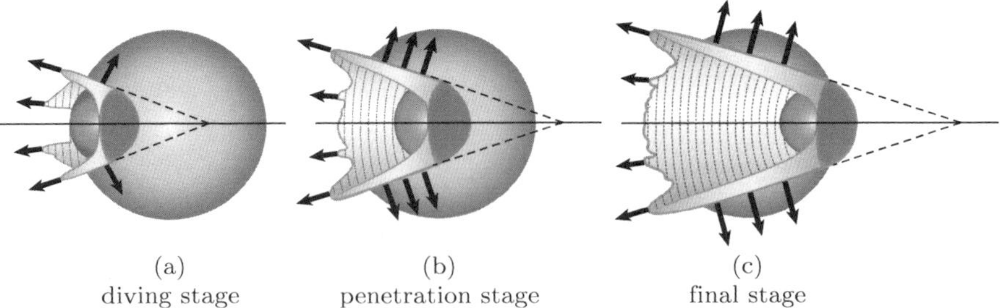



Fig. 1.21   Generation of Mach shock waves in nuclear matter [955] upon collision of a light nucleus (on the left) with a "heavy" one.

EOS models of these superextreme objects. We shall now pass over to a more detailed presentation of models of state in individual portions of phase diagrams [294, 295].

Along with radio pulsars, neutron stars are also sources [477] of powerful X-ray emission (X-ray pulsars), gamma, and X-ray bursts (magnetars), constant X-rays coming from the center of supernova remnants, and very weak optical stars. The study of neutron stars solves and suggests solutions for a number of fundamental problems of high energy density physics. This is in the first place examination of the EOS of superdense matter, $\rho > \rho_0$.

As it ascends the pressure and temperature scale, the substance (its composition, structure, properties, etc.) undergoes radical changes from an ideal gas of noninteracting neutral particles described comprehensively by the classical statistical Boltzmann physics to exotic forms of baryon and quark–gluon matter some of whose manifestations now only begin to be detected in record-parameter experiments on relativistic ion collisions [294, 295] and in interpretation of astronomical observations of so-called "compact" (neutron and "strange") stars [414].

In the latter case we speak of giant pressures $p \approx 10^{25}$ Mbar and $T \approx 10^9 - 10^{11}$ K experienced by our Universe in the course of its evolution within millionth fractions of a second after the Big Bang (Figs. 1.11, 1.14, 1.20 and 1.22) [294, 295]. At that moment the thermodynamic parameters of matter assumed the maximum values conceivable from our today's physical notions. These ultraextreme parameters of matter available for contemporary physical concepts are determined by the so-called Planck quantities [345, 754] which are combinations of fundamental constants, namely, Planck's constant $\hbar$, velocity of light $c$, gravitational constant $G$, and Boltzmann constant $k$:

- length $l_{\mathrm{P}} = \sqrt{\dfrac{\hbar G}{c^3}} = \dfrac{\hbar}{m_{\mathrm{P}} c} \approx 1.62 \cdot 10^{-33}$ cm;

- mass (the so-called "maximon" mass) $m_{\mathrm{P}} = \sqrt{\dfrac{\hbar c}{G}} = 2.18 \cdot 10^{-5}$ g;

- time $t_{\mathrm{P}} = \dfrac{l_{\mathrm{P}}}{c} = \dfrac{\hbar}{m_{\mathrm{P}} c^2} = \sqrt{\dfrac{\hbar G}{c^5}} = 5.39 \cdot 10^{-44}$ s;

- temperature $T_{\mathrm{P}} = \dfrac{m_{\mathrm{P}} c^2}{k} = \sqrt{\dfrac{\hbar c^5}{G k^2}} = 1.42 \cdot 10^{32}$ K;

- energy $W_{\mathrm{P}} = m_{\mathrm{P}} c^2 = \dfrac{\hbar}{t_{\mathrm{P}}} = \sqrt{\dfrac{\hbar c^5}{G}} = 1.96 \cdot 10^9$ J;

- density $\rho_{\mathrm{P}} = \dfrac{m_{\mathrm{P}}}{l_{\mathrm{P}}^3} = \dfrac{\hbar t_{\mathrm{P}}}{l_{\mathrm{P}}^5} = \dfrac{c^5}{\hbar G^2} = 5.16 \cdot 10^{93}$ g/cm$^3$;

- force $F_{\mathrm{P}} = \dfrac{W_{\mathrm{P}}}{l_{\mathrm{P}}} = \dfrac{\hbar}{l_{\mathrm{P}} t_{\mathrm{P}}} = \dfrac{c^4}{G} = 1.21 \cdot 10^{44}$ N;

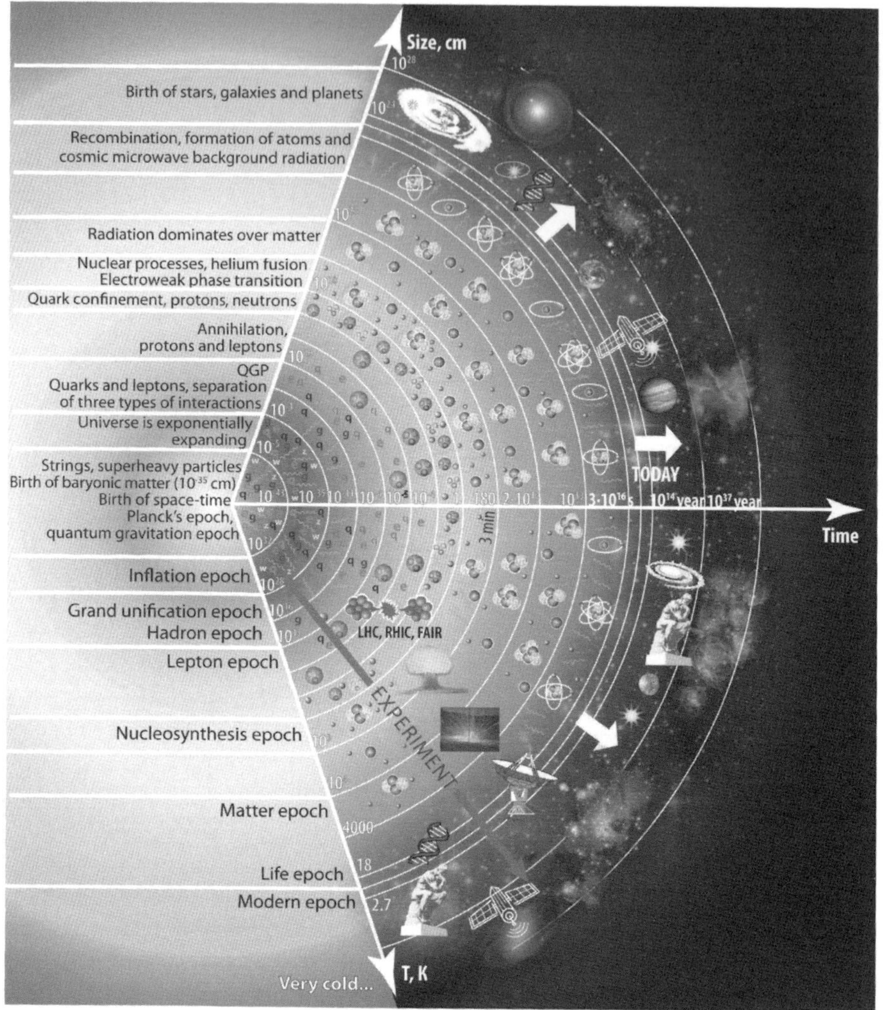


Fig. 1.22    Time scale after the Big Bang.

– pressure $p_P = \frac{F_P}{l_P^2} = \frac{\hbar}{l_P^3 t_P} = \frac{c^7}{\hbar G^2} = 4.63 \cdot 10^{113}\,\text{Pa};$

– charge $q_P = \sqrt{\hbar c 4\pi \varepsilon_0} = 1.78 \cdot 10^{-18}\,\text{C};$

– power $P_P = \frac{W_P}{t_P} = \frac{\hbar}{t_P^2} = \frac{c^5}{G} = 3.63 \cdot 10^{52}\,\text{W};$

– circular frequency $\omega_P = \sqrt{\frac{c^5}{\hbar G}} \approx 1.85 \cdot 10^{43}\,\text{s}^{-1};$

- electrical current $I_P = \frac{q_P}{t_P} = \sqrt{\frac{c^6 4\pi\varepsilon_0}{G}} = 3.48 \cdot 10^{25}$ A;

- voltage $V_P = \frac{W_P}{q_P} = \frac{\hbar}{t_P} = \sqrt{\frac{c^4}{G 4\pi\varepsilon_0}} = 1.05 \cdot 10^{27}$ V;

- impedance $Z_P = \frac{V_P}{I_P} = \frac{\hbar}{q_P^2} = \frac{1}{4\pi\varepsilon_0 c} = \frac{Z_0}{4\pi} = 29.98$ Ohm;

- electric field strength $E_P = \frac{V_P}{l_P} = \frac{1}{G}\sqrt{\frac{c^7}{4\pi\varepsilon_0 \hbar}} = 6.4 \cdot 10^{59}$ V/cm;

- magnetic field strength $H_P = \frac{1}{G}\sqrt{\frac{c^9 4\pi\varepsilon_0}{\hbar}} = 2.19 \cdot 10^{60}$ A/m $= 1.74 \cdot 10^{62}$ Oe.

Such superextreme matter parameters with which the known physical laws supposedly do not work might have existed at the beginning of the Big Bang or in the singularity upon black hole collapse. For this region, physical models are discussed assuming that our space has more than three spatial dimensions, and the ordinary matter resides in a three-dimensional manifold, in the "3-brane world" [866], imbedded into this multidimensional space.

The resource of present-day experiment in high-energy physics which is still very far from these Planck values make it possible to establish the properties of elementary particles at energies up to $0.1-10$ TeV and distances up to $10^{-16}$ cm. An important event was start up in CERN of the Large Hadron Collider (LHC) on which the range of teraelectronvolt energies became accessible.

If we do not overstep [294, 295] the limits of nonrelativistic (relative to nucleons) energies, $mc^2 \approx 1$ GeV, we arrive at the boundary temperature of $10^9$ eV, the energy density of $10^{37}$ erg/cm$^2$, and pressure of the order of $10^{25}$ Mbar, although at early nano- and picosecond stages of Universe evolution more extreme matter states possibly occurred in neutron and quark star cores. Therefore neutron and strange stars, as well as black holes are real astrophysical objects providing although indirect, but still pithy information on the thermodynamics of ultraextreme states needed for the construction of the EOS of matter.

Although very modest, our experimental potentialities allow us to "look" into this range of extreme astrophysical states (Figs. 1.1 and 1.22).

Obviously, the EOS of matter at the inflationary stage of the Universe, $p = -\rho c^2$, under "Planck" conditions is the most "extreme" EOS which can be formulated for the description of superextreme states of matter.

According to the Big Bang theory [294, 295] the instant of the Universe commencement is the era of the classical space–time origination. At the instant of Universe emergence the density $\rho$ and temperature $T$ of matter reached the Planck values $\rho_P \approx 10^{93}\,\mathrm{g/cm^3}$, $T_P \approx 1.3 \cdot 10^{32}\,\mathrm{K}$ when the gravitational interaction is comparable in strength with other (strong, weak, electromagnetic) interactions, and it should be considered with allowance for quantization.

The giant energies implemented by modern accelerators are certainly insufficient for reproduction in laboratory of the conditions corresponding to the range of "Grand Unification" parameters of the order of $10^{15}\,\mathrm{GeV}$ (Fig. 1.22). For this purpose, an accelerator should have the size of the Solar system. Since the instant of Big Bang the Universe has been continuously expanding, the temperature of matter falling and the volume increasing. In the description of the Universe origination the most general ideas about quantum evolution of the Universe are used. This means that the whole Universe can occur without energy consumption with the original geometric size of the order of Planck one.

In the course of expansion, the Universe borrows energy from the gravitational field to produce more matter. The positive energy of matter is exactly balanced by the negative gravitational energy, and so the resultant energy balance is equal to zero.

In $10^{-43} - 10^{-42}\,\mathrm{s}$ after the birth of the classical space–time the Universe inflation stage begins (Fig. 1.22) which lasts till the stage of Grand Unification, i.e., about $10^{-35}\,\mathrm{s}$, when relic radiation appears. Analysis of the inflation stage requires extrapolation of the known physical laws by 30 orders of magnitude on the spatial scale. Beyond this stage obviously lies the boundary of human philosophy and, as the outstanding ancient Christian theologian Tertullianus noticed, it is precisely on such a boundary that theology begins.

The inflationary stage is characterized by exceedingly high negative pressure, $p = -\rho c^2$, under which the very laws of ordinary gravitational physics change. Matter in this state is not a source of attraction. It is a source of repulsion. Possibly, a very early Universe was filled not only with particles, but also with the temporal form of dark energy (inflaton field) with very high density. It is this energy that induced the Universe expansion with great acceleration and then broke out with the formation of high-temperature plasma which then disintegrated into the usual matter and radiation. Only a weak trace of dark matter remained which became significant only in contemporary epoch. During the inflation stage the volume of the Universe increases many orders of magnitude (in some models about $10^{1000}$ times), and so the entire Universe appears to lie in one causally related region. Here, the kinetic energy of Universe expansion and its potential energy become equal. During this stage physical conditions appear which later lead to Hubble-law expansion of the Universe.

Since at this stage the interparticle distance variation rate is proportional to the distance itself and the expansion velocity $R \sim e^{bt}$, this stage is called inflationary, for under inflation the growth of monetary mass obeys the same law of exponential growth.

As distinct from explosion of chemicals, where the scatter of detonation products is related to the pressure gradient, in the case of Big Bang the source of expansion is the negative pressure in a homogeneous medium, i.e., antigravity that existed in the early Universe.

During inflation the gravitational repulsion forces drive away particles so that they further move inertially forming the Hubble expansion law.

Below we shall pass over to a more detailed analysis of different portions of the phase diagram of matter and their thermodynamic description.

# Chapter 2

# Equations of State of Gases and Liquids

Gaseous and liquid states of matter are of greatest importance in everyday life of people and occupy a rather wide portion of the phase diagram (Figs. 2.1 and 2.2) [656]. From a solid body these states of matter are separated by a clearly pronounced interface with melting which, according to paper [598], has no critical point. In the range of high temperatures and pressures, the gas and liquid border upon the plasma emerging as a result of thermal ionization ($T$-ionization) or pressure-induced ionization ($p$-ionization). The boundary is here rather conditional and depends strongly on the quantum-chemical specificity of a particular substance (its ionization potential, electron structure, atomic size, etc.).

The gaseous and liquid states of a substance are characterized by the maximum degree of disorder and the absence of rigidity. Even a small shear stress induces the motion of the liquid and gas that assume the shape of the vessel which they fill and resist a change in the volume rather than the shape [656].

The liquid–gas interface is even more conditional. It is pronounced for the states $T < T_c$, $p < p_c$ below critical. Above them the difference is absent, and the substance here is called a "dense fluid" whose properties will be our main concern in the present chapter.

## 2.1  Virial expansion

While for a low-density (ideal) gas the EOS was found as far back as early in the XVIII Century, the thermodynamically exhaustive description of the liquid state has traditionally been the most difficult task [78].

     *Thermodynamics and Equations of State for Matter*

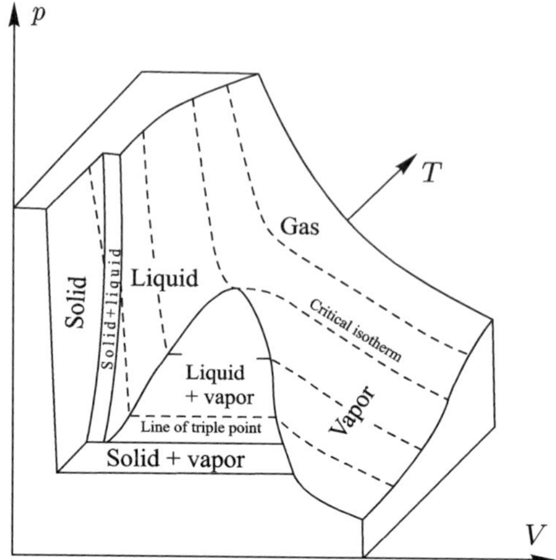


Fig. 2.1  Part of the surface in $(p,\,V,\,T)$ space, which represents the equation of state (EOS) of a pure substance with isotherms given by the dashed line. These regions are called according to the physical state of material.

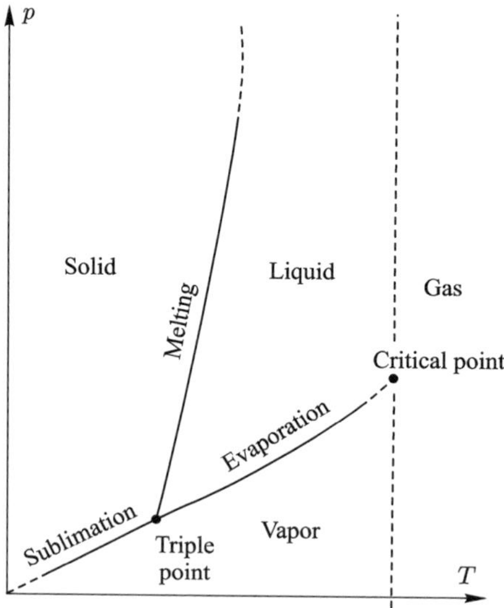


Fig. 2.2  Phase diagram of a pure substance obtained by projection of the $pVT$-surface onto the $(p,\,T)$ plane.

The point is that in a liquid the mean energy of interparticle interaction is comparable with the kinetic energy of particle motion. This fact, along with the structural disorder, makes impossible the use of the perturbation theory methods and obstructs the construction of qualitative models because of a small number of measurements. It is the advance in experimental equipment, as well as the appearance of computer-based parameter-free Monte Carlo and molecular dynamics methods together with the method of integral equations that determine the present stride in the description of thermodynamics and the structure of the liquid state of matter.

The model of an ideal nondegenerate gas is the first and most simple approximation applicable for low densities and temperatures when no processes of thermal dissociation and ionization proceed in the system and the interaction energy of particles is low compared to their kinetic energy. The most wide-spread way to allow for this interaction is to employ the Mayer group expansions [78, 671], which leads to the virial EOS [681]

$$p = \frac{NkT}{V} \sum_i \frac{N^i B_i(T)}{V^i},\tag{2.1}$$

where the virial coefficients $B_i(T)$ take into account the interaction of $i$ particles through the potential $\varphi_{ij}(r)$. The form and parameters of the potentials $\varphi_{ij}$ are selected from the condition of a consistent description of as diversified information as possible, i.e., the data on quantum-mechanical calculations of binding energy and on the diatomic molecular spectrum, as well as experiments with molecular beams and available thermophysical data.

The analysis of the results of such a description of the system carried out for noble gases (for example, with potentials "$\exp - 6$", "$12 - 6$"), allowed finding the numerical parameters of these potentials and their optimal form [966]. The short-range character of the indicated potentials provides convergence of the integrals in Bi, but the practical calculation of higher virial coefficients is restricted by the difficulties of computation of multiple integrals. The first seven virial coefficients for the simplest hard-sphere potential and several lower coefficients for more realistic interaction potentials have been calculated up to now [966]. In particular computations, the character of convergence of the expansion (2.1) worsens sharply with increasing density, which necessitates employing a considerable number of terms of the series (2.1) containing unknown higher virial coefficients. It should be emphasized that the very convergence of the virial series has only been proved for an exceedingly rarefied gas.

In dense schemes the effects of non-additivity of interparticle interaction also become significant, i.e., their account in the framework of the three-body Axilrod-Teller potential (see Ref. [966]),

$$\Delta\varphi_{123} = \nu(1 + 3\cos\theta_{12}\cos\theta_{13}\cos\theta_{23})(r_{12}r_{13}r_{23})^{-3},$$

changes the argon $B_3$ value at the critical point by 40%. These circumstances restrict the applicability limits of the virial EOS on the side of high gas densities. For this case numerous rather complicated empirical dependences were proposed that realize a detailed approximation of experimental data and are convenient for engineer thermodynamic calculations.

## 2.2   Short-range order. Integral equations

In strongly compressed gases the energy of interparticle interaction appears to be comparable with the kinetic energy of particle motion, which practically corresponds to the liquid state [78, 656, 671], although the division into a liquid and a gas is, of course, conditional. Such situations have in recent years been successfully described by physical models based on integral equations and direct numerical simulation (by Monte Carlo and molecular dynamics methods) of strongly compressed classical systems to be considered below (Chapter 5).

The absence of the long-range order typical of a solid body and a strong interparticle interaction obstruct greatly a theoretical description of the liquid state which has long been traditionally regarded as a difficult and the least elaborated area of statistical physics [598, 656]. The attempts to apply gas-related arguments to liquids turned out to be unavailing because of a great (2–3 orders of magnitude) difference in the densities of these phases, which calls for a considerable number of known virial coefficients in an expansion of the form (2.1). Even if the necessary coefficients could be calculated, the question of convergence of the series (2.1) for the densities corresponding to the liquid phase would remain open as the available proofs of virial series convergence refer to the limit $\rho \to 0$ [78, 966].

The density of a liquid differs from that of a crystal by 10–20%, which accounts for the relative success of solid-body approximations as applied to its description.

Being surrounded by neighbors, the atoms in a dense liquid find themselves localized and make chaotic quasi-oscillatory thermal motions around the part of fixed position with a characteristic frequency of $10^{12}$–$10^{13}$ Hz and at the same time change this quasi-fixed position in jumps. The indicated frequencies are close to the characteristic oscillation frequency of soli-body atoms or to frequencies of oscillatory excitations of molecules, and the characteristic times of averagings correspond to the nanosecond range.

This notion of liquid molecule localization in a small space region under the action of the mean field of the neighbors underlies the lattice model or the free volume model [78,248,656]. Naturally, such an approach is the most acceptable for the description of a strongly compressed low-temperature liquid whose kinetic energy makes up only a small fraction of the total energy, that is, under conditions that generally speaking do not correspond to the liquid state when these quantities are comparable. As a result, the lattice model is only applicable to states intimately close to the melting curve, the melting showing up in numerous defects of the crystal lattice as if strongly distorted in the liquid state [947].

The combination of solid and gaseous notions leads to the multistructural model [78, 248] according to which a real liquid is represented as an equilibrium gas–solid mixture whose properties are described in the language of the binary distribution function or by the statistical sum with the empirical parameter providing interpolation between a harmonic crystal and an ideal gas. Similar representations are used in the cell model ([248, 656], Chapter 3) according to which the whole liquid is divided into molecular cells filled according to the thermodynamic conditions. Such kind of models only suggest a schematic description of the properties of a real liquid, but thanks to their simplicity they can be effectively used as an element of semi-empirical models (Chapter 9). We also note that the present-day state of the methods of cell-model computations [195] made it possible to describe in an analytically unified way the properties of a solid and a liquid and the phenomenon of melting (Fig. 2.3).

Improvement of the experimental equipment for diffraction measuring [656, 966] allowed getting important information on the structure of liquid, which contributed to a deeper understanding of its properties. According to the notions based on the experiments on elastic scattering of X-rays and thermal neutrons, realized in the liquid is the short-range order described by the binary distribution function $g(r)$ determining (for a given interparticle interaction potential $\varphi(r)$) all equilibrium properties of the liquid state.

Figure 2.4 of Ref. [656] presents binary correlation functions of oxygen molecules in water (dashed line) and argon (solid curves). Despite the radical distinctions of these substances, their correlation functions have much in common. The clearly pronounced first peak and damping of $g(r)$

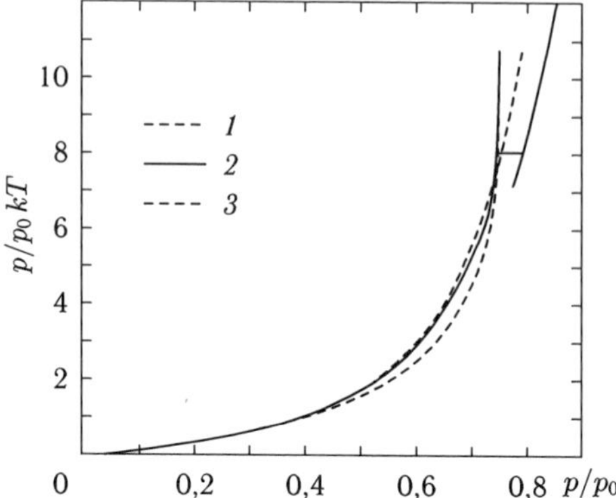

Fig. 2.3　EOS of a set of hard discs [195]. Calculation: *1* self-consistent cell model, *2* with allowance for cluster corrections, *3* Monte Carlo method.

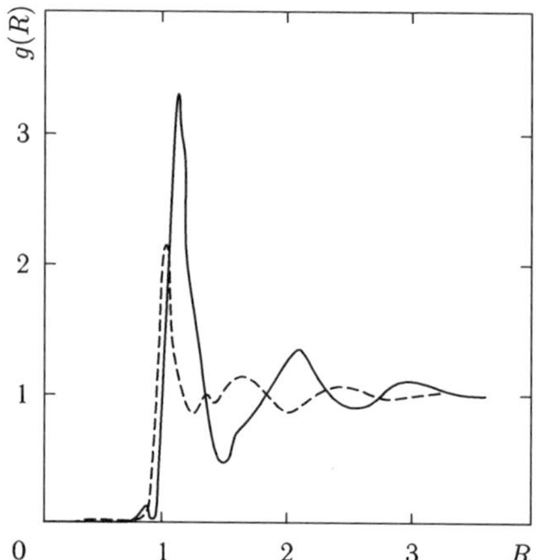

Fig. 2.4　The functions of radial distribution of oxygen atoms in water (dashed line) and argon (solid curves) near the freezing points and at the distance $R = r/\sigma$ scaled by the van der Waals diameter (2.82Å in water and 33.4Å in argon).

oscillations with increasing $r$ testify to the existence of short-range order and the loss of translation symmetry with increasing $r$. The amplitude of the first peak gives the number of nearest neighbors, namely, 10 for argon and approximately 4.5 for oxygen.

It is exceedingly important that in all liquids an "excluded volume" region exists for $0 < r < 1$, where $g \simeq 0$, occupied by a given molecule or atom and inaccessible for other atoms because of the strong interatomic repulsion induced by the Coulomb forces and by the uncertainty principle for electrons bound in atoms. These repulsive forces sharply grow with distance as $r \to 0$ and can be approximated by a hard wall. This leads to a very productive theory of hard incompressible spheres. It is precisely the existence of "excluded volume" that became the key to the current understanding of equilibrium, structural, and transport properties of the liquid state [656].

To predict the thermodynamic behavior of liquid depending on the character and details of intermolecular interaction the statistical theory of liquid state is solving the problem of finding the statistical sum of a many-body system through the so-called pair correlation function $g(r)$ of the system particles and the interaction potential $\varphi(r)$. The function $g(r)$ determines the probability of finding the centers of two particles at a mutual distance $r$ at a given temperature $T$ and particle density $n$ of the system as a whole. The function $g(r)$ describes the inner order in the system and its dependence on the thermodynamic conditions. Known the function $g(r)$, for monoatomic and one-component systems with pair interaction between particles the thermodynamic functions can be found from the well-known exact solutions of equations for pressure, total energy, and compressibility (see equations (2.5)–(2.7)).

Thus, for simple liquids the function $g(r)$ describes simultaneously their structure and thermodynamic properties.

In the theory of liquids, a special place is taken by approaches such as seeking approximate integral equations for $g(r)$ when the interaction potential $\varphi(r)$ is defined.

The statistical theory relates these characteristics through nonlinear integral equations based on different hypotheses concerning the relative role of short- and long-range interactions in the system [45, 78, 195, 656]. In turn, the parameter-free Monte Carlo and molecular dynamics methods and the recent direct measurements of these quantities in "dust" plasma experiments [313, 1012] permit, given the potential, an exact $g(r)$ computation and verification of applicability conditions of different integral equations.

The description of the liquid is thus determined by the choice of an adequate potential $\varphi(r)$ and the employed approximations can be controlled experimentally as the structural factor, i.e., the Fourier-transform of the binary distribution function

$$S(q) = 1 + n \int [g(r) - 1] e^{iqr} dr,$$

is an observable in diffraction X-ray or neutron measurements, and the experiments with "dust" plasma permit a direct measurement $g(r)$ through optical observations. The scheme of description turns out to be closed and comes down to an analysis of the effect of the potential's properties on the structural and equilibrium properties of a disordered system.

The ideological basis of the statistical description of liquids [78,656] is a chain of an infinite number of integro-differential Bogolyubov–Born–Green–Kirkwood–Yvonne (BBGKY) equations [656] relating $s$ and $s+1$ particle correlation functions determining the probability of $s$ particles to reside in the chosen coordinates $\vec{r}_s$ for a given potential $\varphi(\vec{r}_1, \vec{r}_2, \ldots \vec{r}_s)$:

$$-kT\nabla_1(\vec{r}_1, \vec{r}_2, \ldots \vec{r}_s) = g(\vec{r}_1, \vec{r}_2, \ldots \vec{r}_s)\nabla_1\varphi(\vec{r}_1, \vec{r}_2, \ldots \vec{r}_s)$$

$$+ \rho \int g(\vec{r}_1, \vec{r}_2, \ldots \vec{r}_s, \vec{r}_{s+1})\nabla_1\varphi(\vec{r}_1, \vec{r}_{s+1})d(\vec{r}_{s+1}).$$

$$(2.2)$$

Although exact in principle, this infinite chain of equations is cumbersome and of little use for real calculations. Therefore, to "break" this chain the so-called superposition approximation (or the Kirkwood hypothesis) is applied where the three-particle correlation function is represented as the product of two-particle functions [78, 277, 656]:

$$g(\vec{r}_1, \vec{r}_2, \vec{r}_3) = g(\vec{r}_1, \vec{r}_2) \cdot g(\vec{r}_2, \vec{r}_3) \cdot g(\vec{r}_1, \vec{r}_3). \qquad (2.3)$$

This hypothesis radically simplifies the system (2.2) and leads to a family of integro-differential equations for the binary correlations function $g(\vec{r}_1, \vec{r}_2)$ through which the whole thermodynamics of the system is then defined [656]. We shall write this family of equations in the Ornstein–Zernike form [656] for the symmetric potential $\varphi(\vec{r}) = \varphi(r)$:

$$h(r) = C(r) + \int dr' h(|r - r_1|)C(r'), \qquad (2.4)$$

$$\ln g(r) = -\frac{1}{kT}\varphi(r) + h(r) - C(r) + B(r), \qquad (2.5)$$

where $g(r) = h(r) + 1$. The so-called direct distribution function has the form

$$C(r) = \frac{1}{2\pi^2 n} \int_0^\infty \left[\frac{S(q) - 1}{S(q)}\right] q\sin(qr)dq. \qquad (2.6)$$

The approximation

$$B(r) = 0, \qquad (2.7)$$

corresponds to the "hyperchain" approximation.

The Percus–Yevick approximation corresponds to the approximation

$$B(r) = \ln(1 + C(r) - h(r)) - C(r) + h(r). \tag{2.8}$$

The experimentally established fact of weak temperature dependence of the structure of liquid for a fixed density became decisive for creation of modern models of liquid [78, 966]. Similarity of the structures of liquid metals, ion systems, and dielectrics having essentially different attractive forces in turn testifies to an insignificant role of these forces in the formation of equilibrium forces of liquid. These circumstances are indicative of the dominant contribution of the repulsive part of potentials, whereas the influence of the attractive forces and allowance for the temperature effects leads in a dense liquid to corrections which can be taken into account in perturbation theory [78], on the basis of the variational method [651] or in the chaotic phase approximation [45].

For a model description of the repulsive forces it is in turn reasonable to use the simplest potential of hard spheres of radius $\sigma$

$$\varphi(r) = \begin{cases} \infty, & r \le \sigma_2, \\ 0, & r > \sigma, \end{cases} \tag{2.9}$$

characterizing the ultimate state of a strongly compressed and heated liquid. Being the simplest nontrivial liquid-state model the hard-sphere model became a fairly popular object of theoretical analysis for the application of both the technique of integral equations and the numerical methods and has presently been studied most thoroughly [45, 195].

Beginning with the pioneering works [94, 95, 895] it became clear that maximally simplifying the interparticle interaction and reducing the repulsion to an infinite potential at the point of contact the hard sphere model is capable of describing the geometrical specificities of collective behavior in a dense liquid which appeared to be largely responsible for the structural properties and the phase boundary of the liquid state [656].

This model provides the existence of an "excluded volume" (Fig. 2.4), i.e., the volume which cannot already be occupied by the other particles of the liquid. Indeed, a given volume $V$ can only contain a finite number of spherical atoms of radius $\sigma$ whose volume is equal to $V_c$. For a hexagonal close-packed structure and a cubic face-centered structure we have $V_c = \sigma^3/\sqrt{2}$. The ratio of the sphere volume $V_c$ to the total volume $V$ for random packing equals 0.64, which is by 14% less than the packing density equal to 0.7405 [656].

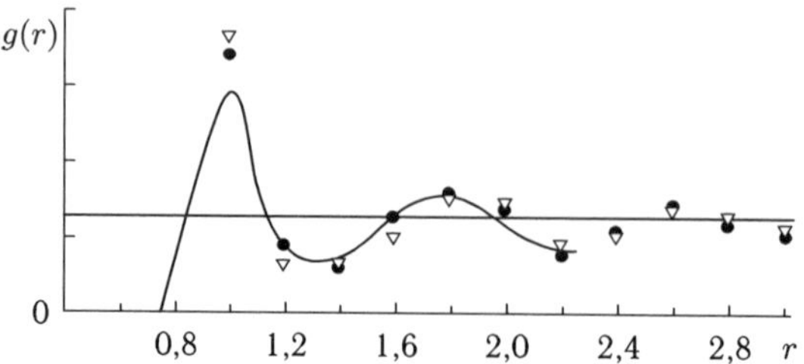

Fig. 2.5   Radial distribution function $g(r)$ depending on the reduced distance $r/\sigma$ for the random packing model [94, 95] (•) and [895] ($\nabla$) with respect to the schematic graph of diffraction data on argon type liquids. For a hard-sphere liquid $g(r) = 0$ for $r/\sigma < 1$ [961].

In Fig. 2.5 the structural data of the models [94, 95, 895] with allowance for thermal effects are compared with argon. The hard-sphere models are seen to reproduce the effect of the "excluded" volume and the short-range order in a dense liquid.

It is of importance that for the potential (2.9) an exact analytical solution of the Percus–Yewick equation was obtained which is the best integral equation for the description of systems with short-range repulsion and is widely used to calculate the properties of real liquids. The existence in the solution of a free parameter, i.e., the packing density $\eta = 4\pi\sigma^2 N/3V$, permits, applying different ways of its definition, a successful description of structural characteristics of simple liquids and liquid metals [45, 78, 195, 656].

The calculations by different methods of systems with potential (2.9) are presently accomplished in the entire range of the liquid state and can be best of all represented analytically by the Carnahan-starling Pade approximation [656]:

$$\frac{F_{\text{TC}}}{NkT} = \frac{4\eta - 3\eta^2}{(1 - \eta)^2},$$

$$\frac{p_{\text{TC}} V}{NkT} = \frac{1 + \eta + \eta^2 + \eta^3}{(1 - \eta)^3}. \tag{2.10}$$

The character of the data description by the approximation (2.10) is shown in Fig. 2.6.

It is interesting that the numerical calculations by the model (2.9) for densities similar to the close packing of spheres, $\eta = \frac{4\pi\delta^3 n}{3} \lesssim 1$, exhibited the presence of structure anomalies and the tendency of the appearance of long-range order in the system, which is associated [834] with crystallization having the geometrical character in this case.

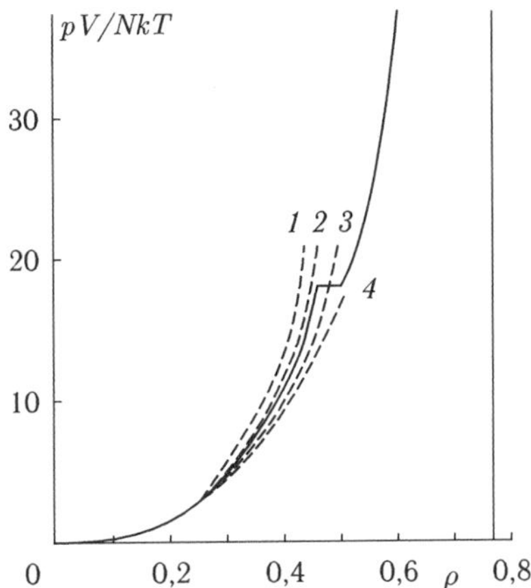



Fig. 2.6  Models for the liquid branch of the hard-sphere system isotherm. The solid line is the result of data simulation; it is well reproduced by the Carnahan–Starling EOS. The dashed line *2* gives the Percus–Yevick EOS of compressibility. The other dashed lines are alternative estimates of the liquid isotherm [656].

The occurrence of "hardening" phase transition in the hard-sphere model of liquid was the most surprising result of the early calculations of an ensemble of solids by the molecular dynamics method [17]. Inasmuch as the internal energy in such a system is purely kinetic, such a transition is of entropic character. The corresponding jump of the volume in the molecular-dynamic computations makes up 5%. The melting curve in the hard-sphere model has a positive slope $dp/dT > 0$ and the entropy jump is $\Delta S = $ constant, which is typical of the subsequent measurements of the entropy jump by the optical methods in the megabar pressure range [575].

We postpone the description of computer-calculational methods of thermodynamics of ensembles of many bodies till Chapter 5, where we shall consider the current quantum-mechanical versions of these approaches. Here we present the particle trajectories [656] computed by the molecular dynamics method for the solid (Fig. 2.7(a)) and liquid (Fig. 2.7(b)) states that visually illustrate the character of particle motion in these phases.

Further progress in the description of the liquid-phase properties was achieved using the soft-sphere model that exploits the repulsive power-law potential

$$\varphi(r) = \begin{cases} \varepsilon \left(\dfrac{\sigma}{r}\right)^{n}, & r \leq \sigma, \\[2mm] 0, & r > \sigma, \end{cases} \qquad (2.11)$$

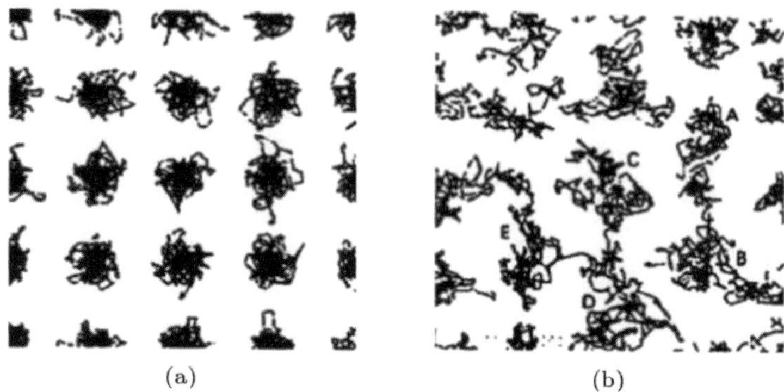

(a)　　　　　　　　　　　　　　(b)

Fig. 2.7　Comparative simulation of the behavior of 32 hard spheres: the behavior of particle trajectories in (a) is in contrast with their behavior in a liquid (b). Pay attention to the particle exchange in (b): A and B, C, D, and E [961].

which also participated in calculations of the equilibrium properties by Monte Carlo and molecular dynamics methods [78,422,445,446]. The results are described by the approximating formula

$$\frac{F_{\mathrm{MC}} - F_0}{Nk_{\mathrm{B}}T} = \frac{1}{6}(n + 4)\eta^{n/9}\left(\frac{\varepsilon}{k_{\mathrm{B}}T}\right)^{\frac{1}{3}},$$

the power $n$ of the potential being an additional free parameter of the model. Similarly to the hard-sphere potential, the repulsive power-law potential leads to the potential for melting [445, 446] whose characteristics depend strongly on the $n$ value. Note in this connection that the numerical simulation of the properties of real systems with the use of more complicated Lennard–Jones type potentials [422], step and rectangular-well potentials [1063, 1064] provided obtaining melting and evaporation curves and a qualitative description of isostructural electron and polymorphic phase transitions in a solid (see Chapter 8).

In what follows, in Chapters 4 and 9, we shall use the "soft"-sphere model to describe the properties of strongly compressed plasma of condensed densities.

The combination of the hard-sphere model (2.9) and the mean field approximation underlies the modified liquid-state model [656,1062] in which free parameters are found from the experimental values of sublimation energy and the position of the first peak of the liquid–metal structural factor at the melting point. This liquid-state model, the same as its predecessor the Van der Waals model, is qualitative and does not yield a detailed description of fine characteristics of the liquid. Coincidence of the

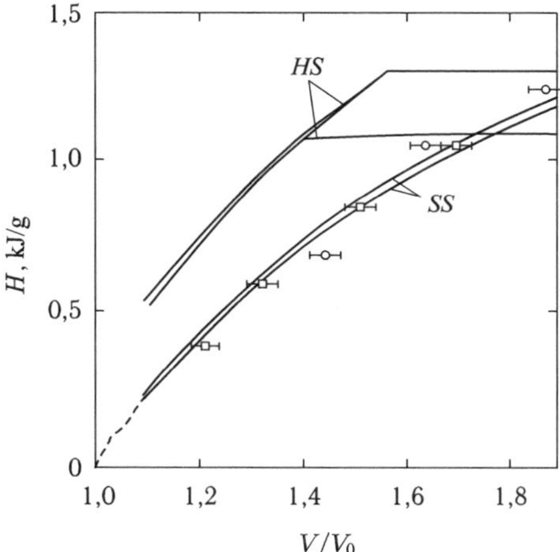


Fig. 2.8  Description of experiments on pressure-induced electric explosion of uranium wires using the hard-sphere (HS) and soft-sphere (SS) model [341, 1061]. Symbols designate experiment, dashed line: statistical data, solid lines: calculations.

theory with experiments can be considerably improved by applying the soft-sphere model (2.11) and introducing additional adjustable parameter into the mean field model [340, 341, 1061]. Such an approximation is exemplified in Fig. 2.8, where the parameters of the EOS were chosen using the results of experiments on pressure-induced electric explosion of uranium wires and the normal-pressure uranium properties. The modified Van der Waals models [340, 1062] are equally applicable for calculation of thermodynamic properties of the dense gas phase and show the asymptotics of an ideal gas, which enables their successful use in the construction of broad-range semi-empirical EOS (Chapter 9).

Despite the exceeding simplifications, the hard- and soft-sphere models take a correct account of the basic qualitative specificities of many-body interactions in a liquid and therefore satisfactorily describe the thermodynamic and structural properties of the liquid phase. These models can thereby be used as zero approximation with a subsequent employment of perturbation theory [45, 195, 656] for account of the details of the real interparticle interaction potential.

The details of the real interparticle interaction potential were taken into account within the thermodynamic perturbation theory, chaotic phase approximation, or by the variational method of perturbation theory. The variational method turns out to be convenient for calculation of the thermodynamics of particular liquids in a wide range of parameters; it is especially effective in the description of the properties of a dense liquid phase. In finding the Helmholtz free energy $F(V,T)$ of a system of particles with an arbitrary interaction potential $\varphi(r)$, this method makes use of the Gibbs-Bogolyubov inequality:

$$F \leqslant F_0(\lambda) + \frac{n}{2} \int [\varphi(r) - \varphi_0(r)] g_0(r, \lambda) d^3 r,$$

where $F_0$ and $g_0$ are the free energy and the pair correlation function of the original system with the interparticle interaction potential $\varphi_0(r)$. The EOS of an arbitrary system is determined by perturbation averaging over the structural characteristics $\lambda$ of the initial approximation according to the condition $(\partial F / \partial \lambda)_{V,T} = 0$. As zero approximation it is appropriate to choose a system of hard spheres for which analytical expressions exist that determine the free energy and the pair correlation function. This makes it possible to calculate in an explicit form the thermodynamic characteristics for the case of arbitrary interparticle interaction potential. The hard-sphere radius $\sigma$ serves in this case as a minimizing parameter $\lambda$. However, for high density the use of the hard-sphere model as original is fraught with considerable errors because of a too strong particle repulsion at small distances. More realistic is here the choice of the "soft" repulsive power-law potential, typically $\varphi(r) = C/r^{12}$, as the zero-approximation model [656]. The use of the solid-body or soft-sphere models as the original approximation allows obtaining to high accuracy the EOS of liquid inert or molecular gases and simple metals.

The development of the variational method of perturbation theory for computation of the EOS of liquid metals has led to the employment as zero approximation of the results obtained by the Monte Carlo method for one-component plasma (OCP). The (OCP) model is based on numerical simulation of the simplest system of point charges with Coulomb interaction, placed for steadiness against a uniform background of compensating charge of the opposite sign. Analytical expressions for $F_0$ exist in this case, the minimizing quantity being the Coulomb nonideality parameter ($\Gamma = e/kTr_{\mathrm{WZ}}$) with the mean interparticle distance $r_{\mathrm{WZ}} = (3/4\pi n)^{1/3}$.

Comparison of computations of the EOS made with the use of the systems of hard or soft spheres and OCP as zero approximation has shown [656] that for liquids with weakly pronounced repulsion (alkali metals, salt melts, alloys) the most acceptable is the one-component plasma model. It should be stressed that in the final expression of the EOS of liquid metals, apart from the configuration term defined by the variational method, summands

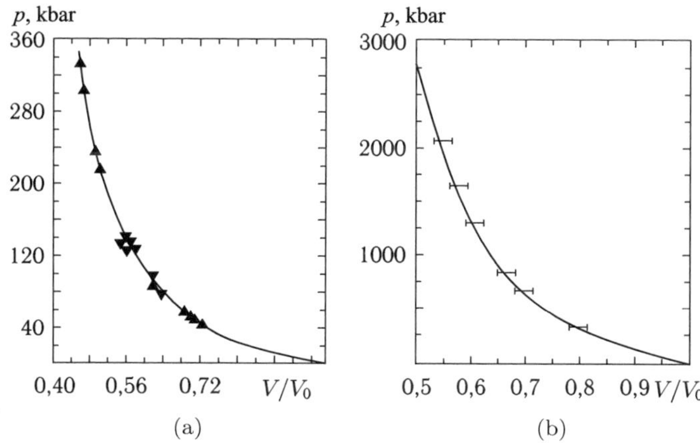

Fig. 2.9 Shock adiabats of sodium (a) and aluminum (b) [195]. Solid lines are calculations by the variational method of perturbation theory and symbols stand for experiment.

should be added to the energy allowing for the Coulomb interaction energy and for the kinetic energy of particles in the system, and the resultant value of these summands becomes substantial at high temperature.

The variational principle of perturbation theory which had first been applied in the description of the simple-liquid properties under near-critical conditions [651] was used on the basis of the hard-sphere system to calculate the thermodynamics of liquid simple metals and alloys [56], OCP [341,1061], and dense partially ionized plasma [858].

This way of description permitted successful calculations of the characteristics of the liquid phase of metals from the normal-pressure state to the state of high-density completely ionized plasma. Figure 2.9 presents the results of shock-wave experiments and calculated shock adiabats of sodium and aluminum. Note that calculations by the variational method not only describe the experimental data, but also demonstrate good coincidence with the results of more rigorous theoretical models in the experimentally unexplored range of superhigh pressures and temperatures [858].

The method was later extended to the case of the use as zero approximation of the soft-sphere model more realistic for substantial compression [78], which made possible to describe successfully the property of liquid gases, namely, xenon [862], argon [859], hydrogen, deuterium [1066], and helium [1065]. The position of melting curves of these elements was determined [1065,1066] by calculations of the thermodynamic characteristics of

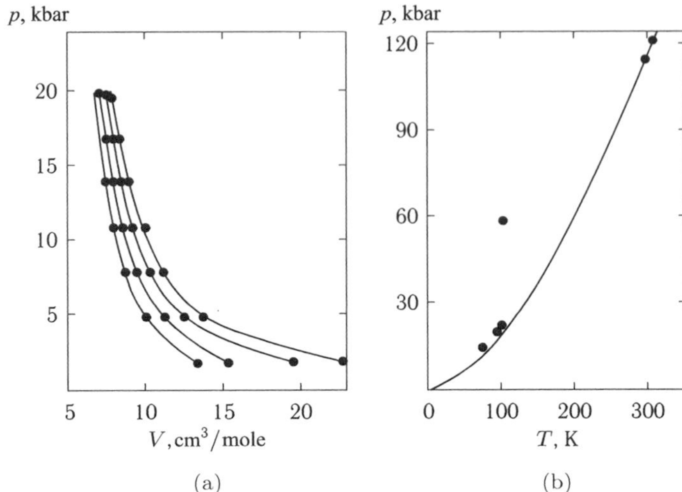

Fig. 2.10   EOS of the liquid phase (a) and melting curve (b) of helium [1065]. The experimental data are given by solid lines (a) and symbols (b); temperatures on the isotherms are equal to 75, 150, 225, and 300 K.

the liquid phase of hydrogen, deuterium, and helium within the variational method and the EOS of the solid phase in quasi-harmonic approximation. In the liquid phase the calculations allowed for quantum corrections and in the solid phase for the anharmonism effects. Figure 2.10 gives comparison of the theoretical and experimental data for helium in the liquid phase and on the melting curve [1065]. The figure demonstrates good agreement between calculations and experiments in the entire investigated range; the case of hydrogen and deuterium is characterized by a similar accuracy of the description [1066].

The development of variational method of perturbation theory for calculation of liquid metal characteristics has led to the use of the results obtained by the Monte Carlo method for OCP as the original approximation of perturbation theory. Comparison of the calculations for different versions of the variational method developed on the basis of the systems of hard and soft spheres and OCP has shown [861] that for simulation of lithium properties the best is the description using the characteristics of OCP. The results of an analogous comparison [707] between the hard-sphere system and one-component plasma as zero approximations in simulation of thermodynamics of liquid sodium and aluminum testify to the fact that more preferable for an optimum description of sodium is also the OCP

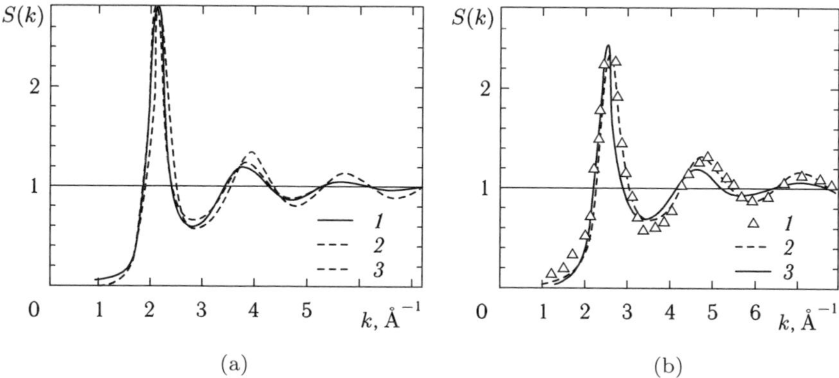

Fig. 2.11   Structural factor of liquid sodium (a) and aluminum (b) [707]. (a) $T = 173\,\mathrm{K}$, experimental data (experiment *1*) and calculation by the variational method on the basis of hard spheres (HS, $\eta = 0.47$ *2*) and one-component plasma (OCP, $\Gamma = 156$ *3*); (b) $T = 933\,\mathrm{K}$, experiment *1*, HS ($\eta = 0.44$) *2*, OCP ($\Gamma = 119$) *3*.

model, whereas for aluminum the best agreement with experiment is given by calculations within the hard-sphere model (Fig. 2.11). This is explained by high rigidity of the aluminum ion core, which is generally inherent in multi-valence metals. The results [707] nevertheless confirm that for soft repulsion systems (alkali metals, salt melts, alloys) more acceptable for the use of variational method of perturbation theory is the OCP model.

The application of the apparatus of perturbation theory in calculation of the properties of the dense liquid phase is considerably obstructed by nonadditivity of the interaction forces and the necessity of allowing for many-body interaction. Since in most cases the exact form of interaction potential is not known in advance, the authors of Refs. [520–522] proposed a peculiar form of perturbation theory exploiting as zero approximation the cold curve of a solid allowing efficiently for nonadditive forces. The energy of the liquid will then be expressed in terms of the variables characterizing the short-range order and the averaging is made over the cold curve taken either from experiment or from computer simulation. The thermodynamic and structural characteristics of liquid for different forms of interaction potentials calculated by the model of corrected hard spheres [520–522] showed good agreement with calculations by Monte Carlo and molecular dynamics methods. The model turned out also effective in the description of thermodynamic properties of real liquids in a wide range of parameters and yields results agreeing with experiment at low [524] and high [216] pressures (Fig. 2.12). Interestingly, the propriety of such an approach [520–522]

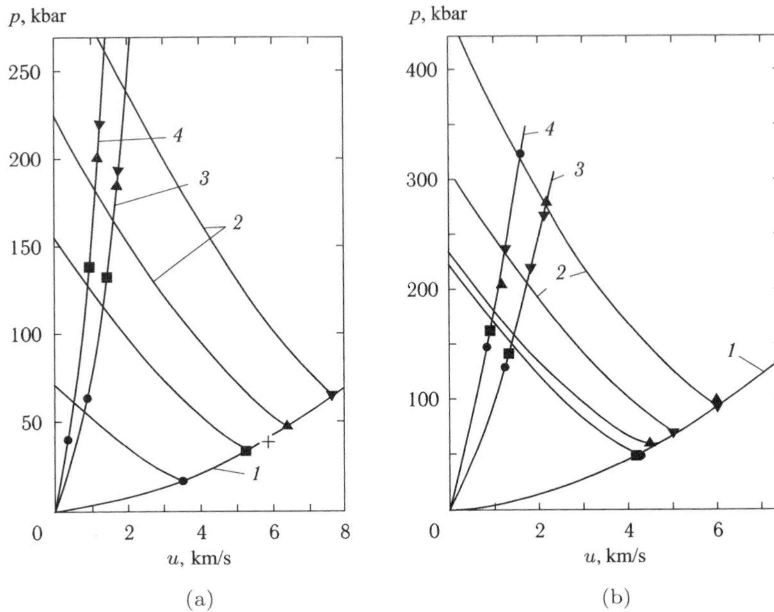

Fig. 2.12  Shock compression of liquid hydrogen (a) and deuterium (b) [216]. *1, 2* calculation of normal and reflected shock waves by the corrected hard sphere model [1010]; *3, 4* shock adiabats of magnesium and aluminum reflecting screens. Symbols stand for experiment.

was independently enunciated by experimental results since the EOS of the liquid and solid phases of alkali metals deduced from experimental data give coincident curves of the ground state [951].

Along with the direct problem of statistical fluid mechanics, i.e., calculation of its equilibrium properties from the defined interparticle interaction potential $\varphi(r)$, the inverse problem was also considered where experimental data with the structural factor $S(q)$ are applied to find an adequate potential [9]. The problem of seeking the potential from the data of diffraction measurements for inert gases and liquid metals is solved by the methods of molecular dynamics and integral equations, the integral approximation giving better results [9].

We have seen that the method of integral equations is attractive for two reasons, first, the possibility of obtaining analytical results regarding the structure and thermodynamic properties of simple liquids and, second, the possibility of solving the inverse problem of reconstruction of the form of intermolecular potential $U(r)$ by the known function $g(r)$. Such an equation has been sought since 1935 when Born, Kirkwood, and Yvonne first proposed a simplified version of integral relation between the space correlation functions $g(r)$ and $\varphi(r)$. Several such approximate equations of different degrees of precision are known at the present time [656]. To retrieve the interparticle interaction pair potential by the known pair correlation function within the method of integral equations, the formal relation

$$g(r, T, n) = e^{-\varphi\left(\frac{r}{T + \omega(r, T, n)}\right)}, \tag{2.12}$$

is used which implies

$$U(r) = -T[\ln g(r, T, n) - \omega(r, T, n)]. \tag{2.13}$$

Thence, if the function $g(r)$ is known and the way exists to calculate the second function $\omega$, the pair potential can be found.

In a rarefied gaseous system the probability to reveal two particles at a distance $r$ between them is proportional to the factor $\exp(-U(r)/T)$. In this case we have $\omega = 0$. When the density of particles increases and the interaction between them becomes substantial, the function $\omega$ is nonzero. In the hyperchain (HC) approximation, used most frequently compared to other approximations, the function $\omega$ is represented in the form $\omega(r) = g(r) - l - c(r)$, where $c(r)$ is a direct correlation function. Another widely used approximation is the Percus–Yevick approximation in which $\exp(\omega(r)) = g(r) - c(r)$. Now widespread is also the approximation proposed by Martynov and Sarkisov: $\omega(r) = g(r) - 1 - c(r) + \alpha\omega^2$ with $\alpha$ as a certain constant. The results obtained by solving the integral equations with such approximations are in some cases in good agreement with the results of numerical simulation of Lennard–Jones systems for hard spheres. On the whole, no unambiguous conclusion now exists in favor of one or another local approximation, and different types of them are used in the literature [9, 199, 332, 656, 666, 880].

The analysis of errors showed however that the inverse problem of finding $\varphi(r)$ from $g(r)$ is incorrect since obtaining a 10% error in $\varphi(r)$ will require the error in the original experimental information on $S(q)$ not above 1%. To increase the precision in seeking the interparticle interaction potential, apart from the structural factor additional information on the EOS and the equilibrium properties of liquid is also used. The characteristics of the potential can as well be found from thermodynamic information alone without the use of diffraction measurement data. Account of the experiments on shock compression of argon and the experiments on molecular beam scattering made it possible [859], for instance, to choose the type and

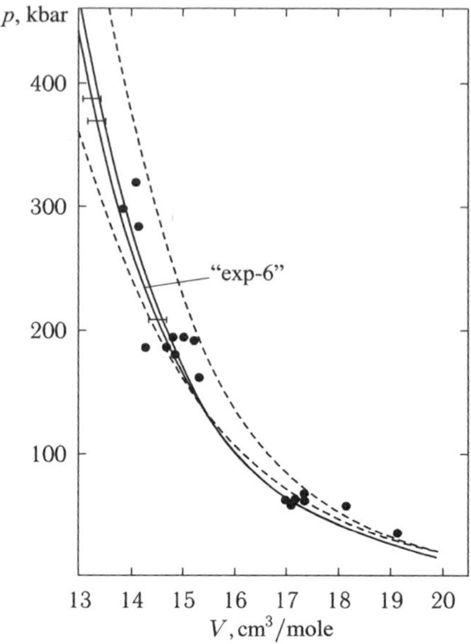



Fig. 2.13 Shock compression of liquid argon [859]. Symbols stand for experiment, curves show the calculation of shock adiabat with different types of potentials (the best description is given by the potential "exp-6").

determine the parameters of the potential describing the experimental data in the best possible way (Fig. 2.13).

## 2.3 Dust plasma for fluid theory

A new interesting trend in statistical physics of gases, fluids, and solids has recently come forth. It is the use of dust plasma experiments for investigation of equilibrium and transport properties of an ensemble of strongly interacting (correlated) particles [1012].

Dust plasma is a partially ionized gas containing negatively charged macroscopic size $(5\text{--}100\,\mu\text{m})$ dust particles. In dust plasma, micron dust particles acquire a considerable negative charge $(q \approx 10^3\text{--}10^5 e)$ and can form quasi-stationary dust plasma structures like a liquid or a solid [181, 316, 626, 682, 972].

Possessing a whole number of unique properties [1012] dust plasma is a good experimental model both for the study of the properties of strongly nonideal systems and from the viewpoint of a deeper understanding of substance self-organization phenomena and processes in nature.

The experimental examination of dust plasma plays an important role in verification of the existing and development of new phenomenological models for strongly nonideal fluid systems. Such models are of great significance because owing to a strong interparticle interaction the fluid theory has no small parameter to be used for an analytical description of the state of its structure and thermodynamic characteristics as is possible in the case of gases.

A unique property of dust plasma is the fact that its heavy particles (dust) scatter visible light and are therefore well seen in the optical range by optical TV cameras or sometimes by the human eye (Fig. 1.3). As distinct from the classical X-ray or neutron methods of finding the binary correlation function $g_2(r)$, this makes it possible to determine by direct observation (at the "kinetic" level) the necessary structural, thermodynamic, and kinetic properties of a system of any order $g_s(r_1, r_2, \ldots, r_s)$, i.e., temperature, pressure, self-correlation functions of EOS, transport factors, diffusion, and thermal conductivity and thus to verify the basic postulates of statistical physics, for instance, the BBGKY equations (2.2), the Kirkwood reciprocity hypotheses (2.3), the Kubo relation, and the integral equations (2.4)–(2.8).

We shall begin with the classical problem of retrieving the interparticle interaction potential from measured structure characteristics of plasma liquid [298]. The experiments were conducted on a high-frequency capacitive discharge setup (Fig. 2.14 [298]). In the course of experiment the vacuum

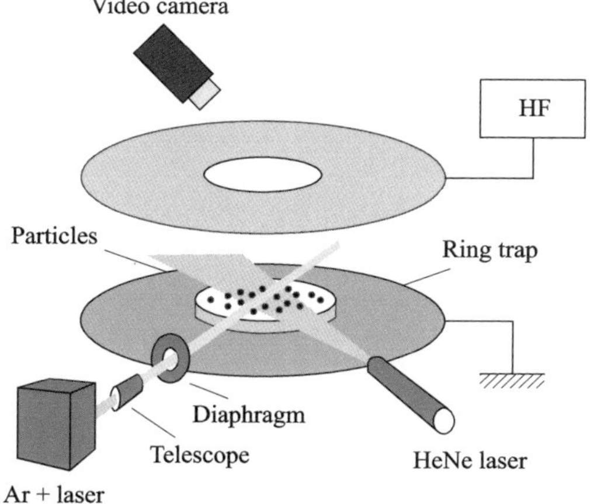


Fig. 2.14   Scheme of experimental setup.

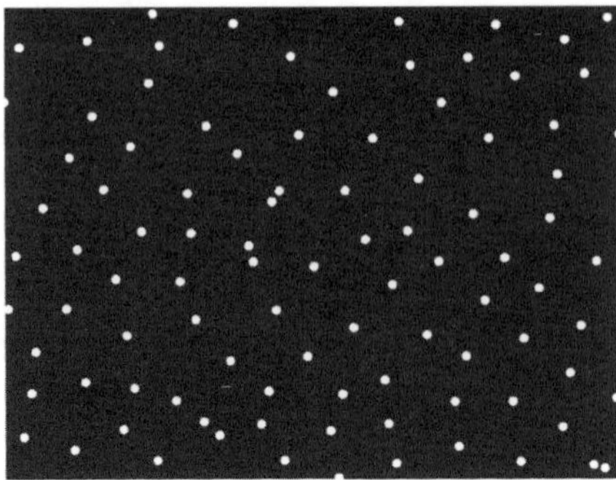

Fig. 2.15    Typical dust structure obtained in experiments.

chamber was filled with argon at a pressure $p = 20-30\,\mathrm{Pa}$, voltage from a high-frequency generator with carrier frequency of $13.56\,\mathrm{MHz}$ was applied to the electrodes generating a glow discharge between them in the argon atmosphere.

Through the opening in the upper electrode, dust particles were thrown in the discharge, which were either plastic (MF) transparent spheres $1.9\,\mu\mathrm{m}$ in diameter or polydisperse ($\mathrm{Al_2O_3}$) spheres $5\,\mu\mathrm{m}$ in diameter. On getting into the discharge and acquiring a negative charge the particles hovered in the near-electrode layer. As a result, the horizontal cross section of the dust cluster was accessible for observation. The position of dust particles was recorder by a video camera (Fig. 2.14).

The typical picture of experimentally obtained dust structures is demonstrated in Fig. 2.15.

The video recording was processed with a special program providing identification of positions of individual particles in the field of vision of the video camera and construction of a binary and a ternary correlation functions for each dust plasma structure. The error in the correlation function measurement is determined by the expression $\delta_N = 1/\sqrt{N}$, where $N$ is the number in the ensemble over which the averaging is taken. In our experiments the error is below 10%.

The potential was reconstructed by the known correlation function $g(r)$ on the basis of the fluid-theory integral equations (2.4)–(2.8) [656]. The basic relation defining the potential $U(r)$ is equation (2.13). The function $\omega$ in formula (2.13) depends on the function $\gamma$:

$$\gamma(r) = h(r) - c(r) + B(r). \tag{2.14}$$

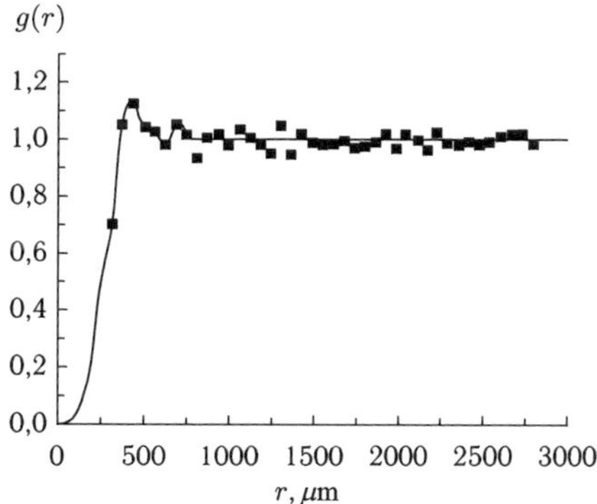

Fig. 2.16  Correlation function for $n = 200\,\mathrm{cm}^{-3}$. Dots correspond to the measurements of [298]. Solid line is the correlation function used in the calculations.

Equation (2.14) involves the difference between the direct correlation function $c(r)$ and the function $h(r) = g(r) - 1$, $B(r)$ is a bridge-functional representing infinite series of irreducible diagrams [665, 666]. The infinite series of the bridge-functional cannot be summed up, and the task of the theory is to seek adequate and physically grounded approximations of the bridge-functional. Passing over to approximate equations is based on replacement of the nonlocal bridge-functional $B(r)$ by the local bridge-function $B(h(r))$ or $B(\gamma(r))$, $B(\omega(r))$. No strict theoretical arguments now give grounds for the possibility of such a replacement.

To calculate the parameters of the potential, two measured correlation functions were chosen which are presented in Figs. 2.16 and 2.17. These correlation functions are seen to have a pronounced first peak and further to oscillate about the theoretical value $g(r) \to 1$.

In paper [298] the potential energy found from the measured correlation functions was approximated by the expression for a screened Coulomb potential with an arbitrary charge and screening radius $R$:

$$\varphi(Z, R) = \frac{(Ze)^2}{r} e^{-r/R}.$$

The parameters $Z$ and $R$ are found from the best correspondence condition of the potentials $U$ and $\varphi$. The potential $U$ was calculated by formulas (2.4)–(2.7), (2.13) and (2.14).

For the correlation function of Fig. 2.16 the dust particle density is low and the gas case is realized for the potential $U(r) = -T \ln g(r)$. The symbols in Fig. 2.19 correspond to the thus found dependence. The solid line of Fig. 2.18 constructed in this way leads to the values $Z = 490$

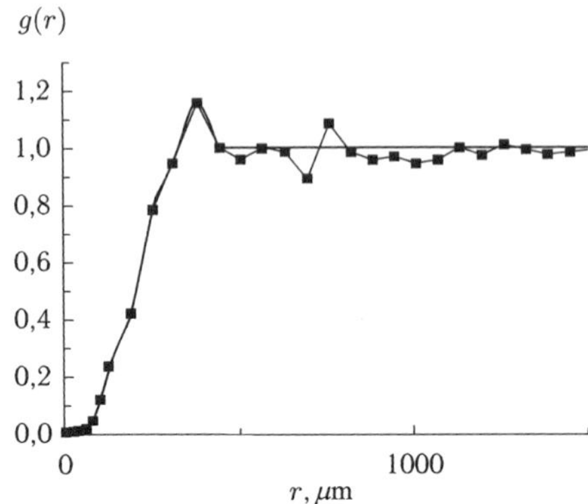

Fig. 2.17   Correlation function for $n = 3 \times 10^4 \, \text{cm}^{-3}$. Dots correspond to the measurements of [298]. Solid line is the correlation function used in the calculations.

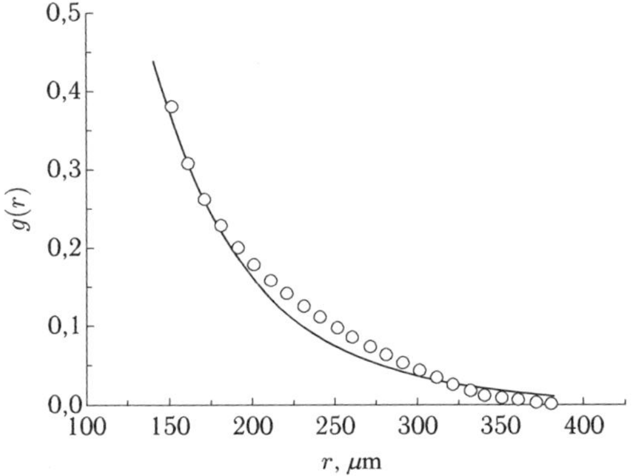

Fig. 2.18   Circles correspond to the dependence $U(r) = -T \ln g(r)$, solid line to the screened potential with $Z = 490$ and $R = 87 \, \mu$m.

and $R = 87 \, \mu$m. Note that the obtained screening radius value is close to the value $R \approx 100$ of the electron Debye radius estimated from the discharge parameters for $n_e \approx 10^9 \, \text{cm}^{-3}$, $T_d \approx 0.3 \, \text{eV}$. The Coulomb non-ideality parameter estimated by the average interparticle distance and by the established charge value $\Gamma = (Ze)^2 n^{1/3}/T \approx 2$. Under these conditions, however, the average interparticle distance $r \approx 1700 \, \mu$m exceeds greatly the Debye radius found above, and therefore the dust particle interaction at average distances is considerably weakened by screening. For the second correlation function (Fig. 2.17) the density correction to the potential

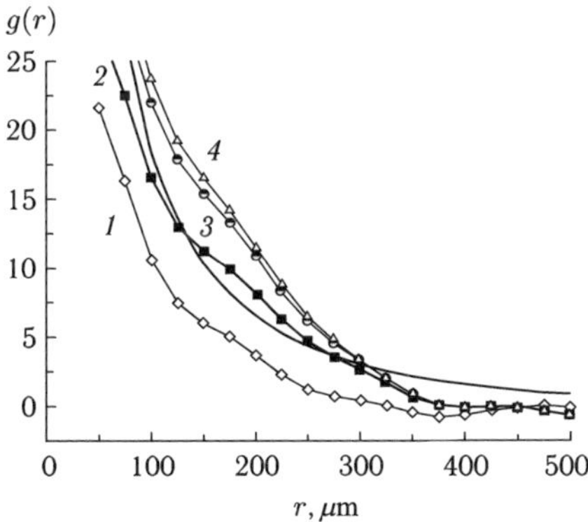

Fig. 2.19   Lines with symbols: *1* $U(r) = -T \ln g(r)$, *2* PY, *3* CHNC, *4* MS. Solid line without symbols shows the screened potential with $Z = 1320$ and $R = 500\,\mu$m.

is substantial, and so the calculation was made with the use of different closures. Figure 2.18 presents the results of calculation.

The lines with symbols in Fig. 2.19 correspond to the approximations [298]: *1* $U(r) = -T \ln(r)$, *2* PY, *3* CHNC, *4* MS. The solid line without symbols corresponds to the screened potential (2.15) with $R = 500\,\mu$m and $Z = 1320$. One can see that it lies most closely to the approximation PY. The Coulomb nonideality parameter for these conditions is approximately equal to 2.6 and the found screening radius corresponds to the following plasma parameters: $n_{\mathrm{e}} \approx 10^9\,\mathrm{cm}^{-3}$ and $T_{\mathrm{d}} \approx 1\,\mathrm{eV}$. The average distance between dust particles is $r \approx 200\,\mu$m and is close in the order of magnitude to the Debye radius.

In the assumption that the dust particle subsystem is in local thermodynamic equilibrium at temperature $T_{\mathrm{d}}$, some thermodynamic parameters of such dust particle system can be found by the measured correlation functions and the found potential. In the first place, such a parameter is the partial pressure of the dust component. Employing the known formula [925], we obtain that the pressure is equal to

$$p = nT - \frac{4\pi n^2}{6} \int_0^{\infty} r^3 \frac{d\varphi}{dr} g(r) dr. \tag{2.15}$$

The reduced isothermal compressibility has the form

$$\chi = T \left( \frac{\partial n}{\partial p} \right)_{\gamma} = 1 + 4\pi n \int_0^{\infty} h(r) r^2 dr. \tag{2.16}$$

Table 2.1    Calculated thermodynamic parameters for the dust subsystem [298].

| N, p/p | $n$, cm$^3$ | $p$, Pa | $R_{\mathrm{d}}$, $\mu$m | $T_{\mathrm{d}}$, eV | $R$, $\mu$m | $Z$ | $\Gamma$ | $p/nT$ | $\Gamma_1$ | $\chi$ |
|---|---|---|---|---|---|---|---|---|---|---|
| 1 | 200 | $\approx 25$–30 | 1.5–5 | $\approx 0.1$ | 87 | 490 | 2 | 1 | 0.03 | 1 |
| 2 | $3 \cdot 10^4$ | 20 | $\approx 20$ | $\approx 5$ | 500 | 1320 | 2.5 | 8.5 | 4.6 | 0.2 |

The total dust particle energy is equal to

$$E = \frac{3T}{2} + 2\pi n \int_0^\infty r^2 \varphi(r) g(r, T) dr. \qquad (2.17)$$

It is a well-known fact that the potential-to-kinetic energy ratio is the system nonideality parameter $\Gamma$. Equation (2.17) implies the definition of this parameter in terms of the correlation function:

$$\Gamma_1 = \frac{4\pi n}{3T} \int_0^\infty \varphi(r) g(r) r^2 dr. \qquad (2.18)$$

The values of the calculated thermodynamic parameters $p/nT$, $\Gamma_1$, and $\chi$ for a dust subsystem are presented in Table 2.1. For an ideal gas the isothermal compressibility is $\chi = 1$. The lower the $\chi$ value compared to unity, the stronger the deviation of the considered state from the ideal gas state. As should be expected, from the thermodynamic point of view the state of the dust system with the correlation function (Fig. 2.16) is fairly close to the ideal gas state.

For the second correlation function (Fig. 2.17) the situation is different. Table 2.1 shows that the reduced isothermal compressibility is $\chi < 1$ and the compressibility factor is $p/nT > 1$. These values may suggest a certain conclusion concerning the thermodynamic state of plasma liquid. We shall apply some conclusions drawn for hard charged spheres with effective radius $d$ and charge $Z$ to a dust particle system. As is shown in Ref. [623], the critical temperature and pressure in such a system can be estimated as follows:

$$T_{\mathrm{c}} \approx 0.05D; \quad p_{\mathrm{c}} \approx 0.01 \frac{T_c}{d}, \qquad (2.19)$$

where $D = Ze^2/d$. Obviously, the critical parameters estimated for a dust plasma liquid have rather low values. Consequently, most of the measured dust plasma parameters lie in the beyond-critical region ($T > T_{\mathrm{c}}$, $p > p_{\mathrm{c}}$). This is indirectly testified by the value of the compressibility factor $\chi > 1$.

For many substances (inert gases, hydrocarbons, water, etc.) the line of the unit factor $p/nT = 1$ drawn in the density-temperature coordinates is straight and lies in the beyond-critical region [52, 664]. The range of states lying under this line corresponds to the values $p/nT < 1$ (in particular, the entire region of biphase states lies under this line). The region with $p/nT > 1$ is always above the lines of phase equilibrium and corresponds to the charged fluid states. From this viewpoint a conclusion can be drawn that obviously the dust plasma for the second correlation function (Fig. 2.17) also corresponds to the beyond-critical state.

Concluding this section we notice that the charge, the screening radius, and the dust particle interaction potential in a gaseous and correlated fluid structure were first found with the help of integral equations of fluid theory from experimentally measured pair correlation functions. These data underlay the estimation of the compressibility factor, the reduced isothermal compressibility, and the internal energy of the system. The analysis of the determined parameters entails the conclusion that one of the regimes of the state of a dust system corresponds to an ideal gas and the other to a beyond-critical fluid.

Making use of this technique of direct visual observations of the dust plasma structure, we shall consider the applicability of the closure (2.3) of the chain of fluid-theory BBGKY integro-differential equations (2.2). We shall begin with the procedure of closing (2.3) of equations (2.2).

As has already been mentioned, the now already classical X-ray and neutron methods of experimental study of the structural properties of liquids allow obtaining information on the binary correlation distribution function $g_2(r)$ only, but are inapplicable to examination of $g_3(r)$ and higher-order correlation functions $g_s(r_n, \ldots r_s)$. For the analysis of three-particle correlation in real liquids indirect methods of diagnostics are typically used, for instance, measurements of the structural factor $S(q)$ for several pressure values of the medium at a constant temperature whence the derivative $\partial g(r)/\partial \rho$ is reconstructed which, in turn, contains information on the three-particle function $g_3(\vec{r}_1, \vec{r}_2, \vec{r}_3)$ [9, 829]. Eduction of such information requires additional data on isothermal compressibility of investigated medium. This drawback is not inherent in the dust plasma experiments [317, 1011] that make it possible to see the real structure of the object (Fig. 2.15) providing the possibility of a thorough structural study of liquid state.

Below we shall discuss the three-particle distribution function $g_3(\vec{r}_1, \vec{r}_2, \vec{r}_3)$ which is not only necessary for verification of the Kirkwood closure condition (2.3), but is important in calculation of physical characteristics of the medium depending on the derivatives of the pair function $g(r)$ with respect to temperature $\partial g(r)/\partial T$ or particle density $\partial g(r)/\partial \rho$ such as entropy, thermal expansion coefficients, etc. The function $g_3(\vec{r}_1, \vec{r}_2, \vec{r}_3)$ determines the probability of a simultaneous detection of three particles near the points $\vec{r}_1$, $\vec{r}_2$, $\vec{r}_3$. As distinct from the binary function, the function $g_3(\vec{r}_1, \vec{r}_2, \vec{r}_3)$ depends on three spatial coordinates and, accordingly, provides additional information on an examined structure, in particular, on the orientation order in an examined particle system.

To approximate the three-particle correlation function the superposition Kirkwood approximation is frequently used:

$$g_3(\vec{r}_1, \vec{r}_2, \vec{r}_3) = g_3^{\mathrm{sp}}(\vec{r}_1, \vec{r}_2, \vec{r}_3) - g(\vec{r}_1 - \vec{r}_2) \cdot g(\vec{r}_2 - \vec{r}_3) \cdot g(\vec{r}_3 - \vec{r}_1).$$

$$(2.20)$$

This approximation is based on disregard of terms of the form $U(\vec{r}_1 - \vec{r}_2, \vec{r}_3 - \vec{r}_1)$ in the Hamiltonian system, which do not come down to pair interactions. Relation (2.20) is often employed in the calculation of integral equations and the kinetics of interacting particles, and also in the reconstruction of interparticle interaction potentials by methods based on the hyperchain approximation or the Percus–Yevick equation [9, 75, 830, 831, 904]. Nevertheless the numerical simulations accomplished for the case of hard-sphere interaction and for particles interacting with Lennard–Jones type potentials show inadequacy of the superposition approximation even for low particle densities [830, 831, 904]. With increasing nonideality of the investigated fluid systems the distinction of the approximation $g_3^{\mathrm{sp}}(\vec{r}_1, \vec{r}_2, \vec{r}_3)$ (2.20) from the exact $g_3(\vec{r}_1, \vec{r}_2, \vec{r}_3)$ value can reach nearly 100%.

The experimental data (Fig. 2.14) with the help of which the position and displacements of macroparticles in the observed structures can be determined were used in the analysis. In all the analyzed cases the structures were quasi-stationary fluid type systems. The average interparticle distance $r_p$ in the analyzed dust structures varied from 260 to 350 $\mu$m. As a result of processing video tapes of the type of Fig. 2.20 pair $g(r)$ and three-particle correlation functions $g_3(\vec{r}_1, \vec{r}_2, \vec{r}_3)$ were obtained and averaged over 2–2.5 s under invariable experimental conditions. The pair correlation functions $g(r/r_p)$ are presented in Fig. 2.21 for different gas pressures $p$ and discharge power $W$. The cross sections of the obtained three-particle correlation functions $g_3(\vec{r}_{12}, \vec{r}_{23}, \vec{r}_{31})$ $(\vec{r}_{ij} = |\vec{r}_i - \vec{r}_j|)$ with a fixed $r_{12}$ value equal to the most probable interparticle distance $r_p^{\mathrm{max}}(r_{12} = r_p^{\mathrm{max}})$ determined by the position of the maximum of the pair correlation function $g(r)$ for the same discharge parameters are presented in Fig. 2.22.

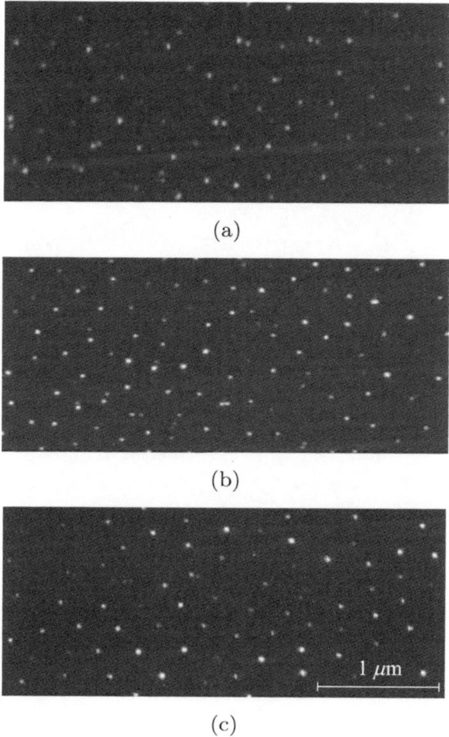



Fig. 2.20   Video pictures of dust cloud particles in the near-electrode layer of discharge for different experiments: (a) $p = 5\,\mathrm{Pa}$, $W = 9\,\mathrm{W}$; (b) $p = 3\,\mathrm{Pa}$, $W = 2\,\mathrm{W}$; (c) $p = 7\,\mathrm{Pa}$, $W = 10\,\mathrm{W}$.

The figure also presents the results of computation of the three-particle function $g_3^{\mathrm{cn}}(\vec{r}_{12}, \vec{r}_{23}, \vec{r}_{31})$ in the framework of the superposition approximation (2.20). To represent the given functions $g_3(\vec{r}_{12}, \vec{r}_{23}, \vec{r}_{31})$, $g_3^{\mathrm{sp}}(\vec{r}_{12}, \vec{r}_{23}, \vec{r}_{31})$ in the visual "two-dimensional" form convenient for the analysis they were normalized to the $g_3(\vec{r}_{12}, \vec{r}_{23}, \vec{r}_{31})$ maximum value: black color corresponds to unity and white color to $g_3 = g_3^{\mathrm{sp}} = 0$. The deviation of the function $g_3^{\mathrm{sp}}(\vec{r}_{12}, \vec{r}_{23}, \vec{r}_{31})$ from the results of $g_3(\vec{r}_{12}, \vec{r}_{23}, \vec{r}_{31})$ computation is indicated in the caption to Fig. 2.22; it was calculated proceeding from the relative root-mean-square error of the superposition approximation:

$$\delta = \left[ \frac{\sum\limits_{i=1}^{N} \left\{ \frac{g_3(\vec{r}_{12}, \vec{r}_{2i}, \vec{r}_{i1}) - g_3^{\mathrm{sp}}(\vec{r}_{12}, \vec{r}_{2i}, \vec{r}_{i1})}{g_3(\vec{r}_{12}, \vec{r}_{2i}, \vec{r}_{i1})} \right\}^2 }{N} \right]^{\frac{1}{2}},$$

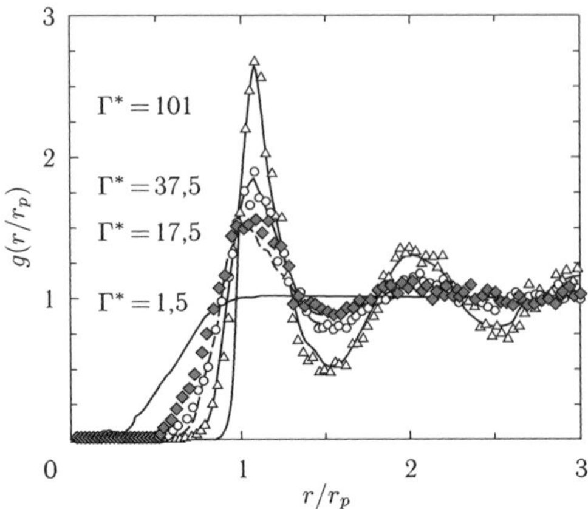



Fig. 2.21  Pair correlation functions $g(r/r_p)$, measured in experiments: ($\bullet$) $p = 5\,\mathrm{Pa}$, $W = 9\,\mathrm{W}$; ($\circ$) $p = 3\,\mathrm{Pa}$, $W = 2\,\mathrm{W}$; ($\Delta$) $p = 7\,\mathrm{Pa}$, $W = 10\,\mathrm{W}$ and obtained in numerical simulations for different $\Gamma^*$ (solid lines) shown in the figure.

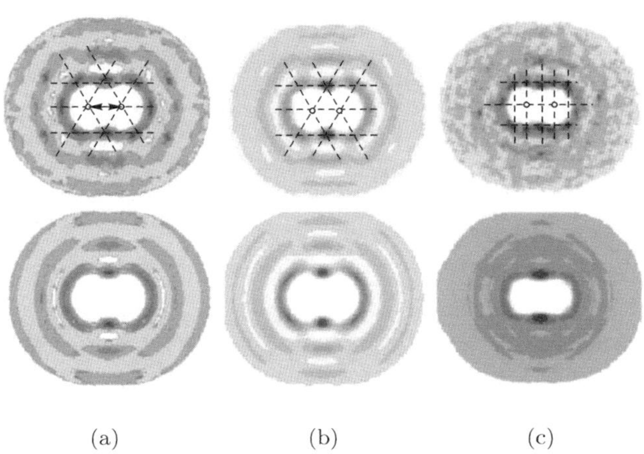

(a)      (b)      (c)

Fig. 2.22  Cross sections of the measured three-particle functions $g_3$ (the upper row) and functions $g_3^{\mathrm{sp}}$ calculated in the superposition approximation (the lower row) for different experiments: (a) $p = 7\,\mathrm{Pa}$, $W = 10\,\mathrm{W}$, $\delta = 0.61$; (b) $p = 3\,\mathrm{Pa}$, $W = 2\,\mathrm{W}$, $\delta = 0.28$; (c) $p = 5\,\mathrm{Pa}$, $W = 9\,\mathrm{W}$, $\delta = 0.3$.

where $N$ is the total number of spatial elements $d\vec{r_i}$ in the neighborhood of the point with coordinate $\vec{r_i}$ into which the analyzed plasma liquid layer was divided. Thus, the error $\delta$ in experiment was varied for strongly correlated structures within the range of 15–20%, and with increasing nonideality approximately from 30 to 60%.

The visual comparison of the obtained results shows that observed in the registered structures is the formation of a short-range orientational order of macroparticles which is reflected in the appearance of $g_3(\vec{r}_{12}, \vec{r}_{23}, \vec{r}_{31})$ peaks in the nodes of hexagonal clusters shown in Figs. 2.22(a) and (b) with dashed lines. With increasing maximum of the pair correlation function (see Fig. 2.21) the value of these maxima located at distances $r$ close to $r_p^{\max}$ grows, and new maxima emerge at distances $r \approx 2r_p^{\max}$. This phenomenon does not show up in the analysis of the superposition approximation $g_3^{\mathrm{sp}}(\vec{r}_{12}, \vec{r}_{23}, \vec{r}_{31})$.

For comparison of the experimentally obtained results with correlation of macroparticles in the systems, numerical simulation was performed. The calculations were carried out for a three-dimensional system by the molecular-dynamics Langevin method with periodic boundary conditions [1011]. This method is based on the solution of the system of ordinary differential equations with the Langevin force $F_{\mathrm{Br}}$ which takes into account random beats of surrounding gas molecules or other random processes leading to setting up the equilibrium (stationary) kinetic temperature $T_{\mathrm{p}}$ of particles characterizing the kinetic energy of their stochastic (thermal) motion. Along with random forces $F_{\mathrm{Br}}$, account was also taken of the forces of pair interparticle interaction $F_{\mathrm{int}}(l) = -eZ_{\mathrm{p}}\frac{d\varphi}{dl}$:

$$m_{\mathrm{p}}\frac{d^2\vec{l}_k}{dt^2} = \sum_j F_{\mathrm{int}}(l)|_{\vec{l}=|\vec{l}_k-\vec{l}_j|} \frac{\vec{l}_k - \vec{l}_j}{|\vec{l}_k - \vec{l}_j|} - m_p\nu_{\mathrm{fr}}\frac{d\vec{l}_k}{dt} + \vec{F}_{\mathrm{Br}}, \qquad (2.21)$$

where $\vec{l} = |\vec{l}_k - \vec{l}_j|$ is the interparticle distance, $m_{\mathrm{p}}$ is the particle mass, $\nu_{\mathrm{fr}}$ is the macroparticle friction factor which for the conditions under consideration can be obtained within the free molecular approximation [620].

The derived pair correlation functions are demonstrated in Fig. 2.21 for different nonideality parameters $\Gamma^*$. One can readily see that the experimental results correspond well to the systems with nonideality parameters $\Gamma^* \approx 100$, 37.5, and 17.5. In the latter case ($\Gamma^* \approx 17.5$) the differences between the calculations and measurements of $g(r)$ are most pronounced because the experimental curve has a wider first maximum related to the small inhomogeneity of the analyzed dust structure. The function $g(r)$ for $\Gamma^* \approx 2.5$ shown in the same figure corresponds to the case when the behavior of the pair correlation function becomes monotonous, i.e., the maxima of $g(r)$ vanish. The cross sections of three-particle correlation functions and their superposition approximations are given in Figs. 2.22 and 2.23 for different $\Gamma^*$ for $r_{12} = r_p^{\max}$. The root-mean-square error $\delta$ for the superposition approximation is indicated in the caption to Fig. 2.23 also for the

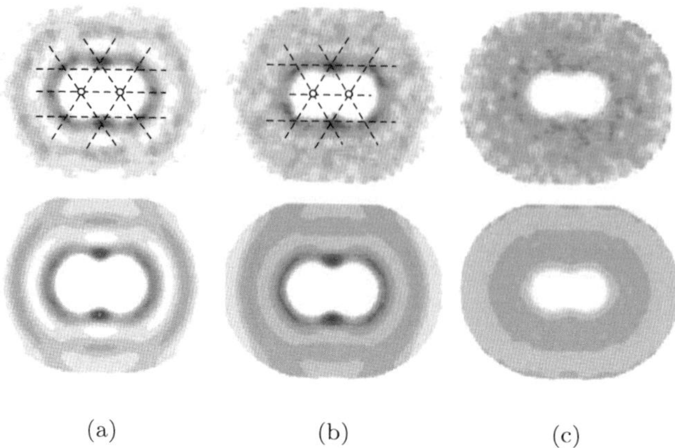

(a)                                    (b)                                    (c)

Fig. 2.23   Cross sections of three-particle functions $g_3$ measured (upper row) and calculated in the superposition approximation $g_3^{\mathrm{sp}}$ (lower row) for different parameters $\Gamma^*$: (a) $\Gamma^* = 37.5$, $\delta = 0.61$ (b) $\Gamma^* = 17.5$, $\delta = 0.28$ (c) $\Gamma^* = 1.5$, $\delta = 0.3$.

above-given numerical $\delta$ values registered in laboratory experiments (see Fig. 2.22).

With increasing $\Gamma^* > 40$–$50$ the form of the three-particle correlation function for numerical experiment begins depending on the orientation of a plane being modeled. The latter circumstance may be due to the fact that with increasing nonideality parameter $\Gamma^* > 40$–$50$ the dynamics of a modeled fluid system becomes analogous to a solid and can be considered within the jump "theory" formulated for the description of strongly correlated molecular liquids [9]. The main point of the theory is that the molecules of such a liquid are in equilibrium ("sedentary") state during the time necessary to deliver them energy (the activation energy) sufficient for breaking the potential bonds with neighboring molecules and going over to the surroundings of other molecules.

Figure 2.24 presents the dependence of the error of superposition approximation in the entire liquid-state range up to system crystallization.

As $\Gamma^*$ increases to 100, the root-mean-square error $\delta$ of the superposition approximation for the analyzed cross section of the three-particle function somewhat rises to make up nearly 70%. As $\Gamma^* \to 100$, a body-centered crystal lattice is formed in the simulated system. Figure 2.25 illustrates the cross section of such a lattice (in the facet of the elementary cubic cell) and also the functions $g_3(\vec{r}_p^{\,\mathrm{max}}, \vec{r}_{23}, \vec{r}_{31})$ and $g_3^{\mathrm{sp}}(\vec{r}_p^{\,\mathrm{max}}, \vec{r}_{23}, \vec{r}_{31})$ for $\Gamma^* \approx 400$. For comparison the figure also shows the results for the hexagonal particle

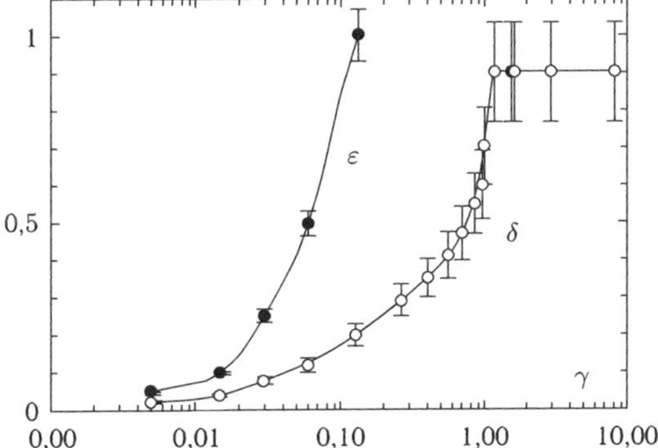

Fig. 2.24   Relative root-mean-square error $\delta$ (o) of the superposition approximation and relative errors $\varepsilon$ (•) of the Percus–Yevick equation depending on the parameter $\gamma$.

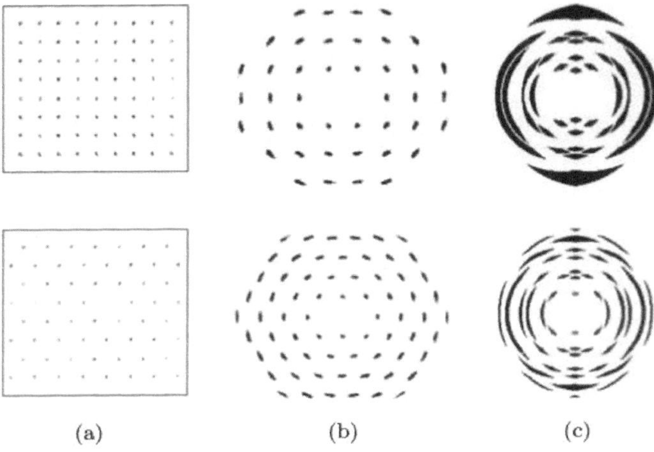

(a)    (b)    (c)

Fig. 2.25   Cross sections of a crystal cubic lattice (a) and the three-particle correlation functions calculated for these planes $g_3$ and for the superposition approximation $g_3^{\mathrm{sp}}$ (b). The upper row is the cross section of the body-centered lattice, the lower row is the cross section of the face-centered lattice ($\delta = 0.36$) corresponding to the hexagonal particle arrangement ($\delta = 0.79$).

arrangement. The error $\delta$ of the superposition approximation is indicated in the caption to the figure.

The analysis of the experimental results showed that the difference of the calculations in the superposition approximation [317] from measurements of the three-particle correlation function makes up about 30–60% for the analyzed cross sections of $g_3(\vec{r}_{12}, \vec{r}_{23}, \vec{r}_{31})$ for $r_{12} = r_p^{\mathrm{max}}$. The formation of regular macroparticle clusters was observed in both the experimental systems and the simulated fluid structures. The numerical calculations of three-particle correlation of macroparticles interacting with a screened potential showed that the formation of such clusters in modeled systems is observed with increasing effective nonideality parameter $\Gamma^*$, which agrees well with the results of modeling reported in Ref. [1013].

We shall now compare the experimental and computational methods of molecular dynamics and integral equations of binary correlation functions to estimate applicability of the method of integral equations for the description of gases and liquids. The results of experiments were used (Table 2.2, Fig. 2.14). The corresponding correlation functions of gas and liquid structures can be seen in Fig. 2.26, where they are compared with computer calculations and the data of integral equations.

Table 2.2   Particle radius ($a_p$), temperature ($T$), average interparticle distance ($r_p$) and parameter $\gamma$ for different measurement series in a hf discharge for power $W$ and pressure $p$.

| N | $W$, W | $p$, Pa | $a_p$, $\mu$m | $T$, eV | $r_p$, $\mu$m | $\gamma = \Gamma/\Gamma_c$ |
|---|---|---|---|---|---|---|
| 1 | 10 | $\approx 7$ | 1.7 | $\approx 0.29$ | 295 | 0.96 |
| 2 | 10 | $\approx 3$ | 1.7 | $\approx 0.57$ | 315 | 0.4 |
| 3 | 3 | $\approx 2$ | 1.7 | $\approx 0.61$ | 270 | 0.13 |
| 4 | 5 | $\approx 5$ | 0.95 | $\approx 0.09$ | 275 | 0.06 |

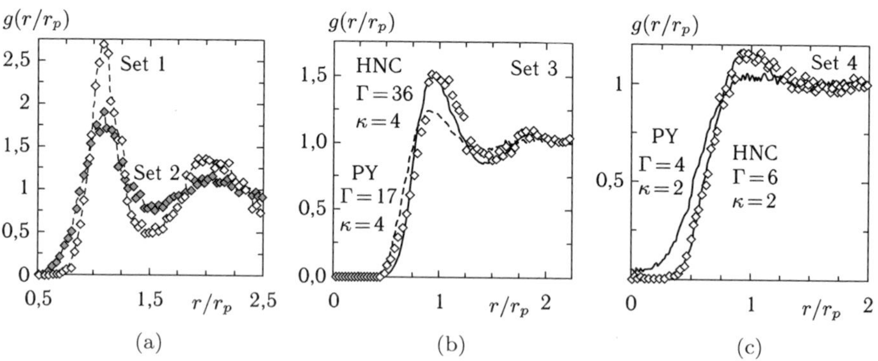

Fig. 2.26   Binary correlation functions $g(\vec{r}/\vec{r}_p)$ obtained in different series of measurements (symbols) and correlation functions (see (b) and (c)) obtained by numerical simulation for different $\Gamma$ and $\kappa$ values for pair potentials reconstructed by the methods of Percus–Yevick approximation (grey lines) and hyperchain equations (black lines).

One can see that the integral equations describe reasonably the behavior of plasma liquid in almost the entire range of liquid state. The hyperchain approximation is more pertinent for the description of plasma liquid states because it allows for long-range correlations better than the other approximations. The Percus–Yevick model obviously has some advantage in the description of short-range order.

# Chapter 3

# Quantum-Mechanical Models of a Solid

## 3.1  Hartree and Hartee–Fock approximations

With lowering temperature of a liquid it passes over to a solid state characterized by translation symmetry of heavy particles (ions, atoms, molecules). And although the solid state does not occupy a broad space on the phase diagram of a substance (see Figs. 2.1 and 2.2), the role of this phase in our life is great. It is the base of numerous technical (materials), biological, geophysical, medical, and other applications.

Huge experimental material on thermophysical properties of solids has been accumulated and sophisticated methods of their calculation created (see Chapter 5 and the references therein). We should emphasize that at moderate (below 5–50 kbar) pressures an important role is played by the strength properties of materials which, when under pressure, resist not only the volume variation (thermodynamics) but, as distinct from gases and liquids, also the shape variation, and the forces themselves and the deformations are of tensor nature. The behavior of a solid under these conditions is described by different endurance models to which vast literature is devoted [294, 295, 508, 509]. Above the ultimate strength $\sigma_{\text{str}}$ the shear stress can be neglected and the plastic model of the motion of medium can be used in the assumption that the pressure and strain tensors are spherical and are described by thermodynamic equation of state (EOS). Since the latent heat, entropy, and density variation under melting normally makes up several percent and decrease along the melting curve $p_{\text{mel}}(T)$, the thermophysical properties and the EOS of a solid and a liquid do not differ strongly in the vicinity of $p_{\text{mel}}(T)$ and, as will be seen below, are described by semi-empirical models (Chapter 9).

85

In its existence region (Figs. 2.1 and 2.2), the behavior of a solid is fairly diverse and shows exceptionally individual properties of each structure element depending on particular features of the electron energy spectrum of atoms, molecules, and ions composing the crystal lattice. A lot of methods of quantum-mechanical descriptions of electronic, thermodynamic, phonon, magnetic, optical, and other characteristics of solid state have been proposed to date, and the methods themselves have reached high perfection.

We shall mainly dwell on considering the results of computation of thermodynamic properties of matter in a possibly wide range of pressures and temperatures and only briefly discuss the particular qualitative characteristics of the methods applied, referring for more details to original papers and specialized monographs and reviews [58, 138, 1096].

In a thermodynamic description of a solid, phonon, and electron components are distinguished [58] which are considered rather independently. This is the so-called adiabatic or Bohr–Oppenheimer approximation. Most band-model calculations of the electron spectra, when a crystal is thought of as an ideal periodic structure with immobile nuclei at the cell sites and each electron is moving in a periodic self-consistent potential created by the ion core and the other electrons, allows expressing the wave function of the whole system in terms of one-electron Bloch functions [1096]. The main distinction of the band structure models lies in the ways of calculation of these one-electron functions inside the elementary atomic cells into which the whole substance is divided and in the techniques of solution sewing at the boundary.

In the general form the quantum-statistical description of thermodynamic properties of a substance requires introduction of the density matrix apparatus and the application of different statistical approximations. The most frequently used is the grand canonical ensemble approximation [76].

Given the density matrix $\hat{W}$ of the system, the mean value of the physical quantity $F$ can be represented as [598]:

$$\langle \hat{F} \rangle = Sp(\hat{W}\hat{F}), \tag{3.1}$$

where $\hat{F}$ is the operator corresponding to the quantity $F$.

The most general variational principle typically applied under equilibrium conditions for the grand canonical ensemble is the requirement of minimum grand thermodynamic potential: $\Omega = E - \mu N - TS$, where $\mu$ is the chemical potential of the system and $T$ is temperature. The joint quantum-mechanical consideration of a system of interacting ions and electrons leads to difficulties that have not yet been overcome and, therefore, normally used is the above-mentioned adiabatic approximation where the ions are considered to be classical particles moving much slower than the electrons because for each new position of ions the electrons are assumed to come quickly into thermodynamic equilibrium. According to equation (3.1) the grand thermodynamic potential of the system of electrons and ions is written down as:

$$\Omega = \langle \hat{\Omega} \rangle = Sp \left[ \hat{W}(\hat{H} - \mu\hat{N} + T \ln \hat{W}) \right]. \tag{3.2}$$

Here, $\hat{H}$ is the Hamiltonian of the system of electrons, $\hat{N}$ is the operator of the number of particles, and the density operator $\hat{W}$ can be written as:

$$\hat{W} = \frac{\exp\left(-\frac{\hat{H} - \mu\hat{N}}{T}\right)}{Sp \exp\left(-\frac{\hat{H} - \mu\hat{N}}{T}\right)}. \tag{3.3}$$

In a nonrelativistic approximation one can define the classical Hamiltonian function as a function of only coordinates and momenta of all the particles of the system. If all the electron spin interactions and all the nuclear effects (for instance, finiteness of nuclear size and mass) are neglected, the Hamiltonian of the electron system will have the form

$$\hat{H} = \sum_i \left[ -\frac{1}{2}\Delta_i + U_a(\mathbf{r}_i) + \frac{1}{2}\sum_{i \neq j} \frac{1}{\mathbf{r}_i - \mathbf{r}_j} \right], \tag{3.4}$$

where $U_a(\mathbf{r}_i)$ is the potential energy of the $i$th electron with coordinates $\mathbf{r}_i$ in a given atomic nucleus potential $V_a(\mathbf{r})$, that is, $U_a(\mathbf{r}_i) = -V_a(\mathbf{r}_i)$. Then the mean value of the operator $\hat{\Omega} = \hat{H} - \mu\hat{N} + \theta \ln \hat{W}$ is calculated by the formula

$$\langle \hat{\Omega} \rangle = Sp(\hat{W}\hat{\Omega}) = \sum_n w_n \int \Psi_n^*(Q)\hat{\Omega}\Psi_n(Q)\,dQ, \tag{3.5}$$

where $w_n = w(E_n, N_n)$ is statistical probability of the state of the system of electrons with a given energy $E = E_n$ and the number of particles $N = N_n$, $\Psi_n(Q)$ is the wave function of such a state, $Q$ is the set of coordinates $q_i$ of the electrons, where $q_i$ includes the spatial $\mathbf{r}_i$ and spin $\sigma_i$ variables of the electron with number $i$. Integration over the entire configuration space $Q \equiv \{q_i\}$ means integration over all the space coordinates $\mathbf{r}_i$ and summation over all the spin variables $\sigma_i$. The mean value of any quantity, as equation (3.5) shows, is calculated in two steps: first, one seeks the mean value of the considered quantity in a state with the wave function $\Psi_n(Q)$ for given energy $E_n$ and the number of particles $N_n$, and then the obtained values are averaged over different states with weight equal to the probability $w_n$ of these states.

The Hartree–Fock equations can be obtained using the following approximations:

(1) The wave functions $\Psi(Q)$ are represented as an antisymmetrized sum of the products of one-particle orthonormal wave functions in occupied states $\nu_i$ (Slater's determinant in the Hartree–Fock approximation), $\Psi(Q) = 1/\sqrt{N!}\det\{\psi_{\nu_i}(q_j)\}$, $i, j = 1, \ldots N$.
(2) The one-particle states $\nu$ are occupied with certain probabilities $n_\nu$ ($0 \leq n_\nu \leq 1$) which can be thought of as independent near the equilibrium state [542].

Then, varying the mean value of the grand thermodynamic potential (3.5), under the condition of normalization $\int |\psi_\nu(q)^2|\, dq = 1$ and orthogonality $\int \psi_\nu^*(q)\psi_\lambda(q)\, dq = 0$ for $\lambda \neq \nu$, one can obtain the Hartree–Fock equations for a substance with a given temperature and density [741]:

$$\hat{H}_0\psi_\nu(q) + \frac{1}{n_\nu}\sum_{\lambda \neq \nu}\Lambda_{\nu\lambda}\psi_\lambda(q) = -\frac{\Lambda_{\nu\nu}}{n_\nu}\psi_\nu(q) \equiv \varepsilon_\nu\psi_\nu(q), \qquad (3.6)$$

$$n_\nu = n(\varepsilon_\nu) = \frac{1}{1 + \exp\dfrac{\varepsilon_\nu - \mu}{\theta}}, \qquad (3.7)$$

where

$$\hat{H}_0\psi_\nu(q) = \left[-\frac{1}{2}\Delta - V_a(\mathbf{r}) + \sum_\lambda n_\lambda \int \frac{|\psi_\lambda(q'|^2\, dq'}{|\mathbf{r} - \mathbf{r}'|}\right]\psi_\nu(q)$$

$$- \sum_\lambda n_\lambda \int \frac{\psi_\lambda^*(q')\psi_\nu(q')\, dq'}{\mathbf{r} - \mathbf{r}'}\psi_\lambda(q). \qquad (3.8)$$

Equations (3.6)–(3.8) were first obtained by Matzubara [670]. The solution of the system of equations (3.6)–(3.8) encounters difficulties due to the last term in the right-hand side of the Hamiltonian (3.8), the so-called exchange interaction. Making use of the quasi-classical approximation one can approximately take into account the exchange effects and obtain the single effective potential for all the electrons. Slater [927] was the first to allow for the exchange effects in local approximation for a free atom at zero temperature. The expressions for a nonzero temperature were derived in Refs. [406, 740]. For example, the interpolation expression for the exchange potential [740, 741] is as follows:

$$V_{\text{ex}} = \frac{\pi\rho(\mathbf{r})}{\theta}\left[1 + 5.7\frac{\rho(\mathbf{r})}{\theta^{3/2}} + \frac{\pi^4}{3}\cdot\frac{\rho^2(\mathbf{r})}{\theta^3}\right]^{-1/3}, \tag{3.9}$$

where $\rho(\mathbf{r})$ is the electron density. The substitution of $V_{\text{ex}}$ from the system of equations (3.6)–(3.8) for the exchange term of the Hamiltonian (3.8) yields the system of Hartree–Fock–Slater equations. Elimination of the exchange term from (3.8) leads to the system of Hartree equations.

The system of Hartree–Fock equations allows, in principle, calculating the one-particle wave functions $\psi_\nu(q)$ for a given position of nuclei and a given value of the chemical potential $\mu$ which is determined from the condition of electrical neutrality of the system of electrons and ions in a substance. It is however impossible to perform in practice the calculations for the original system of equations for the reason that minimization of the grand thermodynamic potential (3.5) is carried out with variation of all the wave functions of the system, which leads to an exponential growth of spatial dimension of the varied parameters [562]. Moreover, for a fairly large number of electrons it is impossible to calculate the wave functions of the system with admissible precision. In paper [562] this phenomenon is referred to as an "exponential barrier". Thus, it is necessary to introduce further simplifications in the solution of the system of Hartree–Fock equations. Traditionally, instead of the potential $V(\mathbf{r})$ with a given position of nuclei one finds the average potential near the examined nucleus by averaging $V(\mathbf{r})$ over different positions of the other nuclei. In the absence of preferred direction, the average potential is considered to be spherically symmetric, $V(r)$, and the wave functions in this potential for a given energy $\varepsilon_\alpha = \varepsilon$ is written as

$$\Psi_{\varepsilon lm}(\mathbf{r}) = \frac{1}{r}R_{\varepsilon l}(r)(-1)^m Y_{lm}(\vartheta, \varphi), \tag{3.10}$$

where the radial function $R_{\varepsilon l}(r)$ satisfies the equation

$$-\frac{1}{2}R''_{\varepsilon l} + \left[-V(r) + \frac{l(l+1)}{2r^2}\right]R_{\varepsilon l}(r) = \varepsilon R_{\varepsilon l}(r) \tag{3.11}$$

and $Y_{lm}(\vartheta, \varphi)$ are spherical functions. Thus, the atom finds itself in a spherical cell of radius $r_0$ and the influence of the other atoms is taken into account with the help of boundary conditions. For $r = 0$ the boundary condition $R_{\varepsilon l} = 0$ should be met for the radial function. The conditions on the cell boundary can be imposed using different approaches. For low-temperature low-density plasma, $R_{\varepsilon l}(r_0) = 0$ is chosen which is the so-called restricted-atom model (Chapter 4). For solid-state densities the most adequate are the Bloch conditions on the spherical cell boundary.

The thermodynamic functions of electrons for Hartree, Hartree–Fock, and Hartree–Fock–Slater models in a spherical cell can be found by differentiating the grand thermodynamic potential $\Omega$:

$$p_e = -\frac{\Omega}{v}, \tag{3.12}$$

$$E_e = -\Omega - T\frac{\partial\Omega}{\partial T} - \mu\frac{\partial\Omega}{\partial\mu}, \tag{3.13}$$

$$S_e = -\frac{\partial\Omega}{\partial T}. \tag{3.14}$$

Here, $v = (4/3)\pi r_0^3$ is the spherical cell volume. The corresponding formulas can be found in monograph [741].

The Hartree–Fock model can be simplified by expanding the electron density in powers of the Planck's constant $\hbar$ [538]. Such a model is called the Thomas–Fermi (TF) model with quantum and exchange corrections (TFC) [505, 538, 545] and reduces to solution of the boundary problem for the self-consistent potential $V(r)$ and the correction $\delta V(r)$ to it. Correction to the potential results in the occurrence of corresponding corrections to the thermodynamic functions [505, 539] which in the TFC model are self-similar. This was the basis of the detailed tables of TFC model calculations [505]. Disregard of quantum and exchange corrections in the TFC models gives the finite-temperature TF model [259]:

$$\frac{1}{r}\frac{d^2}{dr^2}(rV) = \frac{2}{\pi}(2\theta)^{3/2}I_{1/2}\left(\frac{V(r)+\mu}{T}\right), \quad 0 < r < r_0, \tag{3.15}$$

with boundary conditions

$$rV(r)|_{r=0} = Z, \quad V(r_0) = 0, \quad \left.\frac{dV(r)}{dr}\right|_{r=r_0} = 0. \tag{3.16}$$

Quantum-statistical models are widely used to calculate different plasma properties in extreme states [545, 547, 741]. Nevertheless, the approximations used in formulation of the models, in particular, employment of the spherical cell and disregard of correlation effects lead to serious errors in the range of moderate densities and temperatures. In this parameter range, the density functional method is now being applied more frequently (see Chapter 7).

Deferring the description of the property of quasi-classical TF and TFC models and the density functional method till Chapters 6 and 7, we go on discussing quantum-mechanical solid-state models.

While increasing compression solids tend to transform to closely packed highly symmetric crystal structures, which justifies the use in this case of the model of Wigner–Seitz spherical cells. According to this model, a complete self-consistent problem in a crystal is reduced to solution of the wave equations (3.6)–(3.8) in one cell with Bloch boundary conditions [1096]. Provided these conditions are met, regions of allowed states appear instead of discrete energy spectrum for a free atom. The relation of pressure with the wave functions on the cell surface can determine the EOS of the system. Even simplified calculations in the spherical cell model [41, 1027] carried out neglecting the exchange effects (the term $G_{nl}(r)$ is left out of equations (3.6)–(3.8)) revealed individual properties of the elements and allowed determining the structure and the character of the electron energy band occupation. As distinct from the statistical model (see Chapter 6), the band theory could reproduce a number of qualitative effects of the periodic system, for example, difference compressibility and normal density of elements neighboring in the system. The simplest band formation model will be considered in Chapter 4. Allowance for the individual features of energy band occupation causes a lack of evenness or even nonmonotony in the behavior of characteristics of metals under compression. In dynamic experiments this is fixed as a shock compressibility variation [28].

Allowance for exchange effects in the band model leads to substantial computational difficulties due to the necessity to solve systems of integro-differential equations. An approximate allowance for the exchange, when a real nonlocal exchange potential in Hartree–Fock equations is replaced by an averaged local one, was first proposed by Slater and was further developed as the "$X - \alpha$" method [1096]. This method underlays the calculations of thermodynamic characteristics of a condensed phase of a substance at high pressures and temperatures [865]. Such a method of describing the exchange effects was further developed in Ref. [739], where the expression for the effective exchange potential was proposed for the case of arbitrary temperatures and densities.

The Hartree–Fock–Slater calculations [739,915,916] showed a much better coincidence with experimental data and with the results of plasma calculations than those in the quasi-classical model. Typical of all the obtained dependences are oscillations relative to the data of quasi-classical models, the oscillations being observed until the system has shells. It should be

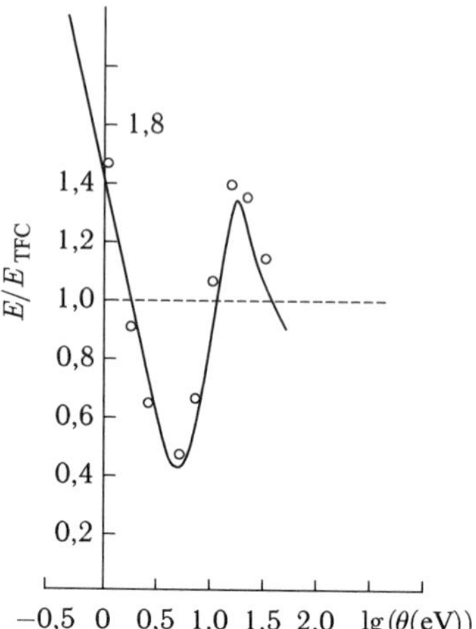



Fig. 3.1 Ratio of the lithium plasma energy by the Hartree–Fock–Slater model [739] to the energy by the TF model [500, 505] at the pressure $p = 1\,\text{kbar}$. The symbols stand for the calculation by the chemical model.

noted that corrections to the shell effect are also substantial where the values of exchange and quantum corrections to the TF model are small. Figure 3.1 shows that the value of the lithium plasma energy at $T \approx 5$ and $16\,\text{eV}$ changes with allowance for the shells by more than a factor of two, although the calculations using the TF model and its modifications are practically coincident. Similarly, the shock adiabats of copper and lead presented in Fig. 3.2 differ noticeably from the quantum-statistical ones at pressures much higher than the traditional boundary value of $300\,\text{Mbar}$ [35], which is indicative of the necessity of taking into account the shell effects up to extreme temperatures and pressures. At the same time, Fig. 3.2 also demonstrates a sharp distinction of the calculated shock adiabats from the experimental ones at lowered ($p \leq 10\,\text{Mbar}$) pressures, which points out an insufficient precision of the Hartree–Fock–Slater model in this region.

## 3.2 Asphericity of cells

For the description of the properties of a solid under normal conditions the spherical cell approximation is too schematic as the elementary cells of real crystals have a much more complicated shape, especially for structures with low coordinate numbers. Furthermore, at low pressures the relative contribution of substances of exchange and correlation terms to the EOS

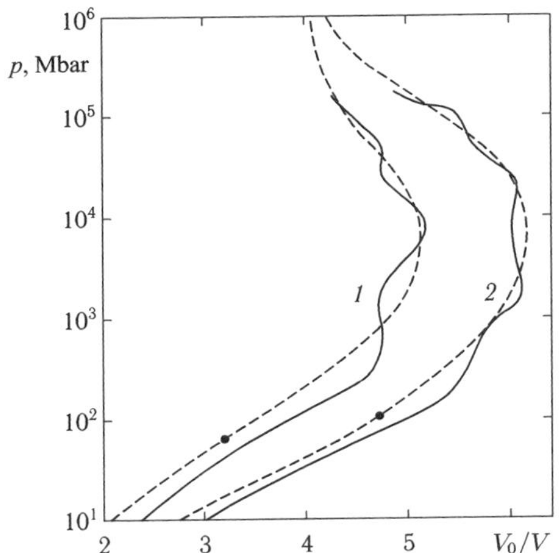

Fig. 3.2   Shock adiabats of copper (*1*) and lead (*2*) [915, 916]. Solid curves stand for calculations by the Hartree–Fock–Slater model and dashed lines show TFC calculations and interpolation [35].

increases, which necessitates their more rigorous account. The Hartree–Fock calculation of intercellular exchange interaction at $T = 0\,\mathrm{K}$ and low compressions, as well as the estimation of the contribution of correlation corrections and intercellular exchange effects showed their substantial influence upon the computational characteristics, which generally impugns the expediency of using spherical approximation for calculations in this region [217]. These circumstances required development of new effective methods of self-consistent calculation of nonspherical structures in the low-pressure range.

In real crystals, ions occupy a limited atomic cell volume, especially for low compression. To describe such a situation, the crystal volume is divided into regions differing in the ways of seeking solutions of Schrödinger equations. The division makes it possible to simultaneously meet the boundary conditions on the surface of a Wigner–Seitz elementary cell of complicated shape and take into account the difference between the lattice potential and the atomic one. This is attained by introducing the so-called MT potential (named after its muffin-tin shape) [928] spherically symmetric in the ion neighborhood, where the wave function is the combination of solutions of Schrödinger equation in the central field, constant in the interstitial space, with solution in the form of linear combination of plane waves. These two solutions are sewed at the boundary sphere inside the cell, which leads to an attached plane wave.

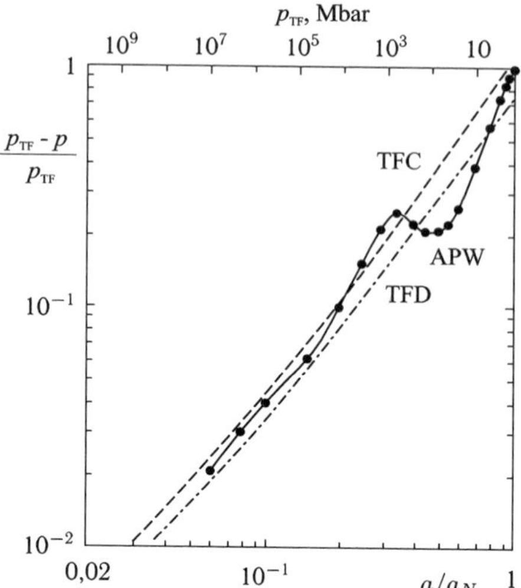



Fig. 3.3   EOS of aluminum ($T = 0\,\mathrm{K}$) [676]. Calculation: TFD is the Thomas–Fermi–Dirac model, TFC is the Thomas–Fermi model with quantum and exchange corrections, APW is the model of attached plane waves, $\alpha$ is the lattice parameter.

Quantum-mechanical calculations by the attached plane wave method are more cumbersome compared to the spherical cell approximation and mainly pertain to structures under normal conditions, whereas the numerical results for high pressures are much more modest here. Figure 3.3 presents comparison of the results of calculations for aluminum within this model with the data of quasi-classical approximation. The characteristic nonmonotonies of the calculation curve are due to transition to the continuum of $L$ and $K$ shells at pressures of 50–750 and $10^4$–$10^6$ Mbar. One can see that the model [676] predicts noticeable distinctions from the simplest versions of the quasi-classical approximation at pressures much higher than the applicability limits of the TF model estimated in paper [35].

The method of the attached plane wave in combination with the "$X$–$\alpha$" method was used in Refs. [673–675] to determine the high-temperature equation of iodine state and to examine the peculiarities of the isostructural phase transition in cesium. In combination with the variational method of the description of a liquid (Chapter 6), this method was applied [862] to calculate the EOS of condensed xenon and to obtain the critical values of the volume and pressure corresponding to its metallization under compression ($V \approx 10\,\mathrm{cm}^2/\mathrm{mole}$, $p \approx 1.5\,\mathrm{Mbar}$). The calculated characteristics showed good agreement with experimental results at low and high pressures, which

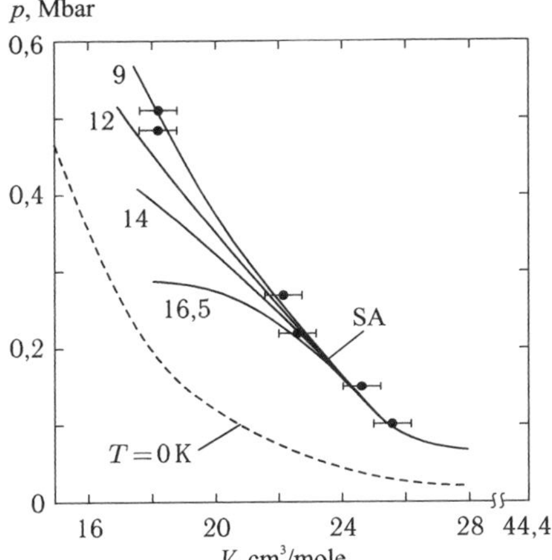



Fig. 3.4   Shock adiabat and zero isotherm of condensed xenon [1096]. Symbols give experimental values, numbers at calculated curves correspond to different xenon metallization volumes under compression.

in particular is illustrated in Fig. 3.4 showing the sensitivity of the run of the calculated shock adiabat of xenon to various values of metallization volume. Independent experiments on measurement of optical radiation absorption in condensed xenon under compression on diamond anvils have recently been carried out [54]. It is of interest that the obtained dependence of the energy gap width on the volume also turned out to be close to the calculated one [862] and predicts metallization for $p \approx 2\,\text{Mbar}$.

Great computational difficulties encountered within the attached plane wave method, especially in the case of nonzero temperatures, have led to a wide use in particular calculations of the method of linear MT orbitals [46] which is less accurate but much more economic in respect of calculations. This way of band calculations, normally used in the case of closely packed structures takes self-consistent account of the exchange-correlation effects in the local density approximation [440, 563]. The application of the linear MT-orbital method permitted a quantitatively correct description for many metals of experimentally observed fine characteristics of the Fermi surface in a broad pressure and temperature range and of a number of peculiarities of the EOS explained by a sequence of electron energy band occupation. Proceeding from first principles and using no input parameters except for the atomic number, the method nevertheless possesses a high precision of description. For example, the calculations of normal density and modulus

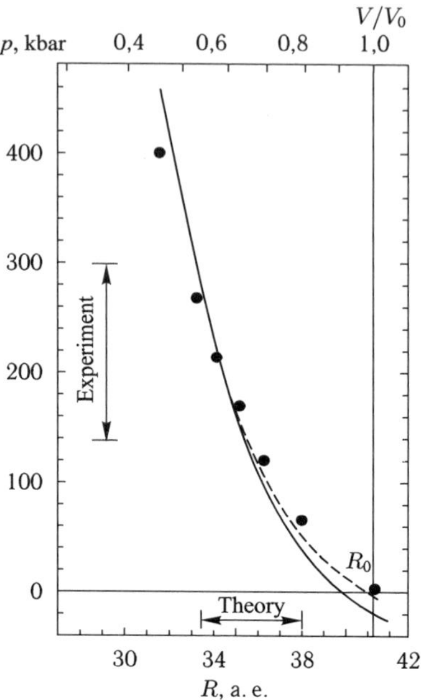


Fig. 3.5 The EOS of calcium [483]. Symbols give the shock-wave experiment, solid line is calculation, dashed line is allowance for nonlocal corrections. Marked are computational and experimental intervals corresponding to transition of calcium to a semimetal.

of dilatation performed within this method yield values coincident with experiment up to respectively 5 and 10%.

The method of linear MT orbitals was used to calculate the EOS of lanthanum [677], thorium [924], calcium [483], and silver [783] that agreed well with the data of static and dynamic experiments from normal conditions to megabar pressures. Figure 3.5 presents comparison of the calculated zero isotherm of calcium [483] with the shock wave data; it should be noted that in addition to a correct description of the EOS the calculation also shows coincidence with experiment of the pressure range corresponding to the transition of calcium to a semimetal. This transition is explained by a change in the electron structure leading for $V = 0.78\,V_0$ to a sharp decrease of the density of states (in the MT method) on the Fermi surface, and it is only an increase in the number $d$-states $V = 0.54\,V_0$ – that restores metallic properties.

The use of the method of linear MT orbitals for calculation of the band structure at normal and heightened pressure provided an

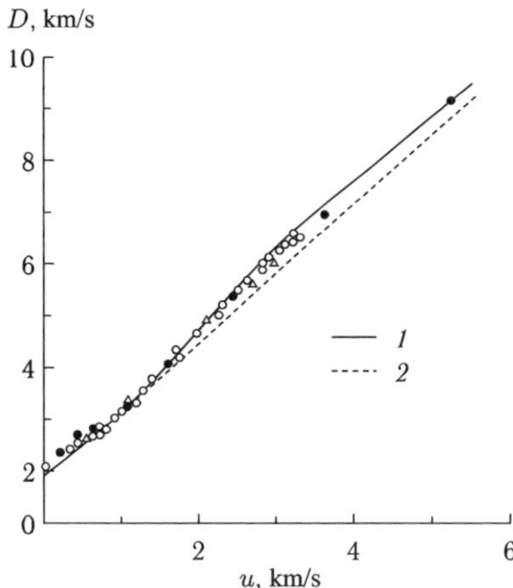

Fig. 3.6  Shock adiabat of lanthanum [781, 782]. Calculation: *1* method of linear MT orbitals allowing for $6s-5d$ electron transition; *2* Mie–Gruneisen EOS ($\gamma/V = $ constant). Symbols stand for experiment.

explanation of electron redistribution regularities and variations of the character of the electron states in cesium [352], in transition metals [811], lanthanides [351, 677] and actinides [922–924]. The thermodynamic consequences of these variations are considered in detail in Chapter 5 and here we only present Fig. 3.6 which shows the experimental data on shock compression of lanthanum and the results of calculations [677] according to which the steep run of the shock adiabat for $u \gtrsim 1\,\mathrm{km/s}$ is explained by termination of $6s - 5d$ electron transition.

All the calculations by the band theory methods are carried out in the statistical lattice approximation, which causes great difficulties in the attempt to overstep the limits of adiabatic approximation and of consistent allowance for crystal atom vibrations. The effect of temperature upon the electron terms in the model is taken into account by computational characteristics averaging over the Fermi distribution [673–675, 677], and the thermal contribution of the lattice atoms is determined from the cold curve [352, 677] in the approximation of different quasi-harmonic models describing the real vibration spectrum in a simplified way. This results in virtually insoluble difficulties in the attempt to describe within the cell model the characteristics of a metal in the vicinity of the phase transition curve (lanthanum [677], cesium [352]), where precisely the fine features of the phonon spectrum, i.e., the loss of stability of one of the vibrational modes, should be taken into account.

## 3.3   Pseudopotential models

In nontransition metals, ions occupy a small part ($\approx 10\%$) of the atomic cell volume, and when moving in a metal electrons spend some little time in the region occupied by ion. Since the characteristics of a metal are practically independent of specificities of conduction electron behavior inside the ion core, it turned out possible to replace the true interaction potential with a multielectron ion by a simplified single-particle one. Such pseudopotential must preserve the scattering properties of the original ion, but inside it can be much weaker than the true potential. Introducing a pseudopotential, one can determine the small parameter of the theory $U_k/\varepsilon_F$ ($U_k$ is the Fourier component of the pseudopotential at the point of reciprocal lattice vector) permitting the application of perturbation theory [433]. Thus, in the pseudopotential theory metal is considered as degenerate dense plasma with phonons as low-frequency collective excitations relative to the ground state corresponding to a regular position of ions.

First progress in the pseudopotential quantitative description of the properties of metals was made in the framework of different one-particle approximations that take an inconsistent account of the multiparticle character of the problem. The shortcoming of such models is allowance for only pair interaction and disregard of substantial indirect interaction of three and more ions through conduction electrons. A consistent pseudopotential theory allowing for the effective two-, three-, etc. -ion interaction is a rigorous series expansion in the electron–ion interaction parameter [138]. The use of multiparticle formalism makes it possible to determine, along with the static properties of metal, also its dynamic characteristics, i.e., the phonon spectrum.

Such an approach to nontransition metals made it possible to describe within a unified scheme the phonon spectrum in the entire phase space region, the elastic constants and the EOS [138,879,898,934–936,999–1001] (Fig. 3.7) and to determine the energy, the lattice parameters, to choose the type of most advantageous crystal structures and to analyze the effect of pressure on lattice anharmonicity [138,999–1001]. Especially successful was the use of the pseudopotential method in the calculation of the EOS of metallic hydrogen [1069] whose ion has no electron shell, which enables calculations up to high degrees of compression. Calculations of hydrogen thermodynamic characteristics were performed up to fourth-order terms of the electron–ion interaction parameter, and the EOS was found for all the structures competing in energy.

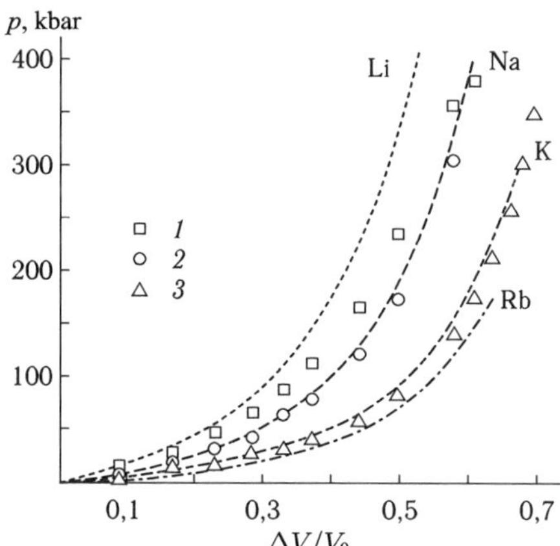


Fig. 3.7   Zero isotherms of alkali metals [999–1001]. Experimental points according to the shock-wave measurements: *1* Li, *2* Na, *3* K.

The application of pseudopotential approach at high pressures is limited by the overlap of ion cores and the lack of necessary experimental data for the formation of compressed substance pseudopotential. According to Refs. [138,999–1001], the upper limit of pseudopotential model applicability for alkaline and alkaline-earth metals lies at the pressure level of hundreds of kilobar, below which the pseudopotential distortion effects are obviously negligible and the effect of compression is reduced to only the electron density variation.

With the help of the pseudopotential model the author of Ref. [206] obtained semiempirical EOS of sodium and aluminum describing the shock data up to pressures $\approx 1$ Mbar. The shock compressibility of alkali metal halides beyond the range of their polymorphic transitions was satisfactorily calculated by choosing the parameters of the potential [999–1001]. It should be added that the EOS of ionic crystals found in the quasi-classical approximation of the Hartree–Fock method and the shock adiabats [1092, 1093] calculated using these equations agree well with the experimental data up to a pressure of the order of $\approx 1$ Mbar above which the anharmonicity and thermal excitation of electrons become significant.

# Chapter 4

# Plasma Thermodynamics

## 4.1 Historical remarks

The thermodynamic properties of a compressed and heated substance in ionized (plasma) state have always been of great interest from the fundamental and applied viewpoint. Dense hot plasma is the most widespread state of matter in the Universe as more than 95% of the visible matter located in stars, planets, and exoplanets falls to its share [294, 295].

With heightening gas temperature the neutral particles dissociate. Such plasma is sometimes called "electromagnetic" (as distinct from "quark–gluon" plasma — Chapter 12) thus laying emphasis on the leading role of Coulomb interparticle interaction in it. The long-range nature of Coulomb interaction shows up in a wide range of parameters of the state of matter. The same feature of the Coulomb potential causes difficulties in the theory [234, 745], for it forbids the application of the ordinary apparatus of statistical gas theory for plasma because of divergence of the corresponding integrals. Regrouping and summation of the most divergent terms of perturbation theory series lead, however, to finite expressions allowing for the screening effects. In dense plasma, the crucial role is also played by quantum effects which are substantial in the region of strong compression and, moreover, provide its stability and regular ideal gas low-temperature asymptotics [234].

The description of thermophysical plasma properties requires correct allowance for a strong collective interparticle interaction, degeneration, a correct separation of discrete and continuous energy spectra, account of the effects of thermal and density ionization and other complicated phenomena in a compressed and heated chemically reacting multicomponent medium. These phenomena determine the behavior of matter in a vast area of phase diagram, occupying the region from solid and liquid up to a neutral gas covering the phase boundaries of melting and boiling of metals, as well as the metal–dielectric transition region. The latter issue is now being intensely investigated in experiments on a repeated (quasi-isentropic) shock-wave compression of dielectrics and their metallization in the megabar pressure range, on pulsed-current electric explosion of conductors, as well as dielectrization of strongly compressed metals [294, 295].

We shall see below that the standard theoretical methods of thermodynamic plasma description become invalid for high densities and strong nonidealities. For this reason, certain assumptions and simplifications have to be used to create thermodynamic models. Such simplifications are not universal since the relative significance of the physical factors responsible for the macroscopic properties of plasma differs in regions with different temperature and density. We often deal with extrapolation of a model beyond its formal applicability limits. Therefore, fairly important in the study of plasma thermodynamics is the role of experiment which we shall rather frequently appeal to in this chapter.

The study of strongly compressed Coulomb systems is now one of the "hottest" and intensely developing fundamental branches of science at the interface of physics of plasma, condensed state, and atomic and molecular physics with a large variety of physical effects stimulated by nonideality and a constantly enlarging set of objects and states where this nonideality plays a decisive role.

Further in this chapter we shall direct our attention toward monograph [229] and only touch upon the latest models of the description of the equation of state (EOS) of plasma. Before turning to the analysis of thermodynamic relations for plasma, we shall make some general remarks [229].

For the general model of a substance as a system formed by nuclei and electrons we are indebted to Rutherford (1911). The potential of a lattice consisting of point charges was first calculated by Ewald (1921) and specified by Wigner (1932). The charge screening theory was introduced by Debye and Huckel (1923) and was later modified by Macke (1949), Mayer (1950), Bohm and Pines (1953), Lindhardt (1954), Mott (1959), Vedenov and Larkin (1959), Kelbg and De Witt (1963), Singwi, Tosi, Land, and Sjolander (1968) and others.

The charged particle theory is undoubtedly the most important part of any plasma theory. However, in the general case plasma contains both neutral and charged particles in different combinations. Two elementary plasma theories consider the limiting cases where either electrically neutral or charged particles only exist. In the former case corresponding to low temperatures and low densities, all the electrons remain in bound state and plasma behaves as an ordinary neutral gas. The elementary gas theory (Chapter 2) goes back to the classical Van der Waals'es works accomplished about a hundred and twenty years ago. In the latter case, the theory is based on the works by Debye and Huckel (1923).

The analysis of partially ionized plasma dates from works of M.N. Saha who published in 1921 in Berlin the paper *"Versuch einer Theorie der physikalischen Erscheinungen bei hohen Temperaturen mit Anwendungen auf die Astrophysik"*.

At high particle densities the Van der Waals and Debye–Huckel theories do not work. As was shown in Chapter 2 the extension of the neutral gas theory to the case of high densities was elaborated by Percus and Yewick (1962), Carnahan and Starling (1969), and also by Ross *et al.* (1980, 1983). The extension of the Percus–Yewick theory to charged rigid spheres is the mean-spherical approximation (MSA) obtained by Weissman and Lebowitz (1972), Trioli, Grygera and Bluhm (1976). The extensions of the mean-spherical approximation are due to Kihara (1978), Parrinello and Tosi (1979), and also to Rosenfeld (1980).

The quantum-statistical theory of thermophysical properties of plasma is based on the methods worked out by Montroll and Ward (1958), Martin and Schwinger (1959), Vedenov and Larkin (1959), Abrikosov, Gor'kov and Dzyaloshinskii (1962), Kadanoff and Baym (1962), Kelbg (1963), De Witt (1963) and March (1968). With the help of these works Kopyshev (1968) and Ebeling (1968) obtained exact results of first corrections for limiting laws. Exact results for lower boundaries and the properties of stability of thermodynamic functions were first obtained by Lyson and Lenard (1967, 1968), Lebowitz and Lieb (1969, 1972), Lieb and Thirring (1975). The solution of bound state problems is closely related with the works by Mott (1959), Larkin (1960), Ebeling (1966, 1968), Kremp and Kraeft (1968, 1972), Kiliman *et al.* (1977), Roepke *et al.* (1978). The random phase approximation goes back to the works by Bohm and Pines (1953), Lindhardt (1954), Nozier and Pines (1958, 1959), Singwi *et al.* (1968), Ichimaru (1973).

The Thomas–Fermi approximation (Chapter 6) and the related density functional approximation are associated with the works of Hohenberg and Kohn (1964), Kirzhnits *et al.* (1975), Lieb and Thirring (1975), Perrot (1982), Darm-Vardan *et al.* (1982).

Another direction is the cluster decomposition method working well in the case of rarefied plasma. This method takes root in the works of Kelbg

(1963, 1964), Ebeling (1968, 1969), Kraeft and Kremp (1968), Rogers (1971, 1974), Solander, Norman and Filinov (1981).

The quantum-statistical Monte Carlo simulation (see Chapter 5) was developed by Ceperly and Older (1980, 1981), Zamalin, Norman and Filinov (1977) and many others. This method is fairly promising but is still at the incipient stage, and we shall consider it in Chapter 5.

The extensive calculations of thermophysical properties of plasma based on the combination of all the above-mentioned techniques were first elaborated by two independent groups, one in Moscow and Chernogolovka [320, 398], the other in Livermore [214, 449, 849, 863].

## 4.2 Hierarchy of models

In the sections to follow, we shall try to present the basic computational methods of thermophysical properties of plasma.

The same as in the description of solids and liquids (Chapters 2 and 3), a rather universal *adiabatic approximation* is used in plasma thermodynamics. It is valid for any density and temperature to an accuracy proportional to the electron-to-nucleus mass ratio $m_e/m$. The main assumption here is that the electron motion can be considered for a motionless nucleus. As a result, the potential energy can be obtained exclusively depending on the nucleus coordinates, which allows one to describe the motion of the nucleus [655].

The next, rather general approximation is the assumption concerning the classical character of nuclear motion (the quantum case is considered in Chapter 5) which is violated at low temperatures and high densities. The applicability condition is the restriction imposed on the ratio $\eta$ of the thermal de Broglie wave length of the nucleus $\Lambda = h(2\pi m k_{\mathrm{B}} T)^{-1/2}$ to the radius of the volume that falls to the share of one nucleus $R_0 = (3m/4\pi\rho)^{1/3}$:

$$\eta = \frac{\Lambda}{R_0} = \left(\frac{2}{9\pi}\right)^{1/6} \frac{h\rho^{1/3}}{m^{5/6}(k_{\mathrm{B}}T)^{1/2}} \ll 1.$$

Here $\rho$ is mass density, $m$ is nucleus mass, and $T$ is temperature.

The electrons in a substance are conditionally divided into bound and free. The latter are also referred to as conduction electrons. Bound electrons are localized near the nuclei owing to a strong electron–nucleus interaction, they move along with the nuclei and always exhibit quantum properties. With heightening temperature and density the bound electrons become free. The value of the quantum effects in a gas consisting of free electrons is determined by the ratio of temperature ($k_BT$) to the Fermi energy $\varepsilon_F = (h^2/2m_e)(3n_e/8\pi)^{2/3}$, where $n_e$ is the electron density.

Condensed matter is characterized by the inequality $k_BT \leqslant \varepsilon_F$. Under this condition, quantum properties of free electrons dominate. The theoretical description can be simplified provided that the thermal energy concentration does not reach the value with which the bound electrons can pass over to an excited state or partially come off. At high temperatures, when $k_BT \gg \varepsilon_F$, the free electrons behave as classical particles.

With lowering density and heightening temperature the condensed matter smoothly passes over to a gaseous state or crosses the two-phase regions (Figs. 2.1 and 2.2). For metals (under normal conditions) the metal–dielectric transition can be thought of as a conditional interface between the gaseous and liquid states. In the plasma state, the region of a relatively weak interaction exists characterized by low values of the nonideality parameter $\Gamma = Ze^2/4\pi\varepsilon_0 R_0 k_B T$ which is the potential-to-kinetic mean-energy ratio.

The plasma–gas interface on the density-temperature plane can be determined by defining the degree of atomic ionization as a certain low value. The gas phase of a substance (Chapter 2) corresponds to a system of neutral weakly interacting classical particles.

The authors of many papers neglect the effect of temperature and consider such a system as strongly degenerate, formally assuming $T = 0$ [677, 685, 1065]. These authors calculate the EOS of a cold substance (e.g., helium) at a high pressure and thoroughly investigate the transition to the metallic state.

The region of dense plasma at intermediate temperatures is the least convenient for examination. The most general method applicable in this region is to consider the nuclear motion on the basis of classical mechanics. Given this, the model is analyzed representing such plasma as an ensemble of electrically neutral weakly interacting cells of volume $V_0$. The area of adequate applicability of this model is strictly speaking unknown, but the limits of its physical groundlessness can be determined with the conception of the Debye screening radius $r_D = (\varepsilon_0 k_B T/e^2 n_e)^{1/2}$ of electron. If the free electrons are unable to screen the ion in a cell of volume $V_0$, no physical grounds exist for the use of such a cell model. In a gas consisting of degenerate electrons, the screening length is determined as $r_{TF} = (\pi/3n_e)^{1/6}(h^2\varepsilon/4\pi m_e e^2)^{1/2}$ and can be used to find the applicability limits of the cell model in this case.

We shall now dwell on the Wigner–Seitz cell approximation which we employed in Chapters 3 and 6. Each cell contains a nucleus with charge $Z$

and the corresponding number of electrons, which provides electric neutrality [455, 501]. The electron motion in the field of the nucleus inside a given cell is described within the chosen model. The influence of electrons and nuclei of neighboring cells is taken into account through the boundary conditions. The contributions of nuclei and electrons are typically described irrespectively of each other in accordance with the adiabatic approximation. A real Wigner–Seitz cell of a complicated geometrical shape is usually replaced in practice by a spherical cell of volume $V_0$.

We shall now say a few words of the physical models describing the electron subsystem. One of the standard approaches is the Green function method [234]. Another widespread approach is represented by the Singwi–Tosi–Land–Sjolander theory [655, 913].

The most consistent and effective methods include the density functional theory (DFT) (see Chapter 7). This method is based on the theorem which says that the free energy of an electron system in an external field can be expressed as an electron density functional. This functional is usually unknown, and the use of any of the approximations gives rise to the so-called quantum-statistical models or quantum-mechanical Hartree type models (see Sec. 6.1). As will be seen in Chapter 7, the employment of DFT approach turned out to be very successful in condensed state physics.

The first among the statistical models was the Thomas–Fermi model (Chapter 7) in which the electron density is calculated in a quasi-classical approximation with employment of quasi-classical potential [547]. When regular and oscillation corrections were taken into consideration, the theoretical results came closer the experimental ones, and shell effects were included in the thermodynamic model.

The range of applicability of Thomas–Fermi models determined by comparison with the results of the dynamics experiments [633] and for condensed densities corresponds to pressures $p \geq 10^2\,\text{GPa}$. A still more consistent description of bound states in plasma is implemented with the help of quantum-mechanical self-consistent field models in which the wave functions of electrons are solutions of the Schrödinger equation with the boundary conditions of the model. The latter represent the translational symmetry of the crystal lattice or an approximate description of this symmetry (see Chapter 3). The first formulations of quantum-mechanical models applied the Thomas–Fermi approximations in combination with the electrostatic theory.

The Hartree–Fock model (ch. 3, Sec. 6.1) allowing for exchange interaction of electrons is the most complete one-electron model, but at the same time the most sophisticated. A simpler description of exchange effects is attained in the Hartree–Fock–Slater approximation making use of the so-called local effective exchange potential. In concrete calculations based on the Hartree–Fock–Slater model [742], the quasi-classical approximation was used for strongly excited states of bound electrons and for free electrons. The other author [865] proposed very simple boundary conditions allowing simulation of the upper and lower boundaries of the energy bands of electrons. Sin'ko [917] closed this model by the approximation for the density of states which has the form typical of free electrons and proposed the so-called self-consistent field model.

As has already been said, the adiabatic approximation allows a separate calculation of the contribution of electrons and the contribution of nuclei to different thermodynamic functions. In the early works the nuclei were regarded as an ideal gas. The model of point ions in a medium with a uniformly distributed negative charge was created taking into account nonideality of nuclear motion [140, 571].

The common deficiency of the above-described cell models is the impossibility of allowing for the interparticle correlations at distances exceeding a unit cell size. The cell models are unable to describe the typical states of plasma characterized by long-range correlations when the Debye sphere contains a large number of particles. The limited atom model combines the ideas of both solid state and plasma; a still simpler approach is the "chemical model" of plasma described below.

## 4.3   Chemical model. Dimensionless parameters

Following paper [229] we shall consider plasma consisting of free electrons with the particle number density $n_e$ and nuclei or ions with the particle number density $n_z$ with positive charges $ze$ ($z = 1, \ldots, Z$). We shall consider the particle motion to be nonrelativistic when the Salpeter parameter is

$$\chi_r \equiv \frac{p_F}{m_e c} \sim \left( \frac{\rho_6 \langle Z \rangle}{A'} \right)^{1/3} \ll 1,$$

$p_{\mathrm{F}} = \hbar k_{\mathrm{F}} = \hbar(3\pi^2 n_e)$, $\rho_6 = \dfrac{\rho}{10^6}$ g/cm$^3$. The opposite, relativistic case will be considered at the end of this section.

We define the resulting positive charge density $n_+$ and the mean charge by the expressions

$$n_+ = \sum_{i=1}^{Z} n_i, \tag{4.1}$$

$$\langle z^p \rangle = n_+^{-1} \sum_{i=1}^{Z} z_i^p n_i. \tag{4.2}$$

Introducing the average interelectron distance

$$d_e = \left( \frac{3}{4\pi n_e} \right)^{1/3}, \tag{4.3}$$

and the distance between positive charges

$$d_+ = \left( \frac{3}{4\pi n_+} \right)^{1/3}, \tag{4.4}$$

we determine two dimensionless correlation parameters:

$$\Gamma_e = \frac{\varepsilon^2}{\theta_e d_e}, \tag{4.5}$$

$$\Gamma_+ = \frac{\varepsilon^2}{\theta_+ d_+} \langle z^{5/3} \rangle \langle z \rangle^{1/3}. \tag{4.6}$$

Here $\theta_e$ and $\theta_+$ are averaged kinetic energies including degeneracy. In the classical case they have the form

$$\theta_e = \theta_+ = k_{\mathrm{B}} T. \tag{4.7}$$

In the quantum case they are determined by the formula

$$\theta_k = \frac{p^{\mathrm{id}}}{n_k}, \tag{4.8}$$

where $p^{\mathrm{id}}$ is the ideal Fermi gas pressure. In other words,

$$\theta_k = \frac{2 k_{\mathrm{B}} T}{n_k \Lambda_k^3} I_{3/2}(\alpha_k), \tag{4.9}$$

$\alpha_k$ is the dimensionless chemical potential, $I_{3/2}(x)$ is the Fermi integral of order $3/2$, and $\Lambda_k$ is the thermal De Broglie wave length. The boundary between the classical and quantum regions can be established with the help of the dimensionless degeneracy parameter

$$\zeta = n_k \Lambda_k^3, \tag{4.10}$$

specifying the ratio of the mean interparticle distance proportional to $n_k^{-1/3}$, to the typical quantum De Broglie wave length $\Lambda_k$. Next, we determine the parameter describing the electron–ion interaction intensity:

$$\xi_{\mathrm{e}} = 2\langle z \rangle \left( \frac{I}{\theta_{\mathrm{e}}} \right)^{1/2}, \tag{4.11}$$

where $I$ is the hydrogen ionization energy. Then we introduce the thermal De Broglie wave length for the relative ion and electron motion:

$$\lambda_{\mathrm{ii}} = \frac{\hbar}{(m_{\mathrm{i}} k_{\mathrm{B}} T)^{1/2}}, \quad \lambda_{\mathrm{ee}} = \frac{\hbar}{(m_{\mathrm{e}} k_{\mathrm{B}} T)^{1/2}}, \quad \lambda_{\mathrm{ie}} = \frac{\hbar}{(2 m_{\mathrm{ie}} k_{\mathrm{B}} T)^{1/2}}, \tag{4.12}$$

where $m_{\mathrm{ie}}$ is the reduced mass for an electron–ion pair.

Using the Landau length $\ell$,

$$\ell = \frac{e^2}{4\pi \varepsilon_0 k_{\mathrm{B}} T}, \tag{4.13}$$

we can determine the interaction parameters

$$\xi_{\mathrm{ee}} = \frac{\ell}{\lambda_{\mathrm{ee}}}, \quad \xi_{\mathrm{ii}} = \frac{z^2 \ell}{\lambda_{\mathrm{ii}}}, \quad \xi_{\mathrm{ie}} = \frac{z \ell}{\lambda_{\mathrm{ei}}}. \tag{4.14}$$

Figures 4.1 and 4.2 present the graph of the temperature dependence of density for single-charge plasma ($Z = 1$). Figure 4.1 corresponds to the case of mass-symmetric plasma (electron–positron plasma, electron–hole plasma, etc.) and Fig. 4.2 to the case of mass-asymmetric plasma (for example, gas plasma). Strong interactions are observed inside the line $\xi_{\mathrm{e}} = 1$ and very strong interactions inside the line $\xi_{\mathrm{e}} = 10$. This will show up, for instance, in the effects of scattering and bound states. In Fig. 4.1 the curves $\Gamma_{\mathrm{e}} = 1$ and $\xi_{\mathrm{e}} = 1$ separate the regions of strong and weak correlations and electron interactions, respectively. The physical meaning of the expression $\Gamma_k \geq 1$ is that the potential energy exceeds the mean kinetic energy and the expression $\xi_k \geq 1$ implies that the interactions are significant for the microscopic scattering processes, that is, the perturbation theory (Bohr approximation) is violated. It is of interest to note that both

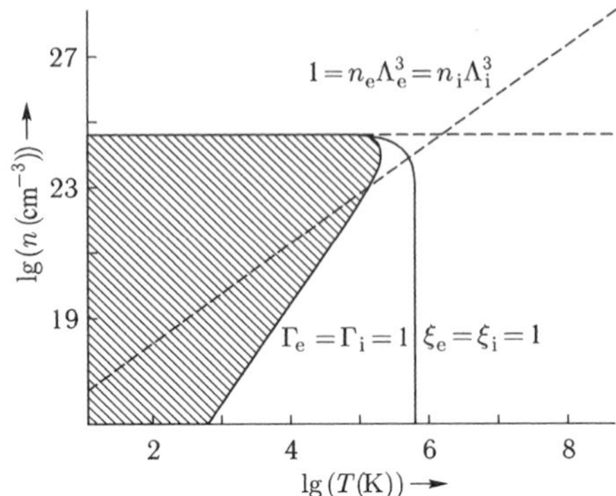


Fig. 4.1   Nonideality region on the graph of temperature dependence of density for symmetric plasma (e.g., electron–positron plasma).

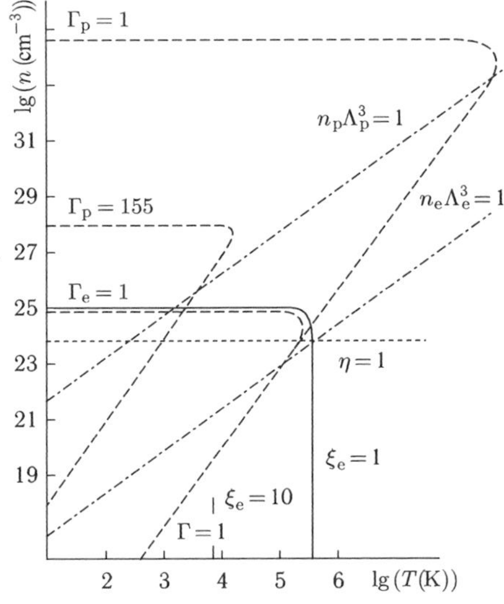


Fig. 4.2   Ranges of electron–proton plasma parameters. Dot-and-dash lines correspond to the boundary of respectively electron and proton degeneracy. The dashed line corresponds to the transition due to Pauli blocking.

the interaction region and the nonideality region are both restricted by the left bottom angle of the $n$–$T$ plane. In this respect high density and high temperatures produce the same qualitative effect: the interactions and nonideality play a smaller role and it is the kinetic energy alone that governs the behavior. This effect is certainly due to the strong increase of Fermi-gas kinetic energy at high temperatures and/or high densities.

The influence of the quantum effects depends strongly on the particle masses. Consequently, the $n$–$T$ plane for mass-symmetric systems (Fig. 4.1) has a much simpler structure than the same plane for mass-asymmetric plasma (Fig. 4.2). Figure 4.2 demonstrates that the high-density behavior of electrons and ions differs radically because of different values of mass $m_\mathrm{p} \gg m_\mathrm{e}$. Lighter electrons reach degeneracy at lower densities. Accordingly, the intermediate region

$$\Lambda_\mathrm{e}^{-3} < n < \Lambda_\mathrm{p}^{-3}, \tag{4.15}$$

exists where the proton's behavior follows the laws of classical physics and the electrons form a degenerate quantum gas. In this region, the occurrence of ion lattice or, in the end, molecular lattice can be expected.

As an example, we shall consider the Sun. The typical parameters of the Sun are:

$$\Gamma_+ < 0.1; \quad 0.01 < \xi < 100. \tag{4.16}$$

This testifies to the fact that the solar matter is in conditions of weak nonideality but strong two-particle interactions.

Another important parameter is a relative occupation of the existing space by bound states:

$$\eta = \frac{4\pi}{3} \sum_b n_b R_b^3, \tag{4.17}$$

where $n_b$ and $R_b$ are the density and the effective radius of the bound state $b$. In the region $\eta > 1$ depicted in Fig. 4.2, the bound states that require a relatively large space decay because of the Pauli blocking effects which make the existence of bound states thermodynamically unfavorable.

The term "Pauli blocking" is used here and below to define the specific class of effects associated with the Pauli exclusion principle [234,581]. From the properties of the wave function under electron rearrangement it follows that each state can be occupied by one electron only. Furthermore, this is precisely the reason why free electrons cannot penetrate into the internal part of the atoms. In spite of the fact that at least in the Bohr's picture the internal part of the atoms is almost empty, the electrons and other particles containing electrons cannot penetrate into the atoms. This is the essence of the Pauli blocking effects playing an important role in hot dense plasma.

The Monte Carlo computations for one-component plasma (see Chapter 5) played a crucial role in the development of the theory of thermophysical properties. This fruitful direction was started by the pioneering studies of Brush, Sahlin and Teller (1966) and was then forwarded by Hansen (1972, 1973), De Witt (1976) and also Zamalin, Norman, and Filinov (1977) and others.

The most thoroughly investigated, although exceedingly simplified model of Coulomb systems is the model of one-component plasma placed for maintaining its stability against the homogeneous background of the compensating charge of opposite sign. In such plasma, recombination is absent and the form of the interaction potential raises no doubts.

This model has no neutrals, and the ion size $r_\mathrm{i} \sim a_\mathrm{B}/Z^{1/3}$ ($a_\mathrm{B}$ is the Bohr radius) can be neglected compared to the average interparticle distance.

This plasma model is the object of studies in the framework of the numerous asymptotical theories [83, 507, 657] based on the expansion in small parameters, regrouping, and random summation of perturbation theory series, as well as approximations exploiting the technique of integral equations borrowed from the theory of fluids [938]. The Monte Carlo simulations [140, 421, 931, 932] based on direct machine computation of the configuration integral serve as standard here, which makes it possible to gain exhaustive information on the classical one-component plasma in an exceedingly wide range of parameters: $\Gamma = 0.05 \div 300$ ($\Gamma = e^2/k_\mathrm{B}Tr_{ws}$), where the mean interparticle distance is $r_{ws} = (3/4\pi n)^{1/3}$, i.e., $\Gamma = (\Gamma_\mathrm{D}^2/3)^{1/3}$), and to construct simple approximation expressions [931, 932]. The background charge screening was taken into account in paper [451] in Monte Carlo simulations by calculating the background permittivity in the linear response approximation. The analytical methods [864, 957] based on the variational principle exploit perturbation theory to describe the ion background, where calculations by the rigid sphere model are taken as zero approximation. The results of such an approach turn out to be close to the Monte Carlo simulations especially for high $\Gamma$ values (Fig. 4.3). It should be noted that the very calculations of the properties of one-component plasma are used in some cases as zero approximation, more realistic than the one in the rigid sphere model, in determination of thermodynamic liquid–metal characteristics (see Chapter 2).

The extrapolation properties of different asymptotic approximations can be radically improved [400] (Fig. 4.4) by additionally imposing on the calculated correlation function the condition of local electroneutrality corresponding to the equality of the polarization plasma cloud charge and the charge of the particle creating this cloud in plasma. This and analogous conditions imposed on the screening of dipole and quadrupole moments [400] do not depend on the interaction strength in the system and follow from

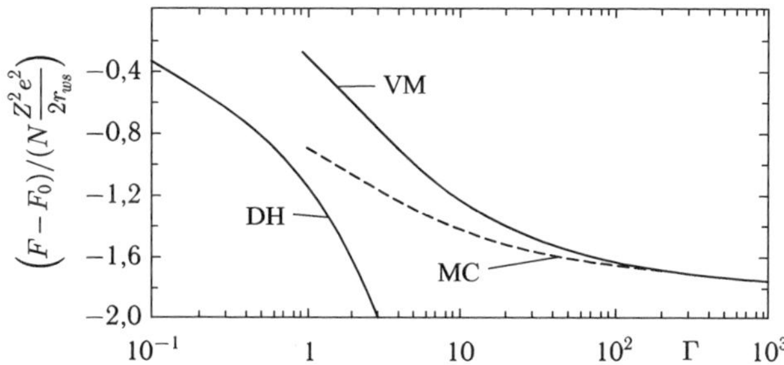

Fig. 4.3 Free energy of one-component plasma [864, 957]. MC are Monte Carlo simulations, VM is variational method (upper boundary), DH is Debye–Huckel law (lower boundary).

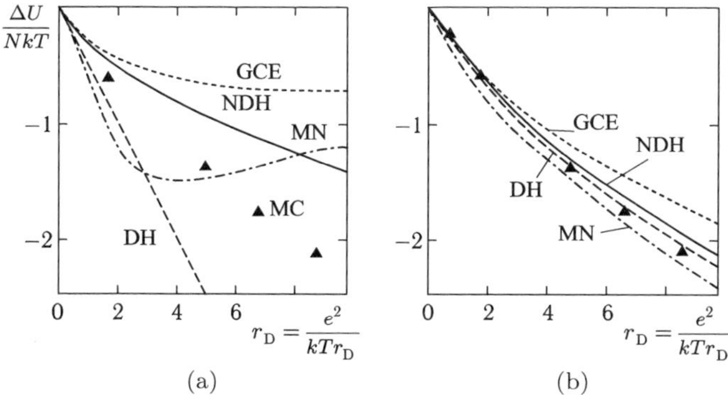

Fig. 4.4 One-component plasma interaction energy without allowance (a) and with allowance (b) for the local electroneutrality condition [400]. MC are Monte Carlo computations, DH is the Debye–Huckel law, NDH is nonlinearized DH, GCE is DH in a grand canonical ensemble, MN: with allowance for the dependence of the screening radius on $\Gamma_D$.

the fundamental fact of the existence of thermodynamical limit of Coulomb systems [607]; allowing for them one can effectively correct approximate correlation functions found from model considerations or constructed by the expansions in small parameters. Note that the solution for one-component plasma of the equations of superentangled chains [938], in which the necessary conditions of positiveness of the correlation function and local electroneutrality are met automatically, yield the results close to Monte Carlo simulations.

The numerical calculations of the properties of one-component classical plasma performed in Refs. [140, 421, 931, 932] testify to the occurrence of thermodynamic anomalies and short-range order under considerable compression, which is interpreted as crystallization [797]. According to the latest results [931, 932] allowing for the dependence of liquid- and solid-phase energy on the number of test particles used in Monte Carlo simulations, in the limit $N \to \infty$ this transition corresponds to the value $\Gamma = e^2 n^{1/3}/kT = 178 \pm 1$.

In real plasma, degeneracy occurs with increasing density, and the interaction is typically described in this case by the dimensionless parameter $r_s = r_{WS}/a_0$ ($a_0 = \hbar/me^2$ is the Bohr radius) characterizing the ratio of the average interparticle distance to the atomic one. In strongly compressed plasma, this parameter is small, which permits the application of perturbation theory and the calculation of several first terms of expansion by the coupling constant [3, 4, 657]. Wigner formulated [1040] the crystal model of rarefied degenerate ($r_s \gg 1$) plasma assuming that as a result of strong Coulomb interaction the electron component goes over into a face-centered cubic lattice. The properties of such a crystal and the conditions of its melting were thoroughly investigated [996] with allowance for zero and thermal vibrations, anharmonism, and exchange effects.

The characteristics of quantum electron plasma were also determined by the Monte Carlo methods [162, 164]; the calculations showed transition from a nonpolarized to a ferromagnetic liquid for $r_s > 26$ and Wigner crystallization for $r_s > 67$.[1] The arguments concerning overestimation of this quantity are presented in Ref. [218], where the local field effects are considered and stability of degenerate electron liquid under charge and spin density wave propagation is analyzedin the language of permittivity. Specific features of phase transitions in Coulomb systems were examined in detail in review [469], where the fact was noticed that the Wigner crystal existence conditions themselves correspond to the region of thermodynamic instability under gas–liquid transition. Note in this connection that no experimental observations of Wegner crystallization in Coulomb systems now exist, and the reports on the experimental discovery of Wigner crystal in experiments involving condensed explosives [192] turned out to be erroneous, as was demonstrated by more detailed studies [314]. We shall return to the variable phase transitions in Chapter 8.

---

[1] The dependences of the ground state energy of different phases on $r_s$ are very close, which leads to a very high sensitivity of the position of transition points to even slightest inaccuracies in the EOS of competing phases [162, 164, 996].

The one-component plasma model is on the hole effective for calculation of the electron gas properties in simple metals, where the main contribution to system's energy is made by direct Coulomb repulsion between ions and their indirect attraction through conduction electrons [364]. Under certain conditions this model describes also the plasma thermodynamics of metal-ammonium solutions, strongly doped semiconductors, the hydrogen–helium plasma of Jupiter and Saturn and the nuclear liquid in dense and superdense stars.

The main shortcoming of the OCP model consists in an exceedingly simplified account of the opposite-sign charge forming a structureless compensating background. More adequate plasma models involve an explicit allowance for the structure and interaction of charges of all signs with an obligatory description of quantum effects under Coulomb interaction leading to convergence of the coordinate part of Gibbs probability upon approaching unlike charges, which in the end provides system's stability. Numerous attempts are known [745] to preserve classical formalism by introducing Coulomb potential cut-off at small distances with induced exclusion of configurations with approached charges. The final result includes the cut-off parameter and, moreover, many models for $\Gamma_D \gtrsim 1$, when the cut-off radius becomes comparable with the interparticle distance, lose thermodynamic stability.

## 4.4 Physical model. Thermodynamic relations

A consistent quantum-mechanical consideration of the problem proceeds from the Hamiltonian containing a complete interaction among all the charges. This corresponds to the physical model of multicomponent plasma in which the contribution of the discrete spectrum appears to be finite and occurs simultaneously with the contribution of the free charge continuum [234]. The physical model is the most general and consistent for real plasma, but practical calculations within this model are rather cumbersome and have not yet become sufficiently widespread [585,847] because its application to the plasma of multielectron elements involves as a constituent a quantum-mechanical calculation of the internal structure of bound states.

The physical model is more consistent, but more complicated for calculation of thermodynamic functions of plasma [234]. For complex systems and, in particular, for plasma containing atoms of several chemical elements the difficulties encountered in realization of the physical model increase tenfold, and therefore a simpler chemical model is normally used in practice of calculations. It is a well-known fact that this model presupposes the presence of a certain kind of particles that have a continuous energy spectrum and the effect of the internal structure of particles (bound states) is allowed for by introducing individual statistical sums. This is the source of certain difficulties associated with the conditional character of the division of all the degrees of freedom into translational and intrinsic. In each particular model these difficulties can be overcome proceeding from these or those physical considerations [234]. Many works can be found in literature devoted to the description of physical aspects of application of the chemical model for calculation of thermodynamic functions and the corresponding mathematical procedures necessary for investigation of nonideal multicomponent plasma.

In rarefied plasma configurations with approached particles are hardly probable, which permits model simplifications. The main one consists in a separate description of states of discrete and continuous spectra respectively determining the internal atomic and ionic structure and the behavior of free particles. Such an approximation underlies the so-called chemical model most popular in plasma physics because the number of different kinds of particles $N_j$ is determined in this case by the condition of chemical equilibrium [229, 598] (4.20) and all the hypotheses concerning the structure of particles and their interaction is contained in the expression for free energy (4.18):

$$F(V, T, \{N_j\}) = F_k + F_d + F_c + F_i.$$

The contribution of the discrete spectrum $F_d$ in this model appears to be singled out and is calculated irrespective of the contribution of the continuum represented by the kinetic part $F_k$; various interparticle interaction corrections are described by the term $F_c$. If the radiation is in local thermodynamic equilibrium with matter, one should also take into account the photon gas energy contribution which becomes substantial under extremely high heating of matter or its strong rarefaction: $F_d = -(4\sigma/3c)VT^4$.

The preliminary remarks made above with respect to the existing methods of calculation of chemical and ionization equilibrium make it possible to analyze the problems encountered in the development of computational methods of thermodynamically nonideal plasma. They are largely due to the necessity of calculating parameters of the EOS and the thermodynamic functions of matter in a broad range of temperatures and densities. Such topical physical problems as the analysis of shock-wave compression of a substance, the interaction of high-power radiant fluxes with a substance need an adequate description of nonideal plasma thermodynamics beginning from the region of condensed state up to extremely high temperatures and pressures corresponding to completely ionized strongly compressed plasma. The plasma composition changes radically upon the temperature and pressure variation in such wide limits. Hence, the supposed method of computation must on the one hand foresee the formation of polyatomic molecules and molecular complexes and on the other hand allow for the processes of dissociation and ionization of arbitrary multiplicity. Furthermore, at high densities, depending on temperature most diverse interparticle interaction mechanisms may occur determining the thermodynamic properties of the system, including the Coulomb interaction between charged particles, the interaction of neutral particles among themselves and charged particles with neutral ones, and at extremely high temperatures and densities — the short-range repulsion of ions because of electron shell overlapping.

Strong interparticle interaction also affects the structure of bound states, which is expressed in the dependence of statistical sums of particles on density and not on temperature only. This restricts the application of ordinary methods of thermodynamic computations of multicomponent systems [23, 452] based on the use of thermodynamic functions of individual substances [453], the region of low densities, where this effect can be neglected. It should also be taken into account that an adequate description of a substance upon a broad-range variation of thermodynamic parameters can be realized by different models beginning from the ideal-gas approximation to the degenerate electron gas model with strong Coulomb interaction of ions. Since the applicability of each of the models is limited, the computational method must be universal enough lest the use of different nonideality models should lead to a substantial change in the mathematical procedure.

We shall now briefly formulate the claims to the method of computation of thermodynamic properties of nonideal multicomponent plasma:

(1) Possibility of including in calculation any kind of particles from complicated polyatomic complexes to ions of arbitrary multiplicity.

(2) Universality with respect to plasma nonideality models allowing for both the interaction of free particles and the effect of this interaction to the structure of the energy spectrum of bound states.

As is known, in the state of thermodynamic equilibrium the free energy of the system reaches its minimum [598]. From this, equations can be derived typically called equilibrium equations relating the numbers of different kinds of particles. For this purpose we shall consider a system of $L$

kinds of particles at a temperature $T$ inside a volume $V$ and containing $K-1$ chemical elements and electrons. The electrons can be considered here as a separate chemical element. According to the chemical model of matter the nuclei and electrons can form compound particles of different kinds, namely, atoms, ions, molecules, etc., and so each kind is characterized by the corresponding number of particles $N_j$. Then the free energy of the system can be written as

$$F(\{N_i\}, V, T) = \sum_{j=1}^{L} N_j k_{\mathrm{B}} T \left( \ln \frac{n_j \Lambda_j^3}{\sigma_j} - 1 \right) + \Delta F_{\mathrm{int}}(\{N_i\}, V, T). \quad (4.18)$$

where $n_j = N_j/V$ is the density (concentration) of the number of particles of sort $j$, $\Lambda_j$ is the de Broglie thermal wave length, $\sigma_j$ is the inner statistical sum in which zero energy corresponds to the state of a complete particle disintegration into nuclei and electrons; $\Delta F_{\mathrm{int}}$ is correction to the ideal Boltzmann part of free energy caused by interparticle interaction and degeneracy. Since the system does not exchange particles with the surrounding medium, the number of nuclei of each of $K$ chemical elements, as well as the number of electrons remain unchanged. Accordingly,

$$\sum_{i=1}^{L} \nu_k^i N_i = \tilde{N}_k; \quad k = 1, 2, \ldots K, \quad (4.19)$$

where $\tilde{N}_k$ is the total number of nuclei of kind $k$ in the system, $\nu_k^i$ is the number of nuclei of kind $k$ in an $i$th kind particle. The condition of extremum of system's free energy inside volume $V$ at a temperature $T$ for constant $\tilde{N}_k$ is expressed as follows:

$$\sum_{j=1}^{L} \left( \frac{\partial F}{\partial N_j} \right)_{V,T,\{N_i\}} \delta N_j = 0. \quad (4.20)$$

Expression (4.19) shows that the differentials $N_j$ are related as

$$\sum_{j=1}^{L} \nu_k^j \delta N_j = 0, \quad k = 1, 2, \ldots K. \quad (4.21)$$

If we agree to designate bear nuclei by the numbers of kinds of mixture from 1 to $K$, then from (4.21) it clearly follows that

$$\delta \tilde{N}_k = \sum_{i=K+1}^{L} \nu_k^i \delta N_i, \quad k = 1, 2, \ldots, K. \quad (4.22)$$

Substituting (4.22) to (4.20) and bearing in mind that the chemical potential is determined by the expression

$$\mu_i = \left(\frac{\partial F}{\partial N_i}\right)_{V,T,\{N_j\}},\tag{4.23}$$

we arrive at the relation

$$\sum_{i=K+1}^{L}\left(\mu_i - \sum_{k=1}^{K}\nu_k^i \mu_k\right)\delta N_i = 0,\tag{4.24}$$

where the differentials $\delta N_i$ are independent. Consequently, from (4.24) we have $L - K$ relations

$$\mu_j = \sum_{k=1}^{K}\nu_k^j \mu_k, \quad j = K+1,\dots,L.\tag{4.25}$$

Obviously, each of relations (4.25) corresponds to a chemical reaction of the form

$$A_{\nu 1}^1 A_{\nu 2}^2 \dots A_{\nu M}^K \Leftrightarrow \nu_1 A^1 + \nu_2 A^2 + \dots + \nu_k A^k,\tag{4.26}$$

which represents the decay of the complex particle $A_{\nu 1}^1 A_{\nu 2}^2,\dots,A_{\nu M}^K$ into nuclei and electrons. Formula (4.18) for free energy implies that the chemical potential has the form:

$$\mu_j = \ln\left(\frac{n_j \Lambda_j^3}{\sigma_j}\right) + \Delta\mu_j^{\mathrm{int}}(\{n_i\},T), \quad j = 1,2,\dots,L.\tag{4.27}$$

Then equation (4.25) assumes the form of ordinary chemical and ionization equilibrium equations

$$\ln n_j - \sum_{k=1}^{K}\nu_k^j \ln n_k = \ln K_n^j(\{n_i\},T),\tag{4.28}$$

whose analogue for an ideal gas is the mass action law and for an ideal plasma — the Saha equations [598]. The right-hand side of (4.28) is the equilibrium constant of the chemical reaction (4.26) which can be written as

$$\ln K_n^j(\{n_j\},T) = \ln\frac{\tilde{\sigma}_j}{\Lambda_j^3} + I_j - \sum_{k=1}^{K}\nu_k^j\left[\ln\frac{\tilde{\sigma}_k}{\Lambda_k^3} + I_k\right]$$
$$- \left(\Delta\mu_j^{\mathrm{int}}(\{n_i\},T) - \sum_{k=1}^{K}\nu_k^j \Delta\mu_k^{\mathrm{int}}(\{n_i\},T)\right),\tag{4.29}$$

where $\Delta\mu_j^{\text{int}}(\{n_i\}, T)$ is correction to the ideal part of chemical-potential responsible for the interparticle interaction, $I_j$ is the energy necessary for the decay of $j$th particle into nuclei and electrons, and the expression in parentheses represents the change in the binding energy of electrons and nuclei in a complex (compound) particle due to the interparticle interaction which, in the case of ionization, corresponds to the "ionization potential lowering" in plasma. Here, the statistical sum

$$\tilde{\sigma}(\{n_j\}, T) = \sum_i \omega_i(\{n_j\}, T) g_i e^{-E_i/k_{\text{B}}T} \tag{4.30}$$

is calculated over all the energy levels of the internal degrees of freedom of the particle, counted from its ground-state energy, the factor $\omega_i$ is generally dependent on both the temperature and the density and is determined by the nonideality model adopted in the calculations. To determine the statistical sum $\tilde{\sigma}(\{n_j\}, T)$ (4.30) one should undertake summation over the number of bound levels, which is generally infinite [1054, 1055]; this leads to the divergence $\tilde{\sigma}(\{n_j\}, T)$ in the bound states of the sum itself. To eliminate this divergence, a large number of restrictions $\tilde{\sigma}$ was proposed [305] based on different physical considerations.

In thermodynamic calculations, the energy computation origin is normally the state of the system in which the whole matter has the form of atoms at the lowest energy level. It is therefore convenient instead of $I_j$ to introduce the quantities $A_j$ equal to the difference between the particle energy and the energy of the particle-constituting infinitely spaced atoms. For example, for a singly charged ion the quantity $A_j$ is equal to the ionization potential and for a molecule — to the dissociation energy with the opposite sign, etc. In what follows, we shall for simplicity discard the tilde above the statistical sum, at the same time bearing in mind that the energy of particles is counted from their ground states. Then the chemical and ionization equilibrium equations (4.28) can be written in the form

$$\ln\left(\frac{n_j\Lambda_j^3}{\sigma_j}\right) + \frac{A_j}{k_{\text{B}}T} + \frac{\Delta\mu_j^{\text{int}}}{k_{\text{B}}T}$$

$$= \sum_{k=1}^{K} \nu_k^j \left[\ln\left(\frac{n_k\Lambda_k^3}{\sigma_k}\right) + \frac{A_k}{k_{\text{B}}T} + \frac{\Delta\mu_k^{\text{int}}}{k_{\text{B}}T}\right]; \quad j = K+1, \ldots, L. \tag{4.31}$$

If the system has a fixed pressure $p_0$ and temperature $T$, equations (4.31) must be supplemented with the EOS

$$\sum_{j=1}^{L} n_j k_{\text{B}}T + \Delta p^{\text{int}}(\{n_i\}, T) = p_0, \tag{4.32}$$

where $\Delta p$ is correction to the ideal part of the pressure, corresponding to the nonideality correction $\Delta F$ in the free energy. When considering a system with given density $\rho_0$ and temperature, instead of the EOS one should use the equation for density:

$$\sum_{j=1}^{L} m_j n_j = \rho_0,\qquad(4.33)$$

where $m_j$ is the $j$th particle mass. In practical calculations, instead of relations (4.19) one takes the ratio of the number of nuclei of a given chemical element to the total number of nuclei of the chemical elements constituting the mixture:

$$\beta_k = \frac{\tilde{N}_k}{\sum_{i=2}^{K} \tilde{N}_i}.$$

If the index 1 designates the electrons, then these ratios can be expressed as

$$\beta_k(\{n_j\}) \equiv \frac{\sum_{j=1}^{L} \nu_k^j n_j}{\sum_{i=2}^{K} \sum_{j=1}^{L} \nu_i^j n_j};\quad k = 2,\ldots,K,\qquad(4.34)$$

and, accordingly, the equations of proportions will have the form

$$\beta_k(\{n_j\}) = b_k;\quad k = 2,\ldots,K.\qquad(4.35)$$

When in equilibrium, plasma must satisfy the electric neutrality condition

$$\sum_{j=1}^{L} z_j^{(+)} n_j = \sum_{j=1}^{L} z_j^{(-)} n_j,\qquad(4.36)$$

where $z_j^{(+)}$ and $z_j^{(-)}$ are charges of positively and negatively charged ions. With allowance for this condition equations (4.35) can be extended by inclusion in the material balance also electrons as an independent "chemical element".

Instead of the quantities $\beta_k$, equivalent quantities $\alpha_k$ are frequently used in calculations:

$$\alpha_k(\{n_j\}) \equiv \frac{\sum_{j=1}^{L} \nu_k^j n_j}{\left[\sum_{i=1}^{K} \left(\sum_{j=1}^{L} \nu_i^j n_j\right)^2\right]^{1/2}};\quad k = 1,\ldots,K.\qquad(4.37)$$

To these quantities there correspond the equations

$$\alpha_k(\{n_j\}) = c_k;\quad k = 1,\ldots,K,\qquad(4.38)$$

which we shall use below.

Thus, we have obtained a complete system of equations necessary for establishing the composition of thermodynamically equilibrium gas system which includes

(1) the EOS

$$\sum_{j=1}^{L} n_j k_{\mathrm{B}} T + \Delta p^{\mathrm{int}}(\{n_i\}, T) = p_0,$$

r a system with given $p_0$ and $T$ or the density equation

$$\sum_{j=1}^{L} m_j n_j = \rho_0,$$

for fixed $\rho_0$ and $T$;

(2) chemical and ionization equilibrium equations

$$\ln\left(\frac{n_j \Lambda_j^3}{\sigma_j}\right) + \frac{A_j}{k_{\mathrm{B}} T} + \frac{\Delta \mu_j^{\mathrm{int}}}{k_{\mathrm{B}} T}$$

$$= \sum_{k=1}^{K} \nu_k^j \left[ \ln\left(\frac{n_k \Lambda_k^3}{\sigma_k}\right) + \frac{A_k}{k_{\mathrm{B}} T} + \frac{\Delta \mu_k^{\mathrm{int}}}{k_{\mathrm{B}} T} \right], \quad j = K+1, \ldots, L;$$

(3) equations of chemical element proportions:

$$\alpha_k(\{n_j\}) = c_k; \quad k = 1, \ldots, K.$$

The system (4.31)–(4.33), (4.38) is the system of transcendent equations in $n_j$ for given $\{c_k\}$ and given $p_0$ and $T$ (or $\rho_0$ and $T$). At the present time two approaches exist to the calculation of the composition of multicomponent equilibrium systems which underlie a lot of concrete realizations. The first approach is based on the solution of the system of equilibrium equations (4.31)–(4.33) and (4.38), the second is based on direct minimization of the corresponding thermodynamic potential. Without claiming generality, we shall discuss the possibilities of each of these methods for nonideal plasma and particular algorithms of calculations.

Considering the use of the approach based on the solution of equilibrium equations, we shall notice the peculiarities of nonideal plasma as an object of thermodynamic calculation. We first of all note that this method is close to the methods [137] based on the mass action law for ideal-gas systems. The difference in our case lies in the appearance of nonideality correction in the EOS and, which is even more important, in the density dependence of equilibrium constants. As has already been mentioned, the latter fact makes impossible the use of the equilibrium constants tabulated in advance (e.g., from Ref. [453]), and it therefore becomes necessary to calculate them in the course of solution of the system of equilibrium equations. To calculate the constants, one should in turn calculate the statistical sum (4.33) either in the energy levels of isolated particles with statistical weights depending on the interaction model chosen, or in the set of levels calculated with allowance for particle interaction in the medium.

In the calculation of thermodynamic properties of strongly compressed nonideal plasma, the following obvious restrictions are imposed. The method should, on the one hand, allow formally obtaining solution for any densities and, on the other hand, be sufficiently flexible in order that different nonideality models [229] could be used. (A detailed description of the applied algorithms can be found in paper [229]).

The presented methods underlie the SAHA-4 program modules for calculation of thermodynamics of multicomponent nonideal plasma containing up to 10 different chemical elements and SAHA-3 for calculation of strongly nonideal one-element plasma. Along with the calculation of the equilibrium composition and thermodynamic functions for given $p$ and $T$, $\rho$ and $T$, the complex of programs for thermodynamic SAHA computations permits calculations of isobars, isochores, isenthalpies, isentropies, shock adiabats, and other thermodynamic curves using various interparticle interaction models.

## 4.5 Thermodynamics of nonideal solar plasma

A separate problem in the calculation of plasma thermodynamics is adequate allowance for bound states [598]. The standard approach consists in the calculation of the statistical sum $\tilde{\sigma}(\{n_j\}, kT)$ by (4.30) which, however, diverges, $\tilde{\sigma}(\{n_j\}, kT) \to \infty$, with allowance for an infinite number of excited states, $j \to \infty$. This needs the introduction of mechanisms of restriction of the number $j$, based on different physical considerations and, therefore, involving some uncertainty. Following paper [941] we shall present here the method free of these shortcomings and permitting a consistent construction of low-temperature weakly nonideal plasma thermodynamics.

The formula for an equilibrium degree of ionization in plasma consisting of neutral atoms, singly charged ions, and free electrons not bound in atoms was first published by the physics and applied mathematics lecturer of Calcutta University M. Saha in 1920 [872] and was presented in more detail in the summary paper [873] which is mainly referred to when mentioning the now well-known Saha formula. Using this formula, one could determine the degree of ionization in solar chromosphere for the explanation of the occurrence of ion lines of various elements with moving away from the solar surface and with increasing plasma temperature. It was derived from the mass action law known for chemical reactions, and its derivation fully corresponded to the derivation presented in the classical book "*Statistical Physics*" [598] by L.D. Landau and E.M. Lifshits. We shall briefly present this derivation keeping to the notation adopted in this book.

We designate the fraction of neutral atoms as

$$c_0 = \frac{n_{\rm a}}{n_{\rm a} + n_{\rm i} + n_{\rm e}}, \tag{4.39}$$

where $n_k$ is the concentration of particles of kind $k$, $k = {\rm a, i, e}$; the fraction of singly charged ions is

$$c_1 = \frac{n_{\rm i}}{n_{\rm a} + n_{\rm i} + n_{\rm e}}, \tag{4.40}$$

the fraction of electrons is

$$c = \frac{n_{\rm e}}{n_{\rm a} + n_{\rm i} + n_{\rm e}}. \tag{4.41}$$

The electroneutrality condition

$$n_{\rm e} = n_{\rm i}, \quad c_1 = c \ {\rm also \ holds}. \tag{4.42}$$

We shall introduce the degree of ionization $\alpha$, equal to

$$\alpha = \frac{n_{\rm e}}{n_{\rm a} + n_{\rm i}} = \frac{n_{\rm e}}{n}, \tag{4.43}$$

where $n = n_{\rm a} + n_{\rm i}$ is the concentration of heavy particles. The quantities $c_0$, $c_1$ and $\alpha$ are related as

$$c_1 = \frac{\alpha}{1 + \alpha}, \quad c_0 = \frac{1 - \alpha}{1 + \alpha}. \tag{4.44}$$

The mass action law for ionization reactions $e + A \rightleftarrows A^+ + e + e'$ (three-particle recombination and the inverse process) gives the relation

$$c_0 = c_1 c p \kappa_p(T). \tag{4.45}$$

In (4.45) $p$ is the resultant particle pressure which in the ideal-gas approximation is equal to

$$p = T(n_{\rm a} + n_{\rm e} + n_{\rm i}) = nT(1 + \alpha), \tag{4.46}$$

$$p_k = n_k T = p c_k, \tag{4.47}$$

Here $\kappa_p(T)$ is the equilibrium constant for the ionization process. Saha himself introduced the equilibrium constant as $\kappa_s = \kappa_p^{-1}$,

$$\kappa_p(T) = e^{\sum_k \nu_k \zeta_k} \, T^{\sum_k c_k} \, p_k^{\nu_k} \, e^{-\sum_k \nu_k \varepsilon_{0,k}/T}. \tag{4.48}$$

Here the designation $\nu_k$ is introduced for the stoichiometric coefficients of the reaction $\sum_k \nu_i A_k = 0$, for which the Gibbs–Duhem chemical equilibrium condition holds:

$$\sum_k \nu_k \mu_k = 0. \tag{4.49}$$

In (4.49) $\mu_k$ is the chemical potential of the $k$th component which is written as

$$\mu_k = T \ln p_k + \chi_k(T). \tag{4.50}$$

The function $\chi_k(T)$ can be written as follows:

$$\chi_k = -T \ln \frac{z_k}{g_k} - C_{p,k} T \ln T - T \ln g_k \left( \frac{m_k}{2\pi\hbar^2} \right)^{3/2}. \tag{4.51}$$

Here $g_k$ is the statistical weight of the ground state of the particle, $m_k$ is the particle mass, $C_{p,k} = 5/2$, $\hbar$ is the Planck constant, and $z_k$ is the statistical sum in the internal states of particles of kind $k$. For atoms ($k = $ a) the value $\varepsilon_{m,\mathrm{a}} = \varepsilon_m < 0$,

$$z_\mathrm{a} = \sum_m e^{-\varepsilon_m/T}. \tag{4.52}$$

At low temperatures $|\varepsilon_{0,k}| \gg T$ (the index "0" means the ground state of the system) the function (4.51) reduces to the form (in the statistical sum one summand is left for the ground state)

$$\chi_k = \varepsilon_{0,k} - C_{p,k} T \ln T - T \zeta_k. \tag{4.53}$$

In (4.53) $\zeta_k$ is the chemical constant of kind $k$ particles, equal to

$$\zeta_k = \ln \left[ g_k \left( \frac{m_k}{2\pi\hbar^2} \right)^{3/2} \right]. \tag{4.54}$$

With these designations, we have for the equilibrium constant $g_\mathrm{e} = 2$:

$$\kappa_p = \frac{g_\mathrm{a}}{2g_1 T} \left( \frac{2\pi\hbar^2}{mT} \right)^{3/2} \cdot e^{I/T} = \frac{g_\mathrm{a}}{2g_1 T} \lambda_\mathrm{e}^3 e^{I/T}. \tag{4.55}$$

Here $m$ is the electron mass, $I = \varepsilon_{0,i} - \varepsilon_{0,a}$ is the atom ionization potential from the ground state, $\lambda_e = \sqrt{\frac{2\pi\hbar^2}{mT}}$ is the thermal de Broglie wave length for an electron. For the ionization degree $\alpha$ we obtain

$$\kappa_s = \kappa_p^{-1} = \frac{\alpha^2}{1-\alpha^2}p. \tag{4.56}$$

In such a form and taking the natural logarithm of both sides of (4.56), Saha wrote his formula for estimation of the degree of ionization. From (4.56) one can obtain

$$\alpha = \frac{1}{\sqrt{1+p\kappa_{\mathrm{p}}}}. \tag{4.57}$$

If we express the degree of ionization $\alpha$ in terms of not pressure, but concentration of heavy particles $n$: $p = n(1+\alpha)T$, then for the degree of ionization we obtain from (4.57)

$$\alpha = \frac{2}{1+\sqrt{1+\frac{g_a}{2g_1}n\lambda_e^3 e^{I/T}}}. \tag{4.58}$$

Making use of formula (4.56), Saha calculated the degree of ionization for the following elements: calcium (in the temperature range of 2000 to 14000 K), barium (from 2 to 12000 K) and strontium (2000–15000 K) and for pressures of $10^{-8}$–10 atm. After the Saha formula had been derived, it was used and extended to various conditions in plasma of various physical systems, including the Sun and others in which the thermodynamic equilibrium condition is met. The most dramatic paradox due to formal application of the chemical potential formulas (4.51)–(4.53) containing the statistical sum over the atomic states consisted in the divergence of the latter with allowance for an infinite number of discrete states degenerate proportionally to the square of the principal quantum number $n$. Then the question arose of whether summation should be taken over the states of the continuum or over the states of scattering in the interacting system consisting of an electron and a positive ion. The latter question is due to the fact that we consider two subsystems that consist of an electron and a positive ion, but we first declare them to be free and then to be interacting.

In connection with the of atomic statistical sum divergence problem in which summation is taken over the discrete states many ways were proposed in scientific literature of restriction summation over principal quantum numbers; for details see books [136, 234, 582, 776]. We note here the way to restrict the statistical sum over the values of the principal quantum number for which the Bohr orbit radius is compared with the characteristic sizes of the system. At temperatures $T$, low compared to the ionization potential $I$ equal to Ry $= 13.6\,\mathrm{eV}$ for the hydrogen atom, this restriction yields the statistical sum value close to the one determined by the contribution of the ground state of the atom (Brillouin [136]). The restriction of statistical sum by such Bohr orbit radii at which they are overlapped for the neighboring atoms appears to be more important from the physical point of view (E. Fermi [255]). With such a restriction the contribution of the ground state is also dominant when the $I/T$ values are low. The most radical way of cutoff was proposed by Planck [794] who restricted the summation in the atomic statistical sum to the values of the principal quantum number for which the energy of ionization (equal to $I/n^2$) from the state with such $n$ value is compared with the temperature of the medium. The corresponding principal quantum number of the cutoff according to Planck is smaller than in the first two ways in weakly nonideal plasma. The techniques also exist of atomic statistical sum cutoff on orbit radii comparable with the Debye radius in plasma [234, 718]. It is now pertinent to mention the method of statistical sum cutoff proposed in Refs. [464, 660]. It consists of restriction of the principal quantum numbers to the values responsible for overlapping of the neighboring spectral lines of the series broadened by the electric microfields and collisions with plasma electrons. Another cutoff method, close in ideology, was proposed by Unsold [991] (see also Ref. [242]). The method consists in restriction of the principal quantum numbers to the values for which the bound states in perturbed atoms vanish under the action of plasma microfields. The above-mentioned methods of statistical sum restriction in weakly nonideal plasma are much weaker than the Planck's technique since the plasma temperature under these conditions is higher than the interaction energies which can be estimated for the described cutoff mechanisms.

It is of importance that Saha himself appealed to optical observations and when deriving his formula allowed for the contribution of bound states to the pressure. From today's point of view the problem is that in the physical model investigating the Hamiltonian of the system of interacting electrons and ions (or for simplicity — protons) the notion of atom is conditional since such an operator is not originally involved in the theory. Atoms are actually understood as a contribution of discrete states due to Coulomb interaction to some physical observed results. For example, the contribution of bound states to pressures is represented by the second virial coefficient described by the convergent sum over the discrete atomic states, which in most papers coincides with the Planck–Larkin statistical sum [603], see also Refs. [228, 570, 573, 583, 584, 590, 845, 981, 1014]; the virial expansion was obtained in Refs. [12–15] (see also Ref. [779]) using the Feynman-Katz continual integrals.

The contribution of bound states to pressure can be written as:

$$\delta p^{\mathrm{BS}} = \zeta_{\mathrm{e}}\zeta_{\mathrm{p}} T \lambda_{\mathrm{ep}}^{3} \sigma^{\mathrm{BS}}, \tag{4.59}$$

Here $\zeta_{\mathrm{e}}$, $\zeta_{\mathrm{p}}$ are the electron and proton activities, $\lambda_{\mathrm{ep}} = \left[2\pi\hbar^2/(\mu_{\mathrm{ep}}T)\right]^{1/2}$, $\mu_{\mathrm{ep}} = m_{\mathrm{e}}m_{\mathrm{p}}/(m_{\mathrm{e}} + m_{\mathrm{p}})$ is the reduced mass. The quantity $\sigma^{\mathrm{BS}}$ in (4.59) can be represented as the statistical sum over the principal quantum number $n$:

$$\sigma^{\mathrm{BS}} = \sum_{n=1}^{\infty} n^2 e^{u_n} F_n(u_n), \tag{4.60}$$

where $u_n = \mathrm{Ry}/(n^2 T)$. In the Planck–Larkin formula

$$F_n(u) = F_{\mathrm{P\text{-}L}}(u) = 1 - e^{-u} - u e^{-u}. \tag{4.61}$$

Making summation over the principal quantum number $n$ without restriction, one obtains $(X = \beta\mathrm{Ry})$:

$$\sigma^{\mathrm{BS}}_{\mathrm{P\,L}} = \sum_{k=2}^{\infty} \zeta(2k - 2)\frac{X^k}{\Gamma(k+1)} = \sum_{n=4,6,8,\ldots} \zeta(n-2)\frac{X^{n/2}}{\Gamma\left(\frac{n}{2}+1\right)}. \tag{4.62}$$

Here and below $\zeta(n)$ is the Riemann zeta function.

At temperatures low compared to the ionization potential the contribution of atoms to the pressure coincides with the estimate by the Saha formula. It should be noted that the Planck–Larkin contribution obtained in the above-cited works by the methods of quantum field theory in statistical physics is actually a resultant contribution of discrete states and partially of the states of continuum — the states of scattering. The converging contribution for the sum both over the discrete states only and separately over the states of scattering was obtained in papers [745,941,946]. In the chemical model of plasma, the latter contribution is normally omitted. Interestingly, the resultant contribution of discrete states and the states of scattering in the main approximation coincides with the Planck–Larkin formula [941]. The expression for the contribution of purely atomic stated to pressure can be extended to the case of allowance for the level broadening effects in plasma with the same accuracy as that in the theory of radiation of atoms in lines [943, 998], whereas a similar extension of the Planck–Larkin formula involving a superposition of contributions of discrete states and the states of scattering seems hardly possible. The same formulas for the contribution of discrete states only were also obtained in Ref. [1026] by diagonalization of the Hamiltonian (in which bound states alone were kept) by the canonical transformation method similar to the well-known N.N. Bogolyubov's method [117].

In the plasma radiation calculation, the contribution of atomic states is represented by the Boltzmann–Saha formulas for atomic populations multiplied by the corresponding spontaneous radiation probabilities and by the energy of emitted photons, and for the obtained sum the restrictions are strictly speaking not necessary as this sum converges because of a sufficiently fast decrease of the Einstein coefficients with increasing principal quantum numbers. No subtractions, like those in Planck–Larkin type formulas, are formally needed. For high values of the principal quantum number it is physically necessary to take into account the mechanisms of Ingles-Teller line merging and the cutoff of highly excited states by plasma microfields by the Unsold mechanism. For the solar photosphere, convergence of the contribution of atomic states to pressure by the Planck–Larkin formula sets in for the values of the principal quantum number less than 6 and the discrete lines of Balmer series are observed up to $n \approx 17$ [709]. This fact emphasizes that the concept of atom is unobservable in a strict sense, and the contribution of bound states to one or another physical phenomenon depends on the problem statement.

Another paradox encountered upon a literal application of the Saha formula is that the neutral-atom concentration in the center of the Sun, where the temperature is by two orders of magnitude higher than the ionization potential but the plasma density is close to 150 g/cm$^3$, makes up nearly 30 percent of the total number of heavy particles. The limit $I/T \ll 1$ cannot strictly speaking be applied in the Saha formula in the derivation of which the inverse relation between temperature and ionization potential was assumed. The use of Planck–Larkin or SRM formula for the solar core makes a contribution of the order of $(I/T)^2 \sim 10^{-4}$ to the pressure from neutral atoms. If the electron and proton charge is formally sent to zero, which must result in the disappearance of bound states, then according to the Saha formula a finite number of neutrals remains whose relative number is determined by the quantity $\sim n_e \lambda_e^3$ which for the solar core is close to unity because the electrons are here is partly degenerate. The Saha formula is obviously inapplicable to degenerate plasma since the relation (4.50)–(4.51) between the chemical potential of the electrons and their concentration gets violated. The use of Planck–Larkin type formulas gives a correct limit, i.e., $n_a$ tends to zero, but it also needs correction with allowance for degeneracy.

The Saha formula written as (4.57)–(4.58) implies that with increasing gas pressure or density the degree of ionization decreases monotonously. The experimentally on measured electroconductivity of air adiabatically compressed to a pressure of 9000 atm [512] showed that the Saha formula holds up to a pressure of the order of 50 atm. At high pressures (densities) the electron concentration increases with increasing density. To explain the observed density dependence of plasma electroconductivity the authors (Karpenko *et al.*) assumed that in a dense plasma the ionization potential lowers leading to the experimentally observed effect. The theoretical estimates of this effect were given in paper [976] which takes into account the contribution of the interaction of charged particles with neutrals through polarization of the latter by the electric field of the charges. These estimates agreed qualitatively with the experimental data. Later an analogous theory was developed by A.A. Likal'ter [621] in application to the experimental data on cesium electroconductivity in transition from high densities, for which the conductivity is close to metallic, to low densities where plasma conductivity is several orders of magnitude lower [19,839]. The theoretical description of these experiments which agrees qualitatively with them in the density dependence of conductivity was given in work [20] in which the electron–neutral interaction is expressed in quantum mechanics in terms of the scattering length and the interaction between positive ions and neutrals in terms of atom polarizability as in B.L. Timan's works. It should be noted that the question of phase transition of a crystalline substance upon density variation (known as metal–dielectric (M–D) or N. Mott's transition [717]) was posed by N. Mott, according to whom in a rarefied substance at zero temperature a forbidden zone occurs because of strong electrostatic repulsion of electrons on one center. This zone may shut with increasing substance density resulting in the dielectric–metal transition. As a criterion of such transition Mott suggested the condition of equality of the radius of Bohr atomic orbit to the Debye radius of plasma.

Ya. B. Zel'dovich and L.D. Landau [1078] discussed the possibility of M–D transition at finite temperatures as a first-order phase transition different from the known liquid–gas and liquid–solid transitions. In that paper they discussed, in particular, the version of M–D transition in the case of mercury in which the critical point of liquid–gas transition is rather low (for later experimental data see papers [327, 536]), and for the metal–dielectric transition a much higher critical temperature was assumed. The existence of a nonconducting liquid phase was hypothesized, and in the M–D transition at temperatures above critical, for the liquid–gas transition a jump-like variation of electroconductivity, volume, and other properties must take place. The hypothesis of a jump-like transition of conductivity in mercury was confirmed neither in the above-mentioned experiments nor in those performed twenty years later: a fairly sharp transition, but without jumps, from metallic to a low plasma conductance is observed in the region of critical mercury density.

Thus, for high plasma densities there is a strong deviation from the Saha formula in its classical version which is associated with the allowance for Coulomb and other interactions that alter the values of chemical potentials of electrons and ions. In the region of strong plasma nonideality, when the Coulomb interaction is characterized by the Debye parameter $\Gamma_{\mathrm{D}} = e^2/(r_{\mathrm{D}}T)$ ($e$ is electron charge, $r_{\mathrm{D}}$ is Debye radius of plasma, and $T$ is plasma temperature) assuming the value of the order of unity, a plasma phase transition (PPT) is possible which was described in detail by G.E. Norman and A.N. Starostin [744, 745] who formulated the general conditions of thermodynamic stability for the chemical model of strongly nonideal plasma (see also Refs. [235, 475]).

In the same papers [744, 745] the authors proposed hypothetic diagrams of phase equilibrium for PPT and the relative mutual position in temperature-density or pressure-density coordinates of the known lines of liquid–gas or liquid–solid type phase transitions and lines of first-order phase transition in nonideal plasma under which the plasma splits into phases differing in density and the degree of ionization; curves of coexistence of phases with two critical points are possible. In most metals in which the critical point of liquid–gas transitions is of the order of $1\,\mathrm{eV}$, the substance in the vicinity of the critical point is strongly ionized and, as distinct from the Van der Waals theory [598], a correct description of the liquid–gas transition requires inclusion of a strong Coulomb interaction. In such systems, a new transition different from the known ones may fail to occur. In substances with a small critical point of liquid–gas transition, where ionization is insignificant, PPT can be observed at high temperatures and densities, where a strong Coulomb interaction occurs. In the same papers [744, 745], the authors point to the erroneous criterion of thermodynamic stability of nonideal plasma and the consequent prediction of the critical point for PPT proposed in books [18, 21]. An analogous phase transition was later rediscovered in paper [881] and was analyzed for the conditions in the Jupiter depth (see also reviews [397, 470, 624]). In experiments on adiabatic compression of deuterium at megabar pressure and densities higher than one gram per cubic centimeter, the characteristic sharp bend of the adiabat in pressure-density coordinates was observed which the authors of associated with PPT [301]. For solar depths the Debye nonideality parameter is small, $\Gamma_D \approx 0.05$, in the center of the Sun and the questions related to Mott type transitions or PPT play no role in the framework of the approach based on the series expansion of thermodynamic potential in the small parameter $\Gamma_D$. The use of different kinds of cutoffs and the disappearance of bound states due to the Debye screening of plasma or the effect of plasma microfields is an uncontrolled increase in precision in this range of nonideality parameters.

The current approaches to the systematic calculations of thermodynamic functions of weakly nonideal plasma are based on the use of the methods of quantum field theory of which the most widespread is the Matsubara technique [5, 668] or the method of nonequilibrium Green functions elaborated in by G. Baum, L. Kadanoff, and L. Keldysh [492, 516, 620]. The Feynman diagrams in the Matsubara technique for the interaction to be considered below are presented in Fig. 4.5.

We shall present the decomposition for the pressure of plasma consisting of electrons and protons:

$$p = p_0 + p_1 + p_{\mathrm{H\ F}} + p_{\mathrm{D\ H}} + p_{\mathrm{BS}} + p_{\mathrm{SS}} + o\big(n^{5/2}\big), \qquad (4.63)$$

where $p_0$ is the pressure of ideal noninteracting gas,

$$p_0 = (\zeta_e + \zeta_p)T, \quad \zeta_p = \frac{g_p e^{\mu_p/T}}{\lambda_p^3}. \qquad (4.64)$$

The electrons can be degenerate (for example, in the center of the Sun $n_e \lambda_e^3 \approx 0.6$), and therefore their activity $\zeta_e$ will be expressed [598] in terms of an ideally gaseous, but in the general case for degenerate particles, concentration $n_e^0$:

$$\zeta_e = n_e^0(\mu_e) = \frac{2}{\lambda_e^3} \frac{2}{\sqrt{\pi}} \int\limits_0^\infty \frac{x^{1/2}\,dx}{\exp\,(x - y) + 1}. \qquad (4.65)$$

Here $\lambda_e = \sqrt{\dfrac{2\pi\hbar^2}{m_e T}}$; $y = \mu_e/T$.

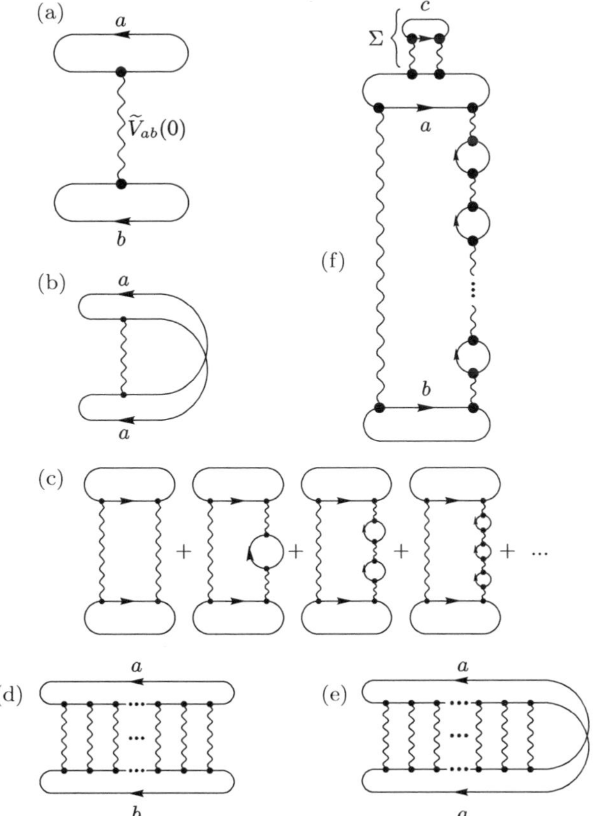

Fig. 4.5   Feynman diagrams in the Matsubara technique [5,668] (the solid line stands for the Green function, the wavy line for the Fourier component of the interaction potential): (a) the Hartree correction, see formulas (4.67), (4.105) (b) the Hartree–Fock approximation (exchange interaction) (4.70) (c) the Debye–Huckel approximation (4.71), (4.75) (d) the ladder diagram (4.85), (4.86), (4.92) (e) the ladder diagram (exchange interaction) (4.139) (f) the diagram describing the influence of the medium (allowance for the energy inset $\Sigma_c$ proper) upon the Debye–Huckel screening (4.84).

Accordingly, the pressure of an ideal gas of degenerate electrons is written in the form [598]

$$p_{0\mathrm{e}} = \frac{4}{3\sqrt{\pi}} T \frac{2}{\lambda_{\mathrm{e}}^3} \int\limits_0^\infty \frac{x^{3/2}\,dx}{\exp{(x - y)} + 1}. \tag{4.66}$$

The first Hartree correction is given in the form

$$p_1 = \frac{4\pi e^2}{\varkappa^2} (\zeta_{\mathrm{e}} - \zeta_{\mathrm{p}})^2, \tag{4.67}$$

regularized with the help of the parameter $\varkappa$ which should be directed to zero in the final results.

To eliminate the divergence, one should require the electroneutrality condition in the form of equality of proton and electron activities. In the general case of multicomponent plasma consisting of electrons and ions with charge $z_k$ the electroneutrality condition in terms of activities has the form

$$\zeta_e = \sum_k z_k \zeta_k, \tag{4.68}$$

and expression (4.67) is generalized in the obvious way:

$$\frac{\delta \Omega_H}{V} = \frac{4\pi e^2}{\varkappa^2} \cdot \left( \zeta_e - \sum_k z_k \zeta_k \right)^2. \tag{4.69}$$

In the next, Hartree–Fock approximation (see Fig. 4.5 (b) one obtains the known [337, 492, 598] final result for the exchange correction at the expense of the electron–electron interaction. In the case of nondrgenerate electrons, for instance, in the first Born approximation we have [598]

$$\frac{\delta \Omega_{exch}}{V} = -\frac{\pi e^2 \hbar^2}{2 m_e T} \zeta_e^2. \tag{4.70}$$

Here for $\zeta_e$ one should take the nondegenerate limit $(\exp(-y) \gg 1)$ in expression (4.65): $\zeta_e = 2\lambda_e^{-3} \exp(\mu_e/T)$.

The ring diagrams, next in the interaction potential [213, 708] see Fig. 4.5, lead to the Debye–Huckel contribution [203] (see, e.g., Refs. [5, 232–234, 570, 582, 583, 603, 1014]):

$$\frac{\delta \Omega_{D\,H}}{V} = -T \frac{\varkappa_D^3}{12\pi}. \tag{4.71}$$

Here $\varkappa_D$ is the inverse Debye radius [5, 326],

$$\varkappa_D^2 = 4\pi e^2 \sum_k z_k^2 \left( \frac{\partial n_k}{\partial \mu_k} \right)_T. \tag{4.72}$$

Expression (4.72) was rigorously derived by Fradkin [326] who showed that in the right-hand side of expression (4.72) derivatives appear of the physical concentrations $n_k$ with respect to the chemical potentials. In perturbation theory we can restrict ourselves by the activity values $n_k^0(\mu_k) = \zeta_k$.

Recall that the physical concentrations are related to the chemical potentials as [598]

$$n_k = -\left(\frac{\partial\left(\frac{\Omega}{V}\right)}{\partial\mu_k}\right)_T. \tag{4.73}$$

For physical concentrations the traditional electroneutrality relation holds:

$$n_e = \sum_k z_k n_k. \tag{4.74}$$

The Debye–Huckel approximation represented in the static limit by the contribution of ring diagrams [5] (see Fig. 4.5(c)) contains several corrections which can be obtained using the technique [492, 516] (the particle spins are assumed for simplicity to be equal to $1/2$):

$$\frac{\Delta\Omega}{V} = -4\sum_{i,j}\int_0^1 \frac{d\lambda}{2\lambda}\lambda^2 \int \frac{d\mathbf{P}\,d\mathbf{q}\,d\mathbf{k}}{(2\pi)^9}\cdot\frac{\exp\left(-\beta(\varepsilon_k^{ij}-\varepsilon_q^{ij})\right)-1}{\varepsilon_q^{ij}-\varepsilon_k^{ij}}$$

$$\times n_i\left(\frac{m_i}{M}\mathbf{P}+\mathbf{q}\right)\left(1-n_i\left(\frac{m_i}{M}\mathbf{P}+\mathbf{k}\right)\right)$$

$$\times n_j\left(\frac{m_j}{M}\mathbf{P}-\mathbf{q}\right)\left(1-n_j\left(\frac{m_j}{M}\mathbf{P}-\mathbf{k}\right)\right)$$

$$\times\frac{16\pi^2 e^4 z_i^2 z_j^2}{(\mathbf{q}-\mathbf{k})^2\left[(\mathbf{q}-\mathbf{k})^2+4\pi e^2\lambda\Pi^R\left(\mathbf{q}-\mathbf{k},\mathbf{V}\cdot(\mathbf{q}-\mathbf{k})\right)\right]}, \tag{4.75}$$

where integration over the parameter $\lambda$ corresponds to the integration over the charge $e^2 \mapsto e^2\lambda$; $\mathbf{P}$ is the total momentum of particles $i, j$; $\mathbf{q}, \mathbf{k}$ are wave vectors of the relative center-of-mass motion respectively before and after the interaction; $n_i(\mathbf{q})$ are occupation numbers for type $i$ particles; $m_i, z_i$ are mass and charge (in elementary units) of type $i$ particle; $M = m_i + m_j$ is the resultant mass; $\varepsilon_k^{ij} = \frac{\hbar^2 k^2}{2\mu}$, $\mu = \frac{m_i m_j}{m_i+m_j}$ is the reduced mass of particles $i$ and $j$; $\mathbf{V} = \mathbf{P}/M$, $\hbar|\mathbf{V}|q \sim \hbar|\mathbf{V}|\varkappa_{\mathrm{D}} < T$.

Here we consider the general case admitting allowance for degeneracy and non-static corrections. The allowance for the contribution of two cross-bars existing in summation of ring diagrams, which is to be subtracted from the ladder diagrams (see Figs. 4.5(d) and 4.5(e)) describing the interaction of a pair of particles in the continuum, corresponds to the replacement of $4\pi e^2\lambda\Pi^R$ (where $\Pi^R$ is the retarding polarization operator [5,492,516,620] determined by the sum of contributions of all the particles participating in the screening of Coulomb interaction, in the denominator of the last factor in (4.75)) by the regularizing parameter $\varkappa^2$.

For the nondegenerate case we obtain

$$\frac{\Delta\Omega}{V} = -\frac{T\varkappa_{\mathrm{D}}^4}{4\pi^2} \int\limits_0^1 \lambda\, d\lambda \int\limits_0^\infty \frac{dq}{q^2 + 4\pi e^2 \lambda \Pi^R(q)}. \tag{4.76}$$

Here $\varkappa_{\mathrm{D}}^2 = 4\pi e^2 \Pi^R(0)$ is the square of the inverse Debye radius, $\Pi^R(q) \equiv \Pi^R(|\mathbf{q}|) \equiv \Pi^R(\mathbf{q}, 0)$; in the nondegenerate case, $\varkappa_{\mathrm{D}}^2 = 4\pi e^2 \beta \sum_i z_i^2 n_i$. In the typically employed approximation consisting in replacement of $\Pi^R(q)$ by its value $\Pi^R(0)$ from (4.76), (4.72), we obtain expression (4.71).

A specification should be made here. In expression (4.76) $\Pi^R(q)$ should be decomposed up to the second derivative with respect to $q$ [940]:

$$q^2 + 4\pi e^2 \lambda \Pi^R(q) \to q^2 \left(1 + 2\pi e^2 \lambda \frac{\partial^2 \Pi^R}{\partial q^2}(0)\right) + 4\pi e^2 \lambda \Pi^R(0). \tag{4.77}$$

We shall introduce the quantity $a = |2\pi e^2 \Pi''(0)|$, where $\Pi'' \equiv \frac{\partial^2 \Pi^R}{\partial q^2}$. For this quantity we obtain [940, 946] the expression

$$2\pi e^2 \Pi''(0) = -\sum_i \frac{\pi \hbar^2 e^2 z_i^2}{m_i} \frac{\partial^2 n_i^0}{\partial \mu_i^2} + \frac{\pi \hbar^2}{9} \sum_i \frac{e^2 z_i^2}{m_i} \frac{\partial^3(\overline{\varepsilon n_i})}{\partial \mu_i^3}. \tag{4.78}$$

Here $n_i^0$ is the ideal-gas concentration, $\overline{\varepsilon n_i}$ is averaging of the kinetic energy over the occupation numbers for sort $i$ particles:

$$\overline{\varepsilon n_i} = \int \frac{d\mathbf{p}}{(2\pi)^3} \frac{P^2}{2m_i} n_i(\mathbf{p}). \tag{4.79}$$

With allowance for this fact, the Debye–Huckel correction assumes the form [946]

$$\frac{\Delta\Omega_{\mathrm{D\,H}}}{V} = -\frac{T\varkappa_{\mathrm{D}}^3}{12\pi} f(a). \tag{4.80}$$

Here

$$f(a) = \frac{3}{2a^{3/2}} \left\{\arcsin \sqrt{a} - \sqrt{a(1-a)}\right\}. \tag{4.81}$$

The estimates for hydrogen plasma give $2\pi e^2 \Pi''(0) = -173\rho T^{-2}$ for the nondegenerate case, here $\rho$ is the matter density in g/cm$^3$ and $T$ is temperature in eV. For the center of the Sun the quantity $f(a)$ differs from unity in the third decimal place, which with allowance for smallness of the Debye–Huckel correction of the order of $10^{-2}$ compared to the ideal-gas approximation makes it possible to restrict the consideration for helioseismological problems by the approximation (4.71) ($f(a) = 1$).

For the "true" radius of charge fluctuation screening $\Gamma_{\mathrm{scr}} = \varkappa^{-1}$ we obtain from (4.77)

$$\varkappa^2 = \frac{\varkappa_{\mathrm{D}}^2}{1 + 2\pi e^2 \Pi''(0)}.\tag{4.82}$$

From the presented estimates for the quantities $\Pi''(0)$ it follows that for the plasma temperature $T \approx 1\,\mathrm{eV}$ the condition $1 + 2\pi e^2 \Pi''(0) = 0$ is satisfied for $n_e \approx 3 \cdot 10^{21}\mathrm{cm}^{-3}$. Under these conditions, however, the nonideality parameter $\gamma = n^{1/3}e^2/T \approx 2$ and expressions (4.78) for $\Pi''(0)$ need to be specified. In the region where $\varkappa^2 < 0$ a charge density wave is formed and the Debye screening vanishes. The parameter by which the corrections $\sim a$ should be taken into account corresponds to $(\lambda_e \varkappa_{\mathrm{D}})^2$ which turns out to be not too small in the center of the Sun (closeness of the function $f(a)$ to unity is due to smallness of the numerical coefficient at this parameter). The ratio of the de Broglie thermal wavelength for an electron to the Debye radius characterizes the quantum effects in the course of Debye screening and the corresponding corrections to expressions (4.71), (4.80) (for $f(a) = 1$) are referred to as diffraction corrections [233]. They can be readily obtained in the nondegenerate case $n_i \ll 1$ from (4.75) in the approximation $4\pi e^2 \lambda \Pi^R (\mathbf{q} - \mathbf{k}, 0) = \lambda \varkappa_{\mathrm{D}}^2$ with allowance for the exponential factors and energy denominators involved in this expression. To a first approximation in the parameter $\lambda \varkappa$ we obtain (see Ref. [233]) we obtain

$$\frac{\Delta \Omega_{\mathrm{dif}}}{V} = \frac{\pi^{3/2}}{4} T \left(\frac{e^2}{T}\right)^2 \left\{ \lambda_{\mathrm{ee}} \zeta_e^2 + 2\zeta_e \sum_k \zeta_k z_k^2 \lambda_{\mathrm{ek}} + {\sum_{kj}}' \zeta_k \zeta_j z_k^2 z_j^2 \lambda_{kj} \right\}.\tag{4.83}$$

Here $\lambda_{\mathrm{ee}} = \dfrac{\hbar}{\sqrt{m_e T}}$; $\lambda_{\mathrm{ek}} = \dfrac{\hbar}{\sqrt{2\mu_{\mathrm{ek}}T}}$; $\lambda_{kj} = \dfrac{\hbar}{\sqrt{2\mu_{kj}T}}$.

Note also another class of corrections to the Debye–Huckel approximation first reported in paper [603] (see also Refs. [233, 583]). Physically these corrections are due to the fact that it is not free particles but particles interacting with the medium ("dressed" particles, see Fig. 4.5(f)) that participate in the screening.

Thus, we arrive at the correction to the Debye–Huckel term (the index "cl" means that the given result is of classical origin)

$$\frac{\delta \Omega_{\mathrm{cl}}}{V} = -\frac{\pi}{3} T \left(\frac{e^2}{T}\right)^3 \left(\sum_i \zeta_i z_i^4\right) \left(\sum_j \zeta_j z_j^2\right).\tag{4.84}$$

In (4.84) summation is taken over all sorts of particles. In paper [603], instead of $\pi/3$ the coefficient $\pi/2$ was used which was further presented in paper [583]. In book [234], this term was written in different places either with the coefficient $\pi/3$ or with $\pi/2$. The independent verification fulfilled in paper [941] showed that the correct value of the coefficient is $\pi/3$. Notice that this correction is small compared to the "big" logarithmic term.

We shall consider the contribution of the quantities $p_{\mathrm{BS}}$ and $p_{\mathrm{SS}}$ involved in formula (4.63) (see Fig. 4.5(d)). In the Matsubara technique [5, 668], for $\delta\Omega/V = -\delta p$ we have (integration over $\lambda$ implies the substitution $e^2 \mapsto e^2\lambda$):

$$\frac{\delta\Omega_L}{V} = \frac{2}{\beta} \sum_{i,\omega} \int_0^1 \frac{d\lambda}{2\lambda} \int \frac{d\mathbf{P}}{(2\pi)^3} G_i(\mathbf{P},\omega)\Sigma_i(\mathbf{P},\omega). \tag{4.85}$$

Here, summation of frequencies $\omega$ (or $p_4$) is performed; for fermions $\omega = \pi T(2n+1)$, the index "$L$" stands for ladder, $G_i(\mathbf{P},\omega)$ is the Green function of sort $i$ particle with momentum $\mathbf{P}$ and frequency $\omega$ in the Matsubara technique [5], and the self-energy part $\Sigma_i(\mathbf{P},\omega)$ can be expressed in terms of the two-particle vertex $\Gamma_{ij}$ obtained in the ladder approximation [337]:

$$\Sigma_i(\mathfrak{p}) = \frac{2}{\beta} \sum_{j,k_4} \int \frac{d\mathbf{k}}{(2\pi)^3} G_j(\mathfrak{k})\Gamma_{ij}\left(\frac{m_j\mathfrak{p} - m_i\mathfrak{k}}{m_i + m_j}; \frac{m_j\mathfrak{p} - m_i\mathfrak{k}}{m_i + m_j}; \mathfrak{p} + \mathfrak{k}\right). \tag{4.86}$$

For example, for the electron–proton interaction $m_i = m_{\mathrm{e}}$, $m_j = m_{\mathrm{p}}$, $\mathfrak{p} = (\mathbf{P}, p_4) \equiv (\mathbf{P}, \omega)$ is the electron momentum, $\mathfrak{k} = (\mathbf{k}, k_4)$ is respectively for the proton. The quantity $\Gamma_{ij}(\mathfrak{q}, \mathfrak{q}'; \mathfrak{P})$ ($\mathfrak{q}$, $\mathfrak{q}'$ is the relative momentum before and after scattering, $\mathfrak{P} = \mathfrak{p} + \mathfrak{k} = (\mathbf{P}, P_4)$ is the resultant momentum) in the ladder approximation for small occupation numbers $n \ll 1$ can be written as [337, 1014]:

$$\Gamma_{ij}(\mathfrak{q}, \mathfrak{q}'; \mathfrak{P}) = (2\pi)^3 \sum_n \left(iP_4 - \frac{\hbar^2 P^2}{2M} - \frac{\hbar^2 q^2}{2\mu} + \mu_i + \mu_j\right)$$

$$\times \frac{\widetilde{\Psi}_n(\mathbf{q})\, \widetilde{\Psi}_n^*(\mathbf{q}')\left(E_n - \frac{\hbar^2 q'^2}{2\mu}\right)}{iP_4 - \frac{\hbar^2 P^2}{2M} - E_n + \mu_i + \mu_j}. \tag{4.87}$$

Here $M = m_i + m_j$, $\mu = \frac{m_i m_j}{M}$; $\mu_i$ and $\mu_j$ are chemical potentials; $E_n$ is the binding energy of the state with the principal quantum number $n$; summation in $n$ is taken in the general case over the discrete states with quantum numbers $\{n\} = (n, l, m)$ characterized by the wave functions of the relative motion of particles $i$ and $j$ (bound states of the electrons in the proton field) $\widetilde{\Psi}_n(\mathbf{q})$ and over the states of the continuum. The sign $\sim$ over the wave function means the Fourier component in momentum space. $P_4 = 2\pi nT$ is the fourth component of the resultant momentum. For the scattering states corresponding to the continuum, instead of the sum over the discrete states $\{n\}$ one takes integration over the momenta $\mathbf{k}$ characterizing the wave function at infinity (in the short-range field it is a plane wave). Using the Schrödinger equation in momentum representation and the completeness theorem for the wave functions ($\widetilde{V}(\mathbf{q})$ is the Coulomb interaction potential in momentum representation)

$$\left(E_n - \frac{\hbar^2 q^2}{2\mu}\right) \widetilde{\Psi}_n(\mathbf{q}) = \int \widetilde{V}\left(\mathbf{q} - \mathbf{q}'\right) \widetilde{\Psi}_n\left(\mathbf{q}'\right) \frac{d\mathbf{q}'}{(2\pi)^3}, \qquad (4.88)$$

one can transform the scattering amplitude to obtain

$$\Gamma_{ij}\left(\mathfrak{q}, \mathfrak{q}'; \mathfrak{P}\right) = \widetilde{V}_{ij}\left(\mathbf{q} - \mathbf{q}'\right) + (2\pi)^3 \sum_n$$

$$\times \frac{\widetilde{\Psi}_n\left(\mathbf{q}\right) \widetilde{\Psi}_n^*\left(\mathbf{q}'\right) \left(E_n - \frac{\hbar^2 q^2}{2\mu}\right)\left(E_n - \frac{\hbar^2 q'^2}{2\mu}\right)}{i P_4 - \frac{\hbar^2 P^2}{2M} - E_n + \mu_i + \mu_j}. \qquad (4.89)$$

Employing (4.85) and (4.86), we shall write $\delta\Omega$ in the form

$$\frac{\delta\Omega}{V} = \sum_{i,j} \frac{4}{(2\pi)^6 \beta^2} \sum_{q_4, P_4} \int_0^1 \frac{d\lambda}{2\lambda} \int d\mathbf{q}\, d\mathbf{P}\, G_i\left(\frac{m_i}{M}\mathfrak{P} + \mathfrak{q}\right)$$

$$\times G_j\left(\frac{m_j}{M}\mathfrak{P} - \mathfrak{q}\right) \Gamma_{ij}\left(\mathfrak{q}, \mathfrak{q}', \mathfrak{P}\right). \qquad (4.90)$$

Performing summation over the frequencies $q_4$ and $P_4$ and integrating over $d\mathbf{P}$, in the nondegenerate case we obtain from (4.90) and (4.87)

$$\frac{\delta\Omega_L}{V} = \sum_{i,j} \zeta_i \zeta_j \lambda_{ij}^3 \int_0^1 \frac{d\lambda}{2\lambda} \int \frac{d\mathbf{q}}{(2\pi)^3} \sum_n \exp\left(-\beta E_n\right)$$

$$\times \left(E_n - \frac{\hbar^2 q^2}{2\mu}\right) \left|\widetilde{\Psi}_n(\mathbf{q})\right|^2 \qquad (4.91)$$

cf. [946]; in what follows we shall also make use of expression (4.89) for $\Gamma_{ij}$). Here, the activities $\zeta_i$ (cf. (4.64)) are employed $\lambda_{ij} = \sqrt{\frac{2\pi\hbar^2}{\mu T}}$. In the discrete case, for the interacting electron and proton $E_n = -\text{Ry}/n^2$, in the continuum $E_k = \hbar^2 k^2/2\mu$. For the two particles $i$ and $j$ interacting through the short-range potential $V = \lambda V_0(\mathbf{r})$, from (4.91) one can obtain the Beth–Uhlenbeck formula [98, 598, 989].

If one uses the representation (4.89) for the quantity $\Gamma_{ij}(\mathbf{q}, \mathfrak{q}; \mathfrak{P})$ involved in (4.90) for the contribution of ladder diagrams, then the contribution $\widetilde{V}_{ij}(0) \sim e^2/\varkappa^2$ can readily be seen to disappear because of the neutrality relation in the activities (4.68). After summation over $q_4$ and $P_4$ in (4.90) with the use of the contribution of the second term of expression (4.89) we come to (see Fig. 4.5(d))

$$\frac{_{\Delta}\Omega_L}{V} = \sum_{i,j} \zeta_i \zeta_j \lambda_{ij}^3 \int_0^1 \frac{d\lambda}{2\lambda} \int \frac{d\mathbf{q}}{(2\pi)^3} \sum_n \left( E_n - \frac{\hbar^2 q^2}{2\mu} \right)$$

$$\times \left| \widetilde{\Psi}_n(\mathbf{q}) \right|^2 \left( \exp\left(-\beta E_n\right) - \exp\left(-\beta \varepsilon_q\right) \right). \qquad (4.92)$$

In formula (4.92) for the e–p interaction, both summation over the discrete spectrum (bound states) and integration over the scattering states characterized by the index $\mathbf{k}$ should be performed. For the e–e and p–p interactions meaningful is only the latter operation.

To calculate the contribution of bound states to BBK we shall use the exact result of Fok [278] for the wave functions of a nonrelativistic hydrogen atom in momentum representation:

$$\frac{1}{(2\pi)^3} \sum_{l,m} \left| \widetilde{\Psi}_{n,l,m}(\mathbf{q}) \right|^2 = \frac{8}{\pi^2 a_0^5 n^3 (q^2 + p_n^2)^4}. \qquad (4.93)$$

Here $p_n = (a_0 n)^{-1}$; $a_0 = \hbar^2/\mu e^2 \lambda$ is the Bohr "radius" with the current charge $e^2\lambda$. Considering that $E_n \equiv -\hbar^2 p_n^2/2\mu$, the part of expression (4.92), which corresponds to bound states, can with allowance for (4.93) be written in the form

$$\frac{\delta\Omega^{\text{BS}}}{V} = -4\zeta_e\zeta_p \frac{\mu^4 \lambda_{ep}^3 e^{10}}{\pi^2 \hbar^8} \sum_{n=1}^{\infty} \frac{1}{n^3} \int_0^1 d\lambda\, \lambda^4 \left[ \exp\left( \frac{\lambda^2 X}{n^2} \right) \int \frac{d\mathbf{q}}{(q^2 + p_n^2)^3} \right.$$

$$\left. - \int \frac{d\mathbf{q}}{(q^2 + p_n^2)^3} \exp\left( -\beta \frac{\hbar^2 q^2}{2\mu} \right) \right]. \qquad (4.94)$$

Here $X = \beta \text{Ry}$.

Making use of (4.94), after integration over $\lambda$ we are led to the expressions:

$$\frac{\delta \Omega^{\mathrm{BS}}}{V} = -\zeta_{\mathrm{e}}\zeta_{\mathrm{p}} T \lambda_{\mathrm{ep}}^3 \sigma^{\mathrm{BS}}. \tag{4.95}$$

$$\sigma^{\mathrm{BS}} = \sigma^{\mathrm{BS}}_{\mathrm{SRM}} = \sum_{n=1}^{\infty} n^2 \mathrm{e}^{u_n} F(u_n), \tag{4.96}$$

$$F(u) = 1 - \mathrm{e}^{-u}\left(4 - \frac{6}{\sqrt{\pi}}u^{1/2} + \frac{4}{\sqrt{\pi}}u^{3/2}\right) \\ + \operatorname{erfc}\sqrt{u} \cdot (3 - 4u + 4u^2), \tag{4.97}$$

where

$$u_n = \frac{\alpha^2}{n^2}, \quad \alpha = \alpha_{\mathrm{ep}} = \sqrt{\frac{\mu e^4}{2\hbar^2 T}} = \sqrt{\beta \mathrm{Ry}} = \sqrt{X}. \tag{4.98}$$

Asymptotically, as $u \to 0$ ($n \gg 1$ or for large $T$) (4.97) implies that

$$F(u) \sim 2u^2. \tag{4.99}$$

This is four times as large as an analogous Planck–Larkin expression

$$F_{\mathrm{P\text{-}L}}(u) = 1 - \mathrm{e}^{-u} - u\mathrm{e}^{-u} \sim \frac{u^2}{2} \text{ at } u \to 0. \tag{4.100}$$

Ultimately, the statistical sum for bound states can be written as

$$\sigma^{\mathrm{BS}}_{\mathrm{SRM}} = \sum_{k=4}^{\infty} \zeta(k-2)\frac{(-\alpha)^k}{\Gamma\left(\frac{k}{2}+1\right)}(k-2)^2 \\ + \sum_{k=1}^{\infty} \zeta(2k+1)\frac{\alpha^{2k+3}}{\Gamma\left(k+\frac{5}{2}\right)}. \tag{4.101}$$

Here $\zeta(n) - \zeta$ is the Riemann function. The corresponding Planck–Larkin formula is of the form

$$\sigma^{\mathrm{BS}}_{\mathrm{P\text{-}L}} = \sum_{k=2}^{\infty} \zeta(2k-2)\frac{\alpha^{2k}}{\Gamma(k+1)} = \sum_{n=4,6,8,\ldots} \zeta(n-2)\frac{\alpha^n}{\Gamma\left(\frac{n}{2}+1\right)}. \tag{4.102}$$

We shall denote the difference between the quantities $\sigma^{\mathrm{BS}}_{\mathrm{P\text{-}L}}$ (4.102) and $\sigma^{\mathrm{BS}}_{\mathrm{SRM}}$ (4.101) as

$$\sigma^{\mathrm{BS}}_{\mathrm{P\text{-}L}} - \sigma^{\mathrm{BS}}_{\mathrm{SRM}} = \sigma_c. \tag{4.103}$$

For $\sigma_c$ we have

$$\sigma_c = \sum_{k=2}^{\infty} \zeta(2k-2)\frac{\alpha^{2k}}{\Gamma(k+1)} - \sum_{k=4}^{\infty} \zeta(k-2)\frac{(-1)^k \alpha^k}{\Gamma\left(\frac{k}{2}+1\right)}(k-2)^2$$
$$- \sum_{k=1}^{\infty} \zeta(2k+1)\frac{\alpha^{2k+3}}{\Gamma\left(\frac{k}{2}+1\right)}. \qquad (4.104)$$

The subtracted term in expression (4.92) can be represented, for example, for the e–p interaction with allowance for the theorem of completeness of functions including discrete and continuous spectra, in the following form (see Fig. 4.5(a))

$$-\zeta_e\zeta_p\lambda_{ep}^3 \int_0^1 \frac{d\lambda}{\lambda} \int \frac{d\mathbf{q}}{(2\pi)^3} \exp\left(-\beta\varepsilon_q\right) \left(\sum_n + \int \frac{d\mathbf{k}}{(2\pi)^3}\right)$$
$$\times \int \widetilde{V}(\mathbf{q}-\mathbf{q}')\,\widetilde{\Psi}^*_{\{{}^n_k\}}(\mathbf{q})\widetilde{\Psi}_{\{{}^n_k\}}(\mathbf{q}')\frac{d\mathbf{q}'}{(2\pi)^3} = -\zeta_e\zeta_p\widetilde{V}_{\mathrm{ep}}(0), \qquad (4.105)$$

that is, the subtracted term in (4.92) exactly compensates the first term in the representation (4.89) for $\Gamma_{\mathrm{ep}}\left(\mathfrak{q},\mathbf{q}';\mathfrak{P}\right)$ which owing to the neutrality in activities (4.68) is nullified after summation in all pair interactions. Thus, the obtained finite statistical sum (4.96)–(4.97) corresponds to the allowance in the counter-term (4.105) for summation over the bound states only. The integral over the scattering states in representation (4.92) also appears to be finite, but only after subtraction of the contribution from the piece of the ladder with two crossbars taken into account in summation of the ring diagram contribution.

In papers [941, 945], the authors investigated the influence of electron degeneracy upon the contribution of bound states to the plasma pressure.

The contribution of bound states with allowance for electron degeneracy in the small population approximation can be obtained in the following form $m_e \ll m_p$:

$$\delta p_{\mathrm{ed}}^{\mathrm{BS}} = \zeta_{\mathrm{P}}\frac{32}{\pi}\mathrm{Ry}\sum_{n=1}^{\infty} 2\int_0^1 \lambda\,\mathrm{d}\lambda \int_0^{\infty} \frac{t^2\,\mathrm{d}t}{(1+t^2)^3}\frac{\exp(\lambda^2 u_n^2(1+t^2))-1}{\exp(\lambda^2 u_n^2 t^2 - y_{\mathrm{e}})+1}.$$
$$(4.106)$$

This expression was obtained disregarding the Pauli blocking in calculation of the scattering amplitude by V.M. Galitskii's method [337].

For moderate quantities $y_e$ it seems that at high $T \gg \mathrm{Ry}$ the integral grows weakly with increasing $y_e$: $\delta p_{\mathrm{ed}}^{\mathrm{BS}} \to p_{\mathrm{as},1} = \zeta_{\mathrm{p}} T (2\pi^2/3)(\mathrm{Ry}/T)^2$, however for large $y_e$ the contribution can be shown to grow sub-exponentially:

$$\delta p_{\mathrm{ed}}^{\mathrm{BS}} \to p_{\mathrm{as},2} = \zeta_{\mathrm{p}} T \left( \frac{2\pi^2}{3} \left( \frac{\mathrm{Ry}}{T} \right)^2 + \frac{64\zeta(3)}{15\pi} \left( \frac{\mathrm{Ry}}{T} \right)^{5/2} \frac{\exp(y_e)}{y_e^{3/2}} \right).$$
$$(4.107)$$

The obtained expressions $X = \mathrm{Ry}/T$ are valid for not strong degeneracy because they were derived from the expression for the scattering amplitude (4.89) valid for small occupation numbers of electrons and ions.

The shifts of the energy levels of atomic and molecular bound states due to the Pauli blocking and exchange effects were estimated within perturbation theory [236]. With increasing plasma density the summary effect related to the electron degeneracy leads to overall ionization of atoms as distinct from the exponential growth of atomic statistical sum with $y_e$ increasing according to (4.107). Qualitatively this means that in strongly degenerate systems of the type of white dwarfs, bound states are absent in contrast to the naive representation obtained using the Saha formula extended to the case of degeneracy.

The authors of paper [942] presented the extension of the expressions for the contribution of bound states with allowance for the plasma effect on the spectral characteristics of atomic states broadened by plasma microfields and collisions with electrons.

Note that expressions of the type of (4.92) are derived by the Keldysh technique (see Refs. [492, 516, 620]):

$$\delta p = \sum_a (2S_a + 1)\hbar \int_0^1 \frac{\mathrm{d}\lambda}{2\lambda} \int \frac{\mathrm{d}\mathbf{P}}{(2\pi)^3}$$

$$\times \int \frac{\mathrm{d}\omega\,\mathrm{d}\omega'}{(2\pi)^2} \frac{\Sigma_a^> (\hbar\mathbf{P}, \hbar\omega) \, G_a^< (\hbar\mathbf{P}, \hbar\omega')}{\omega - \omega'} \left( 1 - e^{-\beta\hbar(\omega - \omega')} \right). \qquad (4.108)$$

Expressing the quantity $\Sigma^>$ symbolically through the imaginary part of the scattering amplitude $\Gamma_{\mathrm{ep}}$ [492]

$$\sum_e^> \sim \int \mathrm{Im}\Gamma_{\mathrm{ep}}(\hbar\omega + \hbar\Omega) \cdot G_i^< (\hbar\mathbf{P}', \hbar\Omega) \frac{\mathrm{d}\mathbf{P}'\,\mathrm{d}\Omega}{(2\pi)^4}, \qquad (4.109)$$

for $\Gamma_{\rm ep}$ we have (cf. (4.89))

$$\operatorname{Im} \Gamma_{\rm ep} \sim \sum_n \left|\widetilde{\Psi}_n(\mathbf{q})\right|^2 (E_n - \varepsilon_q)^2 \delta_\gamma \left(\hbar\omega + \hbar\Omega - E_n - \frac{\hbar^2 P^2}{2M}\right), \quad (4.110)$$

$\delta_\gamma(x)$ is the Lorentz contour going over to the $\delta$ function in the zero-width limit $\gamma \to 0$. From (4.108)–(4.110) one obtains expression (4.92). At the same time such an approach allows one to take into account the effects of atomic state broadening with the same precision with which spontaneous radiation of a set of charged particles can be calculated with allowance for discrete-discrete transitions. In the case of allowance for Stark microfields inducing level shifts and statistical line broadening this approach also contains the effects of destruction of states by microfields [690, 946, 991].

Expression (4.108) can yield the general form of the contribution of bound states (the parametrization $e^2 \mapsto e^2\lambda$ is used)

$$\sigma^{\rm BS} = \beta \int_0^1 \frac{d\lambda}{\lambda} \int \frac{d\mathbf{q}}{(2\pi)^3} \int d\omega \sum_{n=1}^\infty (E_n - \varepsilon_q)^2 \left|\widetilde{\Psi}_n(q)\right|^2$$

$$\times \frac{e^{-\beta\hbar\omega} - e^{-\beta\varepsilon_q}}{\varepsilon_q - \hbar\omega} \, a_n\left(\frac{\omega - E_n}{\hbar}\right), \quad (4.111)$$

where $\mathbf{q}$ is the wave vector of the relative motion of particles, $q = |\mathbf{q}|$, $a_n(\omega)$ is the contour of the atomic state $n$, broadened by plasma ions and electrons [933].

Employing representation (4.93) and considering that

$$E_n = -\frac{\hbar^2 P_n^2}{2\mu}, \quad E_n - \varepsilon_q = -\frac{\hbar^2}{2\mu}(q^2 + P_n^2),$$

$$\varepsilon_q = \frac{\hbar^2 q^2}{2\mu}, \quad \int d\mathbf{q} = 4\pi \int_0^\infty q^2 \, dq, \quad (4.112)$$

and making use of the dimensionless variables

$$x = \frac{\hbar^2 q^2}{2\mu T}, \quad y = \beta\hbar\omega, \quad z = \frac{\hbar^2 P_n^2}{2\mu T} = -\beta E_n = u_n\lambda^2, \quad (4.113)$$

we are finally led to

$$\sigma^{\mathrm{BS}} = \sum_{n=1}^{\infty} G_n = \sum_{n=1}^{\infty} n^2 \int_0^{u_n} \frac{8}{\pi} z^{3/2} \, \mathrm{d}z \int_{-\infty}^{\infty} \mathrm{d}y \cdot \tilde{a}_n(y+z)$$

$$\times \int_0^{\infty} \frac{\sqrt{x}(\mathrm{e}^{-y} - \mathrm{e}^{-x}) \, \mathrm{d}x}{(x+z)^2(x-y)}. \tag{4.114}$$

The integral over $x$ in (4.114) can be calculated analytically:

$$\mathcal{I} = -\frac{\sqrt{\pi} - \pi \tilde{F}(z)\left(1 + \frac{1}{2z}\right)}{z+y}$$

$$+ \begin{cases} \dfrac{\pi \mathrm{e}^{-y}(z-y)}{2\sqrt{z}(z+y)^2} - \dfrac{\pi\left(\tilde{F}(z) - \tilde{G}(y)\right)}{(z+y)^2}, & y > 0 \\[2ex] \dfrac{\pi \mathrm{e}^{-y}(\sqrt{z} - \sqrt{-y})^2}{2\sqrt{z}(z+y)^2} - \dfrac{\pi\left(\tilde{F}(z) - \tilde{F}(-y)\right)}{(z+y)^2}, & y < 0, \end{cases} \tag{4.115}$$

where

$$\tilde{F}(x) = \mathrm{e}^x \sqrt{x} \, \mathrm{erfc}\sqrt{x}, \quad \tilde{G}(x) = \sqrt{x} \, \mathrm{Im}\, w(\sqrt{x}), \tag{4.116}$$

$$\mathrm{erfc}\, z = \frac{2}{\sqrt{\pi}} \int_z^{\infty} \mathrm{e}^{-t^2} \, \mathrm{d}t,$$

$$w(z) = \mathrm{e}^{-z^2}\left(1 + \frac{2i}{\sqrt{\pi}} \int_0^z \mathrm{e}^{t^2} \, \mathrm{d}t\right) = \mathrm{e}^{-z^2} \mathrm{erfc}(-iz), \tag{4.117}$$

and both branches represent one and the same analytical expression.

If broadening is neglected, then $\tilde{a}_n(y) = \delta(y)$ and the integral over $y$ in (4.114) is equal to

$$I_\omega\Big|_{a_n \equiv \delta} = I_{\delta\omega} = \left\{\int_0^{\infty} \frac{\sqrt{x}(\mathrm{e}^{-y} - \mathrm{e}^{-x})}{(x+z)^2(x-y)} \, \mathrm{d}x\right\}\Bigg|_{y=-z}$$

$$= \int_0^{\infty} \frac{\sqrt{x}(\mathrm{e}^z - \mathrm{e}^{-x})}{(x+z)^3} \, \mathrm{d}x. \tag{4.118}$$

An analytical integration over $x$ gives the expression

$$\sigma^{\mathrm{BS}}\Big|_{a_n \equiv \delta} = \sum_{n=1}^{\infty} n^2 \int_0^{u_n} \mathrm{d}z \left( \mathrm{e}^z - \frac{2}{\sqrt{\pi}} \sqrt{z}(1+2z) - \mathrm{e}^z \mathrm{erfc}\sqrt{z}\, \left(1 - 4z - 4z^2\right) \right).$$

$$(4.119)$$

The integration in (4.119) (see Ref. [941]) yields the already known result (4.95)–(4.97).

When the effect of collisional broadening for the atomic state with the principal quantum number $n(n \geq 2)$ is taken into account, the formula [933]

$$a_n(\omega) = \frac{1}{\pi} \int \frac{\mathcal{H}(x)\gamma_n(\omega)\,\mathrm{d}x}{(B_n \mathcal{E}_0 x - \omega)^2 + \gamma_n^2(\omega)}, \qquad x = \frac{\mathcal{E}}{\mathcal{E}_0}, \tag{4.120}$$

$$\mathcal{E}_0 = 2\pi \left(\frac{4}{15}\right)^{2/3} Z e N^{2/3},$$

$$B_n = \left(\frac{3}{8}\right)^{2/3} \frac{\hbar}{Z m_e e}(n^2 - n'^2), \tag{4.121}$$

$$\gamma_n(\omega) = \gamma_{en} \begin{cases} 1, & \omega > 0 \\ \mathrm{e}^{-\beta\hbar|\omega|}, & \omega < 0 \end{cases}, \tag{4.122}$$

$$\gamma_{en} = \frac{32}{3} N \langle v \rangle^{-1} \frac{\hbar^2}{m_e^2}\left(\max\left(1, \ln\frac{R_{\mathrm{D}}}{\rho_0}\right) + 0.215\right) I(n, n'). \tag{4.123}$$

is used. In this formula, $\mathcal{H}(x)$ is the Holtsmark function (see below), $\mathcal{E}$ is the strength of the ion field, $\mathcal{E}_0$ is its characteristic value, $B_n$ is the effective Stark constant for the state $n' = 1$, $N = N_{\mathrm{A}}\rho$; $\gamma_{en}$ is the effective constant of broadening by an electron impact, $\langle v \rangle$ is the mean velocity, $\rho_0$ is the Weisskopf radius, and $R_{\mathrm{D}} = \varkappa_{\mathrm{D}}^{-1}$ is the Debye radius. The ground state $n = 1$ may be thought of as not broadened, $\gamma_1 \to 0$,

$$\langle v \rangle = \frac{2}{\sqrt{\pi}} \sqrt{\frac{2T}{m_e}}, \qquad I(n, n') = \frac{1}{2}(n^4 + n'^4). \tag{4.124}$$

We use the Weisskopf radius as defined in Ref. [933] (formula (22.33)),

$$\rho_0^2 = \frac{2}{3}\left(\frac{e^2}{\hbar \langle v \rangle}\right)^2 I(n, n')a_0^2. \tag{4.125}$$

Taking into consideration the parametrization $e^2 \mapsto e^2\lambda$, we obtain

$$\rho_0 = \frac{1}{\sqrt{3}} \frac{\hbar}{\mu \langle v \rangle} \sqrt{n^4 + 1}. \tag{4.126}$$

The Holtsmark function exploited in (4.120) is given in the form

$$\mathcal{H}(x) = \frac{2}{\pi x} \int_0^\infty t \sin t \, \exp\left[ -\left(\frac{t}{x}\right)^{3/2} \right] \, \mathrm{d}t. \tag{4.127}$$

For the statistical sum represented in the form (4.114) for the case without broadening we have

$$G_n \equiv G_n^{\mathrm{N\ B}} = n^2 \mathrm{e}^{u_n} F(u_n) \sim \frac{2\alpha^4}{n^2} \quad \text{at } n \to \infty. \tag{4.128}$$

The estimates and calculations show that asymptotically the ratio of the terms of the statistical sum with and without allowance for broadening is

$$\frac{G_n^{\mathrm{B}}}{G_n^{\mathrm{N\ B}}} \sim \ln n \quad \text{at} \quad n \to \infty, \tag{4.129}$$

which guarantees convergence of the statistical sum with allowance for broadening. The statistical sum $\sigma^{\mathrm{BS}}$ must actually be limited to the first $n_l$ states for which the dimensionless ionization potential $u_n = \alpha^2/n^2$ is larger than the dimensionless characteristic energy, $B = \beta \hbar B_n \mathcal{E}_0$. Used in the calculations was a smoothed restriction of the number of considered states by introduction of the weight factor of state $w_n$:

$$\sigma_{\mathrm{s}}^{\mathrm{BS}} = \sum_n w_n G_n, \quad w_n = \min(1, \mathrm{e}^{-\tau}), \quad \tau = \frac{B - u_n}{u_n}. \tag{4.130}$$

Note that there exist more correct ways of allowance for the microfield distribution function (see, for instance, the exponential approximation with tunable parameters (APEX) [459]). For applications in helioseismology the Holtsmark approximation is applicable since the nonideality parameter is $\Gamma_{\mathrm{D}} = \varkappa_{\mathrm{D}} e^2/T \ll 1$.

To calculate the contribution of states of continuum to expressions like (4.92), it is necessary to determine Fourier components of the wave functions describing mutual scattering of charged particles. It is convenient to apply a system of Coulomb wave functions as the sum over orbital momenta [10]. In the momentum representation we have

$$\widetilde{\Psi}_k(q) = (2\pi)^{-3/2} \exp\left(\frac{\pi\tilde{\xi}}{2}\right) \Gamma\left(1 - i\tilde{\xi}\right) J_\varkappa, \tag{4.131}$$

$$J_\varkappa = \left(\frac{2\pi\left(1 - i\tilde{\xi}\right)\varkappa}{\left(\frac{(\mathbf{q}-\mathbf{k})^2}{2} + \frac{\varkappa^2}{2}\right)^2} + \frac{2\pi\tilde{\xi}(k + i\varkappa)}{\left(\frac{(\mathbf{q}-\mathbf{k})^2}{2} + \frac{\varkappa^2}{2}\right)\left(\frac{q^2 - k^2 + \varkappa^2}{2} - ik\varkappa\right)}\right)$$

$$\times \exp\left(i\tilde{\xi}\ln\frac{(\mathbf{q} - \mathbf{k})^2 + \varkappa^2}{q^2 - k^2 + \varkappa^2 - 2ik\varkappa}\right). \tag{4.132}$$

Here $\tilde{\xi} = (a_0 k)^{-1} = (\mu e^2 \lambda)/(\hbar^2 k)$. The first term in (4.132) is similar to the regularized three-dimensional $\delta$-function with the use of the parameter $\varkappa$, $\varkappa \to 0$. Such expressions can also be obtained for the "e–e" interaction by the substitution $m_\mathrm{p} \to m_\mathrm{e}$.

Taking into account the subtraction in the factor $\left(e^{-\beta E_k} - e^{-\beta \varepsilon_q}\right)$ in (4.92), for example, for the "e–p" interaction, one can transform expression (4.92) for the scattering states (SS) of the continuum to the form

$$\frac{\delta\Omega_\mathrm{ep}^\mathrm{SS}}{V} = -\frac{\lambda_\mathrm{ep}}{\pi}\zeta_\mathrm{e}\zeta_\mathrm{p}T\left(\frac{e^2}{T}\right)^2 \int\limits_0^1 \lambda\,d\lambda \int\limits_0^\infty\int\limits_0^\infty dx\,dy\,\frac{\pi\tilde{\xi}_\mathrm{ep}}{\sinh\pi\tilde{\xi}_\mathrm{ep}}\exp\left(\pi\tilde{\xi}_\mathrm{ep}\right)$$

$$\times\frac{e^{-y} - e^{-x}}{x - y}\left(\frac{1}{\left(\sqrt{x} - \sqrt{y}\right)^2 + \tilde{\eta}_\mathrm{ep}^2} - \frac{1}{\left(\sqrt{x} + \sqrt{y}\right)^2 + \tilde{\eta}_\mathrm{ep}^2}\right)$$

$$\times\exp\left[-2\tilde{\xi}_\mathrm{ep}\,\mathrm{Im}\ln\left(x - y + \tilde{\eta}_\mathrm{ep}^2 + i2\sqrt{y\tilde{\eta}_\mathrm{ep}^2}\right)\right], \tag{4.133}$$

where $\tilde{\xi}_\mathrm{ep} = \frac{\lambda\alpha_\mathrm{ep}}{\sqrt{y}}$, $\alpha_\mathrm{ep} = \sqrt{\mathrm{Ry}/T}$, $\tilde{\eta}_\mathrm{ep}^2 = \frac{\hbar^2\varkappa^2}{8\mu T}$.

We shall reduce the expression for the counter-term to expression (4.133) at the expense of the contribution from the two ladder crossbars taken into account in summation of a series of ring diagrams (see Ref. [941]; $\varkappa$ is a regularizing parameter which at the end of all calculations should be directed to zero):

$$\frac{\Delta\Omega_\mathrm{ep}^{(2)}}{V} = -8\sqrt{\pi}\zeta_\mathrm{e}\zeta_\mathrm{p}\frac{e^2}{\varkappa^2}\sqrt{X}\int\limits_0^\infty e^{-y}\arctan\left(\frac{1}{2}\sqrt{\frac{\eta_\mathrm{ep}^2}{y}}\right)dy$$

$$= -8\sqrt{\pi}\zeta_\mathrm{e}\zeta_\mathrm{p}\frac{e^2}{\varkappa^2}\sqrt{X}J. \tag{4.134}$$

Here $J$ is the integral over $y$ which is equal to

$$J = \frac{\pi}{2}\left[1 - \exp\left(\frac{\eta_{\text{ep}}^2}{4}\right)\left[1 - \Phi\left(\frac{\eta_{\text{ep}}}{2}\right)\right]\right]. \tag{4.135}$$

The three-dimensional numerical integration gives the following asymptotics as $\varkappa \to 0$ $(a = \text{e}, \text{p})$:

$$\frac{\delta\Omega_{aa}^{\text{SS}}}{V} - \frac{\delta\Omega_{aa}^{(2)}}{V} \to -\zeta_a^2\left\{\frac{e^6}{T^2}\left[\frac{\pi}{3}\ln\left(\frac{\varkappa\lambda_{aa}}{2\sqrt{\pi}}\right) - \frac{\pi}{6}(1 - C)\right]\right.$$
$$\left. + \frac{\lambda_{aa}^3 T}{2}\sigma_{\text{Q}}\left(-\frac{\alpha_a}{2}\right)\right\}, \tag{4.136}$$

where $\delta\Omega_{aa}^{(2)}$ is the portion of the ladder with two crossbars, already included in the Debye–Huckel approximation, $\alpha_a = \sqrt{m_a e^4/\hbar^2 T}$.

$$\frac{\delta\Omega_{\text{ep}}^{\text{SS}}}{V} - \frac{\delta\Omega_{\text{ep}}^{(2)}}{V} \to\to 2\zeta_{\text{e}}\zeta_{\text{p}}\left\{\frac{e^6}{T^2}\left[\frac{\pi}{3}\ln\left(\frac{\varkappa\lambda_{\text{ep}}}{2\sqrt{\pi}}\right) - \frac{\pi}{6}(1 - C)\right]\right.$$
$$\left. + \frac{\lambda_{\text{ep}}^3 T}{2}\left[\sigma_{\text{SRM}}^{\text{BS}} - \sigma_{\text{Q}}(\alpha_{\text{ep}})\right]\right\}, \tag{4.137}$$

$$\sigma_{\text{Q}}(\alpha) = \frac{1}{2}\sum_{n=4}^{\infty}\frac{\zeta(n-2)}{\Gamma\left(\frac{n}{2}+1\right)}\alpha^n$$
$$= -\left(\ln|2\alpha| + \frac{3C}{2} - \frac{4}{3}\right)\frac{2\alpha^3}{3\sqrt{\pi}} + o(|\alpha|^3), \quad \alpha \to -\infty.$$

In [234], the second term in (4.137) contains an erroneous extra summand $\ln 3$ in brackets $(-C - 2\ln 3 + 1)$.

Summation of expressions similar to (4.137) for "p–p", "e–e", and "e–p" interactions with allowance for the fact that $\zeta_{\text{e}} = \zeta_{\text{p}}$ leads to a $\varkappa$-independent expression for the classical part of the second virial coefficient:

$$\frac{\delta\Omega^{\text{cl}}}{V} = \zeta_{\text{e}}^2 T\left(\frac{e^2}{T}\right)^3 \cdot \frac{\pi}{6}\ln\frac{m_{\text{p}}}{4m_{\text{e}}}. \tag{4.138}$$

For the exchange contribution $\delta\Omega_{\text{ee}}^{\text{exch}}$ we obtain the convergent expression

$$\frac{\delta\Omega_{\text{ee}}^{\text{exch}}}{V} = \frac{1}{8\sqrt{\pi}}\zeta_{\text{e}}^2\lambda_{\text{ee}}^3 T E(\alpha_{\text{ee}}). \tag{4.139}$$

For $E(\alpha_{ee})$ one can derive the explicit expression ($\alpha_{ee} = -\alpha_e/2$)

$$E(\alpha) = \alpha + \sqrt{\pi}\ln 2 \cdot \alpha^2 + \frac{\pi^2}{9}\alpha^3 + \sum_{n=4}^{\infty} \frac{\sqrt{\pi}\left(1 - 2^{2-n}\right)}{\Gamma\left(\frac{n}{2} + 1\right)}\zeta(n - 1)\alpha^n.$$

$$(4.140)$$

Formula (4.140) corresponds to work [234] for the exchange contribution and is confirmed by numerical integration.

It should be noted that the expression for the exchange contribution in the limit of low values of the parameter alpha goes over to expression (4.70). For degenerate electrons analogous expressions exist for the exchange contribution to the second virial coefficient [598]. In paper [212], an elegant expression for the exchange contribution is presented which is valid for an arbitrary degree of degeneracy, however account is only taken of the first term of the Born series in the parameter analogous to $\alpha$ but with the chemical potential determining the Fermi energy for the degenerate limit instead of temperature [598].

In the center of the Sun it is necessary to allow for the relativistic corrections in the contribution from the free electron gas to the thermodynamic functions and for the radiation pressure in "transparent" plasma. Using relativism as correction ($T/(mc^2) \lesssim 10^{-3}$) we shall write the relation between the momentum and the kinetic energy of the electron,

$$P = \sqrt{2mE}\left(1 + \frac{E}{4mc^2}\right).$$

$$(4.141)$$

Using (4.141) we arrive at the expression for electron activity (cf. (4.65)):

$$n_e^0(\mu_e) \equiv \zeta_{e,r} = \frac{2}{\sqrt{\pi}}\frac{2}{\lambda_e^3}\int_0^{\infty} \frac{x^{1/2}\left(1 + \frac{5}{4}\frac{T}{mc^2}x\right)}{\exp\left(x - y\right) + 1}\,dx,$$

$$(4.142)$$

and for electron gas pressure (cf. (4.66)) (the index r means that account was taken of relativistic corrections),

$$P_{0e,r} = \frac{4T}{3\sqrt{\pi}}\frac{2}{\lambda_e^3}\int_0^{\infty} \frac{x^{3/2}\left(1 + \frac{3}{4}\frac{T}{mc^2}x\right)}{\exp\left(x - y\right) + 1}.$$

$$(4.143)$$

In (4.142), (4.143), we have $x = \varepsilon/T$, and $y = \mu_e/T$ as before.

Corrections to the energy and to other electron functions can be found in a similar way. To derive the EOS of weakly nonideal hydrogen plasma one should find the relation of activities (or chemical potentials) with concentration with allowance for electron degeneracy

$$n_{\mathrm{e}} + \sum_k n_k = \sum_k (z_k + 1) n_k = -\sum_k{}' \left( \frac{\partial \left( \frac{\Omega}{V} \right)}{\partial \mu_k} \right)_T. \tag{4.144}$$

The thermodynamic potential $\Omega$ is represented as the sum of all the above-mentioned contributions. Here, the prime at summation means that differentiation in (4.144) should be made with account of relations (4.68) and (4.142), (4.143) type expressions for the electron activity and pressure. So, within the simplest model in which along with the contribution of ideal gas the Debye–Huckel contribution is taken into account in the EOS with the use of the general expressions (4.71), (4.72) and the neutrality condition (4.74), we obtain for the simplest case $z_k = 1$

$$2n_{\mathrm{e}} = \zeta_{\mathrm{e}} + \zeta_{\mathrm{p}} + \frac{\Gamma_{\mathrm{D}}}{2} \left( \zeta_{\mathrm{p}} + T^2 \frac{\partial^2 n_{\mathrm{e}}^0}{\partial \mu_{\mathrm{e}}^2} \right). \tag{4.145}$$

For $\delta\Omega_{\mathrm{ep}}^q$ the result can be represented (cf. (4.101), (4.104)) in the form

$$\frac{\delta\Omega_{\mathrm{ep}}^q}{V} = -\zeta_e \zeta_p T \lambda_{\mathrm{ep}}^3 \sigma^{\mathrm{SS}}, \tag{4.146}$$

$$\sigma^{\mathrm{SS}} = \sum_{n=4}^{\infty} \frac{\zeta(n-2)}{\Gamma\left(\frac{n}{2}+1\right)} \left( \tfrac{1}{2} - (-1)^n (n-2)^2 \right) \alpha_{\mathrm{ep}}^n$$

$$- \sum_{k=1}^{\infty} \frac{\zeta(2k+1)}{\Gamma\left(k+\frac{5}{2}\right)} \alpha_{\mathrm{ep}}^{2k+3}. \tag{4.147}$$

Here $\alpha_{\mathrm{ep}} = \sqrt{\beta \mathrm{Ry}}$.

Note that the obtained analytical expressions for the convergent contribution from the bound states (4.95), along with the final expression for the contribution from the scattering states (4.146), (4.147) (obtained, strictly speaking, by guessing and comparison with the results of numerical integration (4.133) with the subtraction of (4.134)) give in total (using the decomposition (4.101)) the expressions

$$\frac{\delta\Omega_{\mathrm{ep}}^q}{V} = \frac{\delta\Omega^{\mathrm{BS}}}{V} + \frac{\delta\Omega_{\mathrm{ep}}^{\mathrm{SS}}}{V}$$

$$= -\zeta_{\mathrm{e}} \zeta_{\mathrm{p}} T \lambda_{\mathrm{ep}}^3 \left( \sigma^{\mathrm{BS}} + \sigma^{\mathrm{SS}} \right) = -\zeta_{\mathrm{e}} \zeta_{\mathrm{p}} T \lambda_{\mathrm{ep}}^3 \sigma^{\mathrm{tot}}. \tag{4.148}$$

Here $\sigma^{\text{tot}}$ is the corresponding sum over the powers of parameter $\alpha_{\text{ep}} = r\sqrt{\beta \text{Ry}}$ coinciding exactly with the expression presented in [234, 582] and obtained from Beth-type formulas written for the summary contribution from the attraction state without division into bound and scattering states

$$\sigma^{\text{tot}} = \frac{1}{2} \sum_{n=4}^{\infty} \frac{\zeta(n-2)}{\Gamma\left(\frac{n}{2}+1\right)} \alpha_{\text{ep}}^n. \tag{4.149}$$

Interestingly, this expression (4.149) in [234, 582] was obtained after discarding the divergent terms $\sim e^2$, $e^4$, $e^6$ with references to the Coulomb screening. Our approach may provide an exact correspondence of expression (4.149) to the sum of convergent contributions from the bound states (requiring no allowance for screening) and the scattering states calculated with the help of three-dimensional integrals (4.133) from which one should subtract the contribution from two crossbars (4.134) already taken into account in obtaining the ring approximation corresponding to the D–H approximation. For high values of the parameter $\alpha_{\text{ep}} \gg 1$ one can estimate the sum in (4.149) replacing summation by integration over $n$ and applying the saddle-point approximation,

$$\sigma^{\text{tot}} \to \exp\left(\frac{I}{T}\right),$$

which corresponds to allowance for only the ground state in the atomic statistical sum as was initially assumed in the Saha formula. The estimates by this formula give the fraction of bound states ("atoms") equal to approximately 30% in the center of the Sun (which is absurd), whereas according to the Planck–Larkin formula this fraction is of the order of $10^{-4}$. Subtraction of the contribution of bound states represented by the Planck–Larkin formula (with $\sigma_{\text{P–L}}^{\text{BS}}$ (4.60), (4.61), (4.62)) from (4.149) may give an analogue of contribution from the scattering states (cf. (4.146), (4.147)):

$$(\sigma^{\text{SS}})' = \sigma^{\text{tot}} - \sigma_{\text{P–L}}^{\text{BS}} = -\frac{1}{2} \sum_{n=4}^{\infty} \frac{(-1)^n \zeta(n-2)}{\Gamma\left(\frac{n}{2}+1\right)} \alpha_{\text{ep}}^n. \tag{4.150}$$

Thus, taking into account (4.103), (4.104), and (4.150), one can derive

$$\sigma^{\text{SS}} = (\sigma^{\text{SS}})' + \sigma_c, \tag{4.151}$$

$$\sigma_{\text{SRM}}^{\text{BS}} = \sigma_{\text{P–L}}^{\text{BS}} - \sigma_c. \tag{4.152}$$

The total contribution (4.149) from the bound states and the scattering states can be represented in two ways

$$\sigma^{\text{tot}} = \sigma^{\text{BS}}_{\text{SRM}} + \sigma^{\text{SS}} = \sigma^{\text{BS}}_{\text{P L}} + (\sigma^{\text{SS}})'. \tag{4.153}$$

In this sense the Planck–Larkin formula is "regular" but contains contributions from the bound states and from the continuum if the contribution from the scattering states is simultaneously determined in an appropriate way. Unlike the Planck–Larkin formula $\sigma^{\text{BS}}_{\text{SRM}}$ describes the contribution of bound states only.

We shall write several first terms of the decompositions (4.101), (4.62), (4.147) and (4.150) $(\delta\Omega/V = -T\zeta_e\zeta_p\lambda^3_{\text{ep}}\sigma)$ in the Born parameter $\sim e^2/(\hbar v_T)$

$$\lambda^3_{\text{ep}}\sigma^{\text{BS}}_{\text{SRM}} = \frac{\pi^4}{3}\left(\frac{e^2}{T}\right)^4\frac{1}{\lambda_{\text{ep}}} - \frac{64\pi^2}{15}\zeta(3)\left(\frac{e^2}{T}\right)^5\frac{1}{\lambda^2_{\text{ep}}}$$

$$+ \frac{8\pi^3}{3}\zeta(4)\left(\frac{e^2}{T}\right)^6\frac{1}{\lambda^3_{\text{ep}}} - \ldots \tag{4.154}$$

$$\lambda^3_{\text{ep}}\sigma^{\text{BS}}_{\text{P L}} = \frac{\pi^4}{12}\left(\frac{e^2}{T}\right)^4\frac{1}{\lambda_{\text{ep}}} - \frac{\pi^3}{6}\zeta(4)\left(\frac{e^2}{T}\right)^6\frac{1}{\lambda^3_{\text{ep}}} + \ldots \tag{4.155}$$

$$\lambda^3_{\text{ep}}\sigma^{\text{SS}} = -\frac{7\pi^4}{24}\left(\frac{e^2}{T}\right)^4\frac{1}{\lambda_{\text{ep}}} + \frac{68\pi^2}{15}\zeta(3)\left(\frac{e^2}{T}\right)^5\frac{1}{\lambda^2_{\text{ep}}}$$

$$- \frac{31\pi^3}{12}\zeta(4)\left(\frac{e^2}{T}\right)^6\frac{1}{\lambda^3_{\text{ep}}} - \ldots \tag{4.156}$$

$$\lambda^3_{\text{ep}}(\sigma^{\text{SS}})' = -\frac{\pi^4}{24}\left(\frac{e^2}{T}\right)^4\frac{1}{\lambda_{\text{ep}}} + \frac{4\pi^2}{15}\zeta(3)\left(\frac{e^2}{T}\right)^5\frac{1}{\lambda^2_{\text{ep}}}$$

$$- \frac{\pi^3}{12}\zeta(4)\left(\frac{e^2}{T}\right)^6\frac{1}{\lambda^3_{\text{ep}}} - \ldots \tag{4.157}$$

The equation of plasma state (the pressure $p(\rho, T)$ and other thermodynamic functions) should be supplemented with the contribution from the equilibrium thermal radiation in plasma. In paper [1088], expressions for radiation intensity in an absorbing medium were presented, from which it follows that for the parameter $\varepsilon''/\varepsilon' \ll 1$ one can use the expression for radiation in a transparent medium. In the center of the Sun, for example, in the framework of the $S$-model [180] we have $\bar{k}_\omega c/\bar{\omega} \sim 10^{-6}$, where $\bar{k}_\omega$ is the mean absorption coefficient (opacity), $c$ is the velocity of light, and $\bar{\omega}$ is the mean frequency of thermal radiation.

We shall write the expression for the energy in transparent dense plasma [600] with allowance for the relation $k = n\omega/c$ ($n$ is the refractive index of plasma and $n = \sqrt{1 - \omega_p^2/\omega^2}$):

$$E_R = 2V \int \frac{\hbar\omega}{\exp\left(\frac{\hbar\omega}{T}\right) - 1} \cdot \frac{d\mathbf{k}}{(2\pi)^3}$$

$$= \frac{\hbar V}{\pi^2 c^3} \int_{\omega_p}^{\infty} \frac{\left(\omega^2 - \omega_p^2\right)^{1/2} \omega^2}{\exp\left(\frac{\hbar\omega}{T}\right) - 1} \, d\omega. \tag{4.158}$$

Here $V$ is the system volume ($V \to \infty$) and $\omega_p$ is the electron plasma frequency $\omega_p^2 = 4\pi e^2 n_e/m_e$. In expression (4.158) it is taken into consideration that radiation with frequencies $\omega < \omega_p$ does not propagate in the medium as free. Expanding the denominator in (4.158) in powers of the exponent $\exp\left(-\hbar\omega/T\right)$, we obtain ($K_n(z)$ are Macdonald functions [366], $z = n\hbar\omega_p/T$)

$$\frac{E_R}{V} = \frac{\hbar\omega_p}{\pi^2 c^3} \frac{T^3}{\hbar^3} \sum_{n=1}^{\infty} \frac{1}{n^3} \left[ z K_0(z) + 2z K_2(z) + (2 + z^2) K_1(z) \right]. \tag{4.159}$$

In the limit $\hbar\omega_p \ll T$ (in the center of the Sun $\hbar\omega_p/T \sim 1/4$) we obtain from (4.159) (cf. [598])

$$\frac{E_R}{V} = \frac{\pi^2 T^4}{15 \hbar^3 c^3} \left[ 1 - \frac{5}{4\pi^2} \left( \frac{\hbar\omega_p}{T} \right)^2 \right]. \tag{4.160}$$

For the free radiation energy we analogously obtain

$$F_R = \frac{TV}{\pi^2 c^3} \int_{\omega_p}^{\infty} \omega \sqrt{\omega^2 - \omega_p^2} \ln\left( 1 - \exp\left( \frac{-\hbar\omega}{T} \right) \right) d\omega$$

$$= -\frac{\hbar V}{3\pi^2 c^3} \int_{\omega_p}^{\infty} \frac{(\omega^2 - \omega_p^2)^{3/2}}{\exp\left(\frac{\hbar\omega}{T}\right) - 1} \, d\omega = -\frac{\hbar V \omega_p^2 T^2}{\pi^2 c^3 \hbar^2} \sum_{n=1}^{\infty} \frac{1}{n^2} K_2(z). \tag{4.161}$$

From expression (4.161) one can obtain an expression for the radiation pressure ($\frac{\partial \omega_{\mathrm{p}}}{\partial V} = -\frac{\omega_{\mathrm{p}}}{2V}$)

$$p_{\mathrm{R}} = -\left(\frac{\partial F_{\mathrm{R}}}{\partial V}\right)_T = \frac{\omega_{\mathrm{p}}^2 T^2}{\hbar \pi^2 c^3} \sum_{n=1}^{\infty} \frac{1}{n^2} K_2(z) + \frac{\omega_{\mathrm{p}}^3 T}{2\pi^2 c^3} \sum_{n=1}^{\infty} \frac{1}{n} K_1(z). \qquad (4.162)$$

In the limit $\hbar\omega_{\mathrm{p}} \ll T$, from (4.162) we have [598]

$$p_{\mathrm{R}}^0 = \frac{T^4 \pi^2}{45 \hbar^3 c^3}. \qquad (4.163)$$

To find the velocity of sound along the solar trajectory we need the values of the quantity $c_V^R$, the radiation heat capacity, $c_V^R = \frac{\partial E_R/V}{\partial T}$, and the derivatives $\left(\frac{\partial p_{\mathrm{R}}}{\partial T}\right)_\rho$, $\left(\frac{\partial p_{\mathrm{R}}}{\partial \rho}\right)_T$. These expressions can readily be obtained from (4.159) and (4.162). We shall not present here the corresponding, rather cumbersome expressions.

Note that equilibrium between radiation and matter implies meeting the requirement $\bar{k}_\omega R \gg 1$. In the periphery, in the photosphere, this condition is already violated [180], i.e., $\rho \approx 4 \cdot 10^{-9}$ g/cm$^3$, $\bar{k} = \varkappa \rho = 8 \cdot 10^{-3} \cdot 4 \cdot 10^{-9} = 3.2 \cdot 10^{-11}$ cm$^{-1}$, $R_\odot \approx 7 \cdot 10^{10}$ cm, $\bar{k} R_\odot \approx 2$.

We shall mention some problems that have to be solved for further progress in this area. In the first place, this refers to successive going beyond the approximation describing the EOS to the second virial coefficient. The account of the contribution to pressure from the states H$^-$, H$_2^+$, H$_2$ even for hydrogen already requires going outside the framework of BBK. The problem of the description of the contribution of multielectron states in neutral and charged particles (HeI, HeII, etc. in all other elements) and multiparticle scattering states was mentioned earlier. The heroic effort [846, 850] made in this direction should not be thought of as closing the problem since with the knowledge of the contribution of simple irreducible diagrams $p_{\mathrm{CD}}$ the very method [850] of allowance for connected diagrams $p_{\mathrm{SD}} = TS(\zeta_k)$ (see also Ref. [234]), consisting in the use of relation

$$p_{\mathrm{CD}} = T \sum_k \frac{\zeta_k}{2} \left(\frac{\partial S}{\partial \zeta_k}\right)^2, \qquad (4.164)$$

results in inaccuracies because of the charge integration in the quantum thermodynamic perturbation theory [5]. Taking as $S(\zeta_k)$ the Debye–Huckel approximation

$$S = \frac{\varkappa_{\mathrm{D}}^3}{12\pi}, \tag{4.165}$$

we obtain from (4.164)

$$P_{\mathrm{CD}} = \frac{\pi}{2}T\left(\frac{e^2}{T}\right)^3 \left(\sum_i \zeta_i z_i^4\right)\left(\sum_j \zeta_j z_j^2\right), \tag{4.166}$$

which differs from the "correct" result (4.84) by the coefficient where because of the charge integration we have $\pi/3$ instead of $\pi/2$ (for the same misprint see Ref. [603]).

At the same time, the approach [846, 850] makes it possible to predict, at least qualitatively, the structure of the convergent statistical sum of a multielectron atom. Note that Planck–Larkin or $\sigma_{\mathrm{SRM}}$ type formulas diverge being literally extended, for instance, to a two-electron (helium-like) atom (ion) because of the presence of doubly excited states. In the limit of a large nuclear charge ($z \gg 1$), when in the principal approximation the electron–electron interaction can be ignored compared to the electron–nucleus interaction, the expression

$$p_{\mathrm{CD}} = \frac{T}{2}\zeta_{\mathrm{p}}\zeta_{\mathrm{e}}^2\lambda_{\mathrm{ep}}^6\left(\sigma_{\mathrm{ep}}^{\mathrm{BS}}\right)^2 \tag{4.167}$$

obtained from (4.164) with allowance for (cf. (4.95)) $S_{\mathrm{SD}} = \zeta_{\mathrm{e}}\zeta_{\mathrm{p}}T\lambda_{\mathrm{ep}}^3\sigma^{\mathrm{BS}}$ seems reasonable at least qualitatively. Substantiation of such type of equations needs employment of Faddeev equations for the three-body problem [250].

We shall also present the contributions from molecules and interactions between charged and neutral particles to the plasma pressure from paper [944].

The pressure of hydrogen molecules can be obtained using the "harmonic oscillator-rigid rotator" approximation [598]

$$p_{\mathrm{m}} = T\zeta_{\mathrm{m}}, \quad \zeta_{\mathrm{m}} = \left(\frac{\zeta_{\mathrm{a}}}{\sigma_{\mathrm{BS}}}\right)^2 \lambda_{\mathrm{aa}}^3\sigma_{\mathrm{m}}, \quad \zeta_{\mathrm{a}} = p_{\mathrm{BS}}/T, \tag{4.168}$$

and the statistical sum for $\sigma_{\mathrm{m}}$ molecules has the form

$$\sigma_{\mathrm{m}} = \sum_{n_1,n_2,J,v} n_1^2 F_{\mathrm{SRM}}(\beta I_{n_1})e^{\beta I_{n_1}} n_2^2 F_{\mathrm{SRM}}(\beta I_{n_2})e^{\beta I_{n_2}}$$

$$\times \left\{\exp\left(\beta\left(D_{n_1,n_2} - \hbar\omega_{n_1,n_2}v - \frac{\hbar^2 J^2}{2I_{n_1,n_2}}\right)\right) - 1\right\}, \tag{4.169}$$

where summation is taken over the atomic states with the principal quantum numbers $n_1$ and $n_2$ of the vibrational $v$ and rotational $J$ states. $I_{n_1}$, $I_{n_2}$ are ionization energies of these atomic states, $D_{n_1,n_2}$ is the dissociation energy of the corresponding molecule, $I_{n_1,n_2}$ is the moment of inertia, and $\beta = T^{-1}$. For our applications a simplified expression was used as the product of statistical sums over the vibrational and rotational states

$$\frac{\sigma_{\mathrm{m}}}{\sigma_{\mathrm{BS}}^2} = \sigma_r \sigma_v = \frac{T}{B_r} \left( \frac{e^{D/T} - 1}{1 - e^{-\hbar\omega/T}} - \frac{D}{\hbar\omega} - \frac{D^2}{2\hbar\omega T} \right), \qquad (4.170)$$

where $D = D_{1,1} \approx 4.52\,\mathrm{eV}$ is the dissociation energy of a hydrogen molecule, $B_r = \hbar^2/2I_{1,1} \approx 0.00736\,\mathrm{eV}$ is the characteristic rotational energy, and $\hbar\omega = \hbar\omega_{1,1} \approx 0.546\,\mathrm{eV}$ is the characteristic vibrational energy.

Corrections for the interaction of charged and neutral particles are introduced in the simple form

$$p_{en} = c_{en}\zeta_e\zeta_n, \quad p_{p,pn} = c_{p,pn}\zeta_p\zeta_n, \quad (n = \mathrm{a,\ m}), \qquad (4.171)$$

where the index $p$ means the allowance for polarization of atoms and molecules by charged particles, and

$$c_{en} = -\frac{2\pi\hbar^2}{\mu_{en}}l_{0,n}, \quad c_{p,pn} = \frac{2\pi e^2\alpha_n}{r_0}, \qquad (4.172)$$

where $l_{0,n}$ is the scattering length for the particle $n$: $l_{0,\mathrm{a}} = 3.125a_0$, $l_{0,\mathrm{m}} = 1.6a_0$, $\alpha_\mathrm{a} = 6.67 \cdot 10^{-25}$, $\alpha_\mathrm{m} = 8.17 \cdot 10^{-25}$ and $a_0 = \hbar^2/(m_e e^2)$, $r_0 = e^2/I$, $I$ being the ionization potential of the neutral component.

In the EOS of low-temperature hydrogen plasma one should also take into account two types of hydrogen ions, namely, the negative (or atomic) ion $\mathrm{H}^- = (ep, e)$ and the molecular ion $\mathrm{H}_2^+ = (ep, p)$. (For unification we use the designations "ae" and "ap", respectively). The contributions to pressure for these ions is calculated as

$$p_{ac} = T\zeta_{ac}, \quad \zeta_{ac} = \zeta_a\zeta_c\lambda_{ac}^3\sigma_{ac}, \quad \sigma_{ac} = e^{I_{ac}/T} - 1, \qquad (4.173)$$

where $c = \mathrm{e,p}$ and $I_{ae} \approx 0.75\,\mathrm{eV}$, $I_{ap} \approx 2.69\,\mathrm{eV}$.

The EOS of weakly nonideal hydrogen plasma, i.e., the dependence of the resultant pressure $p(T)$ along the solar trajectory in the framework of S-model was calculated using relations (4.64), (4.143), (4.71), (4.83), (4.84), (4.138), (4.106), (4.146)–(4.147), (4.136)–(4.137), (4.139)–(4.140), (4.162), (4.168), (4.171)–(4.172), (4.173) [180].

The calculations considering all the above-mentioned contributions refer to the physical picture presented according the thermodynamic perturbation theory to within BBK (up to $\zeta^2$) for hydrogen plasma.

Figure 4.6 presents the partial contributions to the resultant pressure from different components $\delta p/p_{\text{tot}}$ on the temperature scale. One can see that the contribution of the electron and proton pressure which is equal to $\approx 1$ in temperature coordinates falls sharply to $\sim 10^{-5}$ in the periphery where bound states (atoms) prevail whose contribution in the center of the Sun, the same as the contribution of e–p scattering states equals $\sim 10^{-4}$. The contribution of Coulomb interaction $\sim 10^{-2}$ in the center of the Sun reaches its maximum $\sim 10^{-1}$ at $T \sim 5 \cdot 10^4\,\text{K}$ and decreases to $10^{-9}$ in the periphery. The diffraction corrections $\sim 10^{-3}$ in the center of the Sun pass through the maximum $\sim 4 \cdot 10^{-3}$ at $T \sim 5 \cdot 10^4\,\text{K}$, and then, like all the interactions related to charged particles (from the summary logarithm, the scattering states from e–p, p–p, e–e interactions, exchange e–e interaction), fall sharply in the periphery.

In the center of the Sun, the relative contribution from the summary logarithmic term is $\approx 2 \cdot 10^{-4}$ (correspondingly, the screening correction by "dressed" particles is even smaller by an order of magnitude).

The contribution of exchange interaction is approximately equal to $4 \cdot 10^{-3}$ in the center of the Sun and is the next in significance after the Coulomb contribution. The fraction of the radiation pressure in the center

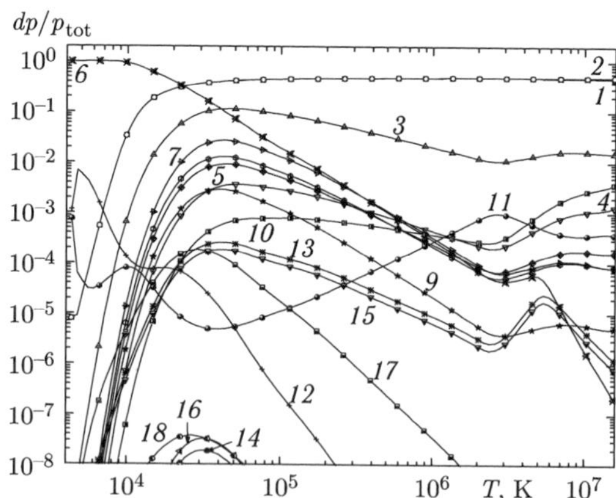

Fig. 4.6　Specific contributions ($\delta p/p_{\text{tot}}$) to the summary pressure depending on temperature along the solar curve from the components: *1* $p_{0,i}$ *2* $p_{0,er}$ *3* $\delta p_{\text{D H}}$ *4* $\Delta\Omega_{\text{dif}}/V$ *5* $\delta\Omega^{\text{cl}}/V$ *6* $\delta p_{\text{ed}}^{\text{BS}}$ *7* $\delta\Omega_{\text{ep}}^{q}/V$ *8* $-\delta\Omega_{\text{pp}}^{q}/V$ *9* $-\delta\Omega_{\text{ee}}^{q}/V$ *10* $-\delta\Omega_{\text{ee}}^{\text{exch}}/V$ *11* $p_{\text{R}}$ *12* $p_{\text{m}}$ *13* $-p_{\text{ea}}$ *14* $-p_{\text{em}}$ *15* $p_{p,\text{pa}}$ *16* $p_{p,\text{pm}}$ *17* $p_{\text{ae}}$ *18* $p_{\text{ap}}$.

of the Sun is $\approx 4 \cdot 10^{-4}$ and the contribution from p–p scattering states there is $\approx 10^{-4}$. The smallest is the contribution $\approx 4 \cdot 10^{-6}$ from the e–e scattering states.

One can also see that many corrections belonging to BBK are fairly significant considering the high precision of the inversion procedure in helioseismological problems.

We should once again emphasize that the contribution from hydrogen plasma only has been considered here in the assumption (certainly erroneous, especially near the center, because of the presence of He and other elements) that it is hydrogen that determines the total density of matter. With all the conditional character of this model, these calculations were aimed at a qualitative determination of the relative fraction of different contributions to the summary plasma pressure. At the same time, the precision of such a model in the usual rather than precession sense, like in helioseismology, is quite adequate in comparison with the data of the complete S-model including He and other elements distributed along the solar trajectory.


## 4.6   Bound atom model

In thermodynamic calculations of rarefied high-temperature gases and plasma the atoms are normally assumed to be an ideal subsystem noninteracting with the surrounding medium. Accordingly, it is assumed that the energy spectrum of bound states is identical to the spectrum of isolated particles and the effect of nonideality on bound states consists within the framework of the chemical model in lowering the ionization potential, the shift of the levels and restriction of the number of upper excited levels in calculation of the statistical sums.

With increasing pressure the interatomic interaction becomes well pronounced upon violation of the condition

$$n_{\mathrm{at}} b_0^3 \ll 1, \tag{4.174}$$

where $b_0 \sim a_0$ is the radius of the atom in the ground state. However, at high pressures and temperatures comparable with the ionization potential a considerable amount of excited atoms appear. The size of the latter can greatly exceed $b_0$ and, therefore, nonisolation and, in particular, the fact that the volume accessible for realization of atomic states is finite affects them in the first place. The influence of this effect was demonstrated in paper [365] where calculations were carried out for hydrogen in the bound-atom approximation. This approximation corresponds to a potential of the form

$$U(r) = \begin{cases} -\dfrac{Ze^2}{r}, & 0 < r < r_c \\ \infty, & r_c \leq r \end{cases}.$$ (4.175)

The results of calculation of the energy and excited states of a hydrogen atom have shown that the dependence of the excited state energy on the atomic volume finiteness increases sharply with increasing principal quantum number. Applying the numerical methods of atomic structure calculation, one can estimate the effect of atomic cell compression to the energy spectrum of multielectron atoms such as argon and cesium. For this purpose the Hartree–Fock method (see Sec. 2.3) was used [429, 1099, 1100] which is our case is reduced to the finite-difference solution [402] of integro-differential equations

$$\left[ \frac{d^2}{dr^2} + V_{nl}(r) - \varepsilon_{nl} \right] f_{nl}(r) + \int_0^{r_c} X_{nl}(r,s) f_{nl}(s) ds = 0,$$ (4.176)

in the radial parts $f_{nl}$ of one-electron functions

$$\psi_{nlms}(r, \theta, \varphi, \delta) = \frac{f_{nl}(r)}{r} Y_l^m(\theta, \varphi) \chi_s(\delta),$$ (4.177)

which obey the normalization conditions,

$$\int_0^{r_c} f_{nl}(r) f_{n'l}(r) dr = \delta_{nn'},$$ (4.178)

and the boundary conditions,

$$f_{nl}(0) = 0, \quad f_{nl}(r_c) = 0,$$ (4.179)

which obviously corresponds to equation (4.175) of the intra-atomic potential. The system of equations (4.176)–(4.178) should be solved for all the electron terms which are possible in the $LS$ coupling approximation. For an adequate description of the electron structure of multielectron atoms it is necessary to take into account the interelectron correlation effects, which for our purpose can be done within the simplest approximation [348]. Figures 4.7 and 4.8 illustrate the calculations performed for argon and cesium atoms and present the schemes of the levels of the ground and excited states depending on the atomic cell radius.

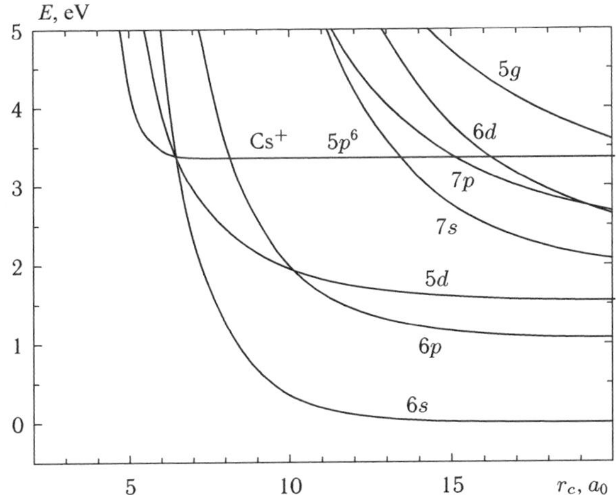

Fig. 4.7   Results of cesium spectrum calculation by the bound-atom model (E, eV).

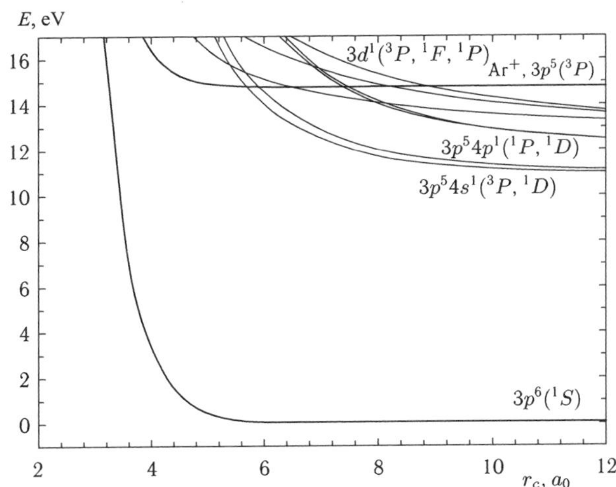

Fig. 4.8   Results of argon spectrum calculation by the bound-atom model (E, eV).

The calculations show that if the ground state of alkali metals starts to experience compression for relatively large cell radii (for cesium this is approximately $12a_0$), then owing to compactness of the electron structure of the ground state of inert gas atoms the energy of their lower level begins notably heightening for much smaller radii. The energy of the ground state of argon for $5.5a_0$ does not shift noticeably, whereas the energy of the first excited state (an external electron in state $4s$) exceeds the ionization limit. The examples considered above show that in spite of the difference in the behavior of compressed-atom spectra a common rule exists, namely, there is a range of atomic cell radii where the energy gap between the ground and excited states increases compared to isolated atoms.

The effect of finite volume on the atomic spectrum of excitations can influence substantially the thermodynamic properties of plasma. It should be noted that, for example, at a pressure of 100 bar and a temperature of $5000\,\mathrm{K}$ the average interparticle distance is equal to about $20a_0$ which only 2–3 times exceeds the "characteristic size" of the ground state of an alkali metal atom and virtually coincides with the size of its first excited states.

Thus, at least for the description of the external electron motion the intra-atomic potential should be effectively changed in order that the presence of plasma surrounding the atom could be taken into account. We shall appeal to the bound-atom approximation to find out how much the deformation of the atomic energy spectrum at the expense of interaction with particles affects the summary thermodynamic characteristics of nonideal plasma.

We shall consider the simplest case of a three-component gas system consisting of atoms, electrons, and single-charged ions. Suppose the atoms have the shape of a sphere with temperature- and density-dependent radius, and the ion and electron sizes will be ignored. The subsystem of finite-size atoms will be represented as a system of hard balls which do not interact when the distance between them exceeds $2r_c$. The free energy of such a model system can be written as

$$F(\{N_j\}, V, T) = F_{\mathrm{id}}(\{N_j\}, V, T)$$

$$+\Delta F_{\mathrm{HS}}(\{N_j\}, V, T) + \Delta F_{\mathrm{Coul}}(\{N_j\}, V, T). \qquad (4.180)$$

The first summand stands for the free energy of an ideal-gas electrically neutral system of atoms, electrons, and ions,

$$F_{\mathrm{id}} = -k_{\mathrm{B}}T\left[N_{\mathrm{a}} \ln \frac{\sigma(r_{\mathrm{c}})}{n_{\mathrm{a}}\Lambda_{\mathrm{a}}^3} + N_{\mathrm{e}} \ln \frac{2\sigma}{n_{\mathrm{e}}^2\Lambda_{\mathrm{i}}^3\Lambda_{\mathrm{e}}^3} + (N_{\mathrm{a}} + 2N_{\mathrm{e}})\right], \qquad (4.181)$$

with the difference that according to the boundary condition of a bounded atom (4.179) the statistical sum of the atom depends on its radius. The second summand in (4.180) is the contribution of repulsion of hard spheres which also depends on the radius $r_{\mathrm{c}}$ through the dimensionless packing parameter $\eta = n_{\mathrm{a}}(4\pi r_{\mathrm{c}}^3/3)$:

$$\Delta F_{\mathrm{HS}} = N_{\mathrm{a}} k_{\mathrm{B}} T f(\eta). \tag{4.182}$$

To describe this contribution, we shall make use of the expression from paper [158]

$$f(\eta) = \eta \frac{3\eta}{(1 - \eta)^2}. \tag{4.183}$$

This expression corresponds to the correction to pressure

$$\frac{\Delta p_{HS}}{n_a k_{\mathrm{B}} T} = \frac{2 - \eta}{(1 - \eta)^3} = \eta f'(\eta), \tag{4.184}$$

and to the chemical potential

$$\frac{\Delta \mu_{HS}}{k_{\mathrm{B}} T} = \eta \frac{3\eta^2 - 9\eta + 8}{(1 - \eta)^3} = f(\eta) + \eta f'(\eta). \tag{4.185}$$

The statistical sum $\sigma_a(r_c)$ was obtained by summation of the energy level contributions calculated in the bound-atom approximation using the Hartree–Fock method. Note that the correction $\Delta p_{\mathrm{HS}}$, the same as the statistical sum $\sigma_a(r_c)$, are monotonically increasing functions of the atomic radius $r_c$, and thereby the dependence of the free energy (4.180) on the atomic cell radius has its minimum. The $r_c$ value for which it is achieved is an equilibrium radius of atoms and can be found from the variational principle of statistical mechanics [598]:

$$\frac{\partial F}{\partial r_c} = 0. \tag{4.186}$$

Making differentiation of the free energy (4.180), we arrive at the system of equilibrium equations

$$k_{\mathrm{B}} T [n_{\mathrm{a}} + 2n_{\mathrm{e}} + n_{\mathrm{a}} \eta f'(\eta)] + \Delta p_{\mathrm{Coul}} = P_0, \tag{4.187}$$

$$r_c \frac{1}{\sigma_{\mathrm{a}}} \frac{\partial \sigma_{\mathrm{a}}}{\partial r_c} = 3\eta f'(\eta), \tag{4.188}$$

$$\frac{n_{\mathrm{a}}}{n_{\mathrm{e}}^2} = \frac{\sigma_{\mathrm{a}}(r) \Lambda_{\mathrm{e}}^3}{2\sigma_i} \exp\left[-f(\eta) - \eta f'(\eta) + \frac{2\Delta \mu_{\mathrm{e\,Coul}}}{k_{\mathrm{B}} T}\right]. \tag{4.189}$$

Solving this system of transcendent equations in $n_{\mathrm{a}}$, $n_{\mathrm{e}}$, and $r_c$ by the above-described methods (Sec. 4.6), one can then find all the thermodynamic functions.

Note that the equilibrium value of atomic radius is pressure- and temperature-dependent. This dependence is individual for each substance and is determined by the form of the bound state spectrum. The equilibrium value of the radius decreases with increasing pressure at a constant temperature, and upon rarefaction accompanied by a monotonous increase of the radius the EOS approaches that of an ideal gas.

## 4.7 Thermodynamic calculations

The calculations within the chemical model of complex-composition plasma in the range of moderate densities and temperatures are one of the important applications of the above-described methods. Such systems normally contain very different sorts of particles, and the range of parameters extends from a gas consisting of polyatomic molecules to plasma with developed ionization and notable Coulomb nonideality. The results of some calculations and their discussion for this range of parameters can be found, in particular, in Ref. [234]. An example of such calculations is given in Fig. 4.9 showing the composition of plasma consisting of the mixture of water and lithium.

At the lower boundary of the temperature range the mixture consists of an ideal gas of $H_2O$, $H_2$, $OH$, $O_2$, $LiOH$ molecules of H and O atoms and at the upper boundary this is a partially ionized plasma with developed Coulomb nonideality consisting of $H^+$, $O^+$, $Li^+$ ions, electrons, and atoms of hydrogen and oxygen.

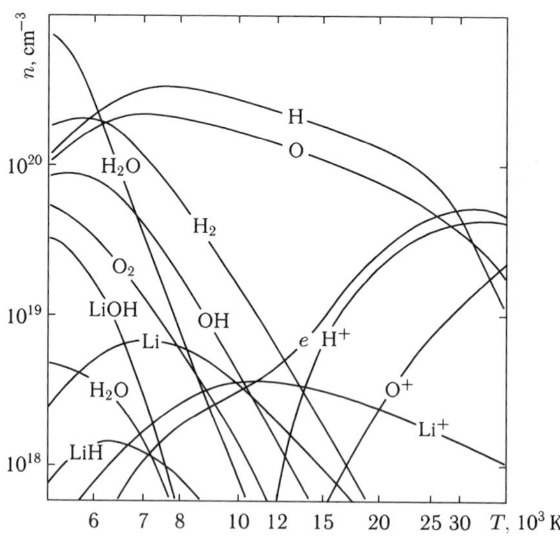


Fig. 4.9  Composition of plasma consisting of a mixture of water (97%) and lithium (3%) for 100 MPa.

The data on shock-compressed cesium [149, 632] and inert gases [97, 307, 404] are the most comprehensive, and therefore we use them for a quantitative comparison with the data of calculations obtained within the above-discussed methods. The most precise data on cesium obtained on the shock tube [149] cover the temperature range from 2600 K to 20000 K and the pressure range from 0.2 to 20 MPa. The chemical model allows one to describe experimental data on the whole correctly with not more than 20% difference in both thermal and caloric EOS. However, in the caloric EOS, the existing difference between the chemical model and the experiments that goes beyond the experimental error cannot be eliminated within the traditional approaches by choosing some of the nonideality Coulomb corrections and the known ways to calculate statistical sums by the energy levels of an isolated atom. At the same time, the experiment and the calculations using the ideal-gas approximation with the statistical sum equal to $g_0$ (the statistical weight of the ground state) showed unexpectedly good coincidence, which has no correct grounds. Nevertheless, such a coincidence suggested the assumption (made in the experimental works [149, 632]) concerning the appearance in the system of additional repulsion disregarded in the traditional approaches [745] and a sharp decrease of the discrete spectrum contribution caused by the energy level deformation at high densities. An analogous behavior of thermodynamic quantities was also revealed in experiments with inert gases, in particular, argon [97, 307, 404], where the temperatures reached 30000 K and the densities had near-critical values.

The above-mentioned Hartree–Fock calculations in the bound-atom approximation showed a high compression sensitivity of the cesium atom, which suggests that at lower densities the discrete spectrum deformation should have a smaller effect on cesium than on argon. This is particularly the case in reality.

The above-mentioned differences of the traditional methods from the experimental data show up in Cs with an interparticle spacing of $20a_0$ and in argon with a spacing less than $10a_0$. The calculations using the bound-atom approximation described above point out that the spectrum deformation effects first of all manifest themselves in the region where the fraction of atoms is notable. This is clearly seen on an example of Cs (Figs. 4.10 and 4.11).

The use of this model improves considerably the agreement with the results in the lower portion of the temperature range attained in the experiment, where the degree of ionization is low. In the region of developed ionization, the Coulomb corrections should be considered in more detail.

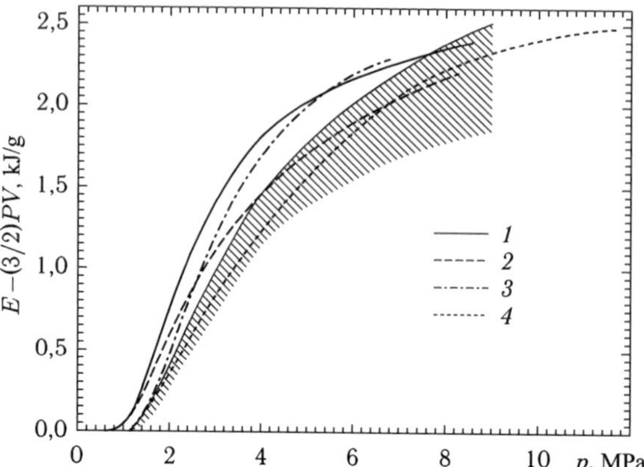

Fig. 4.10   Caloric EOS for cesium on the isochore $V = 200\,\mathrm{cm^3/g}$: shaded is the region (5%) for the smoothed experimental dependence $H(P,V)$ [149]. *1* calculation in the approximation [621], atoms make up an ideal subsystem, *2* calculation in the approximation [468] for charge interaction, *3* bound-atom model with the approximation [621] for charged particles, *4* calculation in the bound-atom approximation with the approximation [468] for a charged subsystem.

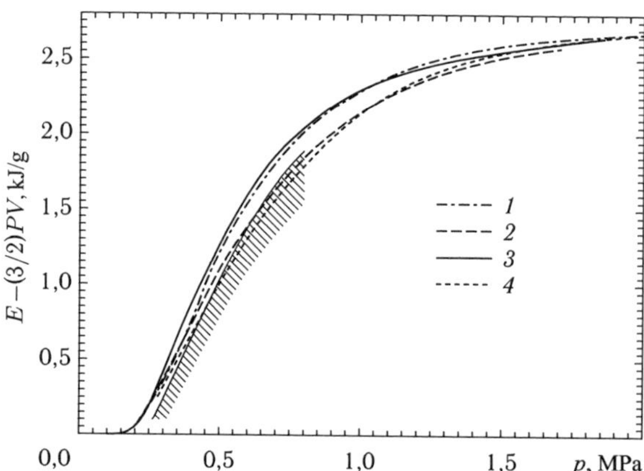

Fig. 4.11   Caloric EOS for cesium on the isochore $V = 1000\,\mathrm{cm^3/g}$. For the notation see Fig. 4.10.

The calculations within the model (4.177)–(4.184) together with the approximation [468] for the description of Coulomb nonideality allowed a satisfactory description of the experiment in the entire investigated region [149, 632]. Along with the agreement in the caloric EOS, the bound-atom model with additional allowance for Coulomb corrections provides

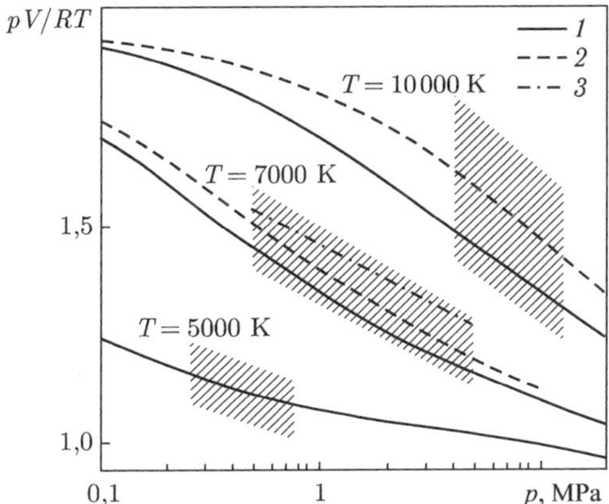


Fig. 4.12 Thermal EOS for cesium plasma. Dashed is the region of experimental error [149]. For the notation see Fig. 4.10.

satisfactory agreement also in the thermal EOS (Fig. 4.12). Note that a distinctive feature of both the bound-atom model and the approximation [468] compared to the traditional calculation of plasma is the effective interparticle repulsion.

Similar conclusions concerning a significant role of interparticle repulsion also follow from the argon-plasma shock-wave compression experiments [404] where the kinographic and temperature measurements were implemented — Figs. 4.13, 4.14, 4.15, 4.16 and 4.17.

We shall consider the hadron hydrodynamics for much larger parameters. The methods presented allow the use of different plasma models taking into consideration all possible plasma effects. This underlies the attempts to extend the chemical plasma model beyond the region of its traditional application and, in particular, to the ultrahigh pressure and temperature range [388, 392]. This appeared to necessitate allowance for such thermodynamic effects as multiple ionization, electron excitation, Coulomb interaction between charged particles, short-range repulsion of atoms and ions, and electron degeneracy. Such states were realized in experiments on superhigh-power shock compression of porous samples of copper [576, 1098] (temperatures to $5 \cdot 10^5$ K, pressures to 20 Mbar, specific energy densities to 75 MJ/g) and aluminum [65, 921] (temperatures to $7 \cdot 10^6$ K, pressures to 400 TPa, specific energy densities to 60 MJ/g).

According to the chemical model plasma consisted of atoms, electrons, and ions. The atoms and ions were considered as Boltzmann particles of

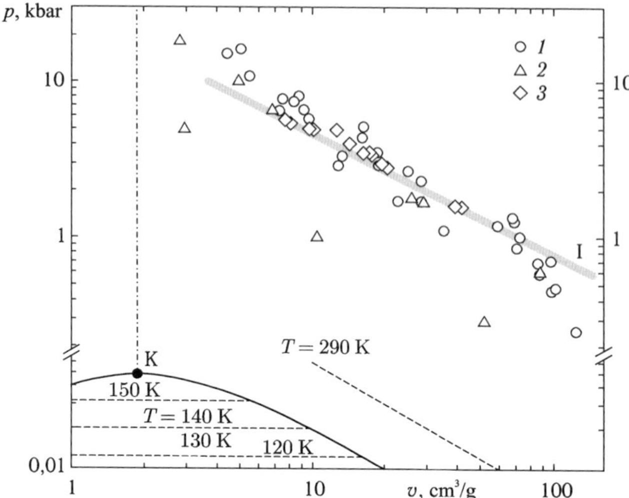

Fig. 4.13   Phase diagram of argon [404]. Marked are the two-phase region boundaries, $K$ is the critical point. *1* experimental data [404], *2* the data of paper [307] corrected with allowance for the EOS of a real gas [1009] before the shock-wave front, *3* [97], the dashed line is the isotherm [1009], the dot-and-dash line is the isochore $v = v_{\mathrm{cr}}$. The grey line is the boundary of single ionization I ($\chi_{\mathrm{Ar}+} = \chi_{\mathrm{Ar}}$, $\chi_{\mathrm{Ar}} = N_{\mathrm{Ar}}/N_{\mathrm{total}}$).

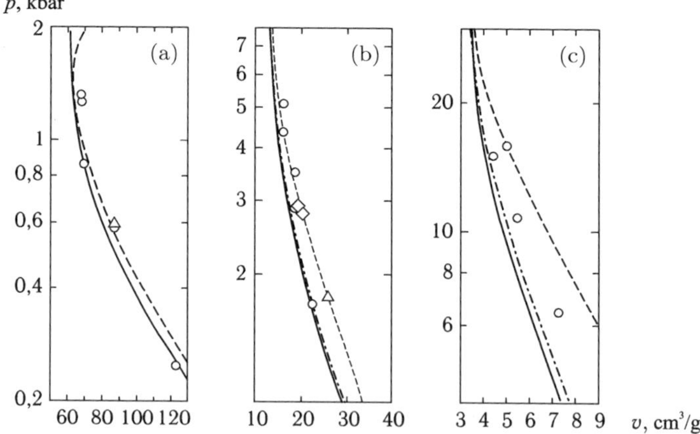

Fig. 4.14   Comparison of the experimental argon shock adiabats with the results of model calculations. The solid curves represent the account of charged particle interaction in the ring (Debye) approximation in the grand canonical ensemble [378, 621], the dot-and-dash curve is additional allowance for the atomic interaction in the approximation of second virial coefficient [436], the dashed lines show the calculation using the "bound-atom" model [230, 398, 404]: (a) $p_0 = 1\,\mathrm{atm}$; (b) $p_0 = 5\,\mathrm{atm}$; (c) $p_0 = 20\,\mathrm{atm}$.

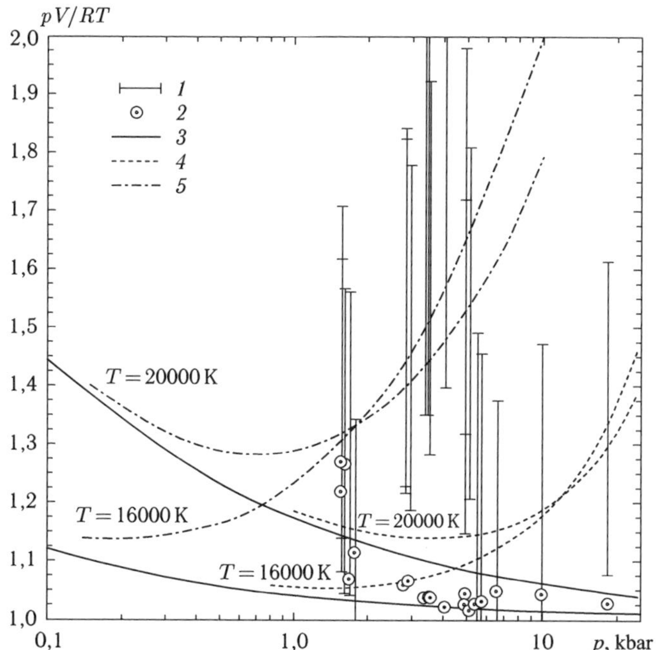



Fig. 4.15  Thermal EOS of argon plasma: $p = 1 \div 20\,\text{kbar}$; $T = 5000 \div 23000\,\text{K}$ [230].
*1* experimental results [97, 307], *2* calculation in the ring (Debye) approximation in the
grand canonical ensemble [378, 621] for charged particles for each experimental point
$p_{\exp}$, $T_{\exp}$, *3*, *4*, *5* calculated isotherms 16000 K and 20000 K in the approximations,
*3* calculation using the pseudopotential model [468] (the pseudopotential depth and
the boundary of intra-atomic states are taken to be equal to $k_{\mathrm{B}}T$), *4* with additional
allowance for the second and third virial coefficients of argon [436], *5* calculation by the
"bound-atom" model [230, 230, 404].

radius $r_i$. The free energy of such a system was given by the sum

$$F \equiv F_{\mathrm{id}}^{\mathrm{e}} + F_{\mathrm{id}}^{\mathrm{i}} + F_{\mathrm{Coul}} + F_{\mathrm{HS}}. \tag{4.190}$$

The first summand is the contribution of an ideal degenerate electron
gas

$$F_{\mathrm{id}}^{\mathrm{e}} = n_{\mathrm{e}} k_{\mathrm{B}} T \left[ \alpha_{\mathrm{e}} - \frac{I_{3/2}(\alpha_{\mathrm{e}})}{I_{1/2}(\alpha_{\mathrm{e}})} \right], \tag{4.191}$$

where $I$ are the Fermi–Dirac functions and the dimensionless chemical
potential $\alpha = \mu / k_{\mathrm{B}} T$ was determined from the relation $I_{1/2}(\alpha_{\mathrm{e}}) = n_{\mathrm{e}} \Lambda_{\mathrm{e}}^3 / 2$.
The contribution of the ideal gas of heavy particles is represented by the

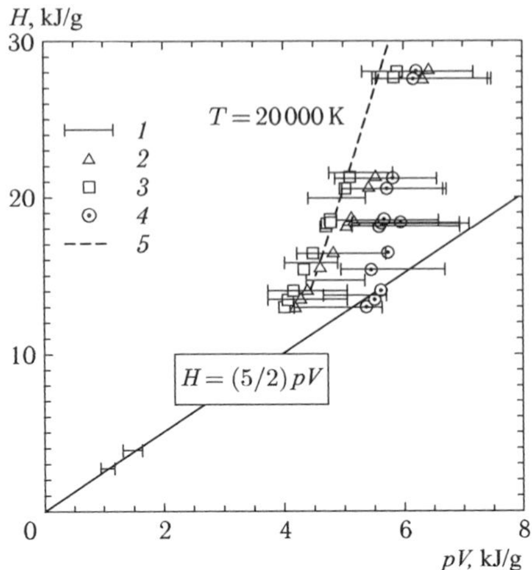

Fig. 4.16   Caloric EOS of argon plasma [230]. *1* the results of experiment [97, 307], *2*, *3*, *4* calculation for each experimental point $V(p_{\mathrm{exp}}, H_{\mathrm{exp}})$ in the approximations: *2* in the pseudopotential model [468] (the pseudopotential depth and the boundary of intra-atomic states are taken to be equal to $k_{\mathrm{B}}T$), *3* in the ring (Debye) approximation in the grand canonical ensemble [378, 621] for charged particles, *4* in the "bound-atom" model [230, 230, 404]; *5* the calculated isotherm $T = 20000\,\mathrm{K}$ within the pseudopotential model [468].

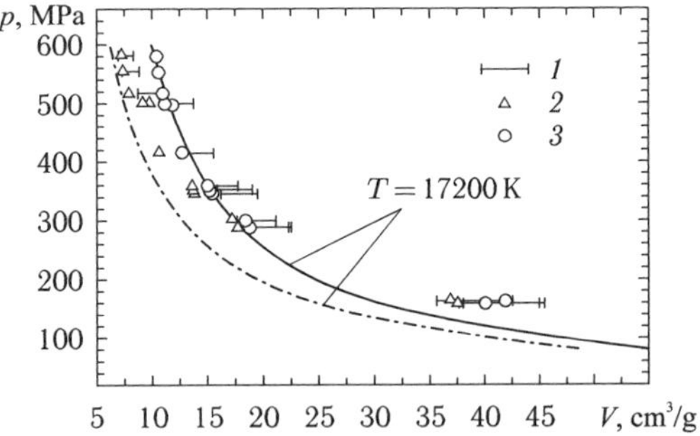

Fig. 4.17   Thermal EOS of shock-compressed argon. *1* experimental results [97], *2*, *3* calculation for each experimental point $V(p_{\mathrm{exp}}, T_{\mathrm{exp}})$ in the approximations: *2* in the ring (Debye) approximation in the grand canonical ensemble [621] for charged particles, *3* calculation in the bound-atom approximation [621] for a charged subsystem. The dot-and-dash and solid curves are calculations of the isotherm in approximations *2* and *3*, respectively.

second summand

$$F_{\text{id}}^{\text{i}} = \sum_j N_j k_{\text{B}} T (\ln \frac{n_j \Lambda_j^3}{\sigma_j} + \frac{A_j}{k_{\text{B}} T} - 1).$$  (4.192)

In (4.192) the statistical sums were defined by the expression [603]

$$\sigma_j = \sum_n q_n^j \left[ \exp \left( \frac{E_n^j}{k_{\text{B}} T} \right) - 1 + \frac{E_n^j}{k_{\text{B}} T} \right].$$  (4.193)

The Coulomb interaction was taken into account in the framework of the Debye approximation in the grand canonical ensemble [621] extended to the case of multiple ionization

$$\frac{\Omega}{V k_{\text{B}} T} \equiv \frac{F - \Sigma N_j \mu_j}{V k_{\text{B}} T} \equiv \frac{p}{k_{\text{B}} T} = \sum_\alpha n_\alpha - \frac{\tilde{\kappa}_{\text{D}}^3}{24\pi}$$

$$= \sum_\alpha \left[ n_\alpha - \frac{\tilde{\Gamma}_{\text{D}}}{6} \frac{n_\alpha z_\alpha^2}{1 + z_\alpha^2 \tilde{\Gamma}_{\text{D}}} \right],$$  (4.194)

where summation was taken over all the charged particles and the parameter $\tilde{\Gamma}_{\text{D}}$ is determined as the root of the equation

$$\tilde{\Gamma}_{\text{D}}^2 = \left( \frac{e^2}{k_{\text{B}} T \tilde{r}_{\text{D}}} \right) = 4\pi \left( \frac{e^2}{k_{\text{B}} T} \right)^3 \sum_\alpha \frac{n_\alpha z_\alpha^2}{\frac{1 + z_\alpha^2 \tilde{\Gamma}_{\text{D}}}{2}}.$$  (4.195)

The short-range repulsion of atoms and ions was described in the hard sphere approximation (4.182)–(4.185). The effective radii $r_c$ were estimated from the preliminary Hartree–Fock calculations for copper and aluminum atoms and ions. For copper these radii were chosen to be equal for all the ions, $r_{\text{i}} = r_c = 1.75 a_0$, which approximately corresponded to the radius of a quintuple ion $Cu^{5+}$. For aluminum, the mean radius for all the ions was determined from the relation

$$r_c = \frac{\sum_{i=1}^L n_i r_i}{\sum_{j=2}^L n_j},$$  (4.196)

and the aluminum ion radii were chosen beginning with $2.5 a_0$ for the atom to 0 for the thirteen-fold $Al^{13+}$ ion in accordance with the results of Hartree–Fock calculation of their structure. The density dependence of $r_c$ determined the corresponding form of the configuration part of the chemical potential

$$\frac{\mu_{\text{i}}^{\text{HS}}}{k_{\text{B}} T} = f_{\text{HS}}(\eta) + \eta f_{\text{HS}}'(\eta) \left[ \frac{3(r_{\text{i}} - r_c)}{r_c} + 1 \right],$$  (4.197)

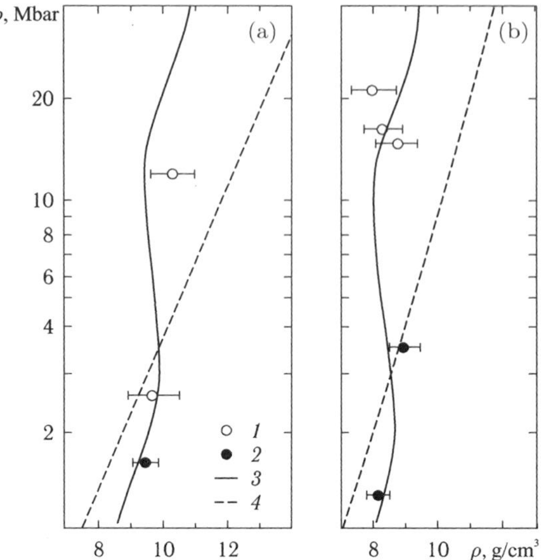


Fig. 4.18    Porous copper compression for $m = \rho_{00}/\rho_0 = 3$ (a) and $m = 4$ (b) by intense shock waves. *1* experiment [576], *2* experiment [1098], *3* Thomas–Fermi model [498], *4* the given model (4.190)–(4.197).

which caused an increase in the degree of ionization compared to the case of similar radius values for all the particles, $r_\mathrm{i} = r_\mathrm{c} = \text{constant}$.

Figure 4.18(a) and (b) presents comparison of the calculated copper shock adiabats with the experimental data [576, 1098] and with calculations by the Thomas–Fermi method with quantum and exchange corrections [498].

The calculations made it possible to find out which of the chemical model modifications has the greatest effect on the results. As was expected, the most significant is the allowance for the short-range repulsion between ions, in the absence of which the calculated densities are much lower than the measured values. Less radical, although substantial, is the allowance for Coulomb nonideality [621] and of the contribution of excited states, which in both cases results in an increase in the calculated density of shock-wave compression. Finally, the effect of allowance for electron degeneracy on the shock adiabat position is the least noticeable. Such a calculation was carried out also for shock-compressed aluminum. It is illustrated in Fig. 4.19 which presents, along with experiment, also the data obtained by other models.

Attempts to extrapolate the chemical model to the region of extreme compressions, i.e., ultrahigh pressures and temperatures were made, apart

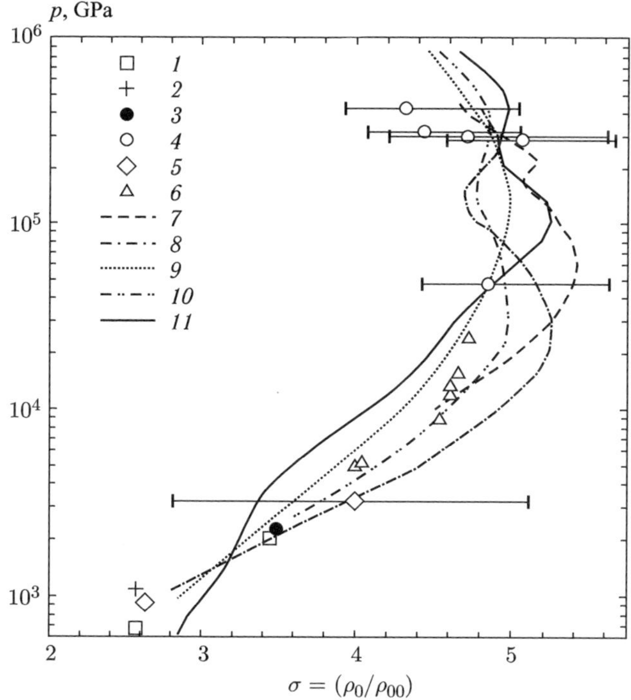



Fig. 4.19  Shock adiabats ($m = 1.0$). *1–6* are experimental data: *1* [140], *2* [598], *3* [598], *4* [881], *5* [921], *6* [65]; *7–11* results of calculations: *7* INFERNO model [1067], *8* [747], *9* TFC [498], *10* [917], *11* calculations within the chemical model according to the approximation (4.223)–(4.231).

from aluminum, also for plasma of iron [395] and nickel [987]. For iron, employed was a modified version of pseudopotential model extended to the case of multiple ionization. Taking into account the high level of densities achieved upon superstrong shock-wave compression of iron the above-mentioned allowance for Coulomb nonideality was combined with restriction of the bound states of multiply ionized iron ions at an average distance between heavy particles. The computational results given in Figs. 4.20, 4.21, and 4.22 demonstrate satisfactory agreement with experimental data of different authors. It should be stressed that in terms of the chemical model the totality of the described states of iron plasma covers the range from strongly to weakly nonideal plasma with a variable degree of ionization, electron degeneracy, and a considerable range of temperatures. This can be seen in Table 4.1 demonstrating the sequence of values

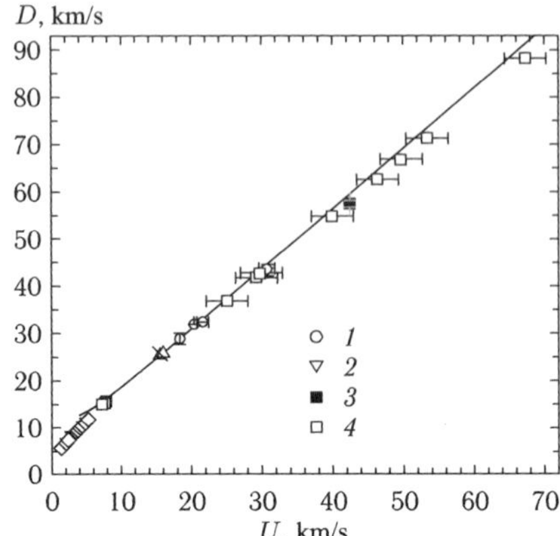

Fig. 4.20   Iron shock adiabat ($m = 1$) in $D$–$U$ variables ($D$ and $U$ are the front velocity and mass velocity [395]). Notation: *1* experimental data [42], *2* experimental data [984], *3* experiment [65]. The solid curve is the pseudopotential model [468] modified with allowance for multiple ionization and electron degeneracy in the plasma EOS (restriction of bound states in atomic and ionic statistical sums at an average distance between heavy particles).

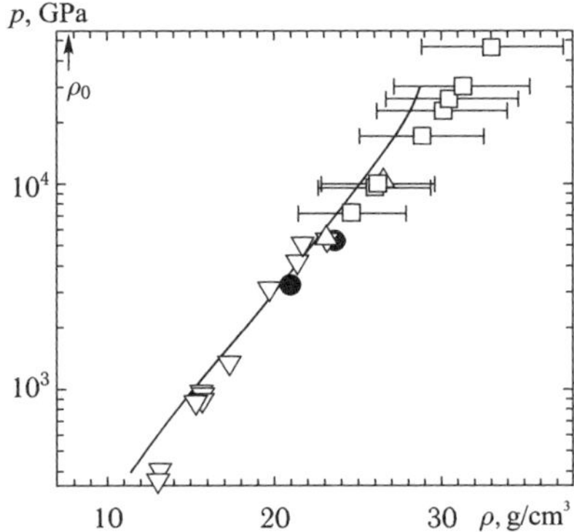

Fig. 4.21   Iron shock adiabat ($m = 1$) in density-pressure variables [395] (the notation is the same as in Fig. 4.20); the arrow is the normal iron density.

of some thermodynamic parameters and dimensionless criteria realized in the Hugoniot adiabat range under consideration.

Note that the presented data on shock-compressed copper and aluminum lie in the region going far beyond the traditional limits of chemical plasma model applicability.

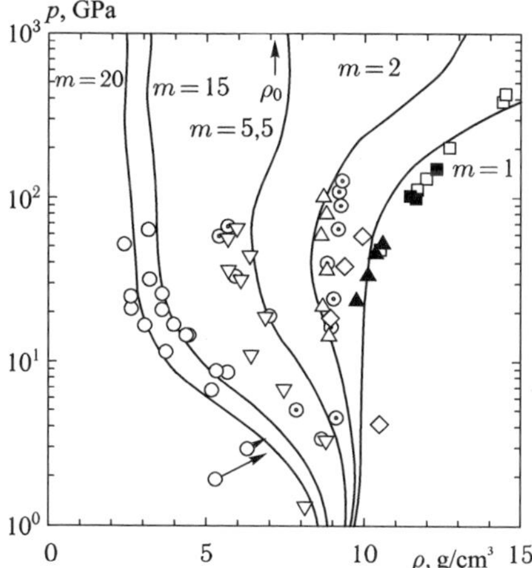



Fig. 4.22   Broad-range properties of the modified chemical model. Shock adiabats of solid and porous nickel [389, 390]). Notation: $m = \rho_0/\rho_{00}$ is nickel sample porosity, the arrow is normal nickel density, the marks are experimental data [982, 987], the solid line stands for calculations by the SAHA-IV code.

Table 4.1   Thermodynamic characteristic of extremely compressed plasma of iron ($\rho_0 = 8.88\,\mathrm{g/cm^3}$; $m = 1$) calculated making use of the modified version of the pseudopotential model [395]: $p$, $\rho$, $H$, $S$ are pressure, temperature, specific enthalpy, and entropy of nickel, $a$ is the velocity of sound, $\alpha$ is the total degree of plasma ionization, $\Gamma_{\mathrm{D}} \equiv e^2/k_{\mathrm{B}}TR_{\mathrm{D}}$ is the nonideality parameter, $\lambda_{\mathrm{e}} \equiv h^2/2\pi m_{\mathrm{e}}k_{\mathrm{B}}T$ is the de Broglie thermal wavelength.

| $p$, GPa | $T$, $10^3$ K | $H$, kJ | $S$, J/(g·K) | $a$, km/s | $\alpha \equiv n_{\mathrm{e}}/(\sum n_j)$ | $\Gamma_{\mathrm{D}}$ | $n_{\mathrm{e}}\lambda_{\mathrm{e}}^3$ |
|---|---|---|---|---|---|---|---|
| $10^3$ | 37 | 89.2 | 2.480 | 13.1 | 1.7 | 48 | 21 |
| $3 \cdot 10^3$ | $10^9$ | 258 | 3.687 | 17.4 | 2.8 | 11 | 6.9 |
| $10^4$ | 350 | 828 | 4.887 | 28 | 4.3 | 3.0 | 2.3 |
| $5 \cdot 10^4$ | 1440 | 4028 | 7.311 | 55.8 | 6.7 | 0.6 | 0.5 |

From this point of view experiments on multiple shock-wave compression seem to be more promising.

## 4.8   Thermodynamics of shock compressed plasma of megabar pressures. Nonideality and degeneracy

The above-presented approach based on the quasi-chemical representation, i.e., the chemical model, (we shall conditionally call it SAHA) has repeatedly been used to describe the thermodynamic plasma properties in very different ranges of parameters — from states close to that of condensed matter to the rarefied solar plasma [66, 387]. In this section we present the results of application of the SAHA approach for the description of hydrogen (deuterium) and inert gases compressed by high-power shock waves. The experiments on shock-wave gas compression [91, 127, 200, 319, 373, 442, 556–558, 715] cover the range of pressures to $100\,\mathrm{GPa}$ and higher and densities to 0.8 $\mathrm{g/cm^3}$ for deuterium and $10\ \mathrm{g/cm^3}$ for xenon. The scope of parameters investigated in the experiments is characterized by extremely complicated and diverse processes which must be reflected in the corresponding physical models. In the first place, as the substance is being compressed the component composition of the medium may change sharply, which is accompanied by the occurrence of strong interparticle interaction: the Coulomb interaction between electrons and ions, the polarization interaction between charges and neutrals, and also the short-range interaction between neutral particles. Since the characteristic interparticle spacing in the medium being considered is comparable with the characteristic size of atoms and ions, the part of phase volume occupied by them becomes inaccessible for other particles, which leads to an increase in their kinetic energy and the corresponding contributions to the free energy of such strongly compressed disordered structures. Furthermore, strong compression induces variation of the energy spectrum of atomic and molecular bound states. With increasing compression one should also take into account the change in the statistics of electrons of the continuum from Boltzmann to Fermi statistics as the degeneracy parameter $n_{\mathrm{e}}\Lambda_{\mathrm{e}}^3$ can increase several times under these conditions.

The thermodynamic plasma parameters at megabar pressures were calculated using the following components of the complete thermodynamic approximation most thoroughly considered in the section for the description of moderate-density nonideal plasma.

- The free energy of a quasi-neutral mixture of electrons, ions, atoms, and molecules can be written in the form of the contribution of the ideal-gas component and the term responsible for interparticle interaction

$$F \equiv F_{\mathrm{i}}^0 + F_{\mathrm{e}}^0 + F_{\mathrm{ii,ie,ee,\ldots}}^{\mathrm{interac}}. \qquad (4.198)$$

- The heavy particles (atoms, ions, and molecules) were assumed to obey the Boltzmann statistics (at densities and temperatures reached in the shock-wave experiment this condition is far from being violated), and their contribution to have the standard form

$$F_i^0 = \sum_j N_j k_{\mathrm{B}} T \left( \ln \frac{n_j \Lambda_j^3}{\sigma_j} + \frac{A_j}{k_{\mathrm{B}} T} - 1 \right), \qquad (4.199)$$

where $\sigma_j$ are statistical sums (4.30) of atoms, molecules, and ions.

- Electrons were regarded as a partially degenerate ideal Fermi gas and the ideal-gas summand $F_{\mathrm{e}}^0$ had the form (4.191) [598].
  We notice that allowance for the effects of electron degeneracy is fairly important in the discussed phase diagram region as the degeneracy parameter $n_{\mathrm{e}} \Lambda_{\mathrm{e}}^3$ can reach several tens. Since the chemical potential correction is positive, the electron degeneracy leads to lower values of the degree of plasma ionization.

- Out of the entire spectrum of interactions, the Coulomb interaction of charged particles and the short-range repulsion of particles at close distances were taken into account.

The Coulomb interaction was described within the version of the pseudopotential model for multiple ionization [401, 468]. The central point of this model is explicit allowance for the nonCoulomb character of free-charge interaction at close distances which under conditions of strong nonideality leads to a considerable positive shift of not only the potential, but also of the mean kinetic energy of free charges. In line with [401, 468] the depth of the electron–ion pseudopotential $\Phi_{\mathrm{ie}}^*(0)$ was related to the boundary separating the free states of each electron–ion pair and the corresponding bound states in the statistical sum. The electron–ion Glauberman pseudopotential had the form (Fig. 4.23)

$$\Phi_{\mathrm{ie}}^*(r) = -\frac{Z_{\mathrm{i}} e^2}{r} \left( 1 - e^{-r/\sigma_{\mathrm{ie}}} \right); \quad \{ \sigma_{\mathrm{ie}} \equiv \sigma_{\mathrm{ie}}(n, T) \},$$

$$\Phi_{\alpha\alpha}^*(r) = -\frac{Z_\alpha Z_\alpha e^2}{r}; \qquad \{ \alpha = i, j \}. \qquad (4.200)$$

In Fig. 4.23(a) this potential is compared with other pseudopotentials proposed for plasma.

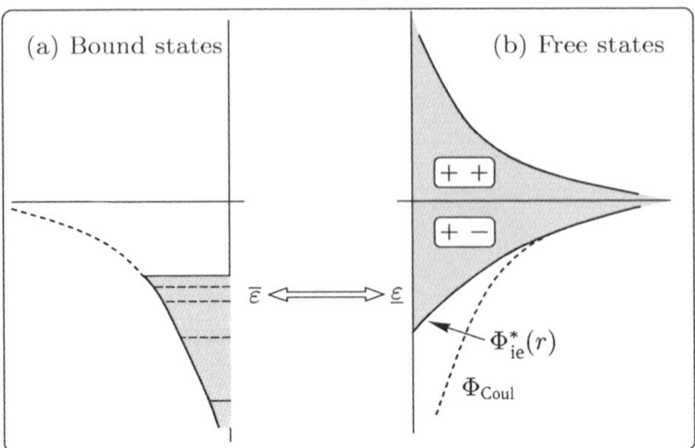


Fig. 4.23   Electron–ion Glauberman pseudopotential: $\underline{\varepsilon}$ and $\bar{\varepsilon}$ is mobile mutual boundary restricting the energies of free and bound states (a) bound states (b) free states.

The parameters of the correlation functions with the potential (4.200) were found from the conditions valid for arbitrary values of the Coulomb nonideality parameter, $\Gamma_{\text{D}} = [4\pi(e^2/k_{\text{B}}T)\sum n_\alpha z_\alpha^2]^{-1/2}$.

1. The local electroneutrality condition:

$$\int \left\{ n_{\text{ie}}[F_{\text{ie}}(r) - 1] + \sum_j n_{ij} Z_j [F_{ij}(r) - 1] \right\} dr = -Z_{\text{i}}. \qquad (4.201)$$

2. The dipole screening condition:

$$\int \left\{ n_{\text{ie}}[F_{\text{ie}}(r) - 1] + \sum_j n_{ij} Z_j [F_{ij}(r) - 1] \right\} \left( \frac{r}{r_{\text{D}}} \right)^2 dr = -3Z_{\text{i}}. \qquad (4.202)$$

3. Nonnegativeness of the correlation functions:

$$F_{ik}(r) \geq 0. \qquad (4.203)$$

4. Relation between the screening cloud amplitude and the depth of the electron–ion pseudopotential:

$$F_{\text{ei}}(0) \equiv 1 + \Psi_{\text{ei}}(0) \approx \beta\Phi^*_{\text{ei}}(0); \quad F_{\text{ii}} \approx 0. \qquad (4.204)$$

### Essential points of the pseudopotential model

The depth of the electron–ion pseudopotential is related to the position of the adopted (generally, arbitrary) boundary $\varepsilon$ separating free and bound states:

$$\Phi_{\mathrm{ie}}^*(r) = \frac{Z_{\mathrm{i}} e^2}{r}\left(1 - e^{-r/\sigma}\right) \equiv -\left(\frac{Z_{\mathrm{i}} e^2}{\sigma}\right)\frac{\left(1 - e^{-r/\sigma}\right)}{\frac{r}{\sigma}}$$

$$\geq -\Phi(0) \sim -\varepsilon; \qquad [\sigma \equiv \sigma(n, T)]. \tag{4.205}$$

The same quantity $\varepsilon$ is simultaneously the boundary of bound states taken into account in the calculation of the statistical sum of atom and ion excitation.

The choice of the meaning and value of $\varepsilon(n, T)$ is not limited to the traditional separation in the relative energy of a coupled pair, but may be arbitrary, in particular, can correspond to the coordinate-wise separation principle, etc. For the ion–ion and electron–electron potentials the Coulomb form was preserved:

$$\Psi_{\mathrm{ee}}^*(r) = \frac{e^2}{r}, \quad \Psi_{\mathrm{ii}}^*(r) = \frac{Z_i Z_j e^2}{r}.$$

After the choice of interparticle interaction pseudopotential, the further approximations necessary for calculation of nonideality corrections are considered in the language of pair correlation functions. Given this, the approach is being developed verified in the Coulomb models when for the indicated pair correlation functions the functional dependence persists which corresponds to the weak nonideality limit:

$$F_{ab}(r) = 1 \pm A\frac{e^{-pr} - e^{qr}}{r} \equiv 1 \pm \Psi_0 e^{-\nu r}\frac{\sinh \omega r}{\omega r}. \tag{4.206}$$

An important advantage of the form (4.206) is the fact that $\nu$ and $\omega$ can be imaginary and the correlation function (4.206) becomes oscillating. At the same time, the parameters of this dependence are chosen from the general conditions independent of the interaction smallness (nonideality). In calculation of nonideality effects, in corrections to the thermodynamic quantities the shift of the mean kinetic energy of free charges, $\Delta U_{\mathrm{kin}} = 3\Delta pv - \Delta U$, following from the virial theorem is explicitly taken into account:

$$U = U_{\mathrm{kin}} + U_{\mathrm{pot}}, \qquad 3pV = 2U_{\mathrm{kin}} + U_{\mathrm{pot}},$$

$$\Delta U_{\mathrm{pot}} = -Vn \int \Phi_{\mathrm{coul}}\left(F_+ - F_-\right) d\mathbf{r}, \tag{4.207}$$

$$\Delta U = -Vn^2 \int \left(F_+ \Phi_{\mathrm{ei}}^* - F_- \Phi_{\mathrm{ij}}^*\right) d\mathbf{r}, \tag{4.208}$$

$$\Delta p V = \frac{1}{3}(2\Delta U - \Delta U_{\mathrm{pot}}) = \frac{1}{3}(2\Delta U_{\mathrm{kin}} + \Delta U_{\mathrm{pot}}), \qquad (4.209)$$

$$\Delta \mu_{\mathrm{i}} = \Delta \mu_{\mathrm{e}} \sim (N_{\mathrm{i}} + N_{\mathrm{e}})^{-1} \Delta U. \qquad (4.210)$$

Here $\Delta U$ and $\Delta U_{\mathrm{pot}}$ are corrections to the total internal and separately to the mean potential plasma energy and $\mu_{\mathrm{i}}$ and $\mu_{\mathrm{e}}$ are chemical potentials of free charges. Relation (4.208) is exact when the pseudopotential (4.205) does not depend on the thermodynamic parameters and (4.209) is a corollary to the virial theorem.

Remaining valid in the weak nonideality region, the widely used relation between corrections to the pressure and internal energy of free charges, $\Delta U = 3\Delta p V$, is no longer preserved in the region of considerable nonideality ($\Gamma \sim I$). This is, first, the result of the explicit account of quantum effects that tell on the contribution of free and bound states and, second, it is a consequence of the very procedure of passing over to the chemical model.

Note that charge interaction corrections obtained from conditions (4.200)–(4.204) are close to the Debye ones for ($\Gamma_{\mathrm{D}} \ll 1$) differing from them by lower values when ($\Gamma_{\mathrm{D}} \gg 1$) [401, 468].

From Fig. 4.24 one may conclude that for equivalent pseudopotentials the results obtained on the basis of the above-given simple relations are in satisfactory agreement with the results of direct numerical Monte Carlo simulations [1083–1086]. This may be regarded as a confirmation of the conclusion made in Refs. [391, 398, 400] that observance of the general relations is the key condition for construction of satisfactory description of thermodynamics of strongly nonideal Coulomb systems.

For the case of partially ionized xenon plasma the calculations [404] were made using the above-mentioned version of the pseudopotential model [468]. In these calculations, the boundary separating free and bound states of atom and ion and the corresponding restriction of the atomic statistical sum were chosen at the binding energy depth of the order of $k_{\mathrm{B}}T$, which virtually coincides with the well-known and frequently recommended procedure of statistical sum calculation by the so-called Brillouin–Planck–Larkin formula (for more details see Ref. [234]). Comparison has shown (Figs. 4.25 and 4.26) that the proposed version of the pseudopotential model [468] allows a qualitative and, allowing for the actual precision and scatter of the experimental data, quantitative description of the parameters of experimentally measured xenon shock adiabats.

Figure 4.25 presents comparison of the experimental data with the results of the given calculations. The shift of shock adiabats obtained in

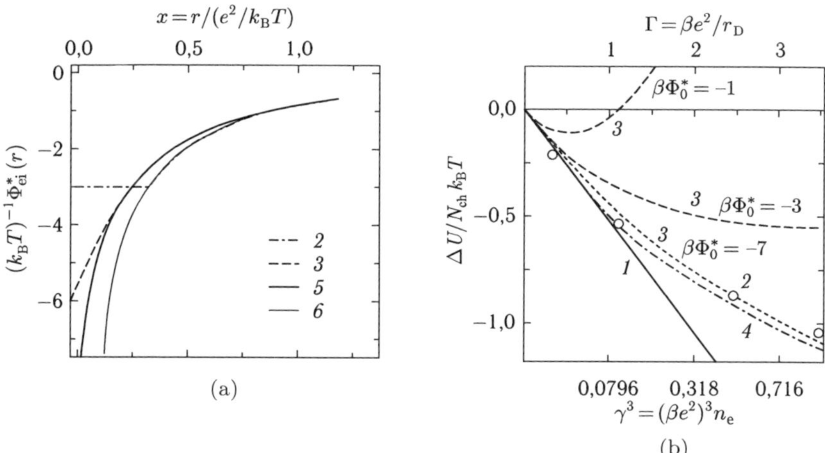

Fig. 4.24 (a) Effective electron–ion pseudopotential $\Phi^*_{ei}(r)$ (divided by $k_BT$) [468]: *2* simplified ("zero model") electron–ion pseudopotential [1083–1086] $\beta\Phi^*_{ei}(0) = -3$, *3* equivalent pseudopotential $\Phi^*_{ei}(r)$ borrowed from [468] $(\beta\Phi^*_{ei}(0) = -6)$, *5* hydrogen electron-ion pseudopotential [1082] at $T = 10^3$ K, *6* Coulomb potential. (b) Dimensionless configuration energy $\Delta U/N_{ch}k_BT$ of the free-charge subsystem: *1* Debye approximation (Debye limit), *2* Norman configuration energy calculated by the Monte Carlo method for a simplified ("zero model") electron-ion pseudopotential with $\beta\Phi^*_{ei}(0) = -3$, *3* linearized version of equivalent pseudopotential $\Phi^*_{ei}(6)$ $\beta\Phi^*_{ei}(0) = -7$, *4* nonlinear approximation [400] for the equivalent pseudopotential $\Phi^*_{ei}(r)$ with $\beta\Phi^*_{ei}(0) = -7$.

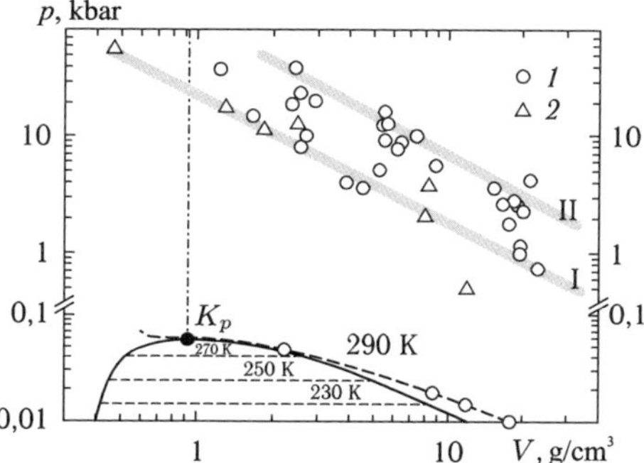

Fig. 4.25 Xenon phase diagram [404]. Marked are the two-phase region boundary and the critical point $K$. *1* experimental data [404], *2* experimental data [307] after correction [1009] of the initial density of unperturbed gas before the shock jump with allowance for the EOS of a real gas; dashed lines are isotherms in the diphasic region [1009], dot-and-dash line shows the critical isochore $(V = V_c)$, grey lines are the boundaries of single ionization: "I" $\{\chi_{Xe^+} = \chi_{Xe}, (\chi_{Xe} = N_{Xe}/N_{total})\}$ and of double ionization "II" $(\chi_{Xe^{++}} = \chi_{Xe^+})$.

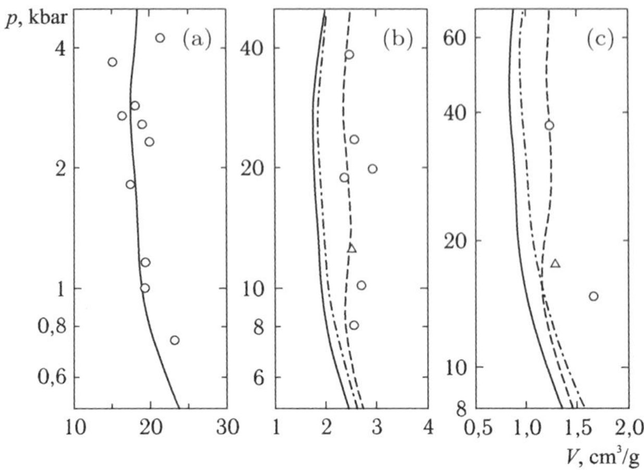

Fig. 4.26 Xenon phase diagram [404]. Comparison of experimental shock adiabats with the results of model calculations. Solid lines the interaction of charged particles was taken into account in the modified Debye ring approximation in the grand canonical ensemble [378,621], dot-and-dash line: added is the atom-atom interaction in the approximation of second virial coefficient [436] dashed line: calculation using the modified pseudopotential model [468] taking into account equality of the depth of free-particle pseudopotential to the upper cutoff boundary of intra-atomic states chosen to be equal to $k_\mathrm{B}T$. The pressure before the shock front is (a) $p_0 = 1\,\mathrm{bar}$; (b) $p_0 = 10\,\mathrm{bar}$; (c) $p_0 = 20\,\mathrm{bar}$.

the model [468] is in this case directly related to the presence in the model of an explicit (positive) correction to the mean kinetic energy of free charges, which in terms of EOS is equivalent, in comparison with the majority of traditional approximations, to the effect of additional repulsion.

An analogous result was obtained in comparative calculations using the model [468] of the parameters of shock-wave compressed cesium plasma [149, 633]. In these experiments, a higher measurement accuracy was achieved and a wider range of pressures and specific volumes and, what is even more important, a wider range of plasma nonideality parameter $\Gamma_\mathrm{D}$ (because of an exceedingly small cesium ionization potential) were involved. Another distinctive feature of the experiments [149, 633] and the accompanying series of computational theoretical papers [309, 633] is obviously one of the first (if not the very first) realization of the well-known idea of Ya B. Zel'dovich [1077] concerning temperature restoration from the results of shock-wave experiment. However, the accuracy of the temperature thus restored and the specific effect of mutual compensation of different uncertainty sources in the model cesium EOS in the given parameter range led to the fact that the finally recalculated "experimental" results on the thermal $p(V,T)$ EOS of cesium turned out to be compatible with virtually all the theoretical approximations proposed afterwards for the description of these experimental data.

The contribution of short-range repulsion of molecules, atoms, and ions was described within the soft-sphere approximation [1061] extended to the case of multicomponent mixture of spheres with different radii:

$$\Delta f_{SS} \equiv \frac{\Delta F_{SS}}{Nk_{\mathrm{B}}T} = C_s y^{s/3}\left(\frac{\varepsilon_{SS}}{k_{\mathrm{B}}T}\right) + \frac{s+4}{6}Qy^{s/9}\left(\frac{\varepsilon_{SS}}{k_{\mathrm{B}}T}\right)^{1/3};$$

$$y = \frac{3Y\sqrt{2}}{\pi}; \quad Y = \frac{4\pi r_c^3}{3} = \frac{\pi\sigma_c^3}{6}, \tag{4.211}$$

where $N = \Sigma N_j$ is the total number of particles described by soft-sphere approximation and $C_s$ is the Madelung constant [1061]:

$$C_s = 6 + \frac{6.669}{s-3} - 1.043(s-4)^{0.389}\exp[0.156(s)],$$

for the potential

$$V(r) = \varepsilon_{SS}\left(\frac{3r_c}{r}\right)^s. \tag{4.212}$$

The mean soft sphere "radius" (the characteristic size) $r_c$ is defined by the expression

$$r_c = \left[\frac{\Sigma n_j r_j^q}{\Sigma n_j}\right]^{1/q}. \tag{4.213}$$

Generally speaking, the degree of repulsion $s$ can also be thought of as the mean quantity if different sorts of particles possess different repulsion "softness", i.e.,

$$s = \left[\frac{\Sigma n_j s_j^l}{\Sigma n_j}\right]^{1/l}, \tag{4.214}$$

where $q$, $\ell$ are the degrees of averaging in (4.213), (4.214) (in the simplest case $q = 3$, $\ell = 1$, which was taken late on).

Then the correction to pressure at the expense of particle repulsion will have the form

$$\frac{\Delta p_{SS}}{nk_{\mathrm{B}}T} = y\Delta f'_{SS}(y) = C_s \frac{s}{3} y^{s/3}\left(\frac{\varepsilon_{SS}}{k_{\mathrm{B}}T}\right)$$

$$+ \frac{s+4}{2}\frac{s}{9}Qy^{s/9}\left(\frac{\varepsilon_{SS}}{k_{\mathrm{B}}T}\right)^{1/3}, \tag{4.215}$$

to the internal energy and

$$\frac{\Delta p_{SS}}{Nk_{\mathrm{B}}T} = \frac{3}{s}\frac{\Delta P_{SS}}{nk_{\mathrm{B}}T} = C_s y^{s/3}\left(\frac{\varepsilon_{SS}}{k_{\mathrm{B}}T}\right)$$

$$+ \frac{s+4}{2}\frac{1}{3}Qy^{s/9}\left(\frac{\varepsilon_{SS}}{k_{\mathrm{B}}T}\right)^{1/3}, \tag{4.216}$$

to the chemical potential

$$\frac{\Delta \mu_i}{k_{\mathrm{B}}T} = \Delta f_{SS} + \left[\frac{\partial \Delta f_{SS}}{\partial N_i}\right]_{V,T}; \tag{4.217}$$

$$\left(\frac{\partial \Delta f}{\partial N_i}\right)_{V,T} = \frac{\partial C_s}{\partial N_i}y^{s/3}\frac{\varepsilon_{SS}}{k_{\mathrm{B}}T} + \left\{\frac{s}{3}\frac{1}{y}\frac{\partial y}{\partial N_i} + \frac{1}{3}\frac{\partial s}{\partial N_i}\ln y\right\}$$

$$\times \left\{C_s y^{s/3}\frac{\varepsilon_{SS}}{k_{\mathrm{B}}T} + \frac{s+4}{2}Qy^{s/9}\frac{1}{3}\left[\frac{\varepsilon_{SS}}{k_{\mathrm{B}}T}\right]^{1/3}\right\}$$

$$+ \frac{1}{2}\frac{\partial s}{\partial N_i}Qy^{s/9}\left[\frac{\varepsilon_{SS}}{k_{\mathrm{B}}T}\right]^{1/3}. \tag{4.218}$$

In formula (4.218), the derivatives in the right-hand side are defined by the following expressions

$$\frac{\partial C_s}{\partial N_i} \equiv \left[\frac{\partial C_s}{\partial N_i}\right]_{V,T} = \frac{\partial s}{\partial N_i}\left\{\frac{6.669}{(s-3)^3} - 1.043(s-4)^{0,389}\right.$$

$$\left.\circ\, e^{0,156(s)}\left[\frac{0.389}{s-4} - 0.156\right]\right\}; \tag{4.219}$$

$$\frac{\partial s}{\partial N_i} \equiv \left[\frac{\partial s}{\partial N_i}\right]_{V,T} = \frac{1}{\ell N}\left[\frac{s_j^\ell}{s^l} - 1\right]; \tag{4.220}$$

$$\frac{\partial y}{\partial N_i} \equiv \left[\frac{\partial y}{\partial N_i}\right]_{V,T} = \frac{y}{N}\left\{1 + \frac{3}{q}\left(\frac{r_j^q}{r_c^q} - 1\right)\right\}. \tag{4.221}$$

Expressions (4.217)–(4.221) imply that the larger the radius of the particle and the degree of its repulsion, the larger the correction to its chemical potential. Hence, with increasing ion charge we obtain for the decreasing sequence of the ion radii also a decreasing series of corrections to the chemical potential (4.217), which causes the corresponding lowering of ionization (dissociation) energy for each type of particles. This is most readily seen for the particular case of particles with identical degrees of repulsion $s_j = s \equiv$ constant and different radii:

$$\frac{\Delta\mu_{jSS}}{k_{\mathrm B}T} = \frac{\Delta F_{SS}}{Nk_{\mathrm B}T} + \frac{\Delta p_{SS}}{nk_{\mathrm B}T}\left[\frac{r_j}{r_c}\right]^3. \tag{4.222}$$

This implies that for particles with different radii the corrections (4.222) are different, which may in turn predetermine a decrease in the ionization (dissociation) energy with increasing matter density. The above-described thermodynamic model has regular asymptotics for low densities of the medium coinciding with the well-known theories of rarified plasma and gases. We shall consider the specificities of calculation of the thermodynamic properties using the approximation (4.198)–(4.222) for inert gases and hydrogen (deuterium).

## 4.9 Gas thermodynamics

In selection of the parameters of the model (4.198)–(4.222) for shock-compressed initially liquefied inert gases, the atomic radii (for a fixed energy constant $\varepsilon_{SS}$) and the degree of repulsion in the soft-sphere mixture approximation (4.211)–(4.222) were chosen proceeding from the condition of the best (within this model) description of computational data of the cold curve ($T = 0\,\mathrm{K}$) for experimental-range densities. The values obtained as a result of such a procedure are presented in the table:

|     | $s$  | $\varepsilon_{SS}$ | $R_{\mathrm{atom}}$ | $Q$ |
|-----|------|--------------------|---------------------|-----|
| He  | 12   | 0.01354            | $2.82a_0$           | 1   |
| Ar  | 12   | 0.0125 eV          | $3.2a_0$            | 1   |
| Kr  | 12   | 0.0171 eV          | $3.54a_0$           | 1   |
| Xe  | 10.5 | 0.0221 eV          | $3.83a_0$           | 1   |

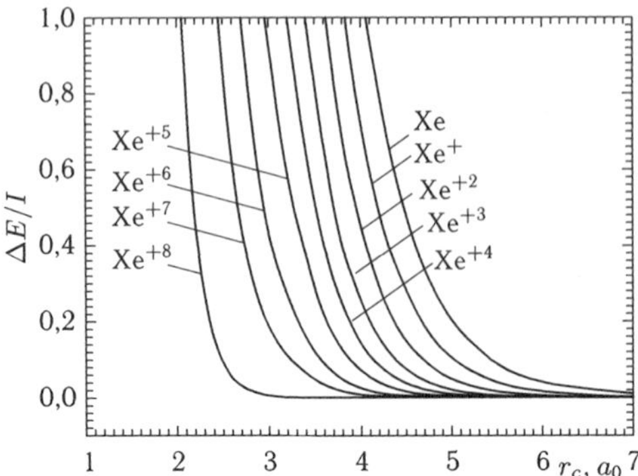



Fig. 4.27   Variation of atomic and ionic energy of Xe upon compression in a spherical cell relative to the ionization potential of an isolated atom.

The relations of the radii of atoms and ions of different multiplicity were found from the calculation of their electron structure in the bound-atom approximation by the Hartree–Fock method (see Sec. 4.6). Figure 4.27 presents the dependence of the xenon atomic and ionic energy on the atomic cell radius.

The ratio of atomic and ionic radii was chosen on the line where the energy shift upon compression was compared with the atom ionization potential, that is, on the line $\Delta E/I = 1$.

The shock adiabats and isotherms of xenon, krypton, and argon were calculated within the above-given model (4.198)–(4.222). In all the three cases the initial state of shock-compressed gases corresponded to a liquid. Figure 4.28 demonstrates the xenon shock adiabat in density-pressure coordinates.

Along with the shock experiments, the data on the multiple shock-wave compression were presented [696]. One can see that on the whole the adopted approximation helps to reach a satisfactory description of the experimental data. The existing difference at low pressures and temperatures can be explained by insufficient precision of the approximation of the liquid-phase xenon states.

This model also permits a satisfactory description of liquid argon and krypton shock adiabats (see Figs. 4.29 and 4.30). The experimental data are borrowed here from papers [354, 372].

Note that a satisfactory agreement can also be reached with the measured values of brightness temperature and the velocity of sound in these substances.

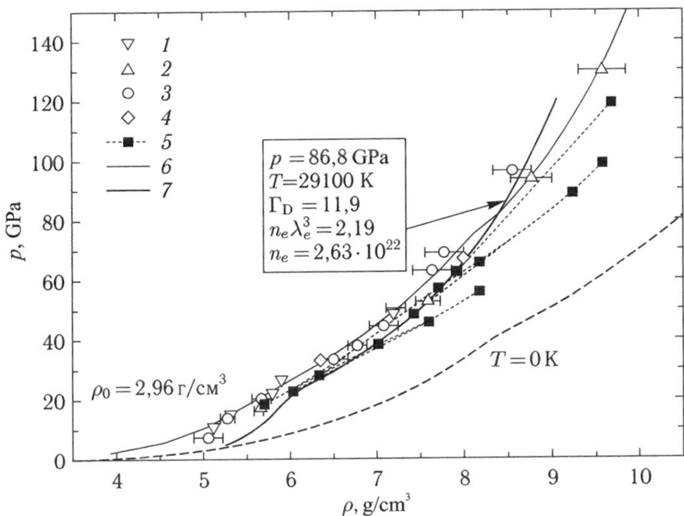

Fig. 4.28   Xenon shock adiabat. Experimental data: *1* [514], *2* [735], *3* [994], *4* [819], *5* [696]. Calculated curves: *6* [994], *7* [229], dashed line is the "cold curve" [994].

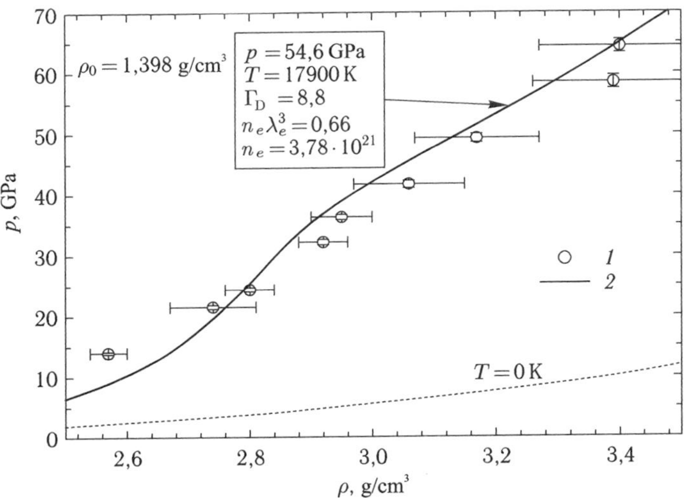

Fig. 4.29   Argon shock adiabat. Experimental data: *1* [372]. Calculated curve: *2* the present work, dashed line is the cold curve.

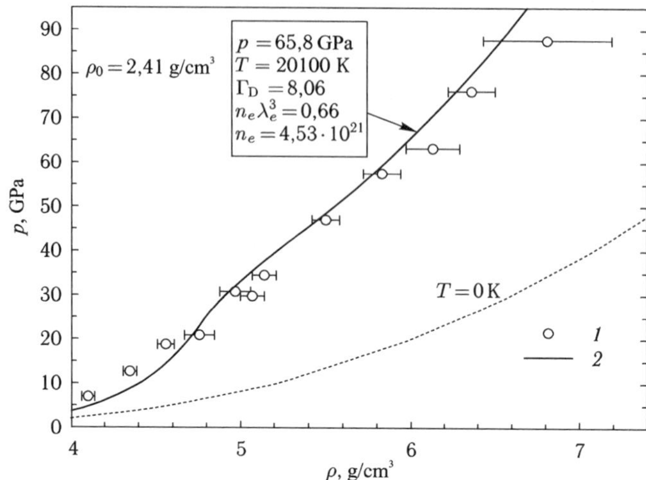



Fig. 4.30   Krypton shock adiabat. Experimental data: *1* [354]. Calculated curve: *2* [401], dashed line is the cold curve.

The above-mentioned comparison of the results of calculations with the experimental data on shock-compressed inert gases shows that the model (4.198)–(4.222) describes qualitatively and in most cases quantitatively the behavior of nonideal plasma in megabar and submegabar pressure range of inert-gas shock compression. We shall emphasize the main physical effects inherent in the model.

(1)  Free electron degeneration was taken into account according to expression (4.191) [598]. The degree of degeneracy ranged from several units to several tens under maximum compression. Obviously, the positive correction to the chemical potential in equations of chemical and ionization equilibrium induced lowering of the degree of plasma ionization.

(2)  By virtue of high values of the Coulomb nonideality parameter, which can be seen in Figs. 4.28–4.30, important was the account of the Coulomb interaction of charged particles (4.200)–(4.204) calculated within the pseudopotential model [401,468] in which attempt was made to avoid the so-called "double account" in singling out the contribution of bound states, when the contribution of one and the same states of the electron–ion pair in the phase space is taken into account as both the contribution among free particles and as the contribution among bound states [401,468]. Corrections to the ideal-gas part of free energy are in this case lower in the absolute value compared to the classical

Debye approximation, which in the range of high nonideality parameters does not lead to the "Debye catastrophe" when the pressure values become negative.

(3) The short-range repulsion (4.211)–(4.222) corresponded to the generalized soft-sphere model. Allowance for this effect is very important in this domain of phase diagram because of high densities of the medium when the interparticle spacing is comparable with the particle size ($\Gamma_\alpha = n_\alpha r_\alpha^3 \approx 1$). Note that with decreasing particle size (for instance, the ion size decreases with increasing charge) the correction to the chemical potential (4.222) decreases, which in turn leads to the fact that the decay of large-radius particles into particles of smaller radius with a gain in volume occupied by particles becomes thermodynamically more advantageous. In the component composition this is expressed in a sharp increase of the degree of ionization (pressure-induced ionization). This will be shown below on an example of the behavior of electroconductivity for high matter densities.

It should be stressed that although the general tendency in the behavior of thermodynamic values remains the same, some regions nevertheless exist (see, e.g., Figs. 4.31, 4.32 and 4.33) where the calculated data obviously deviate from experiment. In particular, this can be observed because of a

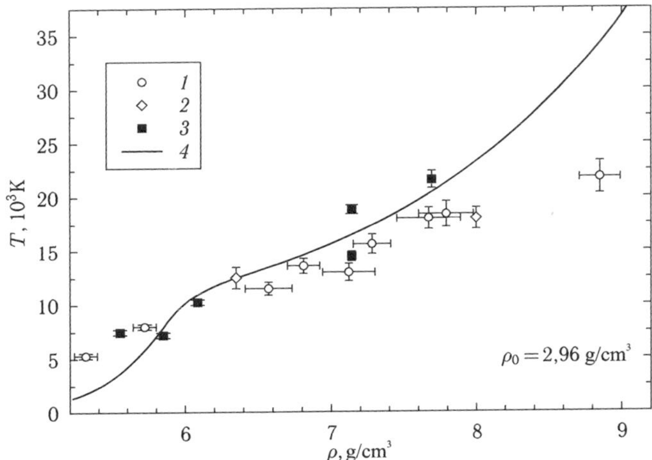

Fig. 4.31 Temperature along the shock adiabat of xenon. Experimental data: *1* [994], *2* [819], *3* [696], *4* calculation by the model (4.198)–(4.222).

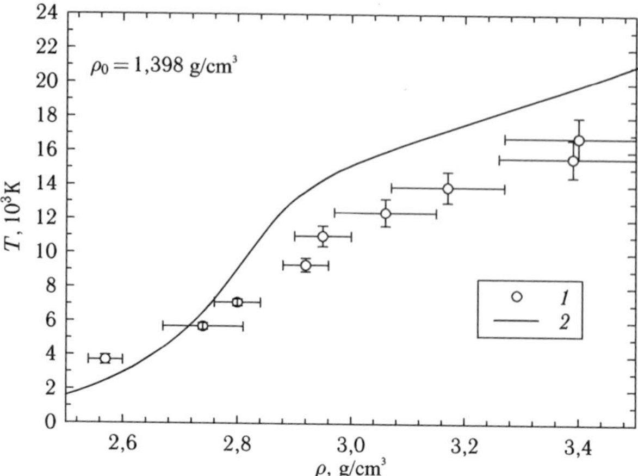

Fig. 4.32   Temperature along the shock adiabat of argon. Experimental data: *1* [372], *2* calculation by the model (4.198)–(4.222).

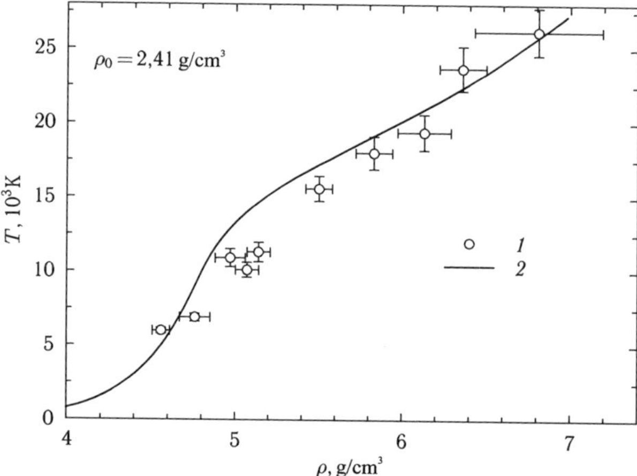

Fig. 4.33   Temperature along the shock adiabat of krypton. Experimental data: *1* [354], *2* calculation by the model (4.198)–(4.222).

simplified definition of repulsion parameters in the soft-sphere model, which may need improvement of this procedure in future.

## 4.10   Hydrogen thermodynamics

The specific feature of hydrogen is the existence of an extensive "monomolecular" region ($\rho \leq 0.3\,\mathrm{g/cm^3}$; $-\mu_\mathrm{H} \geq D(\mathrm{H_2}) \approx 4.5\,\mathrm{eV}$)

where the hydrogen thermodynamics is almost completely determined by the interaction $H_2$–$H_2$. In the framework of the employed generalized soft-sphere model (4.211)–(4.222) the $H_2$–$H_2$ interaction parameters were chosen in this work extremely close to the parameters of the rigorous "nonempirical" atom-atom approximation [1054, 1055]. The noncentral character of the $H_2$–$H_2$ interaction was neglected. The calculations showed that with the use of "soft" repulsion $V(r) \sim 1/r^6$ both the molecular part of the isotherm $T = 0$ ("cold curve") and the results of exact Monte Carlo simulations of $H_2 + H$ mixture thermodynamics [1054, 1055] for an atomic-molecular system.

The main problem in the description of nonideality, including the case of dense hydrogen, is as before a correct definition of the entire set of effective potentials of interaction among all the components of the mixture. This refers to the interactions with participation of both charged and neutral particles and in the first place of $H_2$ and H–H pairs. It is of importance that the effective interaction of free atoms involved in the chemical model differs radically from the singlet (attractive) and triplet (repulsive) branches (given by the rigorous theory) of the total H–H interaction potential since the contribution of H–H pairs interacting along the singlet branch is already taken into account in the class of intramolecular motion. This is the more so in respect of the effective interaction with participation of (free) charged particles since in the chemical model the contributions of free and bound states must be mutually consistent (see the approximation (4.194)–(4.204)). Serious contradiction now exists in the form and parameters of these effective potentials given by different approaches, the parameters of the short-range H–H and $H_2$ repulsion being the main object of the contradictory data beyond the monomolecular region. One of the versions is represented by the results of nonempirical atom-atom approximation [1054, 1055] leading to relatively large "proper volumes" of the hydrogen atom. In terms of the employed modification of the soft-sphere model (4.211)–(4.222) the results correspond almost exactly to the "additive volume" approximation: $v(H_2) \approx 2v(H)$. Such a choice leads for $\rho < \rho^* \approx 0.3\,\text{g/cm}^3$ to the results perfectly coincident with the exact Monte Carlo simulations [1054, 1055] at $T \leq 10$ kK and with the nonanomalous part of the results of quantum Monte Carlo simulations (PIMC [793]) at $T \geq 10\,\text{kK}$. At the same temperatures, the results also agree satisfactorily with the other versions of first-principle approaches, namely, of quantum molecular dynamics (TBMD [190]) and the "wave packets" method (WPMD [551]).

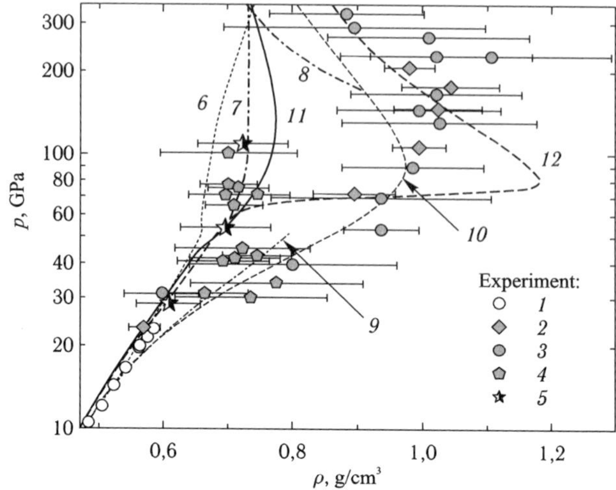

Fig. 4.34　Shock adiabat of deuterium. Experiment: *1* [442], *2* [200], *3* [715], *4* [556–558], *5* [91, 127]. Calculation: *6* [518], *7* [793], *8* results obtained in [229], *9* RACH [101], *10* [860], *11* and *12* calculation according to expressions (4.198)–(4.222).

Figure 4.34 demonstrates all the presently available experimental data on a single shock compression of liquid deuterium.

In the experiments with light-gas guns [442] performed in a direct shock wave, pressures up to 25 GPa (circles *1* in Fig. 4.34) were realized. In the studies of shock wave generation by high-power lasers [200, 715] (*2*, *3*), pressures up to 300 GPa were reached and an anomalously high compressibility of deuterium was revealed at pressures $p > 40\,\mathrm{GPa}$. The recent experiments on Z-pinch [556–558] (*4*) and on spherical explosive systems [91, 127] (*5*) do not confirm the presence of this anomaly up to $p \approx 70\,\mathrm{GPa}$.

The calculated shock adiabats obtained with the use of both the equation of state SESAME [518] (curve *6*) and the semi-empirical EOS [373] do not predict such anomaly in the behavior of shock compressibility. Neither is it expected with the use of "first-principle" approaches: the quantum Monte Carlo method [793] (curve *7*) and molecular dynamics [190]. In paper [860] (curve *10*), the interpolation EOS of deuterium is presented which describes qualitatively the experimental results obtained on lasers.

The approach considered here also fails to reproduce this "dip" towards the unexpectedly high degrees of compression ($\sigma_{\max} \equiv \rho_{\max} \approx 6.5$ against the expected $\sigma_{\max} \approx 4$) in the position of the shock adiabat of deuterium in the region $p \approx 0.5 \div 2\,\mathrm{Mbar}$ (curve *11*) and does not contain phase-transition type anomalies for $\rho \geq 1\,\mathrm{g/cm^3}$. The thermodynamics of compressed hydrogen (deuterium) looks quite different if the H–H (D–D) interaction is described by the potential from [833] commonly accepted in approximate calculations and the H–H$_2$ interaction is described using the standard combination rule. In terms of modification of the soft-sphere model adopted in this work this corresponds to a much smaller ratio of the "proper" volumes H and $H_2 d(H)/d(H_2) \approx 0.4 \to 2v(H)/v(H_2) \approx 0.13$. Such a choice of the "proper" atomic size immediately leads for $\rho \geq 0.3\,\mathrm{mole/cm^3}$ to "pressure-induced dissociation" accompanied by the dip in the deuterium shock adiabat (curve *12*).

The difference in the behavior of deuterium with different ratios of the molecular and atomic radii of deuterium in the first place affects the component composition. Figures 4.35 and 4.36 show the component composition on the shock adiabat of liquid deuterium. One can see that the distinction in deuterium compressibility in two versions of calculation $((d(H)/d(H_2) \approx 0.4$ and $d(H)/d(H_2) \approx 0.8)$ is explained by a faster dissociation in the first case.

New experimental data on shock-compressed deuterium have quite recently been obtained in VNIIEF (All-Russian Scientific Research Institute for Experimental Physics). As distinct from the above-cited experiments presented in Fig. 4.34, the experiments [377] were carried out with gaseous deuterium preliminarily compressed to pressures of 1500 and 2000 bar, which corresponded to the densities of 0.1335 and $0.153\,\mathrm{g/cm^3}$. In this work, temperature was measured simultaneously with the shock adiabat parameters. The model that we had used before when describing the experiments illustrated in Fig. 4.34 was also applied for the description of

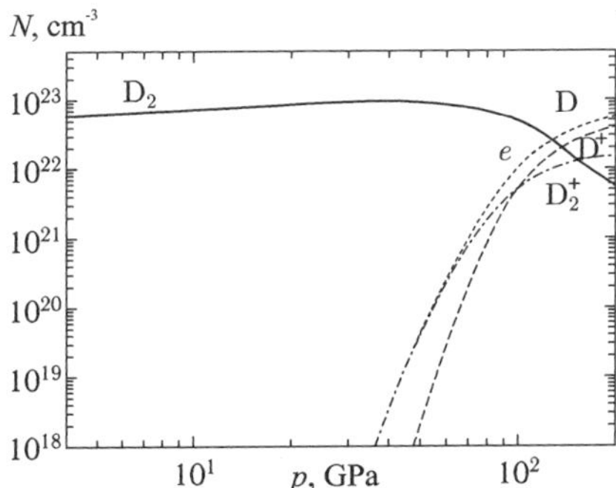


Fig. 4.35 Component composition on the shock adiabat of liquid deuterium with the ratio $d(H)/d(H_2) \approx 0.4$.

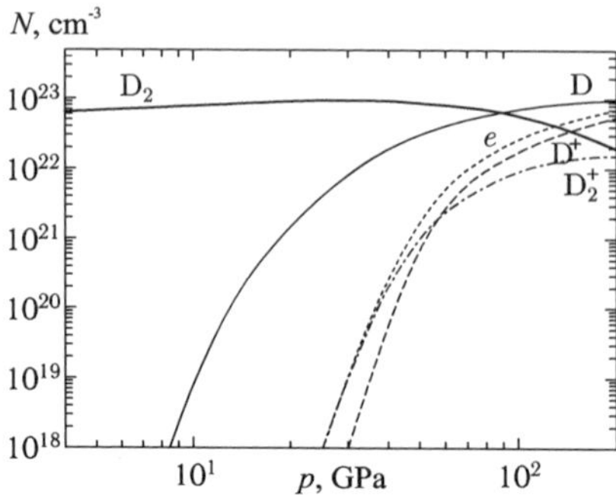

Fig. 4.36   Component composition on the shock adiabat of liquid deuterium with the ratio $d(\mathrm{H})/d(\mathrm{H_2}) \approx 0.8$.

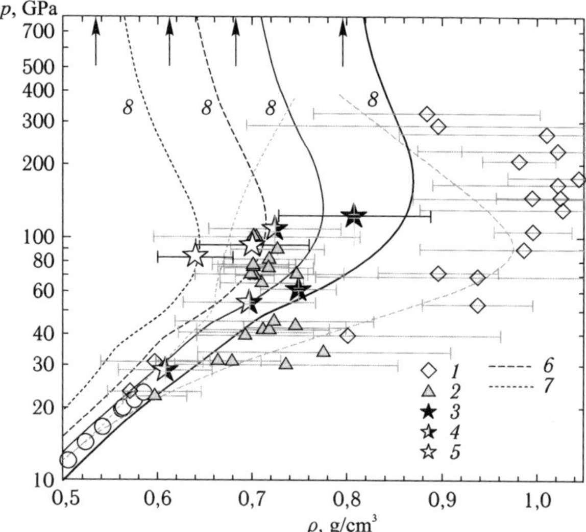

Fig. 4.37   Shock adiabats of gaseous, liquid, and solid deuterium. Experiment: *1* [200, 715], *2* [556–558], *3*, *4* [128], solid and liquid deuterium, respectively, *5* gaseous deuterium [377]. Calculations: *6* [860], *7* [518] *8* calculation by the model (4.198)–(4.213).

the experiments [377]. Figure 4.37 demonstrates the results of calculations within the model (4.198)–(4.222) for four shock adiabats with the initial densities of 0.1335, 0.153, 0.171 (liquid deuterium), and 0.199 (solid deuterium) g/cv$^3$.

We notice that the latest calculations within the model (4.198)–(4.222) illustrated in Figs. 4.37 and 4.38 were carried out with the ratio

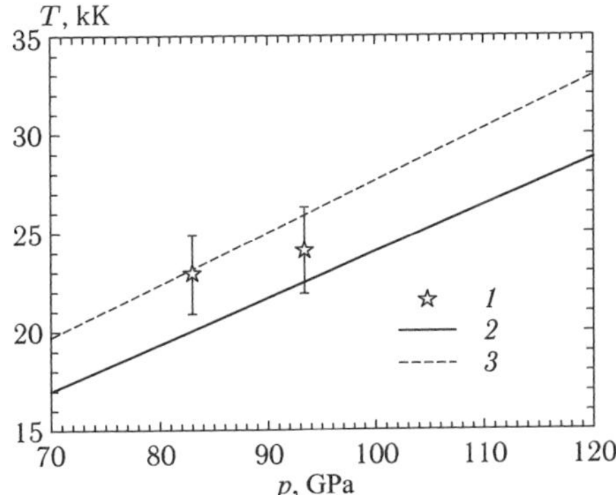



Fig. 4.38  Temperature of shock-compressed gaseous deuterium. Experiment *1* [377], the left point is density $0.1335\,\mathrm{g/cm^3}$, the right point is density $0.153\,\mathrm{g/cm^3}$; *2, 3* calculation by the model (4.198)–(4.222) for the same densities, respectively.

$d(\mathrm{H})/d(\mathrm{H_2}) \approx 0.8$, which corresponds to the results of papers [1054, 1055]. Thus, Figs. 4.37 and 4.38 (Figure 4.38 presents comparison of calculations with experiment in pressure-temperature coordinates) show that the model discussed in the present chapter allows a consistent description of the totality of experiments performed in VNIIEF for different initial densities of deuterium.

## 4.11  Metal plasma thermodynamics

By the present time, the main array of experimental data on the properties of strongly compressed plasma has been obtained by dynamical methods [290] operating with the technique of high-power shock waves for compression and irreversible substance heating. The use of explosive and pneumatic missiles in such experiments made it possible to investigate and analyze the theoretical models of thermodynamic, electrophysical, and optical properties of shock-compressed cesium, inert gases, and hydrogen under strong nonideality conditions. The multiplicity of ionization $\alpha = n_{\mathrm{e}}/(n_\alpha + n_{\mathrm{i}}$ of such a medium did not exceed 1–2. Overstepping the limits of these conditions, i.e., passing over to parameters for which the substance gets multiply ionized with partially degenerate electrons, can be realized through involvement of experimental data on compression of solid and porous metals by shock waves with amplitude pressure of hundreds of thousands or millions of atmospheres. To the present time a large array of experimental data on the dynamic compression of metals has been obtained (see

Refs. [662, 1007, 1095] and the references therein) with the use of shock waves generated by an explosion of chemical [27, 34, 1007] and nuclear [39] BB, pneumatic missiles [490] and, lately, by concentrated laser [50], X-ray [288], and ion [82] flows. The data on shock-wave compression supplemented with the results of recording of shock-compressed metal unloading adiabats underlie the construction of semi-empirical EOS [150] by an optimum choice of constants in functional thermodynamic relations based on simplified thermodynamic models. At the same time, in the course of shock-wave compression already at relatively low (100–200 GPa) pressures melting takes place and then a progressing thermal ionization and pressure-induced ionization of substance. Thus, a dense, disordered multiply ionized system of charged particles, i.e., an electron–ion medium with a complex spectrum of intense collective interactions is realized. For this reason shock-compressed metals seem to be an interesting object for verification of the theoretical models of strongly compressed plasma both from the point of view of search for PPT [745] and for the analysis of different models describing nonideality of strongly compressed plasma for high energy concentrations. Essentially, we speak of the extension of plasma models [321] to the region, unconventional for them, of condensed densities and megabar pressures, where either semi-empirical approximation EOS [321] or far extrapolations of quasi-classical approximations [505] have been applied till recently. Such type of thermodynamic measurements in the "metal–dielectric" transition region would allow, in addition, a verification of the hypothesis [1078] concerning the relation of metallization with first-order phase transitions in disordered media.

The range of parameters of a strongly nonideal system corresponds to the density $\rho_0$ lowered compared to the solid-state values and energies exceeding the atomic and molecular binding energy in a solid (approximately 1 eV per particle). To generate such states of metals, shock-wave compression of fine-dispersed (porous) metals [1094] was used, which allows one to heighten the effects of irreversible energy dissipation at the front of shock-wave discontinuity and to obtain higher substance heating. For a number of metals the porosity value $m = \rho_0/\rho_{00}$ ($\rho_{00}$ is the density of a porous sample) lay in the range $1 \leq m < 30$, and the experimental data covered a considerable region both in the substance density beyond the shock front and in temperature. For nickel the maximum possible porosity is $m = 15$, 20, and 28 and the shock-wave compression pressures are above 80 GPa [982, 987]; for copper we have $m = 10$ [389], for iron $m = 20$ [33], and for aluminum $m = 8$ [986].

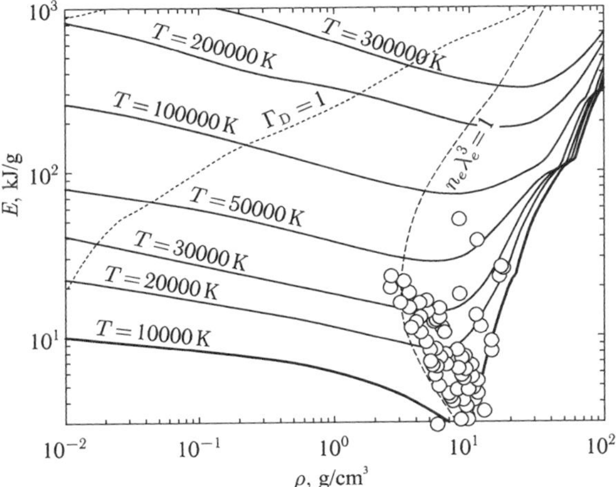


Fig. 4.39  Density-energy diagram for nickel plasma. Plotted are the computational isotherms and the experimental points which are due to shock compression of solid and porous nickel samples [389, 983]. Marked are the lines of constant Coulomb nonideality parameter ($\Gamma_{\mathrm{D}} = [4\pi(e^2/k_{\mathrm{B}}T)^3 n_\alpha z_\alpha^2]^{1/2}$) and electron degeneracy parameter $n_e\lambda_e^3$.

This section is devoted to construction of nonideal plasma models for the description of shock-wave compressed porous metals, to the analysis of the effect of particle interaction on the thermodynamic functions of nonideal plasma in condition of shock-wave loading and to comparison of the computational data with experiment.

To point out the common specific features of the above-mentioned experimental data on shock-wave compression of porous metals we shall consider, in line with Ref. [33], the internal energy ($E$)–density ($\rho$) diagram supplemented with the calculation of substance isotherms using the plasma model of bounded atom. Figures 4.39 and 4.40 present the $E-\rho$ diagram and the diagram of equilibrium composition for nickel [390].

Figure 4.39 demonstrates the fact that the whole phase diagram of the substance (the diagrams of other metals are similar to the one under discussion) falls into two regions of qualitatively different behavior of thermodynamic dependences. A large part is occupied by the region of relatively rarefied ($\rho \ll \rho_0$) "gas" plasma characterized by two distinctive features, namely, a smooth energy lowering upon isothermal compression and a clear manifestation (on large intervals of density variation) of the so-called shell oscillations of all the thermodynamic dependences (for details see Ref. [471]). For $\rho > \rho$ such behavior changes for a sharp growth of energy

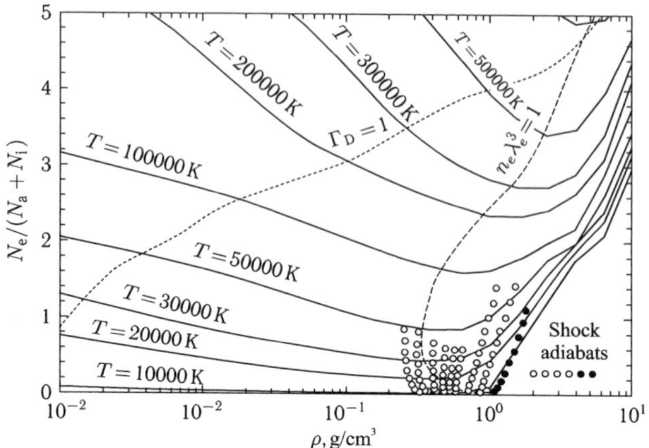


Fig. 4.40 Behavior of the degree of nickel plasma ionization upon compression [390]. The degree of ionization along the isotherms (solid curves) and calculated shock adiabats (dots) of a solid and porous nickel samples corresponding to the experiment [982, 983] is given. The lines of constant Coulomb nonideality parameter ($\Gamma_D = [4\pi(e^2/k_B T)\sum n_a z_a^2]^{1/2}$) and electron degeneracy parameter $n\lambda_e^3 = 1$ in plasma are marked.

and the generalized compressibility factor — $Z(n_{\mathrm{nucl}}, n_e, T) = p/p^{\mathrm{id}}$, which is traditionally interpreted as "pressure-induced ionization". In the limit of very high density, this process ends with an outcome into the range of states well described by the model of a system of mobile nuclei emerged in a weakly nonideal gas of degenerate electrons. For the description of thermodynamics in this region the well-developed apparatus of cell representations can be successfully applied [547, 741, 917]. Between the two above-mentioned ranges of parameters lies a transition region characterized by the minimum values of the internal energy and the compressibility factor and the maximum violation of weak nonideality conditions. The depth and position on the isotherm of the minima corresponding to the thermal and caloric EOS may conditionally be thought of as a center of maximum uncertainty in our knowledge of thermodynamic properties of a compressed and heated substance. Note that the region where the minima are located corresponds to the so-called "nonideality valley". The shock-wave compression of porous targets is unique for it provides information on the behavior of a dense strongly nonideal medium in precisely this most complicated and interesting region.

The description of states of shock-compressed porous metals was based on the quasi-chemical approach (the chemical model) [396, 398, 401] the general features of which were discussed in the previous chapters. Thus, the free energy for a quasi-neutral mixture of electrons, ions, atoms, and molecules is represented as a sum of ideal-gas summands $F^0_{i,e}$ and an ensemble of all the components and terms responsible for different types of interparticle interactions considered separately in the chemical model:

$$F \equiv F^0_i + F^0_e + F^*_{ii,ie,ee,...}. \tag{4.223}$$

The atoms and ions obey the Boltzmann statistics, and their contribution has the form:

$$F^0_i = \sum_j N_j k_B T \left( \ln \frac{n_j \Lambda_j^3}{\sigma_j} + \frac{A_j}{k_B T} - 1 \right). \tag{4.224}$$

Here $k_B$ is the Boltzmann constant, $\sigma_j$ is an individual statistical sum of $j$th sort particle (in this case an atom or an ion) calculated by the energy levels counted from the ground state of the particle, $\Lambda_j$ is the de Broglie thermal wavelength of $j$th sort particles, and $A_j$ is the atomization energy (the difference between the energy of the ground state of a particle and the energy of the ground states of its component atoms).

The electron degeneration effects are important in the considered phase diagram region since the degeneracy parameter for free electrons can reach several units:

$$n_e \Lambda_e^3 \approx 1.$$

As was shown in the preceding sections in the framework of the quasi-chemical representation where electrons are divided into two sorts — free and bound, the degeneracy effect first of all shows up already in the framework of the ideal-gas summand changing qualitatively the density dependence of pressure and the chemical potential. The main effect of electron degeneracy in reconstruction of the nonideality mechanism consists in a gradually weakening participation of electrons (with their increasing degeneracy) in the mechanism of mutual charge screening. The limit of this tendency upon extreme substance compression is the electron "switching-off" from this mechanism and transition of screening in the ion system to the class described by the so-called ion mixture model [83] — a version of the one-component plasma model. In our case the electron degeneracy was mainly considered in the framework of the ideal-gas summand of the free energy, which had the form (4.191). In this case the electron degeneracy led first of all to an effective shift of ionization equilibrium towards lower degrees of ionization and, moreover, to a direct correction in the EOS corresponding to an additional effective repulsion.

The description of nonideality effects was divided in this section into two stages. At one of the stages realized as the SAHA-3 code (see above) and employed repeatedly in applications [234, 389, 392, 404] the "minimum" set of steps was taken which in some cases allowed attaining within the experimental accuracy a satisfactory agreement with the experimental data on the dynamic (shock-wave) compression of different substances: inert gases, cesium, highly porous metal samples, etc. Some simplicity of this approach is necessary for an effective inclusion of the calculation of the EOS in rather cumbersome gas-dynamic calculations that claim rigorous fulfillment of all auxiliary calculations.

Within this reduced approach the following approximations were used.

- Account of electron degeneracy is limited to the ideal-gas summand.
- The Coulomb nonideality effect is taken into account within the so-called Debye (ring) approximation in the grand canonical ensemble [394, 621].
- In calculation of internal statistical sums of atomic and ionic excitations the ground state alone is taken into account.
- Account is taken of intense short-range repulsion of atoms and ions in the framework of approximation of hard-sphere mixture with essentially different sizes of atoms and ions of different multiplicity.
- Atoms and ions are considered to be "penetrable" for electrons, i.e., the latter are not involved in the sphere of activity of the hard-sphere approximation and "do not feel" the effective increase in density because of the presence of atomic and ionic proper volume.
- Additional short-range atomic and ionic attraction is taken into account which provides an effective description of the presence of binding energy of a condensed substance.

From the viewpoint of quantitative calculations one of the advantages of the reduced description under discussion is impossibility of spontaneous loss of system's thermodynamic stability in the employed form of allowance for Coulomb nonideality and statistical sums of excitation (the matrix $\|d\mu_i/dn_j\|$ is positive definite for any degree of plasma compression [234]). The specially performed calculations with separate description of ion nonideality within the so-called ion mixture model [83] and electron nonideality within the interacting electron gas model [458] (the "Twin model OCP" according to the terminology of [468] — the given approximation was exploited in paper [496] to describe thermodynamics of nonideal plasma of hydrogen and helium mixture) have shown that the range of parameters reached upon shock compression of porous samples of most metals falls into the region of existence of phase transition type anomaly ("Van der Waals loops") given by this approximation. The shape and position of these loops depend essentially on the maximum degree of ionization admitted by the adopted computational procedure (instability in the degree of ionization). Such sensitivity of EOS to the choice of approximation used to describe the Coulomb nonideality is typical in precisely this range of parameters of the overwhelming majority of the Coulomb nonideality models proposed in the literature.

We emphasize once again that in this model the Coulomb interaction was taken into account in the framework of the Debye approximation in the grand canonical ensemble [394, 621] extended (see formulas ((4.194)–(4.195)) to the case of multiple ionization: in this case the nonideality parameter $\tilde{\Gamma}_{\mathrm{D}}$ in (4.195) differs from the usual Debye nonideality parameter $\Gamma_{\mathrm{D}}$ ($\Gamma_{\mathrm{D}}^2 \equiv 4\pi[e^2/k_{\mathrm{B}}T]^3\Sigma n_\alpha z_\alpha^2$), and the approximation itself is equivalent to the classical Debye–Huckel approximation in the limit ($\Gamma_{\mathrm{D}} \Rightarrow 0$) and differs from it by noticeably lower values of corresponding corrections in the region of moderate and strong nonideality ($\Gamma \geq 1$).

The overlapping of atomic and ionic electron shells under high degrees of compression leads to intense heavy particle repulsion at short distances. This effect was described in Chapter 2 in the approximation of identical-size hard spheres (the Carnahan–Starling formula [158]) since its application was limited to the region of partial first ionization. In this case, because of the possibility of multiple ionization, this effect was taken into account within the model of different-radii hard-sphere mixture. In direct calculations, the so-called Mansoori formula [650] was used:

$$\frac{\Delta F_{\mathrm{HSM}}}{\sum_i N_i kT} \equiv f_{\mathrm{HSM}}(\nu) = X\frac{\nu}{(1-\nu)^2} + 3Y\frac{\nu}{1-\nu} + (X-1)\ln(1-\nu),$$

$$(4.225)$$

$$\nu = \frac{4\pi}{3}n\overline{r^3}, \quad \overline{r^k} \equiv \sum_i n_i r_i^k \Big/ \sum n_i, \quad k = 1, 2, 3,$$

$$X = \left(\overline{r^2}\right)^3 \left(\overline{r^3}\right)^{-2}, \quad Y = \overline{r^2 r^1}\left(\overline{r^3}\right)^{-1}. \qquad (4.226)$$

The contributions to the pressure, energy, and chemical potential follow from (4.227), (4.228):

$$\frac{\Delta F_{\mathrm{HSM}}}{\sum_i n_i k_{\mathrm{B}}T} = \frac{\partial f_{\mathrm{HSM}}(\nu)}{\partial \nu}, \quad \frac{\Delta \mu_i}{k_{\mathrm{B}}T} = f_{\mathrm{HSM}}(\nu) + \sum_j n_j \frac{\partial f_{\mathrm{HSM}}(\nu)}{\partial n_i},$$

$$\Delta E_{\mathrm{HSM}} \equiv 0. \qquad (4.227)$$

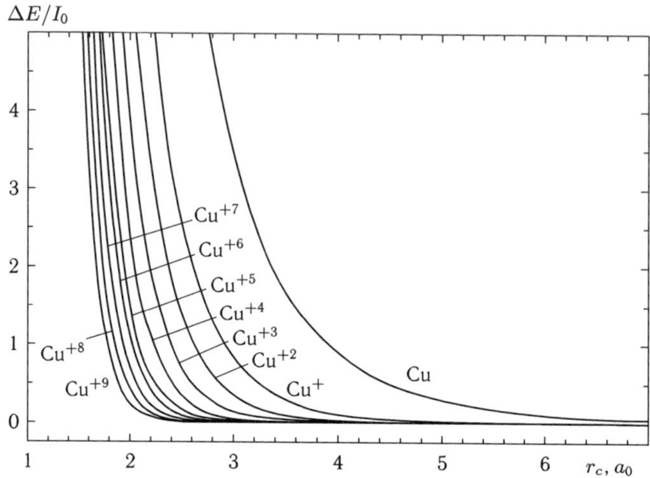


Fig. 4.41 Energy shifts of the ground states of copper atom and its ions. Calculation by the Hartree–Fock method.

Two procedures were used to determine the particle radii $r_j$. In the first procedure, the electron structure of atoms and ions was calculated in the bound-atom approximation (see Sec. 4.6) [234, 404] within which an atom (or an ion) was placed in a spherical cell with hard walls, and the Hartree–Fock method was used to calculate the electron structure of particle with a variable atomic cell radius (see Sec. 4.6). The results of such calculation are given in Fig. 4.41 which presents the energy shifts of the ground state of copper atom and ions depending on the cell radius. The further determination of the "effective" cell radius is based on the relation $\Delta E(r_i) = \text{constant} \cdot I_i$, where $\Delta E(r_i)$ is the ground-state energy shift, $I_i$ is the corresponding ionization potential, and $r_i$ is the atomic cell radius. In this case it was assumed that constant $= 1$.

A simpler procedure [389, 394] uses the assumption of closeness of the atomic structure to the hydrogen-like one. In this case each ion is assigned the corresponding size related to the ionization potential as

$$r_i \approx r_0 \left[ \frac{(z_i + 1)I_0}{I_i} \right], \qquad (4.228)$$

Here $r_0$ and $I_0$ are the atomic radius and ionization potential and $r_i$ and $I_i$ are the same for the $i$th ion, and $z_i$ is the ion charge. In real calculations both procedures were only used to determine the ratio of the radius of the atom and of the other ions. The reference value of atomic radius was determined in line with the so-called Ashcroft–Lekner rule [55] according to which the optimum hard-sphere size corresponding to the normal density is determined from the condition of fixed value of the packing parameter:

$$\nu \equiv 4\pi \sum \frac{n_i r_i^3}{3} \approx 0.45. \tag{4.229}$$

In accordance with Ref. [55] such a choice leads to the best coincidence of the position of first maximum of the pair correlation function of a hard-sphere system with the experimentally observed value for a whole number of simple metals in the liquid state. The calculations in the framework of the approximation (4.223)–(4.230) carried out earlier [390, 394] yield quite acceptable agreement with the experimental shock adiabats of metals for relatively high velocities of shock waves and, correspondingly, high pressures and temperatures.

Allowance for the short-range repulsion is essential for attaining this agreement. However, the performed calculations revealed a range of relatively low pressures [394] corresponding to the low velocities of shock waves where the approximation (4.223)–(4.230) does not, in principle, allow obtaining a solution of the Rankine–Hugoniot shock-adiabat equation [1081] for any set of radii $r_i$. This is explained by the fact that the approximation (4.223)–(4.229) does not contain a mechanism allowing for the binding energy responsible for the existence of the condensed state of matter. To improve the extrapolation properties of the approximation (4.223)–(4.229) in the description of lower regions of the shock adiabats it was supplemented with terms effectively allowing for the above-mentioned binding energy:

$$\Delta F = \Delta U = -A \left( \sum N_i \right)^{1+\delta} \cdot V^{-\delta}; \quad \Delta p = \delta \left( \frac{\Delta U}{V} \right); \quad V^{1+\delta};$$

$$\Delta \mu_i = -A(1+\delta)V^{-\delta} \left( \sum N_i \right)^{\delta}, \quad A, \delta = \text{constant}. \tag{4.230}$$

The corrections (4.230) are temperature-independent. The choice of $\delta = 1$ corresponds to the traditional form of Van der Waals approximation. The choice of $\delta = 1/3$ is considered to correspond better to the "metallic" (plasma) type of condensed matter binding. In this case summation is taken over all heavy particles, and so the corrections (4.230) do not violate the ionization equilibrium. According to the statement of Ref. [622] it is precisely such an exponent that corresponds to the form of effective attraction in "expanded" metals, i.e., metals of intermediate densities corresponding to the density at the critical point of gas–liquid transition.

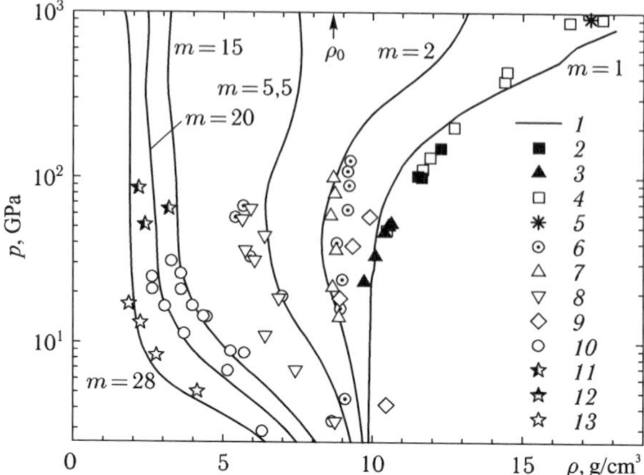



Fig. 4.42  Shock adiabats of porous nickel. Computational data: *1* calculation according to (4.223)–(4.230), summation in (4.230) was taken over all the heavy particles. Experimental data: *2* [678], *3* [1029], *4* [32], *5* [576], *6* [987] ($m = 2$), *7* [987] ($m = 2.32$); *8* [987] ($m = 5.62$), *9* [22], *10* [982], *11* [389], *12* [986] ($m = 20$), *13* [986] ($m = 28$).

The results of shock-adiabat calculations for porous nickel, copper, aluminum, and iron using the above-described model of the EOS (4.223)–(4.230) are presented in the figures to follow.

Figure 4.42 demonstrates the shock adiabats of porous nickel calculated earlier in Refs. [390, 394] and compared there with the experimental data (see Ref. [983]) obtained before. This comparison is supplemented in the present work with the results of additional measurements for porosity $m = 15$ and 20 [389] and with the latest experimental data for high-porosity samples ($m = 20$ and 28) [986]. We pay attention to the fact that the new experimental data agree well with the results of previous calculations. Figure 4.42 also gives the results of comparison for adiabats with low porosity (including the solid adiabat). The goal of this general comparison is to show that the quasi-chemical representation even in the above-described reduced version describes on the whole quite satisfactorily the whole set of experimentally measured shock adiabats of nickel.

Comparison of the calculations with the experimental data for shock adiabats of iron is illustrated in Fig. 4.43. The same as for nickel, one can state for iron that rather satisfactory agreement of theory with experiment for high porosities (and maximum reached degrees of expansion of initially condensed metal) gradually worsens with passing towards the region of increasingly dense plasma in experiments with compression of low-porosity

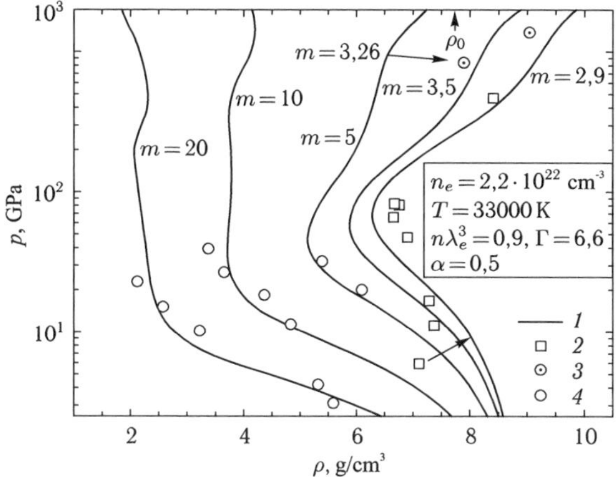


Fig. 4.43 Shock adiabats of porous iron. *1* calculation according to (4.223)–(4.230), summation in (4.230) was taken over all the heavy particles, *2* experimental data [987] ($m = 2.9$), *3* experimental data [987], *4* [402].

Table 4.2  Values of atomic and ionic radii.

| $Z^i$ | 0 | 1 | 2 | 3 | 4 | 5 | 6 | 7 |
|---|---|---|---|---|---|---|---|---|
| Cu | 2.00 | 1.700 | 1.55 | 1.40 | 1.25 | 1.10 | 0.95 | 0.8 |
| Ni | 2.00 | 1.684 | 1.27 | 1.10 | 1.00 | 0.84 | 0.80 | — |
| Fe | 2.05 | 2.00 | 1.58 | 1.18 | 1.075 | 0.978 | 0.904 | — |

samples. Note that the parameters of repulsion (combination of proper sizes of particles) and attraction used in the computational model in the description of new experimental data were chosen according to the same scheme as in the earlier computations [390, 394].

Table 4.2 gives the values of atomic and ionic radii (in atomic units) employed in the calculations of this work for copper, nickel, and iron.

Similar calculations were carried out for porous copper. Comparison of the computational and experimental data for shock-wave compressed copper and nickel is presented in Figs. 4.44, 4.45, and 4.46. In Fig. 4.44, the same as in the previous cases, the earlier results [390, 394] are supplemented with the new experimental data [389, 986]. On the whole, the comparison of the computational and experimental data for nickel, iron, and copper shows that even in an extremely simplified approximation the quasi-chemical representation provides a satisfactory description of the experimental data on

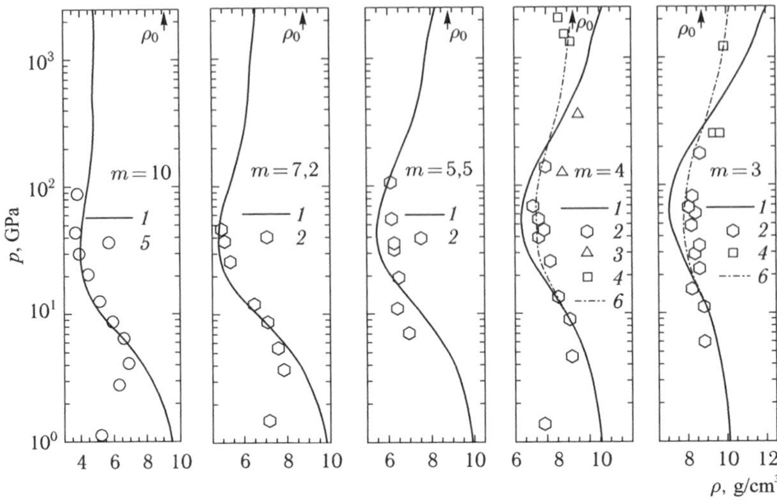

Fig. 4.44　Shock adiabats of porous copper. *1* calculation according to (4.223)–(4.230), summation in (4.230) was taken over all the heavy particles, *2* experiment [987], *3* experiment [576], *4* experiment [1098], *5* experiment [389], *6* calculation [392] by the model (4.223)–(4.230) with altered atomic and ionic radii ($r_\alpha = 2.0\alpha_0$, $r_c(\mathrm{Cu}^{+1}-\mathrm{Cu}^{+3} = 1.75\alpha_0)$).

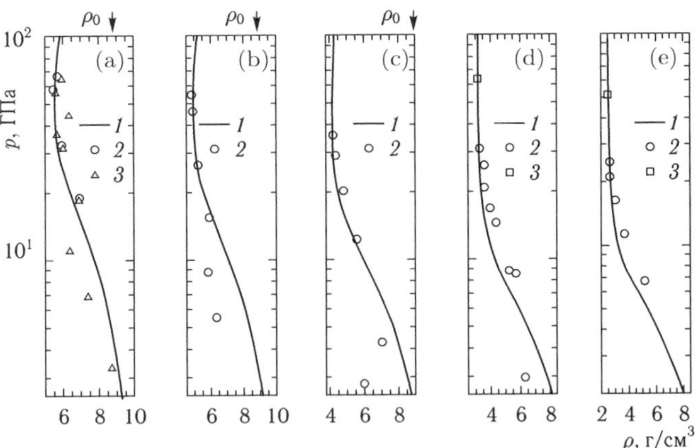

Fig. 4.45　Shock adiabats of high-porosity nickel $m \gg 1$) [390]: (a) $m = 5.45$, (b) $m = 7.2$, (c) $m = 10$, (d) $m = 15$, (e) $m = 20$ ($m = \rho_0/\rho_{00}$ is sample porosity) *1* calculation by the model (4.223)–(4.224), summation in (4.230) was taken over the atoms, *2* experimental data [982, 987], *3*(a) experimental data for $m = 5.62$ [982, 987], *4*(d, e) new experimental points for $m = 15$ and $m = 20$ [389].

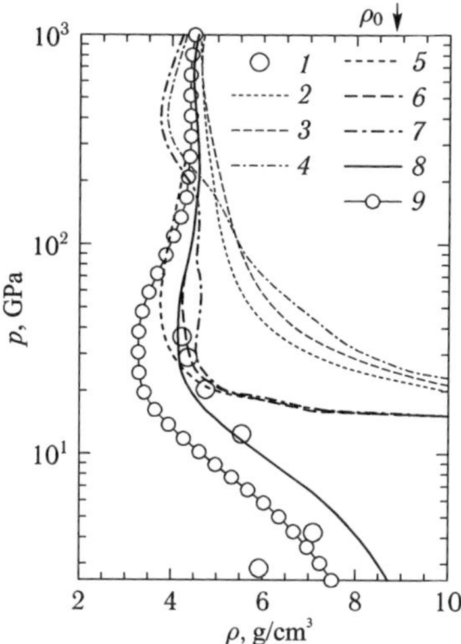


Fig. 4.46  Shock adiabats of porous nickel for $m = \rho_0/\rho_{00} = 10$ [390]. Comparison of computational results in different approximations. *1* experimental data [982, 987], *2* ideal plasma approximation (with allowance for only the ground states in calculating statistical sums), *3* the same as in *2* but with allowance for Coulomb interaction according to (4.225)–(4.227) *4* the same as in *3*, but with Planck–Larkin calculation of the atomic and ionic statistical sums, *5*, *6*, *7* the same as *2*, *3*, *4* but with allowance for short-range repulsion in the hard-sphere mixture approximation (4.225)–(4.229), *8* the same as *6* but with allowance for additional attraction according to (4.223)–(4.230) with $\delta = 1$ (for atoms only), *9* the same as *8* but with enlarged (+20%) atomic and ionic radii.

shock-wave compression of sufficiently high-porosity samples. This is confirmed by the fact that among the results presented in Fig. 4.45, the two experimental points corresponding to the maximum pressures on the adiabats with $m = 15$ and $m = 20$ were published [389, 986] after the main procedure of porous-nickel shock-adiabat calculations had been implemented; the results of the calculation are presented in Figs. 4.42, 4.45, and 4.46.

Good coincidence of the new data with the results of the earlier calculations is an additional proof of correctness of the initial prerequisites underlying the computational procedure based on the generalized chemical model with the purpose of extending its extrapolational resources. It should be stressed that the additionally obtained experimental data discussed here correspond not only to the maximum (for a given range of nickel porosity) pressures ($p \approx 50\,\mathrm{Pa}$) but, according to the present calculations, also to the maximum (among the experimentally attained) temperatures and degrees of ionization (for the shock-compressed nickel plasma). At the same time, as has already been said above, the extrapolational potentialities of the chemical model with the same pressures are gradually getting worse with approaching the range of adiabats of lower porosity and, accordingly, higher densities.

One can assert a generally satisfactory agreement between theory and experiment. At the same time, we see a difference between the computational and experimental data for the case of adiabat with $m = 10$ (Fig. 4.44) in the upper part of the experimentally achieved pressure range. Analyzing the wherefore of this difference, one should take into consideration an obvious simplicity of the theoretical model describing the interaction in the system and an extreme sensitivity of the results obtained with its help to a particular choice of combination of particle proper sizes. This is confirmed by the results of additional calculations carried out with a somewhat varied (enlarged) set of radii also presented in Fig. 4.47. These results show that the earlier obtained distinction between the experimental and computational data lies within the natural uncertainty of the chemical model in the parameter range under consideration. In particular, the authors of papers [392, 394] pointed out the specific behavior of shock data on porous copper adiabats with $m = 3$ and $m = 4$ (Fig. 4.44). To describe this specific behavior, one should use the model EOS with an efficiently higher "rigidity" (curve 6) than that given by the set of proper sizes chosen in Ref. [390,394].

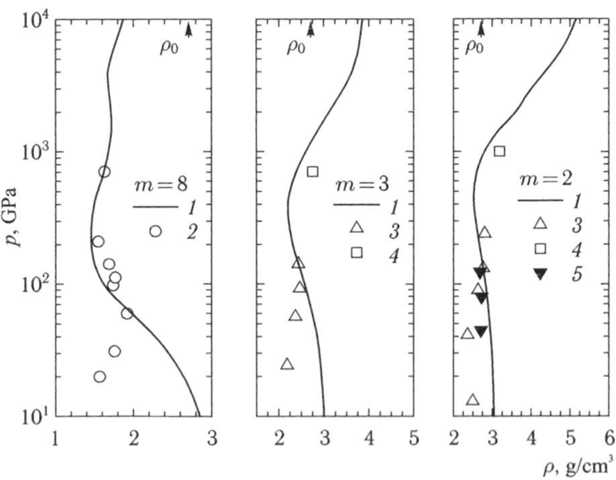


Fig. 4.47 Shock adiabats of porous aluminum ($m \equiv \rho_0/\rho_{00}$ is the original porosity of experimental samples) *1* calculations according to the model (4.223)–(4.231). Summation in (4.230) was taken over the atoms, *2* experiment [986], *3* experiment [71], *4* experiment [576], *5* experiment [1007].

Thus, one may conclude that the model description of thermodynamics of a shock-compressed substance is less universal and requires a higher degree of allowance for the individual features of each particular metal (for example, its electron structure) than the one used in the construction of the simplified method.

Note that the simplified scheme of the choice of the relative atomic and ionic size which gives satisfactory results for the description of shock-compressed porous iron, nickel, and copper does not yield acceptable results for the description of shock adiabats of porous aluminum. For this reason the above-described simplified procedure of determining the ratio of atomic and ionic radii was modified. As the initial data for calculating this ratio the results of the Hartree–Fock calculation of their electron structure were used (see Chapter 3). Calculated was their mean radius over all the electron shells with the use of their occupation numbers $g_{nl}$ and the radial wave functions $f_{nl}(r)$ for the atom and all the ions:

$$\langle r \rangle = \sum_{nl} \int_0^\infty r q_{nl} f_{nl}^2(r) dr. \tag{4.231}$$

The ratios of atomic and ionic radii which were obtained with the help of (4.231) were used as the initial parameters in the thermodynamic model (4.223)–(4.230).

Concluding this chapter we notice that the modern models of nonideal plasma dynamics have now reached a sufficient level of perfection having made accessible the thermodynamic description of states with exceedingly high (concentration, solid-state) densities and ultra–high megabar–gigabar pressures.

# Chapter 5

# Monte Carlo and Molecular Dynamics Methods

## 5.1 Monte Carlo method. Pseudopotentials

The theoretical methods based on perturbation theory become inapplicable in the region of strong interparticle interaction, which necessitates the application of machine Monte Carlo and molecular dynamics (MD) methods [421, 423, 797, 1071].

The numerical Monte Carlo method (MCM) which does not make use of small-parameter expansion is particularly efficient in the case of dense gases and liquids, as well as one-component plasma (OCP), i.e., systems with an exactly known form of the interparticle interaction potential. The method is based on first principles of classical statistical mechanics and on direct machine calculation of mean thermodynamic quantities [1087]:

$$\langle F \rangle = Q^{-1}(N, V, T) \int \ldots \int F_N(q) \exp[-\beta U_N(q)] d^N q, \qquad (5.1)$$

where

$$Q^{-1}(N, V, T) \int \ldots \int F_N(q) \exp[-\beta U_N(q)] d^N q,$$

is the configuration integral, $\beta = 1/kT$, and the interparticle interaction potential is assumed to be defined (and pair in most particular calculations):

$$U_N(q) = \sum_{ij} \Phi(r_{ij}, T), \quad r_{ij} = |q_i - q_j|. \qquad (5.2)$$

By the present time the MCM has been used to perform exhaustive calculations of thermodynamic and structure properties and phase boundaries of a large number of so-called "simple" systems obeying the classical Boltzmann statistics, i.e., ensembles of hard and soft spheres,

211

their mixtures, dense fluids with different interparticle potentials and even nuclear matter [104, 263, 1071], as well as OCP against the background of opposite-sign compensating charge, that is, thermodynamic systems with a given and doubtless interaction potential. Up to the machine-calculation error such calculations are in principle precise and play the role of "computer experiment" to be compared with analytical models of the equations of state.

We will not dwell here on the extensive literature on these problems (part of them was touched upon in Chapter 2), but will discuss the role and the ways of allowance for quantum effects in the framework of the classical Monte Carlo formalism.

The application of Monte Carlo technique to multicomponent plasma encounters specific difficulties of allowance for quantum effects playing a certain role in real plasma and causing the formation of bound states. We recall that it is precisely the introduction of quantum effects that ensures production of atoms, ions, and molecules thus providing stability of the system of negative electrons and positive nuclei for any parameters of the state of matter, but not only where the corresponding dimensionless letter criteria containing Planck constants are not small.

In the pseudo-potential plasma model, quantum effects are taken into account by introducing the effective electron–ion pair potential $\Phi_{\mathrm{ei}}(r, T)$ determined from the condition of equality of the quantum-mechanical probability of particle residence at a given spatial point to the classical Boltzmann exponent [234, 1087]:

$$\lambda_{\mathrm{e}}^3 \sum_a |\Psi_a(r)|^2 \exp(-\beta E_a) \equiv \exp[-\beta \Phi_{\mathrm{ei}}(r, T)], \qquad (5.3)$$

where $\Psi_a$ and $E_a$ are orthonormal wave functions and energy eigenvalues, $\lambda_{\mathrm{e}}$ is the electron thermal wavelength, and summation is carried out over all the states of the discrete and continuous spectra.

For large $r \gg \lambda_{\mathrm{e}}$, the thus determined pseudopotential coincides with the Coulomb one and as $r \to 0$ it has a finite value and depends on a particular electron structure of the element which is determined by a self-consistent solution of the quantum-mechanical many-body problem and cannot already be described by the pair approximation (5.3). A disregard of this fact [1087] can lead to the occurrence of nonphysical complexes because of too large a depth of the pseudopotential (5.3).

A considerable improvement of the pair potential model [1087] was attained by the following decomposition:

$$\exp[-\beta\Phi_{\mathrm{ei}}(r,T)] = S_{\mathrm{ei}}^{\mathrm{b}} + \exp[-\beta\Phi_{\mathrm{ei}}^{*}(r,T)]$$

$$\equiv \lambda_{\mathrm{e}}^{3} \sum_{E_a \lesssim -kT} |\Psi_a(r)|^2 (e^{-\beta E_a} - 1 + \beta E_a) + \lambda_{\mathrm{e}}^{3} \qquad (5.4)$$

$$\times \sum_{E_a \gtrsim -kT} |\Psi_a(r)|^2 \exp(-\beta E_a),$$

where bound states are taken into account by introducing the summand $S_{\mathrm{ei}}^{\mathrm{b}}$ determining the statistical sum of the atom according to Ref. [603], and the continuum is described by the electron–ion interaction pseudopotential $\Phi_{\mathrm{ei}}^{*}$. The thus constructed pseudopotentials in a nonCoulomb region depend weakly on temperature and also (since $\Phi_{\mathrm{ei}}^{*}$ is mainly determined by hydrogen-like states) on the sort of the chemical element. On the basis of similarity in temperature and closeness of pseudopotentials of different chemical elements a simple approximation [1087] was proposed underlying the pseudopotential zero-approximation plasma model:

$$\beta\Phi_{\mathrm{ei}}^{0}(x,T) = \begin{cases} -\varepsilon, & r \leq \sigma, \\ -x^{-1}, & r > \sigma, \end{cases} \quad \sigma = e^2\beta/\varepsilon, \quad x = r/\beta e^2, \qquad (5.5)$$

$$\beta\Phi_{\mathrm{ee}}^{0}(x) = \beta\Phi_{\mathrm{ii}}^{0}(x) = x^{-1},$$

where the numerical parameter $\varepsilon$ of the model is chosen with allowance for experimental data (Fig. 5.1).

On the whole, the applicability region of the pseudopotential model of low-temperature plasma turns out to be restricted as a result of disregard of many-particle interactions and the lack of in information on the discrete energy spectrum in expressions (5.3)–(5.5) which in dense plasma can be distorted because of the strong interaction and generally speaking is not known in advance. In weakly ionized plasma, along with the Coulomb interaction the interaction with participation of neutral particles becomes substantial. An approximate account of mutual repulsion is then realized by the soft- or hard-sphere model allowing for the effects of the proper volume of atoms and leading upon considerable compression to cold pressure ionization.

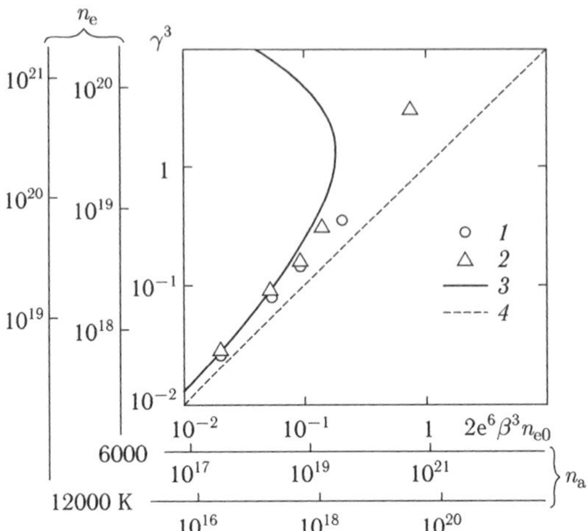

Fig. 5.1 Equation of ionization equilibrium in cesium plasma at $T = 6000$ and $12000$ K [1087]. *1* $\varepsilon = 2$, *2* $\varepsilon = 4$, *3* the Debye–Huckel law, *4* ideal plasma [147].

## 5.2 Quantum MCM. Path integrals

This method has been developed in the framework of the Feynman formulation of quantum mechanics [1087]. The main idea of this formulation consists in representation of the density matrix and, therefore, the thermodynamic quantities in the form of path integrals; the low-temperature density matrix for which no small physical parameters exist will be identically represented as a product of a large number of high-temperature density matrices: $e^{-\beta\hat{H}} = e^{-\Delta\beta\hat{H}} \cdot e^{-\Delta\beta\hat{H}} \ldots e^{-\Delta\beta\hat{H}}$, where $\Delta\beta = \beta/(n+1)$. This leads to the occurrence of intermediate coordinates the effective integration over which is performed by the Monte Carlo met$(n + 1)T)$ density matrices $\rho^{(i)}$ the expressions for which can be obtained by solving the Bloch equations in the high-temperature limit. Thus, each particle in plasma is represented by the point $(n + 1)$ (Fig. 5.2) and the whole configuration of particles is described by the vector

$$\mathbf{q} = \left\{ q_{1,e}^{(0)}, \ldots, q_{1,e}^{(n+1)}, q_{2,e}^{(0)}, \ldots, q_{2,e}^{(n+1)}, \ldots, q_{N_e,e}^{(n+1)}; q_{1,p}^{(0)}, \ldots, q_{N_p,p}^{(n+1)} \right\}$$

with dimension $3\left(N_e + N_p\right)\left(n + 1\right)$.

The spin effects are taken into account by the spin part of the density matrix $S$, and the exchange effects by the permutation operators $\hat{P}_e$ and $\hat{P}_p$ which act on the spatial $q^{(i+1)}$ and spin $\sigma'$ coordinates of electrons and

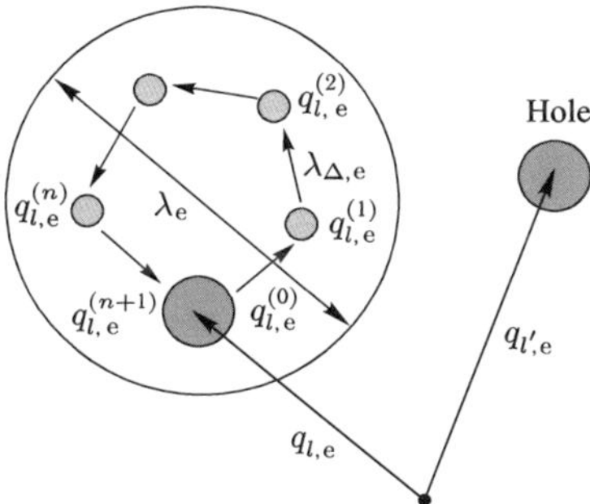


Fig. 5.2 Representation of an electron and a proton (hole) as a cloud of points. Here $\lambda_\mathrm{e}^2 = 2\pi\hbar^2\beta/m_\mathrm{e}$, $\lambda_{\Delta,\mathrm{e}}^2 = 2\pi\hbar^2\Delta\beta/m_\mathrm{e}$, $q^1_{l,\mathrm{e}} = q^0_{l,\mathrm{e}} + \lambda_{\Delta,\mathrm{e}}\xi^1_{l,\mathrm{e}}$, and $\sigma = \sigma'$. The proton has the same representation as a cloud of points, however, $\lambda_\mathrm{p}$ and all the linear scales are $\sqrt{m_\mathrm{p}/m_\mathrm{e}}$ times smaller in this case.

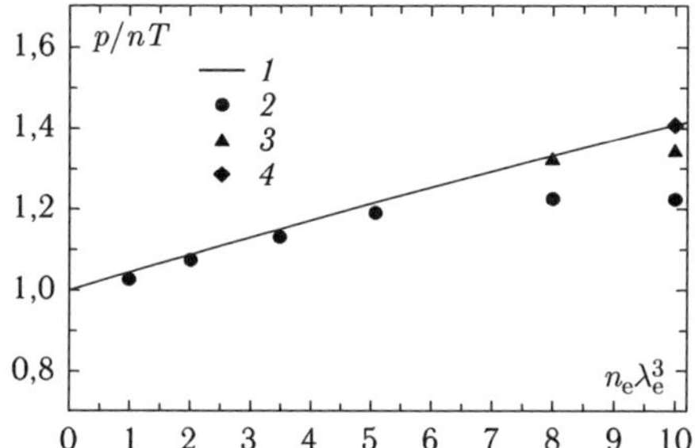


Fig. 5.3 Pressure of ideal plasma of degenerate electrons and classical protons depending on the degeneracy parameter. *1* theoretical dependence; quantum Monte Carlo calculations [1087]: *2* 16 electrons; *3* 32 electrons; *4* 64 electrons.

protons. The sums in expression (5.8) are taken over all the permutations with parities $\kappa_{P_\mathrm{e}}$ and $\kappa_{P_\mathrm{p}}$. The transformation of expression (5.8) to the form in which the sum over all the permutations is replaced by the determinant offers an approximate solution of the "sign problem". As an example, Fig. 5.3 shows the dependence of the calculated pressure for different numbers of particles on the degeneracy parameter $\chi = n_\mathrm{e}\lambda_\mathrm{e}^3$ ($\lambda_\mathrm{e}$ is the electron thermal wavelength, $\lambda_\mathrm{e}^2 = 2\pi\hbar^2\beta/m_\mathrm{e}$), as well as the theoretical dependence. The agreement is seen to be very good up to the value $\chi = 10$; it improves with increasing number of particles.

Figure 5.3 presents the results of quantum MCM calculations in which the exchange effects were only taken into account in the main Monte Carlo cell. The subsequent quantum MCM modification [1087] consisted in a more correct allowance for the degeneracy effects for high densities. For a quantum-particle thermal wavelength sufficiently large compared to the Monte Carlo cell, the exchange effects in the neighboring $3^3-1$, $5^3-1$, etc. cells were additionally taken into account. Comparison of the calculated pressure and energy for an ideal degenerate plasma showed satisfactory agreement with the theoretical dependences up to the value of degeneracy parameter $n_e\lambda_e^3$ of the order of 100. The way of "sign problem" solution presented above allowed the description of thermodynamic properties of ideally strong degenerate hydrogen plasma [266, 271].

Important in nonideal plasma are the interaction effects described by the factors $\rho^{(i)}$. The density matrix asymptotics in the high-temperature limit can be explicitly used as each of the factors $\rho^{(i)}$, $i = 1, \ldots, n+1$ in formula (5.8). Each such $N$-particle high-temperature density matrix can in turn be represented as a product of two-particle density matrices. For a high-temperature two-particle density matrix there exists an analytical solution to the Bloch equation in first-order perturbation theory [515]:

$$
\rho_{a,b}\left(r_a, r_a', r_b, r_b', \beta\right) = \frac{m_a m_b}{(2\pi\hbar\beta)^3} \exp\left[-\frac{m_a}{2\hbar^2\beta}\left(r_a - r_a'\right)^2\right]
$$

$$
\times \exp\left[-\frac{m_b}{2\hbar^2\beta}\left(r_b - r_b'\right)\right] \exp\left[-\beta\Phi^{ab}\right],
$$

where $\Phi^{ab}\left(r_a, r_a', r_b, r_b', \beta\right)$ can be approximated by the half-sum of diagonal pseudopotentials:

$$
\Phi^{ab}\left(|r_{ab}|, \Delta\beta\right) = \frac{e_a e_b}{\lambda_{ab} x_{ab}}\left\{1 - \exp\left(-x_{ab}^2\right)\right.
$$

$$
\left. + \sqrt{\pi} x_{ab}\left[1 - \mathrm{erf}\left(x_{ab}\right)\right]\right\}. \tag{5.6}
$$

Here, $x_{ab} = |r_{ab}|/\lambda_{ab}$, $\mathrm{erf}(x) = 2/\sqrt{\pi}\int_0^x \exp\left(-t^2\right)dt$ is the error function, $\lambda_{ab}^2 = \hbar^2\beta/2\mu_{ab}$, $e_a$, $e_b$ are particle charges, $m_a$ and $m_b$ are particle masses, and $\mu_{ab}^{-1} = m_a^{-1} + m_b^{-1}$ is a reduced mass. Precision of the obtained results is determined by the number of factors $n+1$ in the integrand of formula (5.8), by the temperature $T$, and the degeneracy parameter $\chi = \max\left(n_e\lambda_e^3, n_i\lambda_i^3\right)$:

$$
\varepsilon \sim (\beta\mathrm{Ry})^2 \chi/(n+1).
$$

In paper [693] it is shown that the pseudopotential (5.6) agrees well with the exact quantum potential at temperatures $T > 2\cdot10^5$ K. This means, in particular, that to model the Coulomb system at a temperature of $10^4$ K it suffices to choose $n = 20$; the relative error of the calculations does not exceed 5% in the worst case.

All the thermodynamic functions are expressed in terms of the derivatives of the statistical sum $Z$. In particular, for the pressure and the total system energy the following formulas hold

$$E = -\frac{\beta \partial \ln Z}{\partial \beta},$$

$$\beta P = \frac{\partial \ln Q}{\partial V} = \left[ \frac{\eta}{3V\partial} \ln \frac{Q}{\partial \eta} \right]_{\eta=1}.$$

The multiple integrals in these expressions are calculated by the standard Metropolis method in the cubic computational Monte Carlo cell with the use of periodic boundary conditions [1087].

## 5.3   H and H + He plasmas

The basic Monte Carlo simulations for hydrogen plasma have by the present time been performed in a wide range of degeneracy parameter values, $\chi = n_e \lambda_e^3$, and the nonideality parameter, $\Gamma = \beta e^2 / d$, where $d = (3/4\pi n_e)^{1/3}$ is the mean distance between hod [96, 163, 1087].

The explicit form of each high-temperature factor can be found by perturbation theory with an error inversely proportional at least to the square of the number of high-temperature factors. In this case, the error of the entire product of high-temperature density matrices $\varepsilon$ will be inversely proportional to the number of factors and, therefore, $\varepsilon \to 0$ as $n \to \infty$. Moreover, the error of the method can be estimated by comparison of the results obtained by varying the number $n$ of factors.

Thus, as distinct from the other methods of simulation of degenerate nonideal nonzero-temperature systems this method does not make use of additional physical assumptions and approximations, which allows "*ab initio*" analysis of different physical phenomena. In particular, definite evidence of the existence of plasma phase transition to a metallic state of strongly compressed hydrogen was obtained, and the conditions of the existence of quantum phase transition to a crystalline state of protons in compressed hydrogen and of sufficiently heavy holes in semiconductors were clarified.

The main idea of quantum MCM is presented, for example, in monograph [1087]. For hydrogen plasma at temperatures above $1\,\text{eV}$, protons can be thought of as classical, but for a consideration of the electron–hole or quark–gluon plasma (QGP) this approximation is invalid in the general case. The extension of the quantum MCM to the case where all the plasma components are quantum is discussed below.

The statistical sum of a quantum system at a temperature $T = 1/\beta k$ is expressed via its density matrix by the formula

$$Z\left(N_{\mathrm{e}}, N_{\mathrm{p}}, V, \beta\right) = \frac{1}{N_{\mathrm{e}}! N_{\mathrm{p}}!} \sum_{\sigma} \int_V dq \rho(q, \sigma; \beta), \qquad (5.7)$$

where $q = \{q_{\mathrm{e}}, q_{\mathrm{p}}\}$ and $\sigma = \{\sigma_{\mathrm{e}}, \sigma_{\mathrm{p}}\}$ stand for three-dimensional spatial coordinates and spin degrees of freedom $N_{\mathrm{e}}$ of electrons and $N_{\mathrm{p}}$ of protons (holes) ($N_{\mathrm{e}} = N_{\mathrm{p}}$), i.e., $q_a = \{q_{1,a}, \ldots, q_{l,a}, \ldots q_{N_a,a}\}$ and $\sigma_a = \{\sigma_{1,a}, \ldots, \sigma_{l,a}, \ldots, \sigma_{N_a,a}\}$, $a = \mathrm{e}$, p. The density matrix $\rho$ is unknown in the general case, but it can be calculated when represented in the form of a product of high-temperature factors which in the limit of an infinite number of factors passes over to the path integral:

$$\sum_{\sigma} \int dq^{(0)} \, \rho(q^{(0)}, \sigma; \beta) = \int dq^{(0)} \ldots dq^{(n)} \, \rho^{(1)} \cdot \rho^{(2)} \ldots \rho^{(n)}$$

$$\times \sum_{\sigma} \sum_{P_{\mathrm{e}}} \sum_{P_{\mathrm{p}}} (\pm 1)^{\kappa_{P_{\mathrm{e}}} + \kappa_{P_{\mathrm{p}}}} \, \mathcal{S}(\sigma, \hat{P}_{\mathrm{e}} \hat{P}_{\mathrm{p}} \sigma_a')$$

$$\times \hat{P}_{\mathrm{e}} \hat{P}_{\mathrm{p}} \rho^{(n+1)} \Big|_{q^{(n+1)} = q^{(0)}, \, \sigma' = \sigma}, \qquad (5.8)$$

where $\rho^{(i)} = \rho\left(q^{(i-1)}, q^{(i)}; \Delta\beta\right) = \left\langle q^{(i-1)} \middle| e^{(-\Delta\beta \hat{H})} \middle| q^{(i)} \right\rangle$.

The Hamiltonian $\hat{H} = \hat{K} + \hat{U}$ contains the kinetic energy ($\hat{K}$) and the potential interaction energy ($\hat{U} = \hat{U}_{\mathrm{pp}} + \hat{U}_{\mathrm{ee}} + \hat{U}_{\mathrm{ep}}$). The index $i = 1, \ldots, n+1$ in (5.7) numerates the high-temperature (taken at a temperature electrons with density $n_{\mathrm{e}}$.

Of the main interest is the dashed area $n_{\mathrm{e}} - T$ of the hydrogen plasma diagram (Fig. 5.4) where the correlation and quantum effects are important which cause possible manifestations of interesting physical effects, for instance, the occurrence and decay of bound states, Mott transition, and plasma metallization, proton crystallization, etc. The point $a$ in Fig. 5.4 corresponds to weakly nonideal strongly degenerate plasma when the density and temperature are high. Under conditions of point $b$ the plasma is degenerate and nonideal. The point $c$ corresponds to weak nonideality and weak degeneracy. At the point $d$ nonideality is strong and degeneracy is weak.

To analyze the possibility of the appearance of new physical phenomena in plasma, e.g., such as partial dissociation and ionization, Mott transition,

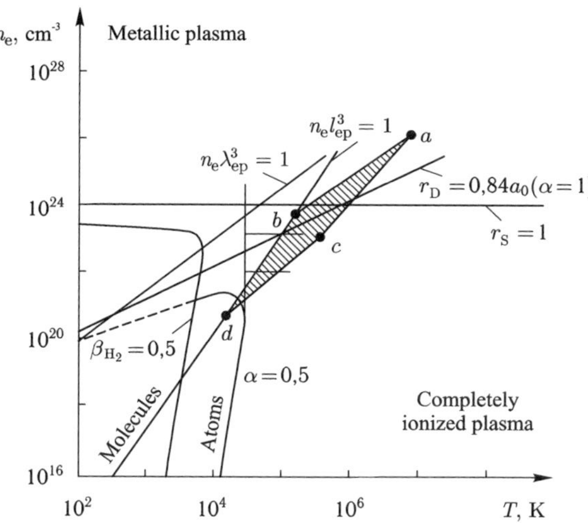


Fig. 5.4   Qualitative $n_{\rm e} - T$ diagram of hydrogen plasma. The parameters $r_{\rm D}$, $a_{\rm B}$, $r_{\rm S}$, $l_{\rm ep}$, $\lambda_{\rm ep}$ denote the Debye radius, the Bohr radius, the Brueckner parameter, the Landau length, and the thermal wavelength of an electron; $\alpha$ and $\beta$ are dissociation and ionization coefficients, respectively.

proton ordering at high density etc., we shall consider the behavior of pair correlation functions (Figs. 5.5 and 5.6). In Fig. 5.5(a) at $T = 20000\,{\rm K}$ and $n = 10^{22}\,{\rm cm}^{-3}$ the proton–proton ($g_{\rm ii}$) and electron–electron ($g_{\rm ee}$) correlation functions have sharp maxima for one and the same coordinate value $1.4\,a_{\rm B}$, which testifies to the existence of hydrogen molecules in these conditions. For this reason, the maximum of the electron–proton ($g_{\rm ie}$) correlation function (multiplied by $r^2$) is displaced towards coordinate values high compared to the Bohr radius $a_{\rm B}$. With density lowering the molecules disappear, and the maximum of $g_{\rm ie}$ appears for $r = a_{\rm B}$. Figure 5.5(b) shows that with increasing temperature to $T = 50000\,{\rm K}$ the maxima of $g_{\rm ii}$ and $g_{\rm ee}$ smear, which testifies to the disappearance of molecules (dissociation).

New physical phenomena are observed as the density increases at the same temperature $T = 50000\,{\rm K}$. Figures 5.5(b), 5.6(a), and 5.6(b) clearly show the character of variation of the proton–proton correlation function. With increasing concentration the curve typical of partially ionized plasma (see Fig. 5.5(b)) is transformed to the one characteristic of liquid (Fig. 5.5(a)) and solid (Fig. 5.6 (b).

Of great interest is also to watch the behavior of the electron–electron correlation function with increasing density: $g_{\rm ee}$ in Fig. 5.5(b) is typical of partially ionized plasma, whereas Fig. 5.6(a) exhibits a sharp increase in

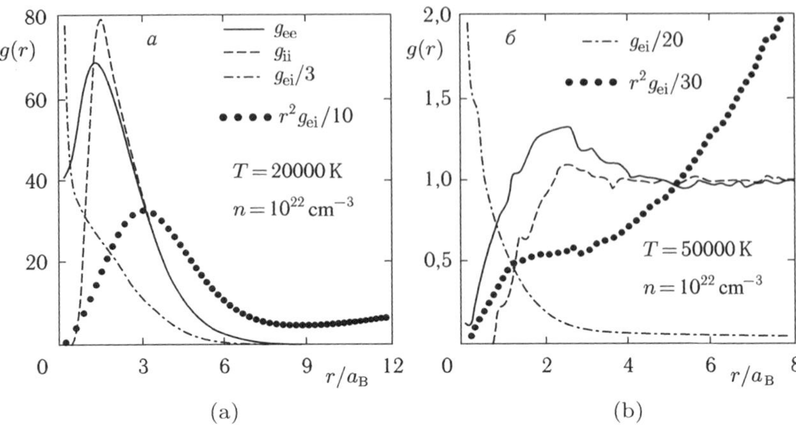

Fig. 5.5　The correlation functions of dense hydrogen plasma: the electron–electron $g_{ee}$, proton–proton $g_{ii}$, and electron–proton $g_{ie}$ plasmas. The values of nonideality, degeneracy, and, Brueckner parameters: (a) $\Gamma = 2.9$, $\chi = 1.46$, $r_s = 5.44$; (b) $\Gamma = 1.16$, $\chi = 0.37$, $r_s = 5.44$.

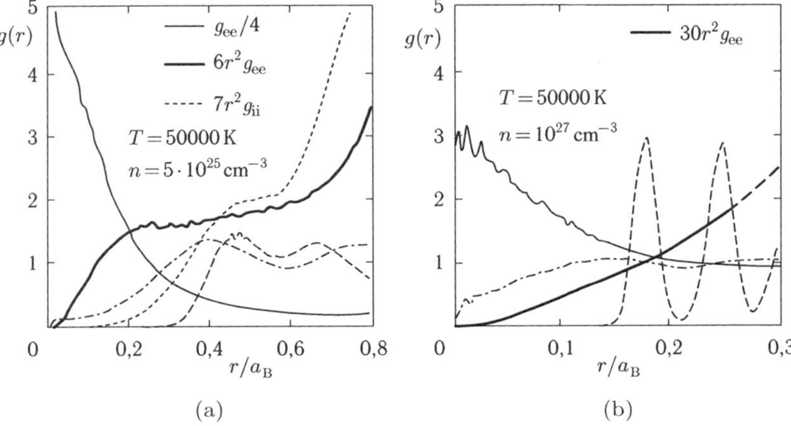

Fig. 5.6　The correlation functions of dense hydrogen plasma [271, 272]. The notation is the same as in Fig. 5.5. The values of nonideality, degeneracy, and Brueckner parameters: (a) $\Gamma = 19.8$, $\chi = 1848$, $r_s = 0.318$; (b) $\Gamma = 53.8$, $\chi = 37000$, $r_s = 0.117$.

the range of low coordinate values. A further increase in the concentration induces a more homogeneous electron distribution in space (Fig. 5.6(b)). To analyze this phenomenon, the functions $r^2 g_{ee}$ and $r^2 g_{ii}$ are also drawn in Fig. 5.6(a). The plateau on the curve $r^2 g_{ii}$ is approximately two times narrower than that on $r^2 g_{ee}$, which means that the most probable spacing between electrons is two times smaller than that between protons. The examination of the electron trajectories for this case shows that we deal with coupling of electrons with opposite spins. The "size" of the electrons is approximately equal to the spacing between protons, which leads to a partial overlapping of coupled electrons.

As an example we shall now consider the equation of hydrogen state at a sufficiently high temperature [266]. Figure 5.7 presents the density dependence of hydrogen plasma pressure at $T = 50000\,\mathrm{K}$. One can readily see that the pressure curve behaves monotonically and the results of calculations agree well with the quantum Monte Carlo simulations with restrictions [693] in the moderate density range.

The situation becomes different as the temperature lowers. Figure 5.8 demonstrates the density dependence of hydrogen plasma pressure at $T = 10000\,\mathrm{K}$. The revealed region of poor convergence to the equilibrium state at densities of $0.1$–$1.5\,\mathrm{g/cm^3}$ is indicative of the existence of phase transition. Since the isotherm $T = 50000\,\mathrm{K}$ exhibits no such anomalies, the phase transition must have a critical point at approximately $T \approx 30000\,\mathrm{K}$. Some other arguments may confirm the existence of phase transition: the calculation by the density functional method in the region of crystalline hydrogen stability with density lowering leads to anomalies in the ion–ion correlation function for $\rho = 0.8\,\mathrm{g/cm^3}$ [1049] (approximately the upper transition boundary in density); the maximum density for which the quantum Monte Carlo simulations with restrictions [693] could be carried out is equal to $\rho = 0.15\,\mathrm{g/cm^3}$ (approximately the lower transition boundary

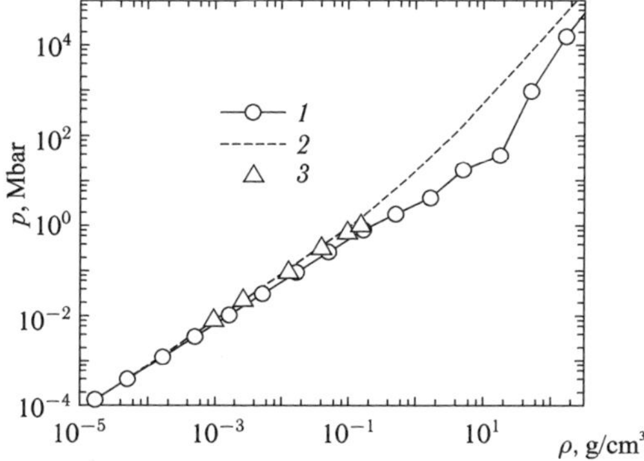


Fig. 5.7   Hydrogen plasma pressure as a function of density at $T = 50\,000\,\mathrm{K}$. *1* Quantum Monte Carlo simulations [266]; *2* theoretical dependence for ideal plasma; *3* quantum Monte Carlo simulations with restrictions [693].

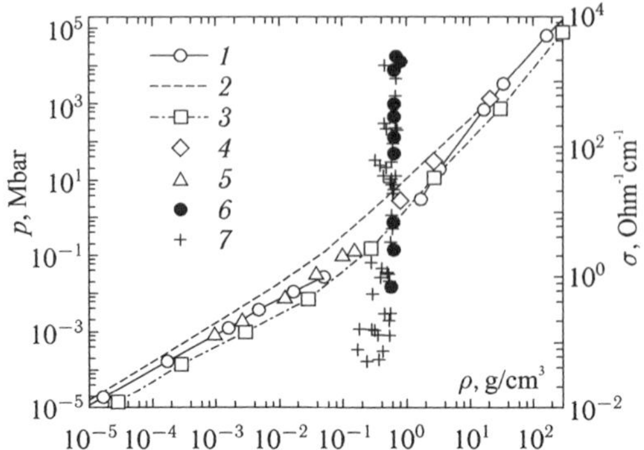

Fig. 5.8  Pressure and electrical conduction of hydrogen plasma as a function of density at $T = 10000$ K. *1* Quantum Monte Carlo simulations  [266]; *2* analytical dependence for ideal plasma; *3* quantum Monte Carlo simulations of helium (33%) and hydrogen (67%) mixture [266]; *4* calculation by the density functional method [1049]; *5* quantum Monte Carlo simulations with restrictions [693] and *6*, *7* experimental electrical conduction of hydrogen plasma, right axis, *6* [1037], *7* Ref. [967].

in density). At the same time, the quantum Monte Carlo simulation for the mixture of 33% of helium and 67% of hydrogen at the same temperature revealed no singular features in the behavior of the equation of state (curve *3* in Fig. 5.8).

A sharp increase in the hydrogen conductance (by 4 to 5 orders of magnitude) was fixed in shock-wave experiments in a very narrow density range, $0.3\text{--}0.5\,\text{g/cm}^3$ at temperatures of 5000–15000 K (symbols *6*, *7* in Fig. 5.8). Finally, in quantum Monte Carlo simulations, the system of particles was observed to split into drops in a Monte Carlo cell.

Of great interest is also the quantum Monte Carlo study of thermodynamic properties of a hydrogen–helium mixture [271]. The results were compared with the chemical plasma model [883,884]. The model makes use of Boltzmann statistics for molecules and ions and Fermi–Dirac statistics for electrons. Furthermore, the model [883,884] takes into account a large number of physical effects, including dissociation and ionization, the interaction between charged particles, atoms and molecules, ion screening at high densities, excitation of molecular electron levels, as well as some other "second-order" phenomena. Because of complicacy of this model the equations of hydrogen and helium state are presented as tables [884]. For this reason, the hydrogen and helium thermodynamic properties were calculated with a relatively simple interpolation program. The hydrogen–helium plasma properties can be obtained within the "linear mixture" method. For example, the energy $E(p, T)$ of the hydrogen–helium mixture with the helium mass concentration $Y = m_{He}/(m_{He} + m_H)$ at a pressure $p$ and a temperature $T$ can be found by the formula $E(p, T) = (1 - Y)E_H(p, T) + YE_{He}(p, T)$, where $E_H(p, T)$ is the internal hydrogen energy and $E_{He}(p, T)$ is the same for helium. Obviously, the linear mixture model works well for weakly interacting mixtures only, when the components do not affect each other noticeably.

Figure 5.9 demonstrates the temperature dependences of pressure and internal energy of hydrogen plasma on the isochores $n_e = 10^{20}$, $10^{21}$, and $10^{22}\,\mathrm{cm}^{-3}$. One can see that the results of quantum Monte Carlo simulations are well consistent with the results of [884] in the entire temperature range. The internal energies calculated by the two models virtually coincide up to temperatures of about $3 \cdot 10^4\,\mathrm{K}$, but for the temperature of $10^4\,\mathrm{K}$ the energy values obtained by the quantum MCM for isochores $10^{20}$, $10^{21}\,\mathrm{cm}^{-3}$ are much higher than the curves obtained by the method of [884]. Figure 5.9(b) also shows the relative concentration of hydrogen atoms as a function of energy. One can see that in the range of $10^4 - 3 \cdot 10^4\,\mathrm{K}$ the plasma consists mostly of atoms. This fact leads to the internal energy lowering. Higher energy values obtained in quantum Monte Carlo simulations can be explained by the presence in the system under these conditions of two characteristic strongly differing scales: intra-atomic and interatomic. To accumulate statistics in such systems, in simulation it is necessary to generate a very large number of configurations which requires a long time

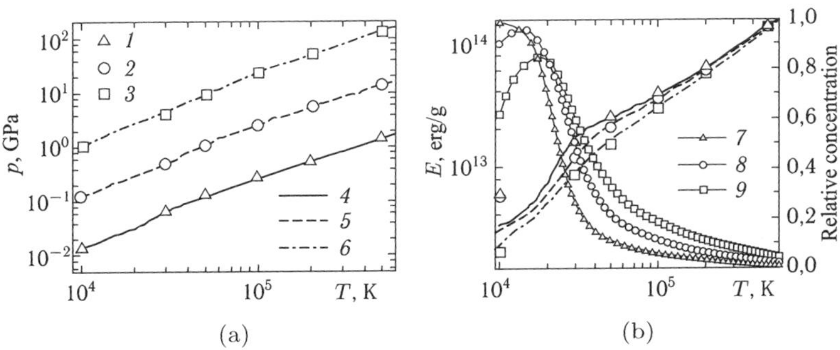

(a)      (b)

Fig. 5.9 Temperature dependence of pressure (a) and internal energy (b) of hydrogen plasma. Quantum Monte Carlo simulations [271]: *1* isochore $10^{20}$, *2* $10^{21}$, *3* $10^{22}\,\mathrm{cm}^{-3}$. Calculations by the chemical model [884]: *4* isochore $10^{20}$, *5* $10^{21}$, *6* $10^{22}\,\mathrm{cm}^{-3}$. Curves *7*, *8*, *9* in Fig. 5.9(b) the fraction of hydrogen atoms relative to the total number of particles in plasma (right axis) depending on temperature on isochores $10^{20}$, $10^{21}$, and $10^{22}\,\mathrm{cm}^{-3}$, respectively [884].

of computation. The decrease in atomic concentration on the $10^{22}\,\mathrm{cm}^{-3}$ isochore at $T = 10^4\,\mathrm{K}$ with a simultaneous decrease in the intermolecular spacing leads to agreement between the internal energy simulated by MCM under these conditions and the calculations by the chemical plasma model [884] (see Fig. 5.9(b)).

The isotherms of hydrogen–helium plasma with two different helium mass concentrations were also calculated. The relatively low helium concentration in the mixture ($Y = 0.234$) adopted in the calculations corresponded to its content in the upper layers of Jupiter atmosphere. Isotherms from $10^4$ to $2\cdot10^5\,\mathrm{K}$ were calculated in the electron concentration range from $10^{20}$ to $3\cdot10^{24}\,\mathrm{cm}^{-3}$. The interest in the thermodynamic properties of a mixture with such helium content was attracted by the experimental work [969] in which the hydrogen–helium plasma conductance was measured under quasi-isentropic compression due to pressure-induced ionization.

Figure 5.10 demonstrates isotherms $4\cdot10^4$, $5\cdot10^4$, $10^5$, and $2\cdot10^5\,\mathrm{K}$. In the figure, good agreement with the model [884] is again observed for pressure and satisfactory agreement for internal energy. The isotherm $T = 10^5\,\mathrm{K}$ of ideal plasma of degenerate electrons and classical protons and $\alpha$-particles is depicted for comparison. This isotherm approaches the one calculated by the chemical model at low and high densities, and in the region of bound state formation the pressure and energy of ideal plasma is higher than those of nonideal plasma. In both models, the isotherms of the dependences of nonideal plasma energy on electron concentration exhibit minima due to atom production.

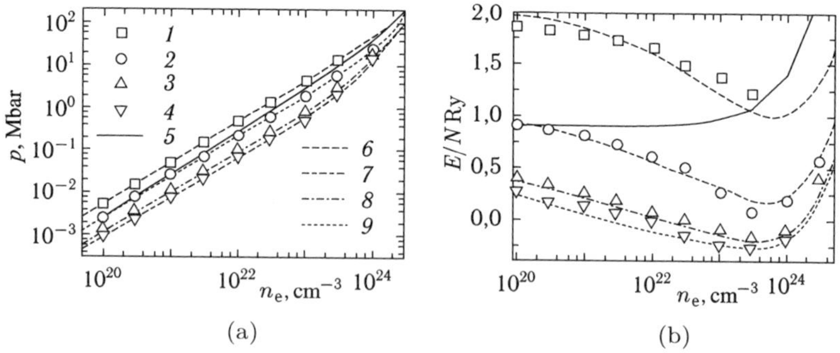

(a)      (b)

Fig. 5.10 Pressure (a) and internal energy per particle in Ry (b) of hydrogen–helium plasma depending on electron concentration with helium mass concentration $Y = 0.234$ [271]: *1* $2\cdot10^5$, *2* $10^5$, *3* $5\cdot10^4$, *4* $4\cdot10^4$ K. Ideal plasma: *5* $10^5$ K. Model [884]: *6* $2\cdot10^5$, *7* $10^5$, *8* $5\cdot10^4$, *9* $4\cdot10^4$ K.

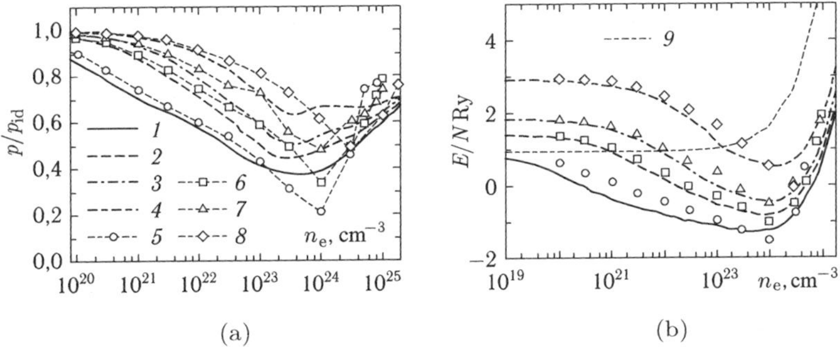

Fig. 5.11 Pressure relative to the ideal plasma pressure (a) and internal energy per particle in Ry (b) of hydrogen–helium plasma depending on the concentration of electrons with helium mass concentration $Y = 0.988$ [271]. Model [884]: *1* 100, *2* 156, *3* 200, *4* 312 kK. MCM: *5* 100, *6* 156, *7* 200, *8* 312 kK. Energy per particle in ideal plasma: *9* $10^5$ kK.

Calculations were also carried out for hydrogen–helium plasma with a high helium mass content $Y = 0.988$. It should be noted that because of the high value of the potential of the first helium ionization the calculations at low temperatures require a large number of high-temperature divisions in the statistical sum representation as the path integral, and therefore the temperatures of modeling for $Y = 0.988$ were chosen to be not lower than $10^5$ K. Figure 5.11(a) presents the pressure referred to the ideal plasma pressure on four isotherms depending on the electron concentration and Fig. 5.11(b) the specific internal energy per particle. One can see that the ionization minima are well reproduced, and the calculated points satisfy the asymptotics for classical ideal plasma (for low concentrations). Good agreement with the model [884] should also be noted.

For temperatures below $4 \cdot 10^4$ K agreement between the results of calculations [269, 271] and by the model [884] for $Y = 0.234$ became worse (Fig. 5.12). This is caused by both poor applicability of the chemical model for nonideal degenerate plasma and by the errors of the linear mixture model. Regions of poor convergence to the equilibrium state were revealed on the isotherms $2 \cdot 10^4$, $1.5 \cdot 10^4$, and $10^4$ K as before for hydrogen plasma [272]. One of these regions in the density range from 0.5 to 5 g/cm$^3$ coincides with the position of the earlier predicted [883, 884, 892] plasma phase transition in hydrogen and hydrogen–helium plasma for low helium concentrations. Moreover, a sharp increase in conductivity at temperatures of $\approx 5 \cdot 10^3$ K is observed in the region of 0.0.83 g/cm$^3$ [969]. However, the isotherms $10^4$ K and $1.5 \cdot 10^4$ K show one more region 0.01–019 g/cm$^3$ in

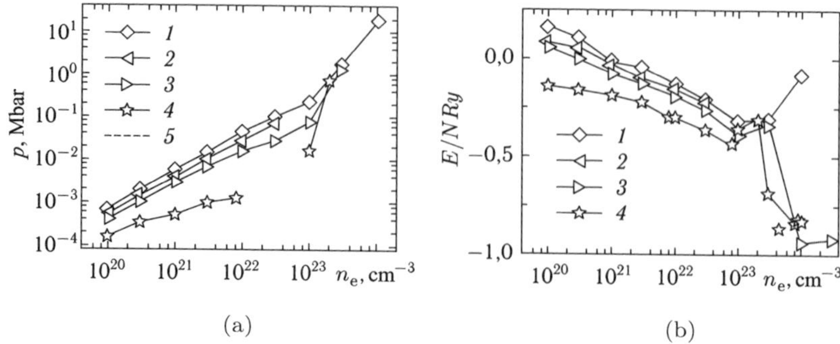

Fig. 5.12   Pressure (a) and internal energy per particle in Ry (b) of hydrogen–helium plasma with helium mass concentration $Y = 0.234$ depending on the electron concentration. Paper [271]: *1* 30, *2* 25, *3* 20, *4* 10 kK. Model [884]: *5*–30 kK.

which the pressure became negative in simulations. Two plasma phase transitions for helium mass concentrations $Y > 0.93$ [892] can exist simultaneously for a hydrogen–helium mixture, but in quantum Monte Carlo simulations the helium content was rather low $Y = 0.234$. Additional examination is needed to explain this phenomenon.

The properties of plasma phase transition in hydrogen–helium plasma are determined considerably by the helium concentration [883, 884, 892]. Such a transition is often associated with a sharp increase in plasma conductance (pressure-induced ionization). The range of temperatures and pressures in which the conductance increases sharply can be estimated from shock-wave experiments. The transition from a low-conducting to a high-conducting state under quasi-isentropic hydrogen compression proceeds at $T \approx 3\text{--}15$ kK and $\rho \approx 0.4\text{--}0.7$ g/cm$^3$ [967, 1037], whereas for helium the transition occurs at $T \approx 140$ kK and $\rho \approx 0.7\text{--}1.25$ g/cm$^3$ [968]. According to the theoretical models the critical point of plasma phase transition makes up $T_c(\mathrm{H}) \approx 12\text{--}19$ kK, $p_c(\mathrm{H}) \approx 0.2\text{--}0.9$ Mbar [234, 883, 892]. According to different data, the critical parameters of plasma phase transition in pure helium are $T_c(\mathrm{He}) \approx 17$ kK or $T_c(\mathrm{He}) \approx 35$ kK and $p_c(\mathrm{He}) \approx 7$ Mbar [892]. For a hydrogen–helium mixture with the helium mass fraction of 0.93 the critical transition temperature was estimated as $T_c(\mathrm{He}) \approx 35$ kK, but the dependence of the critical parameters on the plasma composition has not been studied. According to the quantum-statistical model [892], for the helium mass concentration $Y < 0.93$ and a temperature below the critical temperatures of transitions in pure hydrogen and helium, the properties of hydrogen–helium plasma are largely determined by hydrogen, and only

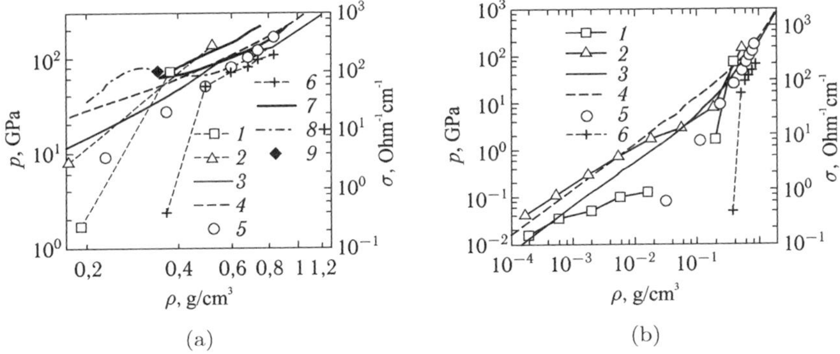

Fig. 5.13   Theoretical and experimental results in the region of assumed plasma phase transition for $Y = 0.234$. Paper [271]: *1* 10 kK, *2* 20 kK. Model [884]: *3* 10 kK, *4* 20 kK. Experiment [969]: *5* quasi-isentrope of hydrogen–helium mixture, $T \approx 5000$ K, *6* electrical conduction of hydrogen–helium mixture on the quasi-isentrope (right axis). (b) *7* boundaries of plasma phase transition in hydrogen plasma [883, 884], *8* subcritical metastable isotherm $T = 12000$ K for $Y = 0.308$ [892], *9* critical point of plasma phase transition in hydrogen–helium plasma for $Y = 0.308$ [892].

one plasma phase transition exists. For high helium mass concentration, $Y > 0.93$, two transitions exist. The Monte Carlo simulations revealed one thermodynamic instability region at $T = 2{\cdot}10^4$ K and two instability regions at $T = 10^4$ K for a low helium concentration, $Y = 0.234$. The results of simulation are given in Fig. 5.13 together with the experimental data and theoretical predictions.

A region of poor convergence to the equilibrium state in the density range from 0.5 to 5 g/cm$^3$ was observed on the isotherm $T = 2{\cdot}10^4$ K. On the isotherms $T = 1.5{\cdot}10^4$ K and $10^4$ K this range extends beginning with 0.38 g/cm$^3$. It is of interest that one more density range showed up in which the pressure became negative in modeling: from 0.015 g/cm$^3$ to 0.19 g/cm$^3$. Physically this phenomenon is due to the formation of many-particle clusters and requires additional study. Figure 5.13 clearly demonstrates that other predictions of plasma phase transition in hydrogen and hydrogen–helium plasma with a low helium content [884, 892] get into the region with poor convergence of the quantum MCM. A rapid increase in the hydrogen–helium plasma conductance along the quasi-isentropy with the initial state $T = 77.4$ K and $p = 8.1 \cdot 10^{-3}$ GPa is also observed experimentally in the density range of 0.5–0.83 g/cm$^3$ [969]. Unfortunately, the quantum MCM cannot yet give reliable results concerning the exact position and properties of plasma phase transition in hydrogen and hydrogen–helium plasma.

The plasma phase transition occurs in different chemical plasma models at approximately the same densities. However, such approaches become more complicated in the region of pressure ionization and dissociation where exceedingly important is a correct allowance for all types of interactions. Furthermore, a possible cluster formation is neglected.

The approach [270, 271] has no such shortcomings. The results of calculation in the region of plasma phase transition show that alternative approaches may exist to explain an exceedingly sharp increase in conductance in this region. The theoretical models predicted earlier that the region of conductance increase appears either for lower densities (the model of hopping conduction in molecular liquids) or for higher densities (free electron gas conduction). However, an increase in the electrical conduction occurs precisely in the region of plasma phase transition, which allows us to take into consideration one more conduction mechanism — the hopping conduction between liquid metallic drops. Obviously, such an effect will only take place between regions where the two other mechanisms dominate.

## 5.4 Shock adiabat of deuterium

The quantum MCM was used to calculate the shock adiabat of deuterium by calculating its thermodynamic properties [270]. The pressure $p$, the specific volume $v$, and the specific internal energy $E$ of matter behind the shock front are related to the initial state $(p_0, v_0, E_0)$ by the Rankine–Hugoniot equation [1081]:

$$E - E_0 + \frac{1}{2}\left(v - v_0\right)\left(p + p_0\right) = 0. \tag{5.9}$$

In line with [692] the parameters corresponding to the state of liquid deuterium under normal conditions were chosen as initial: $p_0 = 0$, $1/v_0 = \rho_0 = 0.171\,\mathrm{g/cm^3}$, and $E_0 = -15.886\,\mathrm{eV}$ per atom. Then calculated were the values of pressure $p_i$ and internal energy $E_i$ at a given temperature $T$ (from $10^4$ to $10^6\,\mathrm{K}$) and three different values of specific volume $v_i$ corresponding to $r_s = 1.7$, 1.86, and 2, where $r_s = \langle r \rangle / a_\mathrm{B}$, $\langle r \rangle = (3/4\pi n_e)^{1/3}$, and $a_\mathrm{B}$ is the Bohr radius. The results of simulation are presented in Table 5.1. The table only shows the pressure and internal energy values for those specific volumes between which the shock adiabat passes at a given temperature. The last two columns give the density $\rho_\mathrm{H}$ and pressure $p_\mathrm{H}$ of deuterium on the shock adiabat.

Table 5.1   Thermodynamic properties and shock adiabat of deuterium.

| $T$, K | $r_s = 1.7$ | | $r_s = 1.86$ | | $r_s = 2$ | | $\rho_H$, g/cm$^3$ | $p_H$, Mbar |
|---|---|---|---|---|---|---|---|---|
| | $p$, Mbar | $E$, eV | $p$, Mbar | $E$, eV | $p$, Mbar | $E$, eV | | |
| 15625 | 2.27 | −19.00 | 1.01 | −9.69 | — | — | 0.854 | 1.113 |
| 31250 | 1.86 | −9.95 | — | — | 1.34 | −6.02 | 0.837 | 1.605 |
| 62500 | — | — | 3.14 | −1.23 | 2.61 | −0.18 | 0.810 | 3.067 |
| $1.25 \cdot 10^5$ | — | — | 7.96 | 17.27 | 6.22 | 17.07 | 0.740 | 7.008 |
| $2.5 \cdot 10^5$ | — | — | 15.97 | 48.22 | 12.38 | 46.85 | 0.720 | 13.305 |
| $5 \cdot 10^5$ | — | — | 32.62 | 112.73 | 26.45 | 114.57 | 0.708 | 27.973 |
| $10^6$ | — | — | 67.66 | 245.99 | 54.40 | 246.45 | 0.698 | 56.722 |

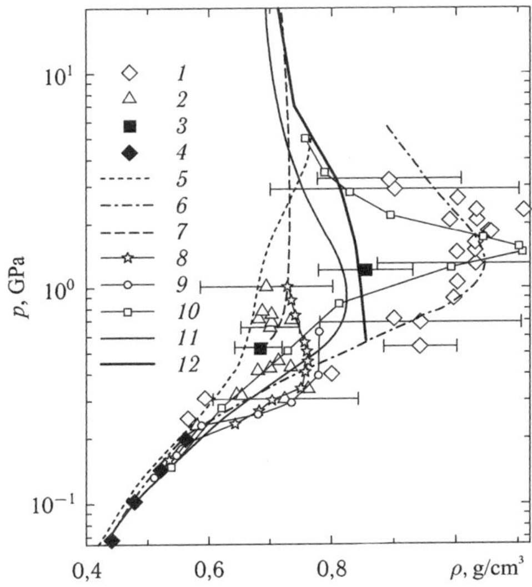

Fig. 5.14   Shock adiabat of deuterium. Experimental data: *1* [188, 200], *2* [556], *3* [92], *4* [736]. Calculation: *5* [899], *6* [860], *7* [692], *8* [208], *9* [122], *10* [552], *11* [103], *12* [270].

Figure 5.14 shows the data of experimental, theoretical, and computational works devoted to shock deuterium compressibility according different authors. The measurements taken on NOVA setup by means of laser-induced shock wave generation [188, 200] testify to the fact that the deuterium density behind the shock front can increase more than six times. However, at high pressures the limiting degree of compression $\rho/\rho_0$ must not exceed 4 for an ideal monatomic gas, and since the deuterium dissociation in a shock wave proceeds at pressures $p \approx 0.5\,\text{Mbar}$ [733], the author of paper [733] arrived at the conclusion that the data of [188, 200] are invalid.

The experiments where the aluminum foil was accelerated by a magnetic field to velocities exceeding 20 km/s [556] exhibit a much lower degree of compression than that in papers [188, 200]. The data of Refs. [188, 200] do not agree with the data of [556] within the experimental error. As distinct from papers [188, 200] and [556], in which the samples were several hundred microns thick, the shock compressibility of solid deuterium on a semispherical exploder [92] was measured in a substance layer nearly 4 mm thick. The experimental point obtained at a pressure of the order of 1 Mbar occupies an intermediate position between the data of Refs. [188, 200] and Ref. [556].

Attempts to calculate the shock adiabat of liquid deuterium using different theoretical approaches have been made in numerous papers published since 1997. The well-known equation of state [899] disregards the dissociation effects, and therefore the degree of deuterium compression behind the shock wave appears to be lower than in calculations by other methods. Allowing for dissociation by the simple model of linear mixing, the semiempirical equation of state [860] seems to overestimate the dissociation effect, which leads to compliance with the experimental data [188, 200]. The use of the so-called "*ab initio*" methods fails to clarify the situation as almost all of them have their shortcomings. In most works where the shock adiabat of deuterium was modeled, the results of "*ab initio*" calculations, in particular, the quantum Monte Carlo simulations with restrictions [692] and calculations by the quantum MD method [122, 208] are consistent with the experimental data [556]. The shock adiabats of deuterium from papers [122, 208] also describe well the measurements carried out on a light-gas gun at relatively low pressures [736]. The MD calculation of paper [552], where the electrons were taken into account as antisymmetrized localized wave packets, also agrees with the data of [736], but as distinct from [122, 208], is in much better compliance with the experiments [188, 200].

The shock adiabat of deuterium calculated by the quantum MCM agrees very well with the calculations [692] at high pressures and has regular asymptotics in the degree of compression in the limit of high pressures, but as the pressure lowers it deviates from the curve [692] towards high pressures. This is caused by the fact that in [692] a specific way of calculation of the sum in all permutations in expression (5.9) was used. It consists in allowance for only positive summands in this sum with a simultaneous decrease of the integration domain. Such a procedure which has no rigorous theoretical grounds introduces additional repulsion to the system, which obviously makes the system less compressible. In paper [262] it is also shown that the use of the approach of [692] for calculation of the density matrix does not allow a correct reproduction of the analytical dependence for the pressure and energy of an ideal Fermi-gas.

Thus, owing to the correct allowance for the contribution of bound states and degeneracy effects, the shock adiabat calculated by the quantum MCM deviates from the curve [692] towards higher densities, passes on the right of the data [556] and quite close to the experimental point [92]. At pressures below 1–2 Mbar the phase transition revealed earlier [266] begins to manifest itself, for which reason the portion of the adiabat below 1 Mbar is not reliable enough. It should be noted that the curve obtained by the classical MCM for reacting mixtures [103] is the closest to the shock adiabat of hydrogen calculated by the quantum MCM. The classical method takes obviously the most exact allowance for the deuterium molecule dissociation effects, which made it possible to obtain even without allowance for ionization a good agreement with the experimental data at low temperatures and pressures [736] and within the experimental error also with the point from paper [92].

## 5.5  Electron–hole plasma of semiconductors

Electron–hole plasma is a rather convenient experimental object for checking theoretical and numerical models of nonideal plasma media because it can rather easily be produced in rare-Earth halogenide crystals by moderate external pressure or by laser excitation of semiconductors. Moreover, in the literature one can find discussions of the possibility of creating high-temperature superconductors and different electronic devices based on electron–hole plasma media.

*Ab initio* simulation of different phenomena in semiconductors does not seem possible at the present time because of an exceedingly complicated energy spectrum (band structure). To simplify this task, the notion of electron–hole plasma is traditionally used. In semiconductors, electrons can rather easily be excited by external energy sources, and so the electrons can be thought of as almost free; an unoccupied vacancy, or a hole, is then formed in the valence band. There is every reason to consider the system of electrons and holes to be plasma medium with particles interacting by the Coulomb law. The main difference from the traditional plasma is that the effective masses of these particles are determined by the energy spectra of the semiconductor and can vary within a rather large range. This means that both the electrons and holes should be interpreted as (finite-size) quantum particles with allowance for their interaction and spins.

By the present time, rather many methods for calculation of the properties of quantum interacting particles have been developed, but most of them have some quite serious drawbacks. For example, the Hartree–Fock–Slater method disregards collective effects in plasma, and the density functional method and the quantum MD method based on it make use of the approximation expressions for the exchange-correlation functional. Furthermore, none of these methods can be used without substantial modifications for modeling two-component quantum systems. Different properties of electron–hole plasma were calculated in [265] by the quantum MCM. This method can be applied for an arbitrarily strong interparticle interaction in degenerate (quantum) systems. It takes into account the spin effects and can be extended to the case of mixture of many quantum particles. The quantum MCM is now the only approach to calculation of thermodynamic properties of quantum systems at nonzero temperatures without simplifying assumptions. Thus, this method is unique for the possibility of predicting new physical phenomena in real rather than model physical media.

The equilibrium electron and hole systems were modeled in Ref. [265] in a wide range of temperatures, electron concentrations, and hole-to-electron effective mass ratios. The calculations showed the production and decay of excitons upon temperature and external pressure variation, the formation of biexcitons and many-particle clusters, as well as hole crystallization at very high densities. The phase diagram for $_{0.45}Te_{0.55}$ was consistently explained on the basis of the results of modeling. Parameter depending crystallization was thoroughly studied and plasma phase diagram was proposed. The study of hole crystallization is of great physical interest in connection with the appearance of photons generated by the hole lattice and, as a consequence, with the possibility of the occurrence of high-temperature superconductivity under these conditions.

On an example of paper [265] we shall consider the calculation of the properties of electron–hole plasma emerging in semiconducting eutectic $_{0.45}Te_{0.55}$ under external pressure. The phase diagram of the electron–hole plasma of this substance is shown in Fig. 5.15.

One can see that with increasing pressure, at a temperature below a certain critical value a so-called excitonic insulator is formed. Such a transition is accompanied by a decrease in conduction because the charge carriers (electrons and holes) are bound into neutral particles called excitons.

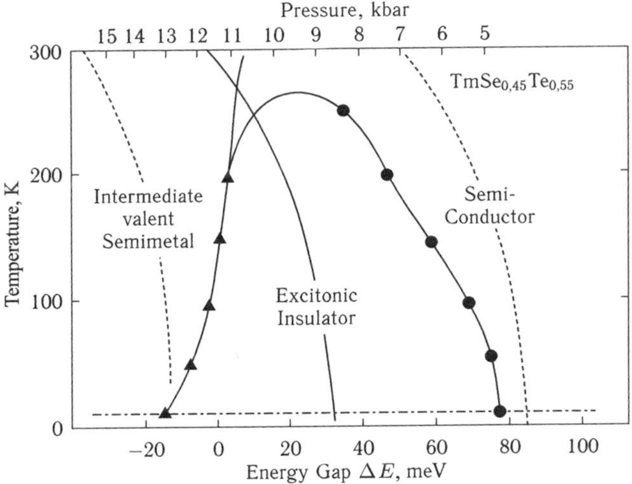


Fig. 5.15   Phase diagram of electron–hole plasma of $_{0.45}$Te$_{0.55}$ [1028].

With a further increase in pressure a Mott transition occurs with increase in conduction ("pressure ionization"). At temperatures above critical, transition to a conducting state is continuous.

Quantum Monte Carlo simulation of different states on the phase diagram was carried out in paper [265]. Electron–hole plasma was represented as a mixture of electrons and holes with constant effective masses $m_{e,ef} = 2.1m_e$, $m_{h,ef} = 80\,m_e$, and $M = m_h/m_e$. The permittivity of the medium was taken to be equal to $\epsilon = 25$ and for some calculations the value $\epsilon = 10$ was also used.

Figure 5.16 demonstrates particle configurations in a Monte Carlo cell. Particles in these configurations make up a cloud of points of size determined by the thermal wavelength of the corresponding particle. The configurations are shown for different values of the Brueckner parameter $r_s = d/a_B$, where $d$ is the mean interparticle distance, $d = [3/4\pi(n_e + n_h)]^{1/3}$, and $a_B$ is the Bohr radius for the medium (with allowance for permittivity).

Figure 5.17 presents pair correlation functions, including those for the configurations of Fig. 5.16. The differences for the cases of low (50 K) and high (200 K) temperatures are well seen. At $T = 50$ K one can clearly see a peak on the function $r^2 g_{eh}$ due to the existence of excitons. The peak is the highest for the lowest density $r_s = 10$. For $r_s = 6$ the peak lowers because of the appearance of biexcitons. For still higher densities $r_s = 4 \div 2$ the pair correlation functions $g_{ee}$ and $g_{hh}$ have maxima for $r = 2a_B$ and

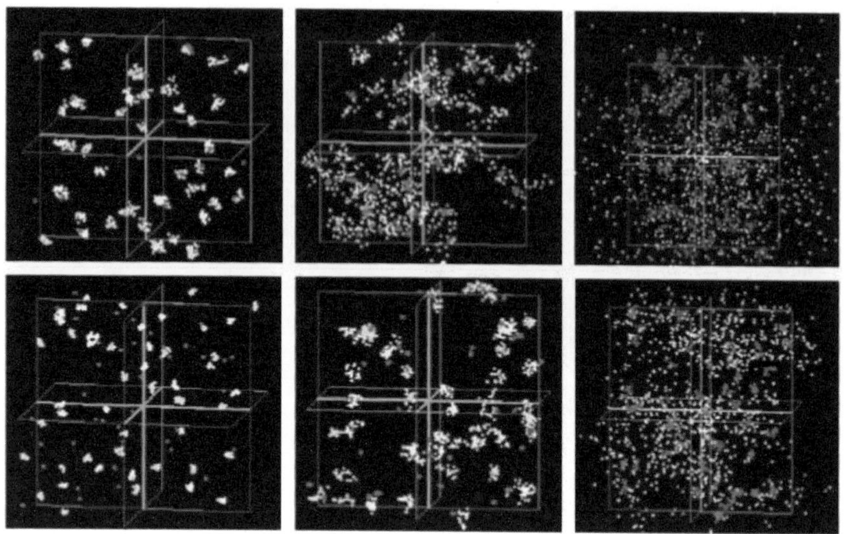

Fig. 5.16   Equilibrium Monte Carlo configurations under different conditions. The upper row: $T = 50\,\text{K}$, $r_\text{s} = 10$, 4, 1. The lower row: $T = 200\,\text{K}$, $r_\text{s} = 10$, 4, 1.

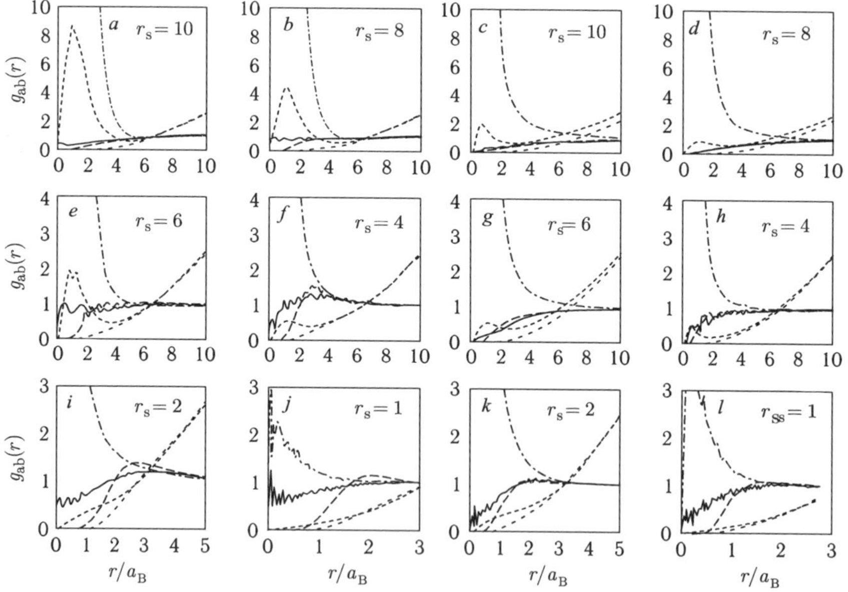

Fig. 5.17   Pair correlation functions of electron–hole plasma under different conditions. The first two columns $T = 50\,\text{K}$, the third and fourth columns $T = 200\,\text{K}$ [1028].

$4a_B$, which testify to a considerable concentration of respectively excitons and biexcitons. Finally, for the highest density $r_s = 1$ the excitons and biexcitons disintegrate because of the many-particle effects associated with pressure ionization and with closeness to the Mott transition.

The qualitative analysis of the correlation functions given in the previous extract can also be grounded by quantitative calculations. For instance, the product $r^2 g_{eh}(r)$ has the meaning of probability of finding an electron at a distance $r$ from a hole. A rough estimation of the fraction of bound states gives the formula

$$\frac{N_b^{eh}}{N_b^{eh} + N_c^{eh}} = \frac{\int\limits_0^{R_b} r^2 \left[ g_{eh}(r) - 1 \right] dr}{\int\limits_0^{R_b} r^2 g_{eh}(r)\, dr},$$

where $R_b$ is the distance at which $g_{eh}(r) - 1 > 0$. The fraction of hole–hole bound states is estimated analogously:

$$\frac{N_b^{hh}}{N_b^{hh} + N_c^{hh}} = \frac{\int\limits_{R_b'}^{R_b} r^2 \left[ g_{hh}(r) - 1 \right] dr}{\int\limits_{R_b'}^{R_b} r^2 g_{hh}(r)\, dr},$$

where $R_b'$ is the distance at which $g_{hh}(r) - 1 > 0$.

The results of calculations are shown in Figs. 5.18 and 5.19. One can readily see that for a low density, $r_s > 6$, and a low temperature ($T < 100\,\mathrm{K}$) practically all the electrons and holes are bound in excitons. An increase in temperature ($T > 100\,\mathrm{K}$) leads to exciton ionization. Irrespective of temperature, for $r_s > 6$ biexcitons and many-particle clusters are practically absent.

For intermediate densities, $2 < r_s < 6$, at a low temperature ($T < 100\,\mathrm{K}$) the fraction of biexcitons and many-particle clusters increases sharply, whereas the exciton concentration decreases. At a high density, $r_s \approx 1$, irrespective of temperature all bound states break because of overlapping of the electron shells, and the Mott transition occurs in the system. Assuming that under transition the number of bound states does not exceed 10% and does not depend on temperature, one can estimate from Fig. 5.18 that this happens for a concentration of $10^{21}\,\mathrm{cm}^{-3}$. Figures 5.18 and 5.19 also imply that at a temperature of $300\,\mathrm{K}$ the bound states disappear completely, and therefore the temperature of $300\,\mathrm{K}$ can be thought of as the

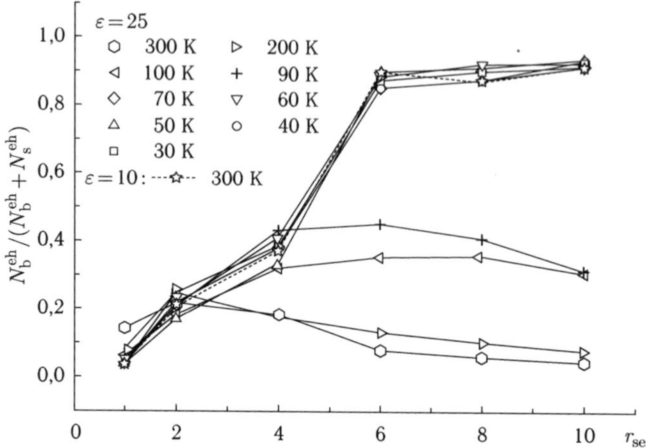

Fig. 5.18   Fraction of electron–hole bound states in $_{0.45}\text{Te}_{0.55}$, $M = 40$ semiconductor depending on the Brueckner parameter.

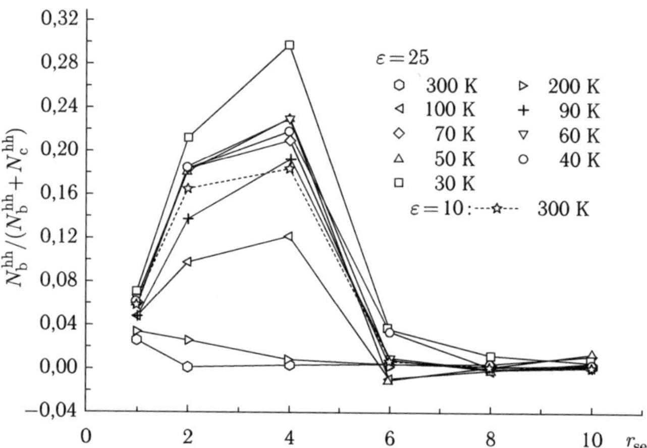

Fig. 5.19   Fraction of the hole–hole bound states in $_{0.45}\text{Te}_{0.55}$, $M = 40$ semiconductor depending on the Brueckner parameter.

upper boundary of the critical transition temperature in the electron–hole plasma under consideration. A more detailed analysis shows that the critical temperature lies in the range of 200–300 K. These conclusions are in qualitative agreement with the phase diagram of Fig. 5.15.

A further increase in density can lead to hole crystallization in electron-hole plasma.

## 5.6 Crystallization of holes

As has already been discussed in Chapter 4, crystallization is one of the fundamental properties of matter. Nearly 80 years ago Wigner [1039] predicted theoretically the possibility of electron crystal formation, and since then this phenomenon has been intensely sought in nature. Electron crystallization was observed on the surface of cold helium drops [376], in semiconductor heterostructures in strong magnetic fields [48], and was also predicted in quantum dots of semiconductors [260]. Crystallization of holes in semiconductors with a high critical temperature has recently been predicted [1]. Meanwhile, Wigner (Coulomb) crystallization is a fundamental property of any system of charged particles; for example, crystallization was experimentally observed in ultracold plasma in traps [478, 1042] and in storage rings [888]. A necessary condition of the existence of crystal in OCP is a considerable excess of the mean potential energy $e^2/r_a$ over the mean kinetic energy ($dkT/2$ for classical plasma or $E_F$ for quantum plasma). Here, $r_a$ is the mean interparticle distance, $r_a \sim n^{-1/d}$ with $d$ as the system dimension. The critical values of nonideality parameter obtained in numerical simulation for the classical OCP was $\Gamma^c = 175$ in the three-dimensional and $\Gamma^c = 137$ in the two-dimensional case [210, 376]. For quantum OCP at zero temperature a crystal is formed at a density determined by the Brueckner parameter, $r_s = r_a/a_B$, where $a_B$ is the Bohr radius of the system with allowance for the permittivity of the medium; in the two-dimensional case the crystallization density corresponds to $r_s = 33$ and in the three-dimensional case to $r_s = 100$ (160) for fermions (bosons) [164, 965].

Most of the Coulomb systems in the Universe are neutral plasma consisting of at least two oppositely charged components (two-component plasma, TCP). Interesting is for instance the question of whether the crystal formation in TCP is possible and under what conditions. Some astrophysical models predict the existence of quantum ionic crystals surrounded by a degenerate electron gas inside white dwarfs and in the neutron star core [896]. Such quantum two-component crystals were not found in laboratory conditions in electron–ion, electron–positron, or electron–hole plasma. On the other hand, examples of classical crystals in TCP are well known, e.g., in colloidal and dust [53, 432, 638, 972], as well as laser-cooled [537, 795] plasma, but such crystals exist at much lower densities, i.e., under conditions when particles obey classical statistics.

The common property for all the above-mentioned classical crystals in TCP is a large mass and charge of certain particles compared to the others.

However, the question remains open of what role is played by these factors for crystal formation in the classical and quantum cases. This question can obviously be answered through computer simulation only. The possibility of quantum crystal formation was convincingly demonstrated and the conditions of its existence were studied by the quantum Monte Carlo simulations. In particular, it was shown that the crystal formation can be expected for a sufficiently large heavy-to-light particle mass ratio, and also for rather low temperatures.

Simulation of crystallization in a Coulomb quantum system is a fairly complicated task: the Coulomb interaction, the bound state formation, as well as quantum and spin effects of the heavy and light components should be taken into account self-consistently. In paper [265], the quantum MCM was used to calculate the mass ratio $M = 1$–$2000$ of heavy and light plasma components for temperatures $T = (0.06 - 0.6)\,T_{\rm b}$ and densities $r_{\rm s} = 0.6$–$13$. Here $kT_{\rm b} = e^2/(8\pi\epsilon_0\epsilon a_{\rm B})$ is the binding energy of an "atom" consisting of a heavy and a light particles of opposite charges (an analogue of the Ridberg energy in a hydrogen atom), $\epsilon$ is the effective permittivity of the system, $a_{\rm B} = 4\pi\epsilon_0\epsilon h^2/\mu e^2$ is the Bohr radius of the system, $\mu$ is the reduced mass, $\mu = m_e/(1 + M)$, $m_{\rm e}$ is the effective electron mass, and $k$ is the Boltzmann constant. In the simulations, positive and negative particles were considered to be quantum fermions with charge $\pm e$ and spin $1/2$. Thus, as a result of modeling one could investigate the properties of different systems, namely, the electron–positron, electron–hole, and hydrogen plasma.

In quantum Monte Carlo simulations, crystallization was actually observed, which is confirmed by Fig. 5.20 showing the Monte Carlo configurations of quantum particles in a computational cell for the same temperature and density, but for different heavy-to-light particle mass ratios. In the quantum Monte Carlo formalism, particles are represented as closed broken lines whose vertices form a cloud of points. The size of this cloud approximately corresponds to the thermal de Broglie wavelength of the corresponding particle, and therefore heavy particles have a smaller size in space than light particles. Figure 5.20 shows vertices of only those broken lines that represent heavy and light particles with different spin directions. For the indicated simulation conditions the electrons form a virtually homogeneous Fermi gas; the electron wave packet exceeds the Monte Carlo cell size; hence, heavy particles belong simultaneously to all the electrons. The position of heavy particles (holes) in space changes drastically: for $M = 1$ the holes form a quasi-gas structure, for $M = 50$ a liquid like one, and for $M = 100$ and $800$ the holes get ordered in space. Thus, crystallization

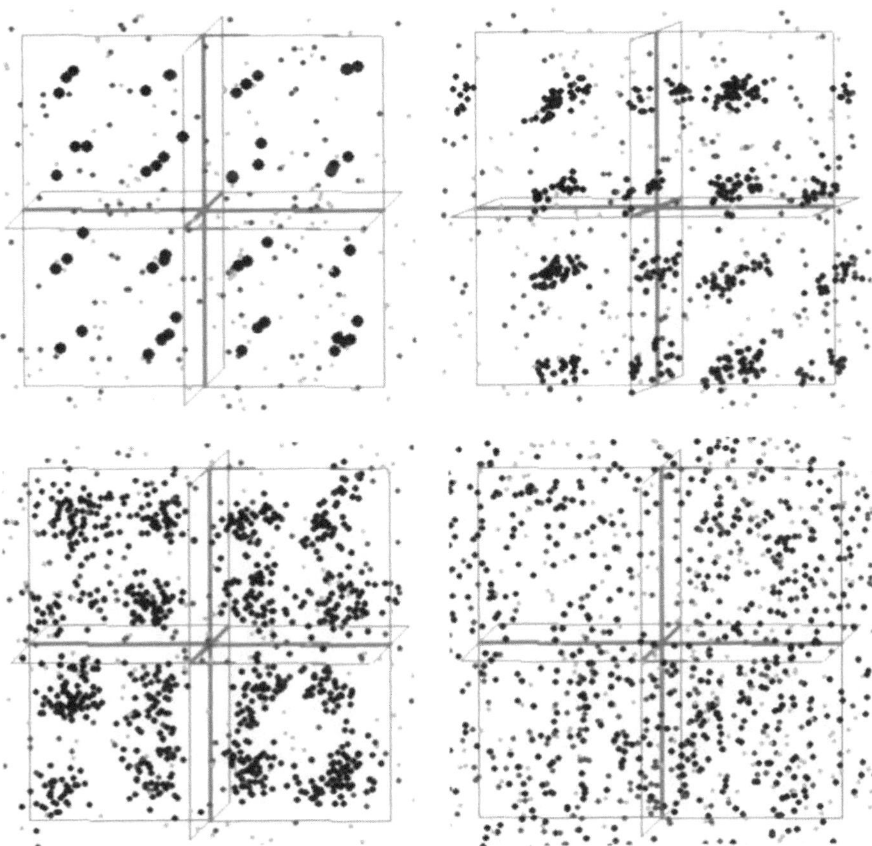

Fig. 5.20    Monte Carlo configurations on an equilibrium domain in electron–hole plasma at $T/T_{\rm b} = 0.096$ and $r_{\rm s} = 0.63$. $M = 800$ (upper left), $M = 100$ (upper right), $M = 50$ (lower left), $M = 1$ (lower right). Points are representations of particles in the form of path integrals (only vertices of closed broken lines are presented). Black and dark-grey clouds of points are holes, grey and light-grey points are electrons with opposite spin direction. The lines indicate Monte Carlo cell boundaries.

proceeds between $M = 50$ and $100$. The figure clearly shows the mechanism of this phase transition: with increasing $M$ the wave packets of individual holes diminish, and crystallization sets in when the hole wave packet reaches a certain critical size ($M \approx 80$).

For a quantitative illustration of the crystallization process, Fig. 5.21 shows the pair correlation functions $g_{\rm ab}(r)$. The formation of a long-range order with $M$ increasing from 1 to 100 is clearly seen in Figs. 5.21(a)–(d). In Fig. 5.21(c), the hole-hole correlation function has only one pronounced

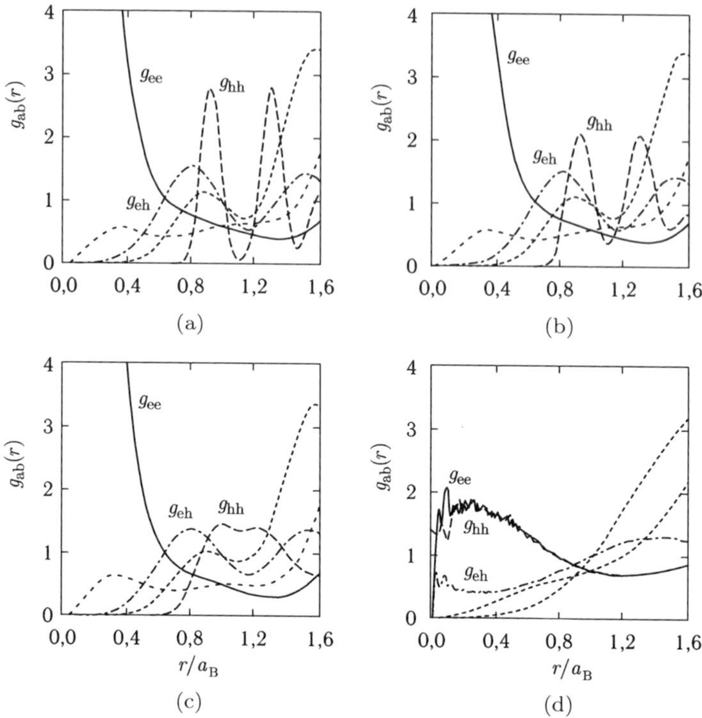

Fig. 5.21    Pair correlation functions of electron–hole plasma, $T = 0.064T_{\mathrm{b}}$, $r_{\mathrm{s}} = 0.63$. The $M$ values and the arrangement of figures correspond to Fig. 5.20.

peak as in the case of a liquid. For $M = 800$ and $100$ (Fig. 5.21(a) and (b)) $g_{\mathrm{hh}}(r)$ exhibits periodic oscillations with a large ratio of values in the maximum and minimum, which is typical of crystals. A crystal exists in a limited temperature range since with a two-fold temperature heightening the crystal melts. In conditions of a hole crystal, the correlation functions $g_{\mathrm{ee}}$ have maxima at small distances, which is indicative of band structure formation. It is therefore qualitatively clear that a crystal exists at temperatures sufficiently low compared to the binding energy in a limited density range and for the heavy-to-light particle mass ratio $M > 60$–$80$.

The relative fluctuation of the mean distance $R$ between the holes can be analyzed. In the crystal phase, this parameter is small because of hole localization, and upon melting it increases rather sharply (the Lindemann criterion). Figure 5.22 shows the dependence of $R$ on the mass ratio $M$ in TCP for $r_{\mathrm{s}} = 0.63$ and $T = 0.064\,T_{\mathrm{b}}$. For the large mass ratio ($M = 2000$, which is close to hydrogen plasma) $R$ is much smaller than

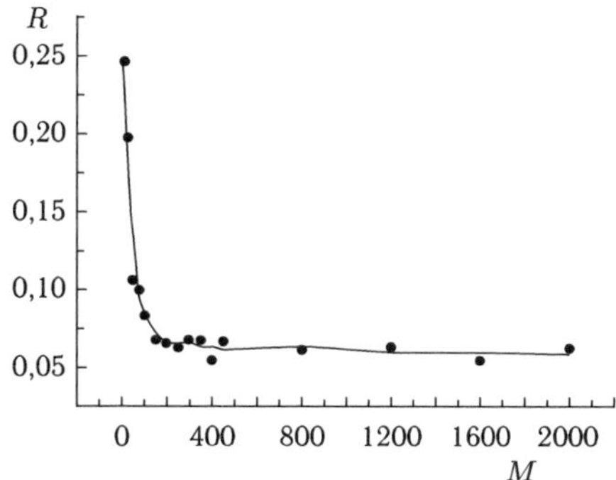

Fig. 5.22   Fluctuation of the distance between holes relative to the mean interparticle distance depending on the hole-to-electron mass ratio.

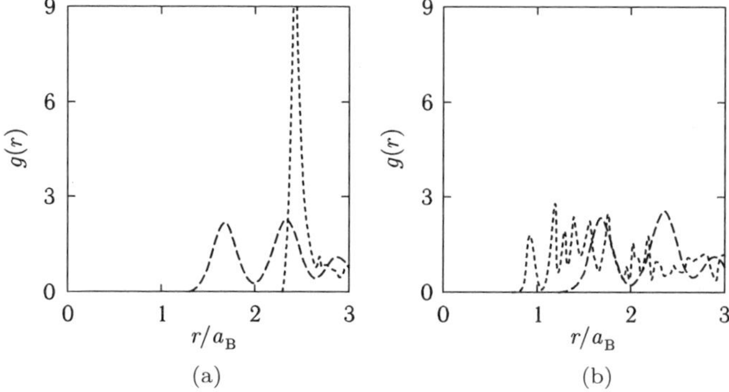

Fig. 5.23   Pair correlation functions of electron–hole plasma, $T = 0.002T_\mathrm{b}$, $r_\mathrm{s} = 0.5$. The dashed lines show the hole–hole pair correlation function for $M = 400$: (a) the dotted line corresponds to the hole–hole pair correlation function with parallel spins, (b) the same with antiparallel spins for $M = 100$.

0.1; this corresponds to the crystal state. The $R$ value remains practically unchanged up to $M = 150$ and then rather rapidly increases approximately 3 times. Thus, the customary ideas of melting are quite applicable also to Coulomb quantum crystals; in particular, the crystal melts with rising temperature.

An analysis of the correlation functions of holes with different spin directions reveals several interesting properties of the hole crystal. Figure 5.23 illustrates the hole–hole pair correlation functions; the dotted line in Fig. 5.23(a) shows the correlation function in the construction of which the holes with parallel spins only were taken into account, and the holes with antiparallel spins only are demonstrated in Fig. 5.23(b). The correlation function disregarding spin is given in both cases for comparison. Figure 5.23 clearly shows that all the holes with parallel spins are located at approximately the same distance from one another, while holes with antiparallel spins are arranged rather chaotically in the cell volume. This signifies that holes form two embedded lattices each of which contains particles with the same spin projections. The physical reason for this phenomenon lies in a strong Fermi repulsion of particles with identical spins.

The existence criterion for a crystal in TCP can be obtained using the results of simulation for OCP. For example, in the classical OCP a crystal exists if the nonideality parameter exceeds a certain critical value $\Gamma^c$. On the other hand, in quantum OCP, a crystal is formed at densities below a certain critical value characterized by the Brueckner parameter $r_s^c$, $r_s > r_s^c$. Furthermore, bound states must be absent from the system, which at low temperatures (quantum system) is determined by the pressure ionization condition ($r_s < r_s^{\text{Mott}} \approx 1.2$) and with rising temperature by the thermal ionization condition (bound states vanish at $T_e \approx 1$, where $T_e$ is dimensionless temperature measured in ionization energies). The condition of bound state nonexistence will be written as follows: $r_s < r_s^{\text{Mott}}(T_e)$ in the assumption that the density at which bound states occur decreases with rising temperature. It can be shown that the above-mentioned conditions of crystal existence are reduced to the following:

$$M > M^c(Z, T_e) = \frac{r_s^c}{Z^{4/3} r_s^{\text{Mott}}(T_e)} - 1, \qquad (5.10)$$

where $M$ is the heavy-to-light particle mass ratio and $Z$ is an analogous charge ratio; the density range is defined by the formulas

$$n^{(1)}(T_e) = \frac{3}{4\pi} \left[ \frac{1}{r_s^{\text{Mott}}(T_e)} \right]^3, \quad n^{(2)}(T_e) = n^{(1)}(T_e) K^3,$$

$$K = \frac{(M+1)}{(M^c+1)}, \qquad (5.11)$$

and the temperature must be below $T^*$,

$$T^* = 6 \frac{Z^2 \Theta (M+1)}{\Gamma^c r_s^c}, \qquad (5.12)$$

where $\Theta$ — the light-to-heavy particle temperature ratio.

The crystal existence region in TCP is qualitatively characterized by the temperature-density phase diagram given in Fig. 5.24. The crystal exists in the absence of bound states, i.e., at sufficiently high densities. On the other

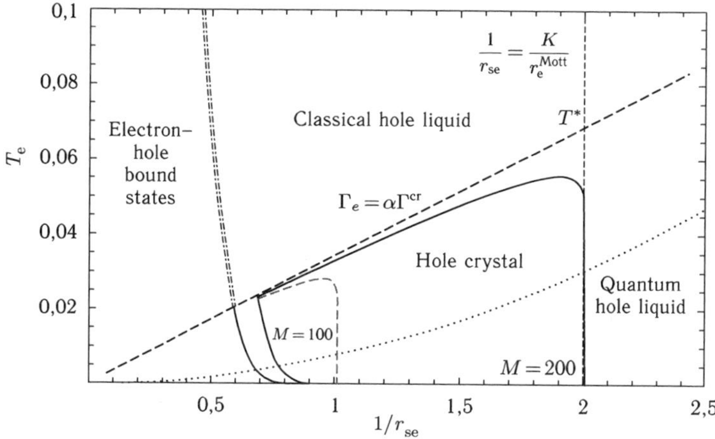


Fig. 5.24   Estimation of the phase diagram of two-component Coulomb plasma [1028].

hand, as was shown by the calculations for OCP, the crystal melts at very high densities. The classical crystal in OCP only exists for fairly high bonding parameter values. In addition, a crystal can form for a sufficiently high heavy-to-light particle mass ratio. Figure 5.24 shows the crystal existence region for $M = 200$ and $M = 100$; for still lower $M$ values a crystal cannot be formed. The dotted line shows schematically the boundary between the quantum and classical systems.

## 5.7   Electron–hole plasma of germanium

The existence of phase transition in degenerate nonideal hydrogen plasma can hardly be verified experimentally. However, the phase transition in a Coulomb system of strongly interacting quantum particles can under certain conditions be observed in experiment. When a thin germanium film cooled to the liquid helium temperature is exposed to laser radiation, electrons and holes are excited, and the obtained plasma can exist for long enough time for observation [486]. For low electron concentrations excitons (electron–hole pairs, (EHP)) are produced and such plasma possesses low conductance. As the concentration increases at temperatures below critical, the system falls into electron–hole drops accessible to visual observation. With a further increase of the concentration plasma again becomes homogeneous and its photoconduction increases by 4–5 orders of magnitude [486]. Because of an exceedingly complicated band structure of germanium the properties of the electron–hole plasma can be described satisfactorily by a degenerate system of holes and electrons with constant effective masses only at temperatures rather low compared to critical. The effective masses of electrons and holes then appear to be close, and hence in simulations for holes and for electrons the exchange effects should necessarily be taken into account. Figure 5.25 presents the configurations of electron–hole plasma for $T = 0.1\text{Ry}$ at different densities.

The quantum MCM predicts correctly the sizes of electrons and holes and reproduces the electron–hole drop formation. Calculating the isotherms of germanium electron–hole plasma at temperatures below critical, $T_\text{c} = 6.5\,\text{K}$, one could determine the boundaries of the region of poor convergence to the equilibrium state. Figure 5.26 illustrates the results

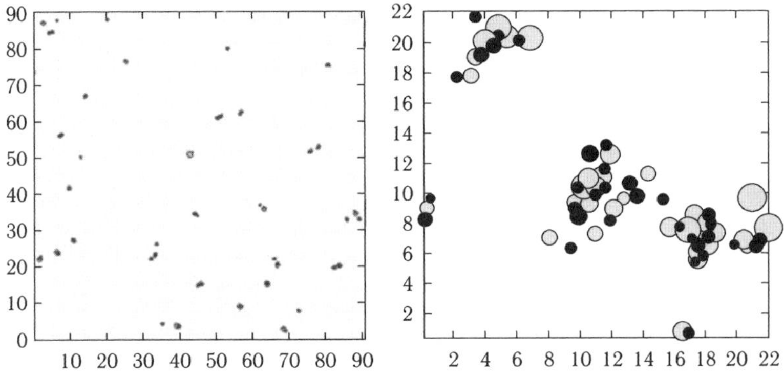

Fig. 5.25   Configurations of EHP in quantum Monte Carlo 2D-simulation at a temperature below critical [126]. Light points are electrons and dark ones are holes. (a) $n = 1.54 \cdot 10^{14}\,\text{cm}^{-2}$, (b) $n = 2.58 \cdot 10^{15}\,\text{cm}^{-2}$.

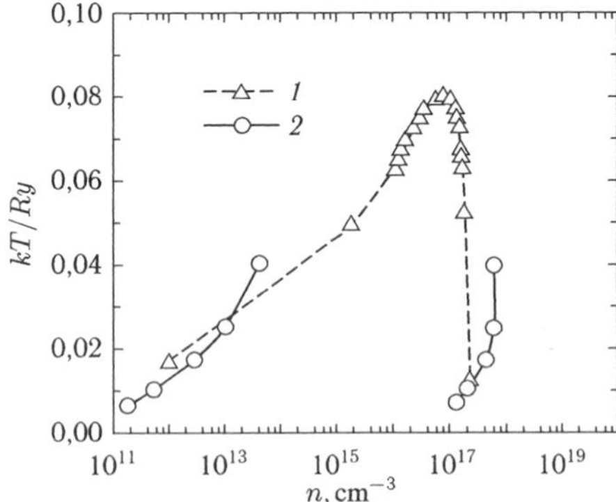

Fig. 5.26   Phase diagram of germanium electron–hole plasma *1* experiment [971], *2* MCM [126].

of simulation. The points on curves *2* at a given temperature correspond to the concentrations limiting the phase transition region. One can see that the calculated phase boundary agrees well with the experimental data [971].

## 5.8 Quantum MD

In Chapter 2 we touched upon the MD method for the description of classical systems of the type of gases and simple liquids. At the same time, the application of this method to quantum structures requires the development of this approach, which is to be briefly discussed in this section.

The method of quantum MD first proposed in paper [157] and its further modifications are widely used to calculate different properties of strongly nonideal media. This became possible because the forces acting on ions are calculated on the basis of the solution of many-electron quantum-mechanical problem.

Let us consider a system of classical ions with coordinates $\{\mathbf{R}_I\}$, masses $M_I$, and interaction energy $E[\{\mathbf{R}_I\}]$ and write for them the Newton equations of motion:

$$M_I\ddot{\mathbf{R}}_I = -\frac{\partial E}{\partial \mathbf{R}_I} = \mathbf{F}_I[\{\mathbf{R}_J\}]. \tag{5.13}$$

This system of equations is solved numerically, for example, with the help of Verlet algorithm [24]. The key property of this and similar algorithms is that the particle system energy is conserved with good accuracy during the entire simulation time in condition that the integration step $\Delta t$ is adequately chosen. The forces acting on ions are determined by the position of both ions and electrons. In classical MD these forces are determined by effective interaction potentials. For complex systems the effective potentials can contain a lot of parameters determined from experiment or from comparison with more accurate quantum-mechanical calculations.

The progress achieved in the methods of calculation of the electron structure of matter opens up the prospects of calculating the forces acting on ions on the side of electrons without using additional approximations. Such computational methods called in the literature *"ab initio"* are based on the Born–Oppenheimer approximation. In the framework of this approximation the electron density is calculated for motionless ions, and ions move driven by the forces from the side of electrons when the electrons are motionless.

The authors of paper [157] were the first to unify the problems of finding the electron density for moving ions. This made it possible to seek equilibrium structures, to study the thermal motion of ions, to simulate phase transitions, etc. On the other hand, the algorithm of Ref. [157] does not describe the dynamics of electrons but is meant for seeking the ground state of a many-electron system for moving ions. To this end, along with the electron interaction energy and the ion kinetic energy the Lagrangian of the system will also involve the fictitious kinetic energy of electrons:

$$\mathcal{L} = \sum_{i=1}^{N} \frac{1}{2}(2\mu) \int d\mathbf{r}|\dot{\psi}_i(\mathbf{r})|^2 + \sum_I \frac{1}{2}M_I\dot{\mathbf{R}}_I^2 - E[\psi_i, \mathbf{R}_I]$$

$$+ \sum_{ij} \Lambda_{ij} \left[ \int d\mathbf{r}\psi_i^*(\mathbf{r})\psi_j(\mathbf{r}) - \delta_{ij} \right]. \tag{5.14}$$

The last term in equation (5.14) is significant for allowance for the electron wave function normalization condition. The Lagrangian (5.14) implies the equations of motion for both the electrons and the ions:

$$\mu\ddot{\psi}_i(\mathbf{r}, t) = -\frac{\delta E}{\delta\psi_i^*(\mathbf{r})} + \sum_k \Lambda_{ik}\psi_k(\mathbf{r}, t)$$

$$= -H\psi_i(\mathbf{r}, t) + \sum_k \Lambda_{ik}\psi_k(\mathbf{r}, t), \tag{5.15}$$

$$M_I\ddot{\mathbf{R}}_I = \mathbf{F}_I = -\frac{\partial E}{\partial\mathbf{R}_I}. \tag{5.16}$$

Equations (5.15) and (5.16) are analogues of the Newton equations and are solved similarly to the equations of the classical MD method [24]. The ion masses coincide with their physical masses and the electron "masses" $\mu$ are chosen from the condition of rapid convergence to the ground state.

The approach [157] displays several potential drawbacks. The main of them is a very small time step $\Delta t$ necessary in most cases for integration of equations (5.15) and (5.16). It should be noted in addition that the introduction of the fictitious kinetic energy of electrons to the Lagrangian (5.14) leads in some cases to a nonphysical energy transfer to the electron degrees of freedom. For this reason, much more widespread is another formulation of the quantum MD method.

The equations of motion for ions (5.13) can be solved if the forces acting on ions are known. For this purpose the Gell-Mann–Feynman theorem [257] and the fast algorithms [93, 775] specially worked out thereto are usually employed. As distinct from the method [157], algorithms for calculation of the trajectories of ion motion and finding the electron density are much oftener applied at the present time. Such separation has its advantages and shortcomings. A rather large MD time step can be chosen for integration of the equations of motion for ions, but at the same time the electron density distribution for given ion positions should be calculated much more accurately than in the algorithm [157]. Nevertheless, the now existing program packets most often use precisely this approach.

The thermodynamic functions of ions, in particular, the energy and pressure, are calculated by the method [24], standard for classical MD, through averaging over a rather large number of configurations. The energy of each configuration is determined by the sum of energies of pair interactions of ions and pressure is found by the virial theorem.

The method of quantum MD can be applied to simulate disordered systems, including liquids and plasmas. Rather interesting results were obtained in modeling the melting of metals: sodium [828], aluminum [131], and iron [90]. Such modeling is most often realized using the single-phase approach in which melting of initially crystalline matter is registered in some way, but attempts are made to carry out also two-phase modeling in which the liquid and solid phases are brought to equilibrium at a given pressure [131]. As an example, Fig. 5.27 presents the aluminum melting curve in comparison with experimental data.

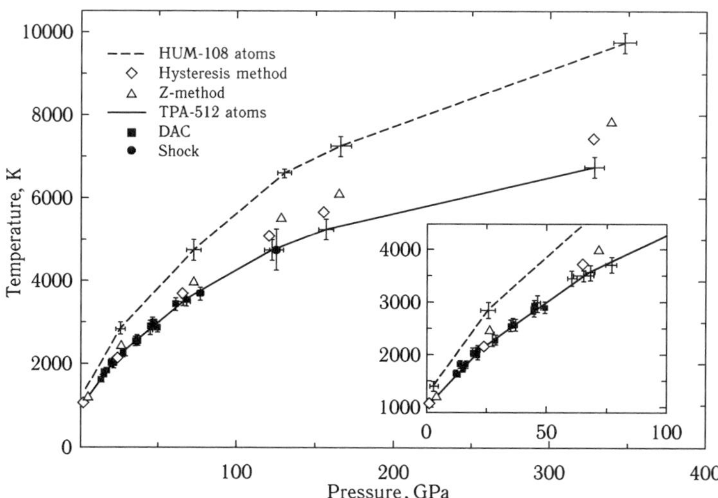


Fig. 5.27 Aluminum melting curves. The solid curve shows two-phase simulation (512 particles), the dashed line gives single-phase simulation. The experimental data: heated diamond anvils [116, 424], shock-wave melting [900].

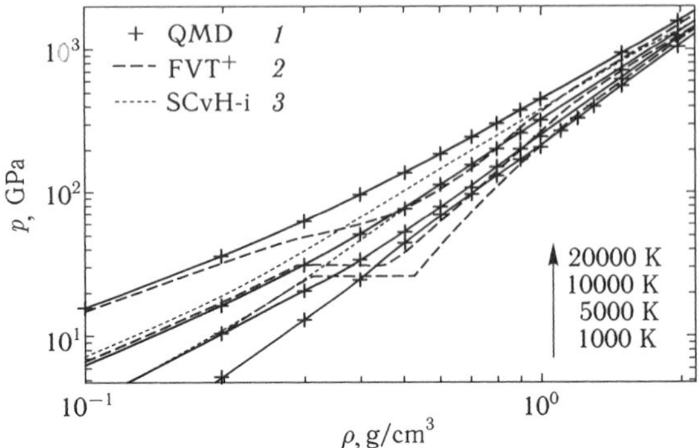



Fig. 5.28   Isotherms of hydrogen plasma. *1* quantum molecular dynamics, *2* (dashed line) variational fluid theory [443], *3* chemical plasma model [884].

A lot of results in simulation of thermodynamic properties of different substances were obtained by the method of quantum MD in the region of liquid and plasma. In particular, the shock adiabats of hydrogen [444], deuterium [122, 208], and helium [534, 691], quartz [554], and hydrogen [555] in solid and liquid phases were calculated, the phase diagram of water was investigated at high temperatures and pressures [669], and the isochors of plasmas of different metals were calculated [836–838]. As an example, Fig. 5.28 presents isotherms of hydrogen plasma at different temperatures.

## 5.9   QGP simulations

Examination of the properties of QGP is an exceedingly sophisticated problem of high energy density physics from both theoretical and experimental points of view. Postponing a detailed discussion of the properties of this exotic state of matter to Chapter 12, we shall now only touch upon the application of MCM for a numerical simulation of QGP.

The properties of QGP were examined experimentally on the relativistic heavy-ion collider (RHIC) in Brookhaven. The most interesting result obtained from the analysis of experimental data is that quark–gluon matter is an ideal liquid rather than an ideal gas, which could be expected from the theory of asymptotical freedom [909]. The spatial correlation effects in QGP can possibly lead to the formation of ordered structures similar to those in dense liquid. Such structures were observed earlier in electromagnetic plasma.

QGP are being studied using different approaches with their advantages and shortcomings. The most straightforward way for examining the properties of strongly nonideal QGP is the numerical calculation of field integrals which appear in the framework of quantum chromodynamics (QCD) [89, 197, 280]. The interpretation of these exceedingly complicated calculations requires employment of different simplified models and, moreover, QCD calculations are only possible for low values of baryon chemical potential and under thermodynamic equilibrium.

The quasi-classical approach introducing the conception of quasi-particles was proposed in [628–631]. It is expected that the main properties of nonabelian plasma can be described within the quasi-classical approximation, and the difficulties inherent in quantum-mechanical calculations within quantum field theory are not encountered in this case. Similar ideas were used within the classical method of MD [438]. This approach has recently been developed in the series of papers [177, 178, 226, 343, 970].

These concepts underlay the formulation of the classical nonrelativistic model of QGP with color Coulomb interaction [343] which was analyzed in MD calculations. The quantum effects were ignored there or were taken into account phenomenologically by means of potential of short-range quasi-particle repulsion. This rather rough approach may yield erroneous results at high densities. For the temperatures and densities considered in paper [343] these effects are very important as the thermal wavelength of quasi-particles is comparable in the order of magnitude with the mean interparticle distance.

Of considerable interest is the extension of the classical model [343] to the case of quantum quasi-particles [261]. The approach is based on the MCM using the path integral (PIMC) with consistent allowance for Fermi (Bose) statistics for quarks (gluons) and quantum effects in quasi-particle interaction.

We shall consider the model based on paper [343]. Its main points can be formulated as follows.

(1) All color quasi-particles are heavy ($m > T$), where $m$ is the quasi-particle mass and $T$ is the temperature, and therefore the relativistic effects can be disregarded. This assumption is based on the analysis of QCD calculations [614, 788].

(2) Since the particles are nonrelativistic, the interparticle interaction is described by the color Coulomb potential. The effects due to the magnetic field are ignored.

(3) The color operators are treated in the quasi-classical approximation, i.e., are replaced by classical color vectors whose time evolution is described by the Wong equations [1048].

The validity of the exploited approximations and their restrictions were discussed in paper [343]. This model requires the following quantities as input parameters: (1) the quasi-particle mass $m$ and (2) the coupling constant $g^2$. All the input parameters should be determined from QCD lattice calculations or from other models based on these data.

Thus, we are dealing with three-component QGP consisting of $N_{\mathrm{q}}$ heavy (surrounded by gluons) quarks, $N_{\bar{\mathrm{q}}}$ antiquarks, and $N_{\mathrm{g}}$ gluons. In thermodynamic equilibrium, the mean values of the numbers of particles can be found in the large canonical ensemble with a temperature-dependent Hamiltonian and can be written as $\hat{H} = \hat{K} + \hat{U}$. The kinetic and potential interaction energies of quasi-particles have the following form:

$$\hat{K} = \sum_i \left[ m_i(T, \mu_{\mathrm{q}}) + \frac{p_i^2}{2m(T, \mu_{\mathrm{q}})} \right],$$

$$\hat{U} = \frac{1}{2} \sum_{ij} \frac{g^2(|r_i - r_j|, T, \mu_{\mathrm{q}}) < Q_i|Q_i >}{2\pi |r_i - r_j|}.$$

Here, $Q_i$ stands for the classical color Wong vector, $T$ is the temperature, and $\mu_{\mathrm{q}}$ is the quark chemical potential. The quasi-particle mass and the coupling constant obtained from lattice models are in fact functions of $T$ and, in the general case, $\mu_{\mathrm{q}}$. Furthermore, $g^2$ is a function of the distance $r$, which results in the linear potential growth for large $r$ [840]. At a given temperature $T$, chemical potential $\mu_{\mathrm{q}}$, and volume $V$, the thermodynamic properties in a large canonical ensemble are completely described by the statistical sum:

$$Z(\mu, \beta, V) = \sum_{N_{\mathrm{q}}, N_{\bar{\mathrm{q}}}, N_{\mathrm{g}}} \frac{\exp(\mu_{\mathrm{q}}(N_{\mathrm{q}} - N_{\bar{\mathrm{q}}})/T)}{N_{\mathrm{q}}! N_{\bar{\mathrm{q}}}! N_{\mathrm{g}}!}$$

$$\times \sum_{\sigma} \int_v dr dQ \rho(r, Q, \sigma, N_{\mathrm{q}}, N_{\bar{\mathrm{q}}}, N_{\mathrm{g}}, \beta),$$

where $\rho(r, Q, \sigma, N_{\mathrm{q}}, N_{\bar{\mathrm{q}}}, N_{\mathrm{g}}, \beta)$ stand for the diagonal matrix elements of the density operator $\rho = \exp(-\beta \hat{H})$. Here $\sigma$, $r$, and $Q$ are spin, coordinate, and color vector of quarks, antiquarks, and gluons, respectively. In order to calculate the thermodynamic functions, in particular, the internal energy or the mean density $n_a = \langle N_a \rangle / V$, the statistical sum logarithm should be differentiated with respect to the corresponding thermodynamic parameters.

The same as for electromagnetic plasma, an exact expression for the density matrix of a nonideal quantum system can be obtained using the path integrals [258, 1087] on the basis of the operator identity $\exp(-\beta\hat{H}) = \exp(-\Delta\beta\hat{H})\ldots\exp(-\Delta\beta\hat{H})$, where the right-hand side contains $(n+1)$ identical factors, $\Delta\beta = \beta(n+1)$, which makes it possible to rewrite the expression, as before, in the form of a path integral. The quasi-particle spins are taken into account in the spin part of the density matrix, and the exchange effects are described by the permutation operators acting on the spatial, spin, and color coordinates of the quark, antiquark, and gluon.

Accordingly, each particle is represented by a set of spatial, spin, and color coordinates. The dominating contributions to the statistical sum are made by the configurations of quasi-particles in which the "size" of the cloud of points representing a quasi-particle is comparable in the order of magnitude with the thermal wavelength; the characteristic distances between points in the cloud are of the order of the thermal wavelength at a temperature $(n+1)$ times higher than a given value.

The MCM calculations were performed with zero baryon chemical potential $\mu_{\mathrm{q}} = 0$, and, moreover, additional simplifications were introduced to the model [343].

(1) The canonical ensemble is used instead of the large canonical ensemble. In the canonical ensemble, at given temperature $T$ and volume $V$ the thermodynamic properties are completely described by the density operator $\hat{\rho} = \exp(-\beta\hat{H})$ which, in turn, determines the statistical sum of the system:

$$Z(N_{\mathrm{q}}, N_{\bar{\mathrm{q}}}, N_{\mathrm{g}}, \beta, V) = \frac{\sum\limits_{\sigma}\int\limits_{V} drdQ\rho(r, Q, \sigma, N_{\mathrm{q}}, N_{\bar{\mathrm{q}}}, N_{\mathrm{g}}, \beta)}{N_{\mathrm{q}}!N_{\bar{\mathrm{q}}}!N_{\mathrm{g}}!}.$$

(2) Since the masses of quarks of different flavors, determined in QCD lattice calculations, are very close, the masses of all the quarks are considered to be identical. Furthermore, the quark and gluon masses are thought of as identical according to the results of lattice calculations [614, 788].

(3) Since the masses of quarks, antiquarks, and gluons are identical and all the quasi-particles have an almost equal number of degrees of freedom, the number of quasi-particles of each sort is assumed to be the same: $N_{\mathrm{q}} \sim N_{\bar{\mathrm{q}}} \sim N_{\mathrm{g}}$.

(4) To simplify the calculations, the color group SU(2) is used instead of SU(3).

Thus, to close the model one more quantity is needed, namely, the quasi-particle density $(N_q + N_{\bar{q}} + N_g)/V = n(T)$ as a function of temperature.

The most consistent way of finding the parameters of the model consists in employing the data of QCD lattice calculations, but at the present time it seems hardly possible. Therefore, the Monte Carlo calculations employ approximate dependences for input data. According to papers [343,614], the quasi-particle mass parametrization can be chosen as follows: $m(T)/T_c = \frac{0.9}{(T/T_c - 1)} + 3.45 + 0.4(T/T_c)$, where $T_c = 175\,\text{MeV}$ is the critical temperature. This parametrization agrees with the quark mass at two temperature values obtained in the lattice calculations [788]. According to [788] the quasi-particle masses are fairly large: $m_q/T \approx 4$ and $m_g/T \approx 3.5$. These values are much higher than those needed for the quasi-classical description [479, 785] of lattice calculations of thermodynamic QGP properties: $m_q/T \approx 1 - 2$ and $m_g/T \approx 1.3$. Furthermore, in the recent paper [513] the quark mass $m_q/T \approx 0.8$ was obtained from lattice calculations. Nevertheless, this work made use of the parametrization [343,614] for comparison with the MD data of paper [343]. The temperature dependence of the quasi-particle mass is shown in Fig. 5.29 (on the left).

The quasi-particle density necessary for calculations in a canonical ensemble was chosen with the help of the dependence $n(T) = 0.24T^3$. At first glance this as a rather low density. For example, in the classical calculations [343] the density was determined by the formula $n(T)/T^3 = 6.3$, which

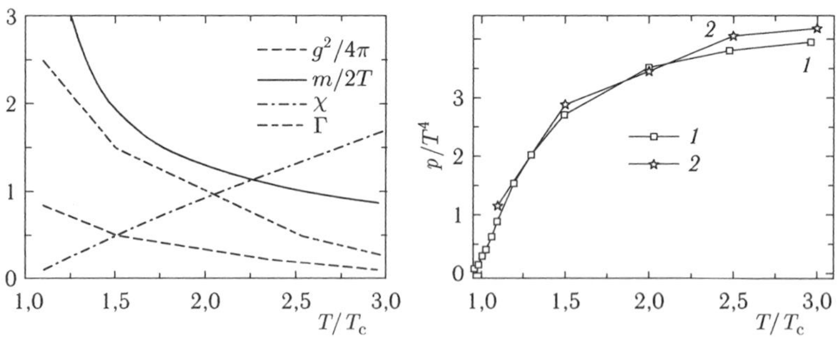

Fig. 5.29   On the left: temperature dependence of the input model parameters and nonideality QGP parameter. On the right: the equation of QGP state (temperature dependence of pressure) obtained from the results of PIMC simulation in comparison with the data of lattice calculations [89, 197].

corresponds to the density of an ideal gas of massless quarks, antiquarks, and gluons. Since the quasi-particle masses are very large in the considered model (the same as in paper [343]), the latter dependence greatly overestimates the density. Even in the quasi-particle models [479, 785] with lower masses, the densities are lower, $n(T)/T^3 \approx 1.4$. The above-mentioned parametrization of $m(T)$ gives even larger masses than those of [479, 785] — the chosen dependence $n(T)$ is not unrealistic. The dependence $n(T)$ was defined using the mean interparticle distance $r_s(T)$ in units $\sigma = 1/T_c = 1.1\,\mathrm{Fermi}$ (the Wigner–Seitz radius). It should be noted that the chosen dependence for $n(T)$ corresponds to the relation $r_s(T) = 1$.

The coupling constant employed in the Monte Carlo calculations was chosen from the condition of best agreement between the calculated pressure and the corresponding results of lattice calculations (see Fig. 5.29 (on the right)). It is also presented in Fig. 5.29. Such $g^2$ values are approximately consistent with the data of lattice calculations [513]. Hence, comparison with the data of lattice calculations by the equation of state (Fig. 5.29 on the right) was applied to optimize the model parameters for the purpose of predicting different QGP properties, including the internal structure and later on the nonequilibrium properties.

Note that although the dependences of the model parameters are not undeniable, they can be called self-consistent. The temperature dependence of density $n(T)$ is planned to be further on excluded from the list of parameters by passing over to the large canonical ensemble.

The nonideality parameter $\Gamma$ defined as the ratio of the mean potential energy to the mean kinetic energy and calculated for the above-described temperature dependences of the model parameters is also shown in Fig. 5.29 on the left. Its value appeared to be of the order of unity, which confirms that QGP is a nonideal Coulomb liquid rather than a gas. In the investigated temperature range, $1 < T/T_c < 3$, QGP is also degenerate because the degeneracy parameter $\chi = n\lambda^3$ (where $\lambda$ is the thermal wavelength of quasi-particles) varies from 0.1 to 1.7 (Fig. 5.29 on the left).

The details of Monte Carlo simulations with the use of path integrals were discussed above. The main idea consists in generation of the Markovian process in which a subsequent configuration is obtained from a previous one through a change in the particle coordinates. For QGP, in configuration generation the color variables $Q$ of all the particles are also used according to the group measure. The length of configuration sequence is so chosen as to provide convergence of thermodynamic quantities. A cubic cell with periodic boundary conditions is used in the simulations. The number of particles was chosen to be equal to $(N_\mathrm{q} + N_\mathrm{\bar q} + N_\mathrm{g}) = 40 + 40 + 40 = 120$ and the number of high-temperature factors in the operator identity to $n = 20$.

The study of the pair correlation functions $g_\mathrm{ab}(r)$ may provide insight into the spatial distribution of quasi-particles in QGP. These correlation functions determine the probability density of finding a pair of particles of types a and b at a given distance $r$ and can be written as follows:

$$g_\mathrm{ab}(R_1 - R_2) = \frac{1}{Z N_\mathrm{q}! N_\mathrm{\bar q}! N_\mathrm{g}!} \sum_\sigma \int dr dQ \delta(R_1 - r_1^\mathrm{a})$$

$$\times \delta(R_2 - r_2^\mathrm{b}) \rho(r, Q, \sigma, \beta).$$

The pair correlation functions only depend on the difference of quasi-particle coordinates because of system isotropy. In the classical system of noninteracting particles we have $g_\mathrm{ab}(r) \approx 1$, whereas the interaction and the effects of quantum statistics lead to a redistribution of particle position. The upper row of Fig. 5.30 shows the results for pair correlation functions at a temperature $T/T_\mathrm{c} = 3$. The pair correlation functions of identical particles

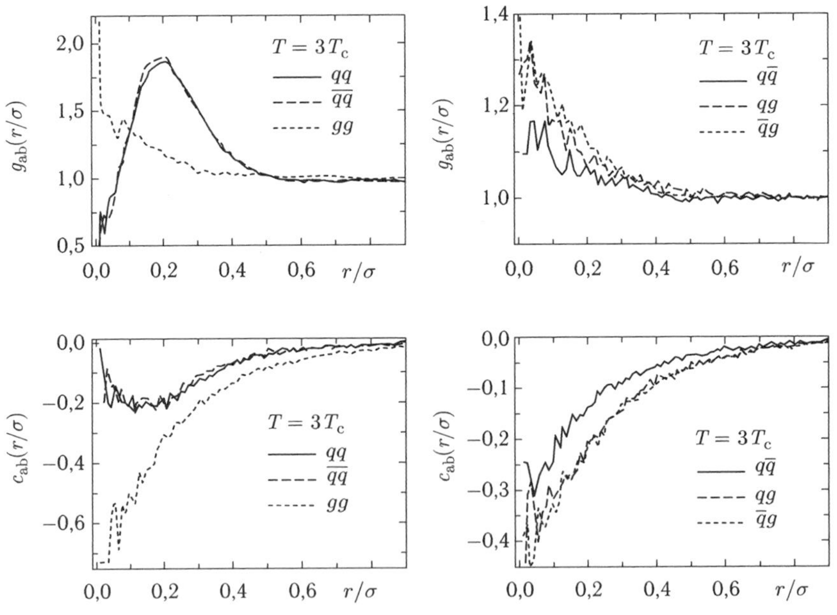

Fig. 5.30   Pair correlation functions (upper row) and color pair correlation functions (lower row) of identical (left column) and different (right column) quasi-particles at a temperature $T = 3T_\mathrm{c}$.

are presented in the upper left angle of Fig. 5.30. At large distances $r > 0.5\,\sigma$ all pair correlation functions coincide and tend to unity.

A great difference in the behavior of pair correlation functions of quarks and gluons (the pair correlation function of antiquarks is identical to that of quarks) is encountered at small distances. If the pair correlation function of gluons increases monotonously as the distance reduces to zero, the pair correlation function of quarks (and antiquarks) has a broad maximum. In simulations, the thermal wavelength $\lambda$ is approximately equal to $0.37\sigma$, i.e., the difference becomes obvious at distances of the order of $\lambda$.

The increase in the population of pair states of gluons at small distances is due to the effects of Bose statistics and attraction described by the color Coulomb potential. On the contrary, the decrease of the pair correlation function of quarks at small distances is a consequence of the Pauli principle. In an ideal Fermi gas, $g(r)$ is equal to zero for particles with the same spin projection and values of the color variable, whereas for particles with different values of the color variable and/or opposite spins the pair correlation function is equal to unity as $r \to 0$. As a consequence, the spin and color averaging of the pair correlation function gives its value less than unity for $r = 0$. At small distances such behavior was also observed in nonideal dense astrophysical electron–ion plasma and nonideal electron–hole plasma in semiconductors [124, 265]. The low probability of finding a pair of quarks (or antiquarks) at small distances causes an increase in the number of pairs of these particles at intermediate distances, which results in the occurrence of maximum on the corresponding pair correlation function.

All pair correlation functions of different types of quasi-particles (the upper right angle in Fig. 5.30) have similar behavior. At small distances, $r < 0.3\sigma$, a rapid growth is observed similar to the behavior of the gluon-gluon pair correlation function. This growth of the pair correlation functions at small distances is indicative of effective pair attraction of quarks and antiquarks, as well as quarks (antiquarks) and gluons.

Thus, the color vectors of these pairs of quasi-particles are antiparallel. This assertion can be verified by calculating the color pair correlation functions defined as

$$c_{\mathrm{ab}}(R_1 - R_2) = \frac{1}{Z N_{\mathrm{q}}! N_{\bar{\mathrm{q}}}! N_{\mathrm{g}}!}$$

$$\times \sum_{\sigma} \int dr\, dQ\, \langle Q_1^{\mathrm{a}} | Q_2^{\mathrm{a}} \rangle \delta(R_1 - r_1^{\mathrm{a}}) \delta(R_2 - r_2^{\mathrm{b}}) \rho(r, Q, \sigma, \beta).$$

The color pair correlation functions are shown in the lower row of Fig. 5.30. All of them assume negative values at small distances, which testifies to an antiparallel orientation of the color vectors of quasi-particle pairs. The minimum of $c_{qq}$ for $r = 0.2\sigma$ corresponds to the maximum of $g_{qq}$. The deep minimum on the gluon color correlation function at small distances is a consequence of Bose statistics and corresponds to the $g_{qq}$ maximum. Thus, for $T/T_c = 3$ the indiciations of spatial ordering are observed, for instance, the maximum of the quark–quark correlation function for $r = (0.1-0.2)\,\sigma$ which can be associated with the onset of liquid-like behavior of QGP. The QGP energy lowers because of the decrease in the color Coulomb interaction energy through a spontaneous "antiferromagnetic" ordering of color vectors. This causes clusterization of quarks, antiquarks, and gluons.

Figure 5.31 presents the pair correlation functions of identical particles for two temperatures $T = 1.1\,T_c$ and $T = 2\,T_c$ (the upper and middle rows). The contribution to the correlation functions is made by both bound and "free" states of quasi-particles. It should be noted however that a rigorous criterion of the division of states into bound and free cannot be introduced because of overlapping of quasi-particle clouds and a strong effect of the surrounding plasma. Nevertheless, a rough estimate of the fraction of bound states can be obtained from the following considerations. The product $r^2 g_{ab}(r)$ characterizes the probability of finding a pair of quasi-particles at a distance $r$ from each other. On the other hand, the corresponding quantum-mechanical probability is the product of $r^2$ and the two-particle Slater sum

$$\sum_{ab} = 8\pi^{3/2}\lambda_{ab}^3 \sum_a |\Psi_a|^2 \exp(-\beta E_a) = \Sigma_{ab}^d + \Sigma_{ab}^c,$$

where $E_a$ and $\Psi_a$ is the energy (with subtracted center-of-mass energy) and the wave function of the pair of quasi-particles, respectively. $\Sigma_{ab}$ determines the diagonal part of the corresponding density matrix. Summation in it is carried out over all possible states $f$ of both discrete ($\Sigma_{ab}^d$) and continuous ($\Sigma_{ab}^c$) spectrum.

At temperatures below the binding energy and at distances less than or of the order of the interaction radius the main contribution to the Slater sum is made by bound states. In electromagnetic plasma the product $r^2 \Sigma_{ab}^d$ has a sharp maximum at distances of the order of Bohr radius. Similarly, at low temperatures $r^2 g_{q\bar{q}}$ forms a clearly pronounced maximum near $r = 0.2\,\mathrm{Fermi}$ which can be interpreted as the size of the bound pair $q\bar{q}$.

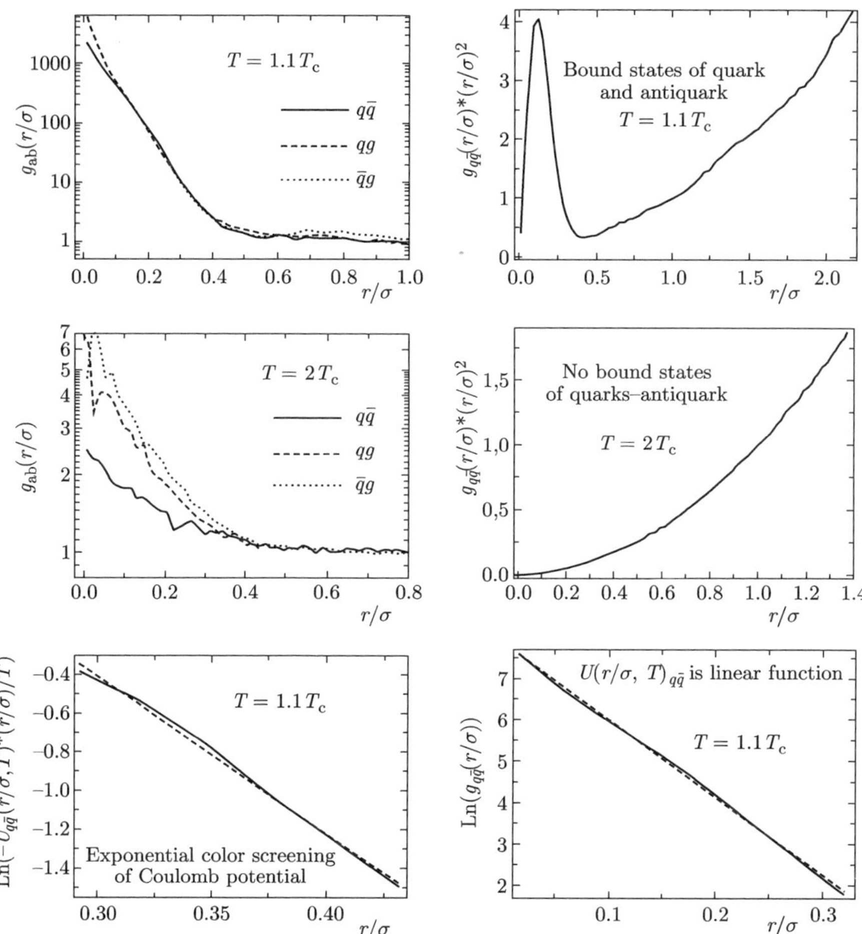

Fig. 5.31 The upper and middle rows: pair correlation functions at two different temperatures $T$ (on the left) and quark–antiquark correlation functions multiplied by the distance squared (on the right). The lower row: the middle force potential at small distances at $T = 1.1\,T_{\rm c}$ (on the right) and at large distances (on the left).

The calculations confirm the existence of bound states of heavy quarks and gluons proposed in papers [139,912] at intermediate temperatures near $T_{\rm c}$. With rising temperature these bound states break up much faster than was assumed earlier [139,912], which is consistent with the analysis of [559]. At a temperature $T = 2\,T_{\rm c}$ the bound states disappear completely, and $r^2 g_{\rm ab}$ behaves like $r^2 \Sigma_{\rm ab}^{\rm c}$.

Figure 5.31 also presents the mean force potential defined as the logarithm of the corresponding pair correlation function, $U_{ab}(r, T) = -T \ln g_{ab}(r, T)$, as depending on the distance. Such a definition agrees with the virial expansion of $g_{ab}(r, T)$ in the low density limit. Near the phase transition in QGP (at a temperature of the order of $T_c$) the mean force potential (the lower right angle) is a linear function at distances smaller than the bound-state interaction radius. This means that the quasi-particle interaction force does not depend on the distance. At large distances the mean force potential (the lower left angle) can, to a high accuracy be approximated by a screened Coulomb potential (Yukawa potential).

Chapter 6

# Statistical Substance Model

In the previous chapters, we saw that the description of thermodynamic, structural, and transport properties of real substances requires a correct quantum-mechanical calculation of the wave functions and the energy spectra of many-electron systems, which in the general case encounters great computational difficulties in solution of the quantum-mechanical many-body problem. This necessitates the introduction of simplifying assumptions underlying the modern condensed state models considered in the two chapters to follow. To begin, we shall review some history which will be presented in line with the perfect work [801].

## 6.1 On the quantum-mechanical many-electron structure calculations

The quantum-mechanical description of thermodynamic, structural, and electron properties of atoms, molecules of solids and liquids has long history [186, 562, 648, 801].

The relation between the positive charge of a nucleus, atomic number, and the atom position in the Periodic system was revealed in 1913. This relation was quantitatively grounded by quantum-mechanical methods.

In 1926, Heisenberg proposed matrix mechanics and Schrödinger — the basic nonrelativistic wave equation describing the motion of nuclei and electrons in molecules:

$$H\Psi = E\Psi, \tag{6.1}$$

which is a differential equation for the eigenvalues of energy $E$ and wave function $\Psi$ of a given state. Here, $H$ is the Hamiltonian operator and $\Psi$ depends on Cartesian and spin coordinates of particles. The only additional restriction is that the wave function $\Psi$ must possess a certain symmetry under particle permutations (i.e., must be antisymmetric for fermions, e.g., electrons and symmetric for bosons). The relativistic generalization of this equation was proposed shortly after by Dirac.

The Schrödinger equation is readily solved for the hydrogen atom and, as it became clear, the results thus obtained are identical to the earlier results due to Bohr. When the Dirac equation is used, allowance for the relativistic corrections results in virtually perfect compliance with experimental spectroscopic data. However, one could not find solution for any other system, which led to the widely known Dirac's 1929 remark: "The fundamental laws necessary for the mathematical treatment of a large part of physics and the whole of chemistry are thus completely known, and the difficulty lies only in the fact that application of these laws leads to equations that are too complex to be solved."

For the majority of researchers-physicists this meant that the fundamental discoveries in chemistry had ended, but the grand mathematical problem of their realization remained. Retrospectively, this assertion, considering its final character, looks exceedingly daring. Then, in 1929, only one — preliminary and approximate — quantum-mechanical hydrogen-molecule calculation existed which had been carried out by Heitler and London. The binding energy obtained in this calculation made up only about 70% of the experimental value. But in the 1930s, most physicists passed over to the study of the internal nuclear structure. Obviously, this standpoint was formally justified, for no serious shortcomings have yet been revealed in the complete Schrödinger–Dirac theory.

At that time, the main attempts were made to calculate the chemical structures of multielectron atoms [801]. Following paper [801], we shall first present calculations of isolated models that constitute the subject of quantum chemistry and then go over to consideration of condensed state.

If the attempts to obtain an exact solution are hopeless, then how can approximate mathematical procedures be worked out such that (a) will permit a qualitative interpretation of chemical phenomena and (b) will be capable of prediction?

In the 1930s, most of the papers were qualitative and assumed that electrons move along independent molecular orbitals. The basic principles of orbital theory for multielectron systems were laid by Hartree, Fock, and Slater. Let in a molecule with closed electron shells $2n$ electrons be distributed over $n$ orbitals $\psi_i$, $(i = 1, \ldots, n)$. Then the corresponding multielectron wave function is written as

$$\Psi = (n!)^{-1/2} \det[(\psi_1\alpha)(\psi_1\beta)(\psi_2\alpha)\ldots]. \tag{6.2}$$

Here $\psi_i$ are orthonormal and $\alpha$ and $\beta$ are spin functions. This single-configuration wave function is typically called Slater determinant.

If the molecular orbitals $\psi_i$ are varied so as to minimize the energy calculated as the mean value of the complete Hamiltonian $H$,

$$E = \langle\psi|H|\psi\rangle, \tag{6.3}$$

then the energy $E$ appears to be completely defined and, according to the variational principle, it is the upper boundary for the exact Schrödinger energy implied by the general wave equation (6.1). Such a procedure leads to the system of coupled wave equations for $\psi_i$ first derived by Fock. This method is called the Hartree–Fock approximation; its first application (to atoms) was due to Hartree.

Great success was achieved in 1951 when the Roothaan equations [855] were proposed in Chicago. Roothaan considered molecular orbitals in the form of linear combinations of a certain set of given three-dimensional one-electron functions $\chi_\mu$ ($\mu = 1, 2, \ldots, N$, where $N > n$). Thus,

$$\psi_i = \sum_{\mu=1}^{N} c_{\mu i}\chi_\mu. \tag{6.4}$$

Then the total energy (6.3) is varied with respect to the coefficients $c_{\mu i}$. This gives a system of algebraic equations which can be written in the matrix form (here, real functions and atomic units are used everywhere):

$$FC = SCE, \tag{6.5}$$

where

$$F_{\mu\nu} = H_{\mu\nu} + \sum_{\lambda\sigma} P_{\lambda\sigma}\left[(\mu\nu|\lambda\sigma) - \frac{(\mu\lambda|\nu\sigma)}{2}\right], \tag{6.6}$$

$$H_{\mu\nu} = \int \chi_\mu H \chi_\nu d\tau, \tag{6.7}$$

$$S_{\mu\nu} = \int \chi_\mu \chi_\nu d\tau, \tag{6.8}$$

$$E_{ij} = \varepsilon_i \delta_{ij}, \tag{6.9}$$

$$P_{\mu\nu} = 2 \sum_{i}^{n} c_{\mu i} c_{\nu i}, \tag{6.10}$$

$$(\mu\nu|\lambda\sigma) = \iint \chi_\mu(1)\chi_\nu(1)\frac{1}{r_{12}}\chi_\lambda(2)\chi_\sigma(2)d\tau_1 d\tau_2. \tag{6.11}$$

These and consequent equations make use of Latin indices for the molecular orbitals $\psi$ and Greek indices for the functions $\chi$ in which the expansion is carried out. Here $H$ is the Hamiltonian of the core describing the motion of one electron in the nuclear field only. The $\varepsilon_i$ eigenvalues are one-electron Fock energies, $n$ lower eigenvalues corresponding to the occupied molecular orbitals with indices $1, 2, \ldots, n$.

If the given functions $\chi_\mu$ are unambiguously determined by the positions of nuclei, then the above-presented nonlinear equations represent a self-consistent mathematical model. These are typically called self-consistent field (SCF) equations. In the early versions of molecular orbital method, the atomic orbitals of molecule-constituting atoms were chosen as $\chi_\mu$; this method is known as "a linear combination of atomic orbitals" (LCAO) or LCAOSCF. In the general case, the set of functions $\{\chi_\mu\}$ is referred to as the basis set. Customarily chosen are basis functions centered at nuclei and are only dependent on the atomic number (positive charge) of these nuclei.

Roothaan type equations can be extended to the case of electron configurations such that some orbitals are occupied by two electrons and the others by one electron. Another extension consists in the fact that electrons with spins $\alpha$ and $\beta$ refer to different molecular orbitals $\psi^\alpha$ and $\psi^\beta$; such approximation is usually called a spin-unrestricted configuration. In this case, two sets of coefficients, $c_{\mu i}^{\alpha}$ and $c_{\mu i}^{\beta}$, occur. The corresponding extension of the Roothaan equations was published in 1954 [803]. This approximation is called the unrestricted Hartree–Fock (UHF) method, and the approximation admitting a double or single orbital occupation is called the restricted open shell Hartree–Fock (ROHF) method.

The introduction of expansion in the basis set played a great role in the development of quantum chemistry. It transformed the mathematical problem of numerical solution of the system of coupled differential equations (like in the Hartree atomic calculations) into the double problem of calculation of three- or six-dimensional integrals (6.7), (6.8), and (6.11) and a subsequent solution of the system of algebraic SCF equations (6.5). If these integrals were calculated analytically, the model could appear to be exact in the sense that good arithmetical accuracy would be attained even in the case when the basic approximations of the model (the use of a single-configuration determinant and a finite basis) remain unsatisfactory.

In the 1950s, the calculation of these integrals was considered to be the main obstacle for progress. The atomic Slater-type orbitals (STO) that have exponential radial parts analogous to those in atomic hydrogen orbitals seemed to be the best set of basis functions in LCAOSCF theory. The one- and two-electron integrals (6.7), (6.8), and (6.11) in the two-center case can be calculated analytically. However, in three- and four-center cases great difficulties were encountered. At that time, this catch situation was known as "integration nightmare" [801].

Two ways were proposed to eliminate the above-mentioned difficulties with integration. One of them is to introduce approximations for the most complicated integrals and to introduce some parameters for the other integrals, as well as to find the values of these parameters empirically by way of fitting them to the experimental data. Such methods were then called semi-empirical. As concerns an alternative approach, without introduction of approximations or empirical parametrization, its range of applicability was then inevitably limited to only very small molecules; this approach was referred to as non-empirical or *ab initio*. The most wide-spread were the semi-empirical methods based on the zero differential overlap (ZDO) approximation, when in the majority of integrals the products of different atomic orbitals $\chi_\mu \chi_\nu$ were neglected. In application to $\pi$-electrons of organic molecules with conjugate bonds this approximation was called the Pariser–Parr–Pople (PPP) theory [770–772, 800]. Later, in 1964–1966, the generalizations of this theory describing all valence electrons, i.e., the methods of complete and intermediate neglect of differential overlap (CNDO, INDO) were obtained [804, 805], and then M.J.S. Dewar with colleagues reduced the PPP method to a more empirical version. The CNDO/INDO methods were real chemical models in the sense that they could be employed to examine many molecules, to vary the structure for the purpose of determining equilibrium geometry, and to construct potential surfaces. However, their resources were restricted by the errors due to numerous approximations of integrals and to a large number of empirical parameters.

The proposition to use Gaussian type functions as basis ones appeared to be a truly great achievement for the "*ab initio*" community. In 1950 S.F. Boys showed [132] that if the radial parts have the form $P(x, y, z) \exp(-r^2)$, where $P(x, y, z)$ is any polynomial of Cartesian coordinates $x$, $y$, and $z$, then all the integrals in SCF theory can be taken analytically. In its original form this proposition did not obviously arouse great interest because separate

Gaussian functions are poor approximation to atomic orbitals. It was clear, however, that later on it can be improved if a large number of such basis functions are used. The competition between the assertors of Slater and Gaussian functions lasted several years.

Computers were used in quantum mechanics beginning also with the 1950s. In 1959 several groups already existed that developed the programs of non-empirical calculations underlain by both Slater and Gaussian bases. In the early 1960s, other programs of general character were worked out. Among them, POLY ATOM and IBMOL program packets based on Gaussian functions should be specially mentioned. These programs made it possible to carry out a number of original calculations of molecular orbitals in LCAO or minimal basis.

A new algorithm was worked out for calculation of integrals, which allowed heightening the efficiency of calculations on basis sets of grouped Gaussian functions by more than two orders of magnitude [802].

This underlay the GAUSSIAN70 program for general access.

The models based on the Hartee–Fock approximation work much better if one set of ungrouped d-functions per each heavy atom (i.e., except hydrogen atoms) is included in the basis. Such a basis is denoted as $31G^*$ or $31G(d)$ [329, 425]. If, in addition, one set of ungrouped p-functions per each hydrogen atom is included in the basis, the latter will be denoted as $31G^{**}$ or $31G(d, p)$. These additional basis functions are called polarization functions. The complete model with basis $31G^*$ is accordingly denoted as $HF/31G^*$. The other important examples of basis set extension are as follows: the introduction of polarization functions with still higher angular momenta (e.g., the basis $31G(2df, p)$ which contains two sets of d-functions and a set of f-functions per each heavy atom, as well as one set of p-functions per each hydrogen atom); the use of diffuse functions that are especially useful for anions and excited electron states. In this case the basis is marked with an additional sign "+", for instance, $31G + (d)$.

A significant hidden defect of the Hartree–Fock method consists in neglect of the electron correlation under the motion of electrons with antiparallel spins ($\alpha\beta$-correlation). Already on the pioneering days of quantum chemistry, the theoreticians understood that when the correlation is neglected, the bond dissociation (i.e., breakage) energy becomes strongly underestimated. This can be qualitatively perceived by considering the process of complete hemolytic dissociation of bond when one electron remains on one center and the second electron on the other center. If the motion of these two electrons is uncorrelated, a nonzero probability exists that they both will remain on one and the same center.

The use of single-determinant wave functions implies neglect of the electron $\alpha\beta$-correlation; the work with refined functions inevitably means employment of several determinants. In the most practical approaches allowing for correlation, the first to be constructed is the Hartree–Fock determinant and then its linear combinations with other determinants. It is especially convenient to construct additional determinants of unoccupied or virtual molecular orbitals which are eigenfunctions of the Fock operator which correspond to higher energies. If a certain finite basis is used for a problem with $2n$ electrons and $N$ Cartesian basis functions, then $N - n$ virtual orbitals are obtained which can be occupied by spin $\alpha$ or $\beta$ electrons.

It is convenient to somehow change the notation and pass over to other basis functions, i.e., spin orbitals which are products of Cartesian basis functions and the spin functions $\alpha$ or $\beta$. Now $N$ is the size of the basis of spin orbitals (which is twice as large as the number of previous basis functions) and $n$ is the total number of electrons. This notation allows the description of both spin-restricted and spin-unrestricted cases. We shall use the indices $i, j, k, \ldots$ for occupied spin-orbitals and $a, b, c, \ldots$ for virtual ones. In this case, the single-determinant functions constructed on Fock orbitals can be divided into initial (i.e., Hartree–Fock) $\Psi_0$, singly excited $\psi_i^a$, doubly excited $\Psi_{ij}^{ab}$, etc. Then the full multi-determinant function can be written as

$$\Psi = a_0\Psi_0 + \sum_{ia} a_i^a \Psi_i^a + \sum_{ijab} a_{ij}^{ab}\Psi_{ij}^{ab} + \ldots \qquad (6.12)$$

The coefficients $a$ can be determined by the variational method minimizing the energy to be calculated. This is the configuration interaction (CI) method. If it is only singly excited determinants that are added, the energy does not lower because the occupied orbitals are already optimized. In the simplest effective form of the CI method, formula (6.12) only takes into account doubly excited determinants. This method is typically determined as CID. If singly excited determinants are also taken into account, the method is called CISD. The above-described versions of the CI method were first realized in the form of iteration schemes somewhere around 1970 and have been frequently used in practical calculations. If the expansion involves all possible excitations, which leads to a large but finite basis (if a finite basis set is used), this is the method of full configuration interaction (FCI). Although it is in principle desirable to exploit the FCI procedure, it usually needs too much computational efforts except for the case of very small systems.

In assignment of a standard basis, the CID and CISD models are well defined, and nevertheless they have some serious shortcomings. This is due to dimensional consistency. If a certain method, for example, CID is applied to two completely isolated systems, the resultant energy is not equal to the sum of energies obtained by applying the same method to each system separately. If the CID method is applied to, say, two isolated helium atoms, the wave function disregards a simultaneous excitation of pairs in each of the atoms since such excitation is strictly fourfold. This deficiency of the CID and CISD models must obviously result in unsatisfactory description of large molecules and interacting systems.

The second main method of allowance for the electron correlation makes use of perturbation theory. Suppose, we defined the excited Hamiltonian as

$$H(\lambda) = F_0 + \lambda\{H - F_0\}, \tag{6.13}$$

where $F_0$ is the Fock Hamiltonian (for which the functions single-determinant in expression (6.12) are exact eigenfunctions). Then $\Psi_0$ is the wave function for $\lambda = 0$ and the exact $\Psi$ (FCI) function is obtained for $\lambda = 1$. Next, the following procedure is applied: the energy under calculation is expanded in powers of $\lambda$:

$$E(\lambda) = E_0 + \lambda E_1 + \lambda^2 E_2 + \lambda^3 E_3 + \ldots, \tag{6.14}$$

the obtained series is cut off at a certain summand and then assume $\lambda = 1$. This method of perturbation theory was first proposed by Moller and Plesset [703]; the method often employs the notation MP$n$ which means that the series is cut off at a term of the order of $n$. The energy $E_0 + E_1$ in the approximation MP1 is identically equal to the Hartree–Fock value. MP2 is the simplest approximation used in practice to take into account the electron correlation; it only includes effects of double excitations. In the third order, MP3 also allows double excitations only. In the fourth order, the approximation MP4 describes (indirect) effects of single excitations, the main part of triple and to a certain extent some quadruple excitations.

If calculations are performed completely up to any given order of magnitude, the Moller–Plesset theory is dimensionally consistent. The problem is that with increasing order of magnitude the expressions for restrained terms turn out to be algebraically complicated and their calculation becomes increasingly complicated. Indeed, calculations by the Hartree–Fock method (without introduction of approximations for integrals) take time proportional to $N^4$, i.e., $N^5$ for MP2, $N^6$ for MP3, and $N^7$ for MP4. Allowance for the contributions of triple excitations to the energy in MP4 is the most cumbersome, and the applicability of the Moller–Plesset theory is normally limited to this level. The MP2, MP3, and MP4 models realized by several groups in the 1970s were included in the GAUSSIAN program [587,588].

More details concerning the state of the art in this area can be found in the brilliant review of the Nobel Prize winner J.A. Pople [801].

Models have been developed in recent years that allow calculation of the chemical system energy to an accuracy approaching the one reached in good experimental works. The description of the models given above shows that the models have two basic characteristics, namely, the basis set and the degree of allowance for correlation.

A detailed comparison of different models with experimental data was performed. The verification was performed with a large number (299) of experimental energy differences for molecules reaching the size of benzene (42 electrons). This data set included 148 formation heat values obtained from atomization heats, 85 ionization potentials, 56 values of electron affinity, and 8 values of proton affinity. All these experimental results are considered to be known with an error not higher than $1\,\text{kcal·mole}^{-1}$. The obtained mean deviation of calculation from experiment amounts to $1.02\,\text{kcal·mole}^{-1}$, which is close to the measurement error.

The worst results should also be mentioned. The largest absolute deviations are equal to $4.9\,\text{kcal·mole}^{-1}$ for the formation heat ($C_2F_4$), $7.0\,\text{kcal·mole}^{-1}$ for the ionization potentials ($B_2F_4$), $4.2\,\text{kcal·mole}^{-1}$ for electron affinity (NH) and $1.8\,\text{kcal·mole}^{-1}$ for proton affinity ($PH_3$ and $SH_3$).

Characterizing the modern state of non-empirical quantum-mechanical models one can say that great progress is reached on the way of approach of their predictive force to the experimental accuracy. For small molecules containing up to 50 electrons the computational error is not very far from the measurement error equal to $\Delta \approx 1\,\text{kcal·mole}^{-1}$.

## 6.2  Statistical model

As we have seen, the application of direct quantum-mechanical methods to multielectron systems is associated with the numerical solution of the non-relativistic Schrödinger equation for the multielectron ($N$) wave function

$\psi(\mathbf{r}_1, \mathbf{r}_2, \dots \mathbf{r}_n)$:

$$\left( -\frac{\hbar^2}{2m_e} \sum_j \nabla_j^2 - \sum_{j,i} \frac{Z_i e^2}{|\mathbf{r}_j - \mathbf{R}_i|} + \frac{1}{2} \sum_{j \neq j'} \frac{e^2}{|\mathbf{r}_j - \mathbf{r}_{j'}|} + E \right) \Psi = 0, \quad (6.15)$$

where $\mathbf{r}_j$ are positions of electrons and $R_i$ and $Z_i$ are positions and atomic numbers of nuclei; $\hbar$, $m_e$, and $e$ are the known fundamental constants, and $E$ is energy. This equation is written in the Born–Oppenheimer approximation in which (with the purpose of studying the electron dynamics) much heavier nuclei are considered to be motionless. The elaborated methods of numerical solution of this equation are cumbersome and require a lot of calculations growing exponentially together with the number of electrons in a system being considered. In his Nobel lecture, W. Kohn [562] called this effect the "exponential wall", and his Nobel Prize colleague J. A. Pople, when discussing the difficulties with numerical quantum-mechanical calculations, called them the "integration nightmare" [801]. This necessitates the search for simplified approaches to the problems some of which are to be considered in the chapters to follow. One of such models is the statistical model (the SCF method) which was proposed by Thomas [973] and Fermi [256] in 1927 and was widespread owing to its clearness and simplicity (see the reviews in Refs. [414, 547]).

The statistical model of matter (the Thomas–Fermi model, or method, TFM) [147, 547] makes the basis of a rather popular approximate approach widely applied to describe the properties of matter at its different hierarchic levels (atomic nucleus, atom, molecule, solid, etc.). Particularly developed are TFM applications to the theory of extreme states of matter occurring under the action of high pressures, high temperatures, or strong external fields. The corresponding fields of physics and related sciences (astrophysics, quantum chemistry, a number of applied disciplines) just make up the range of TFM applicability. The TFM popularity is due to its simplicity, obviousness, and universality. The latter means that the result of TFM calculations refers simultaneously to all chemical elements: passing over from element to element is realized by a simple scale transformation. These TFM features make it an exceedingly convenient tool of a qualitative and in many cases quantitative analysis. They are certainly due to the approximate character of TFM which is only capable of an exact description of reality in some extreme situations. Namely, to apply TFM as a quantitative theory the matter density or its temperature should be sufficiently high.

In this chapter, we shall consider the TFM theory as applied to ordinary electron–nuclear systems (atom, solid, plasma) without touching upon the TFM application to the atomic nucleus theory (for more details see Chapter 13 [100, 360, 414, 544, 566, 567]). The point of TFM application is the electron component of matter for a given state of nuclei. Under conditions of TFM applicability the phonon effects can to a considerable extent be described independently (in this connection see Refs. [3, 4, 545]).

The description of bound states of electrons in the range of high densities $\rho \geqslant \rho_0$ at extremely high pressures $p \gg e^2/a_0^4 \approx 300\,\mathrm{Mbar}$ or temperatures $T \gg \mathrm{Ry} \approx 10^5\,\mathrm{K}$ is radically simplified when the electron shells get crushed and their properties are described by a quasi-classical approximation to the SCF method, i.e., the Thomas–Fermi theory. In this model, the description of a quantum-mechanical system in the language of wave functions and energy eigenvalues is replaced by a simplified statistical representation in terms of the mean electron concentration $n(x)$ and energy $\mu$ at the Fermi distribution boundary for which there hold relations for a quasi-homogeneous degenerate electron gas [545][1]:

$$n(\mathbf{x}) = \frac{p_{\mathrm{F}}^3(\mathbf{x})}{3\pi^2},\tag{6.16}$$

$$\mu = \frac{p_{\mathrm{F}}^2(\mathbf{x})}{2} + U(\mathbf{x}),\tag{6.17}$$

where $p_{\mathrm{F}}$ is momentum and $U(\mathbf{x})$ is potential energy of an electron in the field of the rest of the electrons and external sources (nuclei).

Allowing for the Poisson equation we arrive at

$$-\Delta U = \frac{\Delta p_{\mathrm{F}}^2}{2} = 4\pi n + \text{external source};\tag{6.18}$$

and then one can readily pass over to the nonlinear differential equation for $p_{\mathrm{F}}^2$ or $U$ (the Thomas–Fermi equation).

Numerical integration of the nonlinear differential Thomas–Fermi equation allows finding the electron density $n(\mathbf{x})$ by which all the thermodynamic functions of the electron gas of an atomic cell are reconstructed.

The local characteristics of a "cold" electron system are described by the distribution function

$$f(\mathbf{x}, \mathbf{p}) = \theta(p_{\mathrm{F}}^2(\mathbf{x}) - p^2),\tag{6.19}$$

where $\theta(x) = 1$ $(x > 0)$, $\theta(x) = 0$ $(x < 0)$. In particular, the densities in coordinate and momentum space are given by the formulas $(d^3p \equiv d\mathbf{p}/(8\pi^3))$

$$n(\mathbf{x}) = 2 \int d^3p\, f(\mathbf{x}, \mathbf{p}), \quad n(\mathbf{p}) = 2 \int d\mathbf{x}\, f(\mathbf{x}, \mathbf{p})\tag{6.20}$$

---

[1] Here and below, the atomic units $l = \hbar = m = k = 1$ are used.

(the first of these leads to (6.16)). The thermodynamic properties of the system are described by the expression for its energy: $E = E_k + E_e + E_i$, where

$$E_k = \frac{1}{10\pi^2} \int d\mathbf{x} p_F^5(\mathbf{x}) = \frac{3(3\pi^2)^{2/3}}{10} \int d\mathbf{x} n^{5/3}(\mathbf{x}),$$

$$E_e = \int d\mathbf{x} n(\mathbf{x}) U_e(\mathbf{x}), \quad E_i = 1/2 \int d\mathbf{x} d\mathbf{x}' \frac{n(\mathbf{x})n(\mathbf{x}')}{|\mathbf{x} - \mathbf{x}'|}. \tag{6.21}$$

To generalize the TFM to the case of finite temperatures it is necessary to replace (6.19) by the corresponding quantum-statistical expression

$$f(\mathbf{x}, \mathbf{p}) = \left[ \exp\left( \frac{p^2 - p_F^2(\mathbf{x})}{2T} \right) + 1 \right]^{-1}. \tag{6.22}$$

Then (6.20) gives

$$n(\mathbf{x}) = \frac{\sqrt{2}}{\pi^2} T^{3/2} I_{1/2}(\lambda(\mathbf{x})), \tag{6.23}$$

where $I_n$ is the special Fermi–Dirac function [547], $\lambda = p_F^2/2T$. The expression (6.23) replaces (6.16); as regards (6.17)–(6.18), these equalities remain valid but by $\mu$ we now mean the chemical potential of the "hot" system. Having available the solution of the Thomas–Fermi equation for $T \neq 0$ and the expression for the free energy,

$$F = \frac{\sqrt{2}}{\pi^2} T^{5/2} \int d\mathbf{x} \left( \lambda I_{1/2}(\lambda) - \frac{2}{3} I_{3/2}(\lambda) \right) + E_e + E_i, \tag{6.24}$$

we can obtain a complete thermodynamic description of the "hot" system. In particular, the chemical potential and pressure are obtained from (6.24) by differentiation with respect to the number of particles and the volume of the system, respectively.

This calculation is substantially simplified if $U|_S = 0$ at the boundary of the system. In this case,

$$\mu = \frac{1}{2} p_F^2 \Big|_S, \quad P = \frac{2\sqrt{2}}{3\pi^2} T^{5/2} I_{3/2}\left( \frac{\mu}{T} \right). \tag{6.25}$$

The expression for the pressure $P$ can be regarded as the result of applying the principle of quasi-uniformity and the virial theorem $(P = (2/3)E_k/V)$ at points where the electron gas is locally ideal. This is precisely the situation in the widely used model of spherical Wigner–Seitz cells (see e.g., [361, 544]). In this model a substance is divided into an assembly of spherical cells, each of which contains one nucleus and is electrically neutral as a whole; the latter condition determines the radius $R$ of the cell, $\int d\mathbf{x}\, n(\mathbf{x}) = Z$ and the behavior of $U(x)$ near the boundary of the cell:

$$U(x) \sim (R - r)^2. \tag{6.26}$$

Use of the cell model makes it possible to go over from a many-center problem to a spherically symmetric single-center problem, and this makes it very much simpler to solve the Thomas–Fermi equation.

An explicit solution of a many-center problem is possible only in that region of high pressures or temperatures in which the TFM differs little from the model of an ideal uniform gas [544] (in the following we shall call this region the region of uniformity). In the zeroth approximation (the ideal gas) the relations (6.23)–(6.25) give:

$$n = \frac{\sqrt{2}}{\pi^2} T^{3/2} I_{1/2}(\lambda), \quad F = \frac{\sqrt{2}}{\pi^2} V T^{5/2}\left(\lambda I_{1/2}(\lambda) - \frac{2}{3} I_{3/2}(\lambda)\right),$$

$$\tag{6.27}$$

$$P = \frac{2\sqrt{2}}{3\pi^2} T^{5/2} I_{3/2}(\lambda), \qquad \lambda = \frac{\mu}{T} = \frac{p_{\mathrm{F}}^2}{2T}.$$

In the next approximation, which takes into account the small effects of the Coulomb interaction,

$$\delta p_{\mathrm{F}}^2(\mathbf{x}) = 2\delta\mu - 2n \int \frac{d\mathbf{x}'}{|\mathbf{x} - \mathbf{x}'|} - 2U_e(\mathbf{x}),$$

$$\delta F = \frac{n^2}{2} \int \frac{d\mathbf{x}\, d\mathbf{x}'}{|\mathbf{x} - \mathbf{x}'|} + n \int dx\, U_e(\mathbf{x}) + E_0,$$

where $E_0$ is the energy of the Coulomb interaction of the external sources with each other. Taking into account the electrical neutrality of the system over scales $\sim (Z/n)^{1/3}3$, we have

$$\delta F = -\alpha Z^{2/3} n^{4/3} V, \tag{6.28}$$

where $\alpha$ is the analog of the Madelung constant and depends on the concrete structure of the short-range order in the system; in the cell model, $\alpha = (9/10)(4\pi/3)^{1/3}$.

Correspondingly,

$$\delta\mu = -\frac{4}{3}\alpha Z^{2/3} n^{1/3}, \quad \delta P = -\frac{\alpha}{3} Z^{2/3} n^{4/3}. \tag{6.29}$$

The TFM can also be easily generalized to the relativistic case. However, the corresponding effects would be appreciable only under conditions (near the nuclei, at ultra-high pressures or temperatures) such that the TFM is either completely inapplicable or differs little from the ideal-gas model.

The TFM possesses the important property of self-similarity: the atomic number $Z$ appears in any physical relation only in the form of the combinations:

$$xZ^{1/3}, \quad \omega Z^{-1}, \quad pZ^{-2/3}, \quad EZ^{-4/3}, \quad TZ^{-4/3}, \quad PZ^{-10/3}, \quad nZ^{-2},$$

$$(6.30)$$

where $x$ is the length, $\omega$ the frequency, $p$ the momentum, $E$ the energy (free energy) per particle, $T$ the temperature, $P$ the pressure, and $n$ the density. This is associated with the convenient property of TFM universality with respect to the nuclear charge $Z$.

The derivation of the TFM was based on the assumption of a small degree of nonuniformity of the system. Below we shall need a quantitative measure of this quantity. For this we use a characteristic "nonuniformity length" $L$, over which the characteristics of the system change noticeably, In the region of applicability of the TFM this quantity can be found directly from equation (6.18) by omitting the term with the external sources and replacing $\Delta$ by $1/L^2$:

$$L \sim \frac{p_{\mathrm{F}}}{\sqrt{n}} \sim r_{\mathrm{D}}. \qquad (6.31)$$

This quantity coincides with the Debye radius $r_{\mathrm{D}} \sim v/\omega$ for the electron subsystem, where $v$ is the characteristic velocity and $\omega$ is the characteristic (plasma) frequency.

To obtain resultant thermodynamic characteristics of the model, the electron terms should be taken into account together with the motion of nuclei typically described in the ideal-gas or quasi-harmonic [333] approximation. At $T \gg 10^7$ K and solid-state densities one should also take into account the contribution of equilibrium radiation.

The TFM is a quasi-classical limit $\lambda \to 0$ with respect to the equations of self-consistent Hartree field, and hence the modifications of this model involve more detailed allowance for the correlation, quantum-mechanical, and relativistic effects [545]. The presence of correlation corrections is explained by the difference between the self-consistent Hartree field and the true field inside the atomic cell. These corrections are due to antisymmetry of the wave functions of electrons and are interpreted as exchange correlation effects. In addition, because of inaccuracy of the picture of independent particles adopted in the model the dynamic correlation effects occur.

## 6.3  Oscillation effects

Quantum-mechanical corrections are due to the use of quasi-classical formalism and are divided into the part regular in $\hbar^2$ (called quantum part) reflecting the existence of the nonlocal coupling $n(x)$ with potential $U(x)$ because of the uncertainty principle and the irregular correction reflecting nonmonotony of the physical quantities due to the discrete energy spectrum [335]. It is important that the introduction of the oscillation correction characterizes the most up-to-date TFM modifications [545], whereas the allowance for the exchange, correlation, and quantum corrections [334, 408] is traditional for high energy density physics. The relative value of the correlation and quantum effects is controlled by the dimensionless parameters [545], $\delta_{\mathrm{cor}} \sim \delta_0^\nu$ and $\delta_{\mathrm{quant}} \sim \delta_{\mathrm{exch}} \sim n/p_{\mathrm{F}}^4$, which in the domain of degeneracy ($n^{2/3} \gg T$, $p_{\mathrm{F}} \sim n^{1/3}$, $\delta_0 \sim n^{-1/3}$, $\nu = 2$) are equal to $\delta_{\mathrm{cor}} \sim n^{-2/3}$, $\delta_{\mathrm{exch}} \sim n^{-1/3}$, and in the classical region ($n^{2/3} \ll T$, $p_{\mathrm{F}} \sim T^{1/2}$, $\delta_0 \sim n^{1/3}/T$, $\nu = 3/2$) $\delta_{\mathrm{cor}} \sim n^{1/2}/T^{3/2}$, $\delta_{\mathrm{exch}} \sim n/T^2$.

As is shown in paper [547], the shell effects reflect irregularities of the physical quantities caused by the discrete energy spectrum; however, in the case of continuum these effects can result from de Broglie wave interference. In the quasi-classical region ($\xi \ll 1$), the shell effects are described by the expressions that are periodic functions of $1/\xi$. The quasi-classicality parameter $\xi$ is equal to $\xi = \frac{1}{2\pi}\left|\frac{d\lambda}{d\mathbf{r}}\right| \ll 1$ with $\lambda(\mathbf{r})$ as the particle wavelength and $\mathbf{r}$ as the particle radius vector. In the three-dimensional case we have $\xi \sim N_{\mathrm{e}}^{-1/3}$ [547].

Confining to the case $T = 0$, we shall represent the shell correction to the density in the form

$$n' = n_1' + n_2' + n_3'. \tag{6.32}$$

The quantity $n_1'$ describes the irregularities that appear when the individual energy levels cross the Fermi level. The second term of (6.32) describes not only "energy" shell oscillations but also spatial shell oscillations. Finally the quantity $n_3'$ being a normalizing term, satisfies the equality

$$\int dx(n_2' + n_3') = 0. \tag{6.33}$$

The shell oscillations of thermodynamics quantities are determined, according to (6.33), entirely by the term $n_1'$. If, however, we are interested in the shell oscillations of the density itself, only the term $n_2'$ is important for these.

We first give, following Ref. [547], the complete solution of the problem for the one-dimensional case, considering a finite motion with turning points $R$ and $R'$. We introduce the notation

$$S = \int_R^x dx\, p_F, \quad \tau = \int_R^x \frac{dx}{p_F},$$

for the action function and time of motion on the Fermi boundary (the same symbols with the subscript 0 correspond to integration from $R$ to $R'$). In addition, we introduce the periodic function

$$[f(S_0)]^* = f(S_0) \left( -\frac{\pi}{2} < S_0 < \frac{\pi}{2} \right),$$

with a periodic continuation outside this region; the discontinuities of this function, as is clear from the quantization condition, correspond exactly to coincidence of an energy level with the Fermi boundary. In this notation

$$n_1' = -\frac{2}{\pi p_F \tau_0}[S_0]^*, \qquad \int_R^{R'} dx\, n_1' = -\frac{2}{\pi}[S_0]^*. \tag{6.34}$$

In combination with the expression $n = 2p_F/\pi$ (the one-dimensional TFM), this leads to the result that the total number of particles changes by two as each energy level passes through. Furthermore

$$n_2' = -\frac{\csc\left(\frac{\pi\tau}{\tau_0}\right)}{p_F \tau_0} \cos\left(2S - 2\frac{\tau}{\tau_0}[S_0]^*\right). \tag{6.35}$$

On going over to semi-infinite motion ($R' \longrightarrow \infty$) $\tau_0 \longrightarrow \infty$ and

$$n_1' = 0, \quad n_2' = -\frac{1}{\pi p_F \tau} \cos(2S). \tag{6.36}$$

The three-dimensional spherically symmetric problem, in which there is in addition a summation over the orbital quantum number, leading to partial smoothing of the oscillations (e.g., in place of the discontinuities of the function $n_1'$ itself, discontinuities only of its derivative appear), is solved analogously. We introduce the notation,[2]

$$S = \int_R^r dr \sqrt{p_F^2 - \frac{1}{4r^2}} \approx \int_0^r dr\, p_F + \frac{\pi}{2}, \quad \tau = \int_0^r \frac{dr}{p_F}, \quad \delta = \oint_0^r \frac{dr}{r^2 p_F}.$$

---

[2]In the case (to which the formulas cited pertain) when attractive Coulomb forces act at the fcoordinate origin, $R = 0$ and the symbol $\oint$ means that the quantity $\int_0^r dr\, r^{-3/2}\sqrt{2Z}$ is subtracted from the integral.

for the action function, time of motion and derivative of the angle of rotation with respect to the angular momentum in the state with zero orbital angular momentum. Then the three-dimensional analogs of (6.34) and (6.36) take the form

$$\int d\mathbf{x} n_1' = -\frac{2}{\pi \delta_0} [S_0^2]^*,$$
(6.37)

$$n_2' = -\frac{1}{4\pi^2 r^2 p_{\mathrm{F}} \tau \delta} \sin\left( 2 \int_0^r dr p_{\mathrm{F}} \right),$$
(6.38)

(for more-general relations see Refs. [548, 907]).

There is also another method for describing shell effects, which is more transparent physically and is applicable in the general case of nonseparable variables. It is based on an analysis of the semi-classical Green function, which, in the one-dimensional case, has the form

$$G(x, x', \varepsilon) = -\frac{1}{\sqrt{P_\varepsilon(x) P_\varepsilon(x')}} \sum_j \exp\left[ i \left( \int_{\Gamma_j} ds |p_e| - \frac{\pi}{2} n_j \right) \right], \quad (6.39)$$

$$P_\varepsilon = \sqrt{2(\varepsilon - U)},$$

here, $\varepsilon$ is the particle energy in the Hartree method, $\Gamma_j$ is the trajectory of the classical motion from $x$ to $x'$; there exists, generally speaking, a whole set of such trajectories, of which one ($j = 0$) joins these points directly, as it were, and the others join them by varying numbers of reflections at the turning points. The number of such reflections is equal to $n_j$, and the last term in the phase of (6.39) corresponds to the change of phase in the reflections.

The density is expressed in terms of the Green function at coincident points. The trajectory with $j = 0$ then degenerates to a point, and the corresponding term of (6.39) loses its oscillatory character and leads to the TFM result. The remaining terms of (6.39) are responsible for the shell effects. In the case of semi-infinite motion (Fig. 6.1(a)) there are two terms ($j = 0, 1$), with $n_0 = 0$ and $n_1 = 1$, in the sum (6.39). For finite motion, however, the number of terms is infinite. These include, first of all, terms corresponding to a complete cycle ($n_j$ is an even number); the sum over such cyclic trajectories (Fig. 6.1(c)) forms the term $n_1'$. In addition, there are "semi-cyclic" trajectories (Fig. 6.1(b)), for which $n_j$ is an odd number; the sum over such trajectories gives the term $n_2$ describing spatial oscillations of the density. Thus, the latter arise as a result of interference

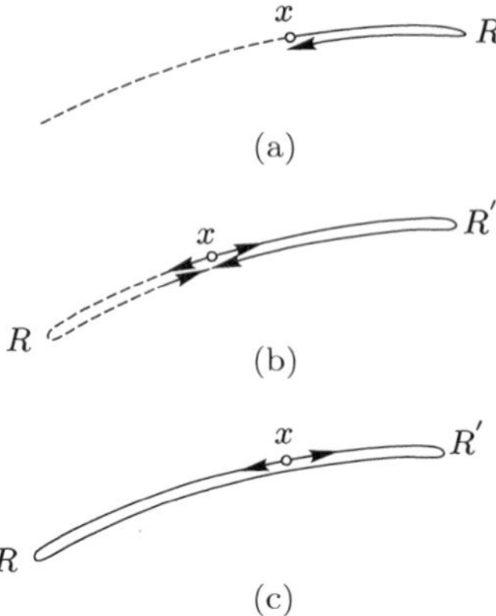

Fig. 6.1   Trajectories corresponding to shell effects, (a) infinite motion in one direction, (b) finite motion–semicyclic trajectory, (c) finite motion–cyclic trajectory.

of the incident wave and the wave reflected at the turning point (a standing de Broglie wave).

In the three-dimensional case the qualitative picture remains as before, although the corresponding expressions are appreciably more complicated. We note, first of all, that for the semi-cyclic trajectories it is necessary to take linear trajectories, on which the particle returns from the turning point along the same path as in its motion to this point. This corresponds to a true turning point (Fig. 6.1), at which all components of the momentum vanish (the distinct character of an s-wave is connected with precisely this; see above). Furthermore, the Coulomb center of attraction, which, because of the well-known quantum-mechanical effects, does not allow particles to pass through itself, must be counted as one of the turning points. However, the inapplicability of the semi-classical description at this point leads to the result that the phase change in the reflection is opposite in sign to the usual phase change; one can verify this by using the exact expression for the Green function in a Coulomb field [447] and matching it with the semi-classical expression in the same way as is done in the case of an ordinary turning point [358, 359].

We note that shell effects violate the self-similarity of the TFM in a substantially more radical way than do the other effects. After the change to the variables (6.30) there remains a complicated irregular dependence on $Z$, which corresponds to the actual oscillations of the physical quantities with variation of the composition of the substance.

In line with Ref. [547] we will consider, shell effects at nonzero temperatures, and also their contribution to thermodynamic quantities. We note immediately that, in a "hot" system, a number of factors leading to smoothing of the oscillations appear. Our formulation of the problem makes it possible to take the chief of these into account, namely, the smearing-out of the distribution of electrons over the levels. It is found that in the region of interest to us this smoothing is described simply by a temperature-dependent coefficient:

$$n_2'(T) = k_n n_2'(0),$$

$$\delta F(T) = k_{\mathrm{F}} \delta F(0) \equiv k_{\mathrm{F}} \delta E. \tag{6.40}$$

In the region of pressures and temperatures under consideration, matter is "metallized"; accordingly, at any temperature the electrons occupy parts of both the continuous and the discrete spectrum (Fig. 6.2). The boundary separating these parts corresponds, according to (6.26), to the energy value

$$\bar{\mu} = 0, \tag{6.41}$$

in reality, it is somewhat lower because of the smearing-out of the levels into bands and because of their overlap. However, when the condition for semi-classicality at the level $\bar{\mu}$, which has the form $(Z/n)^{1/2}U^{1/2} \gg 1$ or

$$n \ll Z^4, \tag{6.42}$$

is fulfilled, these effects are small and, to logarithmic accuracy, do not affect the results (for more detail, see Ref. [549]).

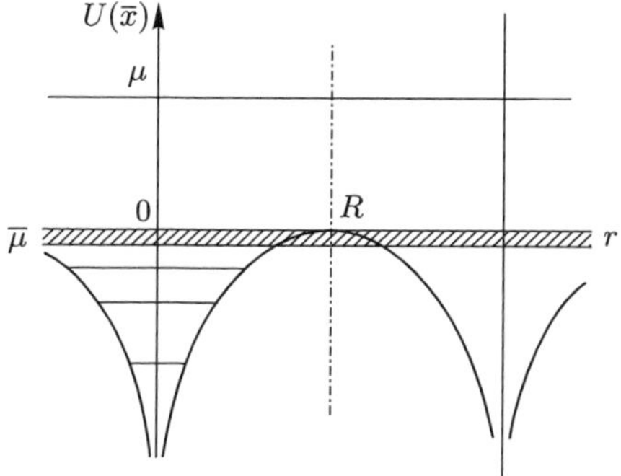


Fig. 6.2 Scheme of the levels of highly compressed matter. The region of overlap of bands is shaded.

The shell corrections to the density are found from (see also Ref. [743]):

$$n'(T,\mu) = \int\limits_{-\infty}^{\infty} d\varepsilon \, \frac{f(\mu - \varepsilon)\partial n'(0,\varepsilon)}{\partial \varepsilon}. \tag{6.43}$$

Using (6.38), we find

$$k_n = \frac{2\pi\tau T}{\operatorname{sh}(2\pi\tau T)}. \tag{6.44}$$

Since $\tau \sim Z^{1/3} n^{-2/3}$, the density oscillations become exponentially small for $T > n^{2/3} Z^{-1/3}$ (in the case of relatively weakly compressed matter — for $T > Z$). But if the temperature is less than the above value, the coefficient (6.44) is close to unity and the dependence of $n'$ on $T$ is determined by the second factor $n'(0,\mu)$ (see (6.40)). With increasing temperature the chemical potential becomes negative and increases in magnitude.

Therefore, an increase of $T$ leads to disappearance of the oscillations corresponding to deeper and deeper shells for which the potential energy becomes comparable with the quantity $|\mu|$. These conclusions are confirmed by a comparison with the results of an exact calculation by the Hartree–Fock method (see Refs. [865, 1097]).

The corrections to the free energy are determined by the quantity $n'_1$, which is nonzero only in the region of the discrete spectrum (see (6.36), (6.38)). Therefore, using (6.43), it is necessary to put the upper limit equal to $\bar{\mu} = 0$ (see (6.41)). Analogously, it is easy to see that the expression for is itself equal to $n'(0,\bar{\mu})$ for $\varepsilon > \bar{\mu}$. Estimation of the correction to free energy with the use of (6.43) gives

$$\delta F(T) = -T \ln\left[\exp\left(\frac{\mu}{T}\right) + 1\right] \int dx n'_1(0) - \int\limits_{-\infty}^{0} d\varepsilon f(\mu - \varepsilon) \int dx n'_1(0).$$

Integrating the second term by parts, it is easy to see that it makes a relative contribution of the order of $1/\mu\tau_0$ (degenerate case) and $1/T\tau_0$ (classical case). When we take into account that $\tau_0 \sim n^{-1/2}$, these parameters are found to be small throughout the region of applicability of the TFM. Therefore, the problem reduces to the first term of $\delta F(T)$. In particular, for $T = 0$ (6.37) gives

$$\delta E = -\mu_0 \int dx n'_1(0) = \frac{2}{\pi}\frac{\mu_0}{\delta_0}[S_0^2]^*, \tag{6.45}$$

where $\mu_0$ is the "cold" chemical potential and the other quantities in (6.45) are referred to the boundary $\bar{\mu}$. Hence,

$$k_{\mathrm{F}} = \frac{1}{\mu_0} T \ln\left[\exp\left(\frac{\mu}{T}\right) + 1\right].\tag{6.46}$$

This coefficient also falls off with temperature, but substantially more slowly than (6.44): it can be seen from (6.27) that (6.46) decreases in a power-law fashion, becoming negligibly small only when $T \gg \mu_0$.

The physical nature of the shell oscillations of the free energy consists in the fact that, on variation of the density of matter or its composition, energy levels are "squeezed out" of the discrete spectrum into the continuous spectrum, or vice versa, whenever the value of the action in (6.45) becomes equal to an odd multiple of $\pi/2$ (see Fig. 6.2).[3] It is this which leads to the irregularities in the behavior of the physical quantities. Leaving a more detailed discussion of these questions, we give here explicit expression valid in the uniformity regime and pertaining to the case $T = 0$.

In place of the density we shall use the cell volum $v = Z/n$, and denote $\Im = 5P/(3\pi^2)^{2/3}$. Then, we have

$$S_0 = kZ^{1/2}v^{1/6} - \frac{\pi}{2}, \quad k = 3\left(\frac{3}{4\pi}\right)^{1/6} \ln\frac{1+\sqrt{3}}{\sqrt{2}} \approx 1.56.\tag{6.47}$$

Analogously [549]

$$\delta_0 = \frac{(36\pi)^{1/6}}{36} \ln(Z^3 v) Z^{-1/2} v^{-1/6}.$$

Hence, for the shell correction to the energy we have

$$\delta E = 36\left(\frac{9\pi}{4}\right)^{1/6} \frac{Z^{7/6}v^{-1/2}}{\ln(Z^3 v)}[S_0^2]^*,\tag{6.48}$$

this quantity, like (6.45), refers to the volume of one cell.

It is important to note that, in reality, (6.48) corresponds to a situation that is unstable near the points of irregularity

$$v_n = Z^{-3}\left(\frac{\pi n}{k}\right); \quad n = 1, 2 \ldots.\tag{6.49}$$

This can be seen from the expression for the shell correction to the pressure:

$$\delta P = -12\left(\frac{9\pi}{4}\right)^{1/6} k\frac{Z^{5/3}v^{4/3}}{\ln(Z^3 v)}[S_0]^*,\tag{6.50}$$

---

[3]The shell effects disappear after all the energy levels have moved out into the continuous spectrum; this corresponds to $n > Z^4$ (see (6.42)).

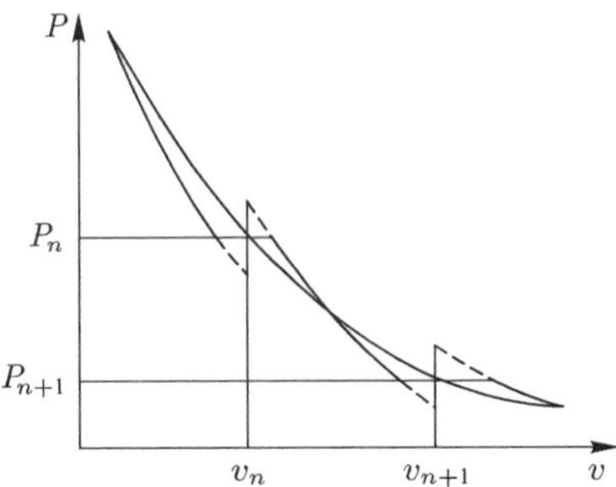

Fig. 6.3   Dependence of the pressure on the volume (shell effects).

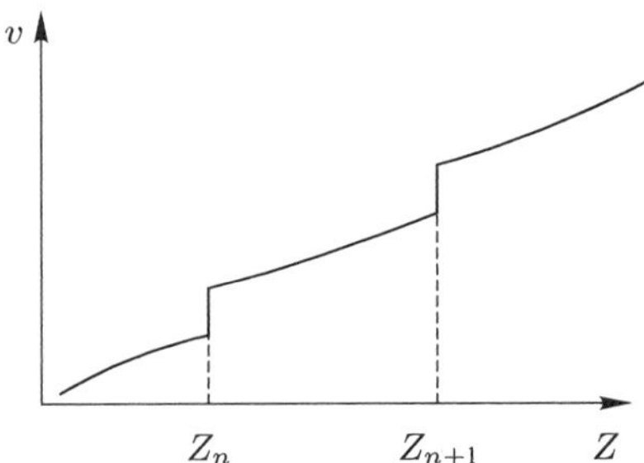

Fig. 6.4   Dependence of the atomic volume on the atomic number (shell effects).

(see the dashed lines in Fig. 6.3) which is obtained from (6.48) by differentiating its last, most rapidly-varying factor with respect to the volume. It is well known that, because of the negative value of the bulk modulus, such a state is unstable and rearranges, as a result of which the pattern depicted in Fig. 6.3 by the thick solid curve arises. This corresponds to a first-order phase transition between a phase in which the $n$th level lies in the discrete spectrum and another phase in which the $n$th level lies in the continuous spectrum. The pattern of oscillations of the volume on variation of the pressure or atomic number is depicted by Fig. 6.3 (rotated through $90°$) and Fig. 6.4. The corresponding analytic expression has the form

$$\delta v / v = -36 \left( \frac{9\pi}{4} \right)^{1/6} \cdot \frac{k Z^{1/3} [S_0]^*}{(3\pi^2)^{2/3} \Im^{1/5} \ln \left( \frac{Z^{20/3}}{\Im} \right)}, \qquad (6.51)$$

where

$$S_0 = \frac{k Z^{2/3}}{\Im^{1/10}} - \frac{\pi}{2},$$

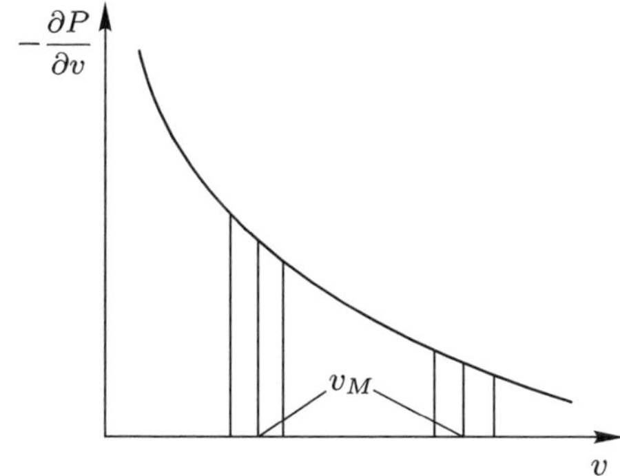



Fig. 6.5   Dependence of the compressibility on the volume (shell effects).

and the points of irregularity (phase-transition points) are

$$\Im_n = Z^{20/3} \left(\frac{k}{\pi n}\right)^{10}, \quad Z_n = \Im^{3/20} \left(\frac{\pi n}{k}\right)^{3/2}. \tag{6.52}$$

The large value of the numerical coefficient in (6.51) is worth special mention, and makes the shell oscillations entirely noticeable even under conditions when they are essentially small.

The shell effects in quantities obtained from the pressure by differentiating it further with respect to the volume are particularly clearly manifested. Thus, the modulus of elasticity (Fig. 6.5) is found to be always equal to zero in the region of coexistence of the phases, while the Gruneisen constant for the electronic component, corresponding to a further differentiation with respect to the volume, even becomes negative.

The numerous TFM modifications corresponding to the exchange, correlation, quantum, and shell effects were obtained in the assumption that their contribution to the physical quantities is relatively small. In fact, when we speak of the first terms of expansion in the corresponding parameters $\delta$, smallness of the first term of expansion allows discarding all the other terms. The corresponding conditions expressed in the language of density and temperature are shown schematically in Fig. 6.6.

However, meeting these conditions does not yet guarantee applicability of the theory to all physical characteristics of a substance. This is because of the above-mentioned substantial inapplicability of TFM in the neighborhood of the nucleus (and on the periphery of an isolated atom). In this region, the quantum effects are so substantial that it devalues not only the TFM itself, but also its generalizations. This manifests itself in the

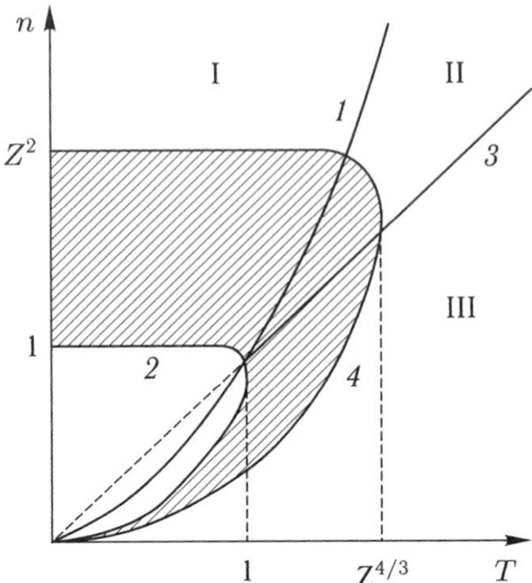

Fig. 6.6   Limits of applicability of the TFM. *1* degeneracy curve, *2* boundary of applicability of the TFM, *3* curve along which the exchange, quantum-mechanical and correlation effects are equal, *4* boundary of the uniformity regime. I region in which exchange and quantum-mechanical effects dominate for a degenerate gas, II the same for a Boltzmann gas, III region in which correlation effects are dominant. The region in which the use of the TFM is justified is shaded.

divergence of the quantum correction to energy at small distances. For this reason, the physical quantities for which the discussed region is significant (the total energy, the neutral atom polarizability, etc.) must be described in a special way.

## 6.4   Quantum and exchange corrections

Bearing in mind first of all the region in the vicinity of the nucleus (and, accordingly, quantum effects) we will note two ways of TFM modification. The first one is the consideration of the quantum correction equivalently with the zeroth term corresponding to TFM [503, 504, 538, 541]. A series expansion in the parameter $\delta_{\text{quant}}$ is not made in this case. The process of successive approximations is as follows. At the first stage the TFM itself is considered and at the next stage it is TFM and equivalently the first quantum correction, at the third stage — the same plus the second-order quantum correction in $\delta_{\text{quant}}$, etc., with no additional assumptions at each stage. Such an approach (sometimes referred to as quantum-statistical model, QSM) proves to be fairly effective. The process of successive approximations rapidly converges, and a rather good description is already attained at the second state. At any rate, the corresponding result is an immeasurably better starting point for subsequent improvements (if necessary) than the TFM itself.

For zero temperature the QSM program can be realized by considering the minimum of the functional $E(n) = E - \mu \int dx n$. This can be done either by the variational method [538, 541] or by solving the Lagrange–Euler equation [503, 504]:

$$[\zeta\Delta + p_F^2(x)]\psi = (3\pi^2)^{2/3}\psi^{7/3}, \tag{6.53}$$

where $\psi^2 \equiv n$, $\zeta = 1/9$. Equation (6.53) resembles the Schrödinger equation but differs from it by the nonlinear term and by the quantity $\zeta$.

The success of the QSM program is due to the following simple circumstance. TFM is inapplicable at small distances because it leads to incorrect behavior of density in this region: as equation (6.53) shows, for $\zeta = 0$ the density is singular at zero, $\psi \sim r^{-3/4}$, $n \sim r^{-3/2}$. Meanwhile, in an exact quantum-mechanical consideration the density is described by a regular function. It turns out that already at the second QSM stage the behavior of the density is radically improved. Equation (6.53) shows that its solution at zero has the form $n \sim 1 - (Zr/\zeta)$. Hence, the next QSM stages only yield insignificant quantitative specifications of the result.

The other way to improve TFM at small distances from the nucleus is based on the fact that in this region the potential created by electrons is practically constant and can be put equal to the constant $C$. At the same time the problem with the potential $U' = -Z/r + C$ can be solved in an explicit form. From this, without any additional assumptions one can come to the expression for the distribution function:

$$f = f_1 - f_2 + f_3, \tag{6.54}$$

here $f_1$ is an exact quantum-mechanical expression of $f$ for the potential $U'$ which can be expressed in terms of the Whittaker functions [447];

$$f_2 = \theta[2(\mu - U') - p^2],$$

is the quasi-classical expression for the same potential;

$$f_3 = \theta[2(\mu - U') - p^2],$$

is the ordinary FTM result. At small distances, where the potential is close to $U'$, the first term alone remains, while at relatively large distances, where the motion becomes quasi-classical, the TFM expression is reconstructed (see Ref. [905]).

Other methods also exist that improve TFM at small distances. We shall only mention paper [325] suggesting the introduction of a special "cutoff" of the energy spectrum of electrons from below, which leads to a sharp decrease in the density in the neighborhood of the nucleus. However, the QSM is now the most thoroughly elaborated and effective method.

## 6.5  Application of TFM to an isolated atom

We shall consider the application of TFM to a neutral isolated atom with a sufficiently high $Z$ value. In the framework of TFM, this corresponds to

$$\mu = 0, \quad p_{\mathrm{F}}^2(x) = \frac{2Z}{r(1+\xi)^2}, \quad \xi = \frac{r}{\alpha}, \quad \alpha = \left(\frac{9}{2Z}\right)^{1/3}, \tag{6.55}$$

where for $p_{\mathrm{F}}^2$ we have used the approximate expression of Ref. [975], which will be applied below.

We begin with the question of the total energy of the electron cloud of an atom. The TFM gives for this quantity the expression $E(Z) = -0.769Z^{7/3}$, which is considerably greater in magnitude than the results of experiment and of quantum-mechanical calculations. This happens because of the inapplicability of the semiclassical description at short distances from the nucleus. In addition, the formula given does not reflect the weak, but perfectly observable, oscillations of the function $E(Z)$, that are associated with shell effects. Figure 6.7 shows the results of calculating the energy of an atom by the TFM, by the TFD model, by the QSM (with inclusion of an exchange correction), and by the Hartree–Fock method. It can be seen that allowance for quantum effects considerably improves the result: e.g., for $Z \approx 20$–$30$ the disagreement with the results of the quantum-mechanical calculation falls from $\approx 25$ to $\approx 5\%$.

The shell contribution to the atomic energy can be estimated assuming that the main role is played by low $\mu$ values and, accordingly, high $r$ values, where quasi-classic is inapplicable. Therefore, we can only count on obtaining qualitative results. The first of these is an explanation of the extremely small amplitude of the oscillations: its relative magnitude $\delta E/E$ is substantially smaller than the parametric estimate $\delta_{\mathrm{quant}} \sim Z^{-2/3}$. The explanation lies precisely in the unsuitability of the semi-classical method for describing the oscillations under consideration. In the region of importance for these oscillations, i.e., the region of the outer shells of the atom, where the charge of the nucleus is practically completely screened, the amplitude of

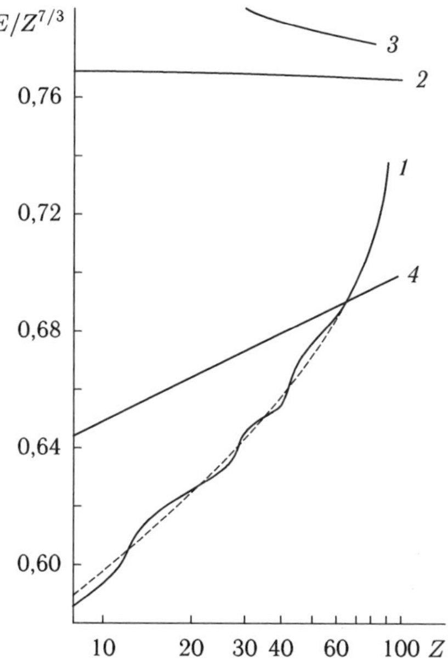



Fig. 6.7  Energy of the electron cloud of an atom. *1* from the Hartree model [430], *2* from the TFM, *3* from the Thomas–Fermi–Dirac (TFD) model, *4* from the QSM.

the oscillations does not depend on $Z$. Correspondingly,

$$\frac{\delta E}{E} \sim Z^{-7/3}, \tag{6.56}$$

which is in agreement with the curve of Fig. 6.7.

The second result concerns the type of oscillations. It is easy to see (6.37) that the oscillations of the energy are smoothed compared with the oscillations of the density. Namely,

$$\delta E \sim \left[ S_0^2 - \frac{\pi^2}{4} S_0 \right]', \tag{6.57}$$

which corresponds to wave-like behavior with a discontinuous second derivative, rather than with a discontinuous first derivative as in $\delta n$. This also agrees qualitatively with Fig. 6.7. The points of discontinuity correspond to the "magic" values $Z_{\text{mag}} = \dots 19, 37, 55 \dots$ (the alkali metals), at which a level with quantum numbers $n$ and $l = 0$ emerges into the continuous spectrum, and $S_0 = \pi[n - (1/2)]$ for $E = 0$. It can be seen from (6.57) that the nodes of the oscillations are located both at the values $Z_{\text{mag}}$ and in the intervals between them. This property corresponds qualitatively to Fig. 6.7, in which the nodes correspond to $Z \approx 21, 29, 37, 53, 59 \dots$.

We turn now to the description of the density distribution in the core of the atom (for $Z^{-1} \ll r \ll 1$). In this region the quantum effects are small, while the shell effects are quite noticeable because of the large numerical coefficient in formula (6.38). Use of this formula and (6.55) gives

$$\delta n \sim \sin\left(\sqrt{8Z\alpha}\,\arccos\frac{\alpha - r}{\alpha + r}\right), \qquad (6.58)$$

(for simplicity, the nonoscillating factor has been omitted) and leads to good (about 5%) agreement with the quantum-mechanical calculations for atoms of average and large atomic number (Fig. 6.8) [548]. What is measured directly in experiment is not the density, but the atomic form factor $F(\mathbf{q}) = \int d\mathbf{x}\, n(\mathbf{x}) e^{-i\mathbf{q}\mathbf{x}}$, where $\mathbf{q}$ is the momentum transfer in scattering of fast electrons or photons. Using (6.38), (6.55) and proceeding by the method of stationary phase, it is possible to arrive at the expression [906]

$$\delta F(q) \sim \cos[1.82 Z^{1/3}(\pi - x)], \qquad (6.59)$$

where $x = q/1.1 Z^{2/3}$. The inclusion of this correction appreciably improves the behavior of the function $F(q)$ (Fig. 6.9).

The next quantity with which we shall be concerned is the potential $p_{\mathrm{F}}^2$ of the atom. With quantum effects taken into account, it was calculated in the paper by Kompaneets and Pavlovskii [568]. These results were used to calculate the polarizability and magnetic susceptibility of the atom and

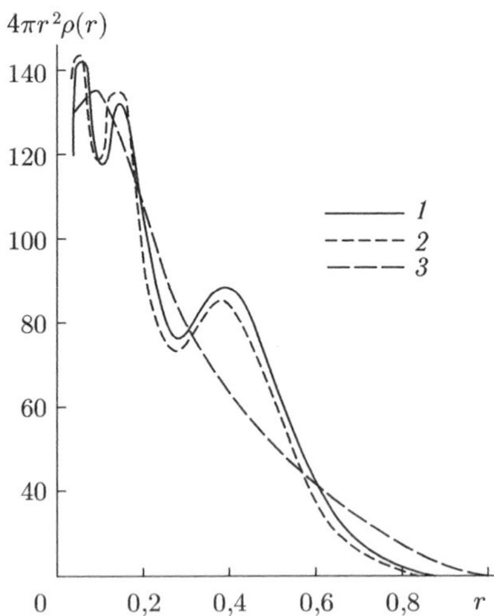


Fig. 6.8   Radial electron density in the mercury atom. *1* from the Hartree model [430], *2* from the TFM, *3* from the TFM with shell corrections.

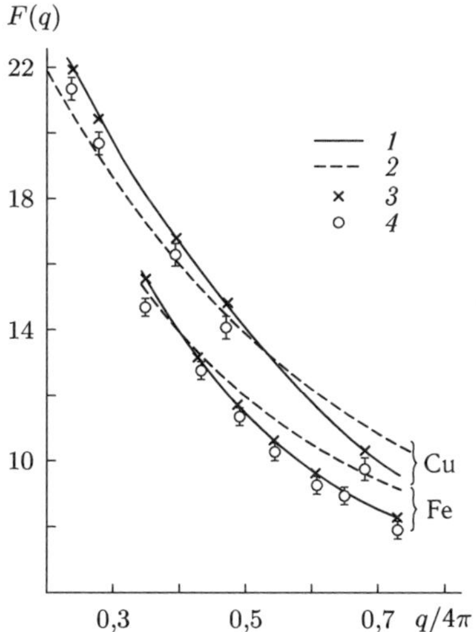

Fig. 6.9   Atomic form factor for iron and copper. *1* from the Hartree–Fock model [641], *2* from the TFD model [974], *3* the same with a shell correction, *4* experiment [1024].

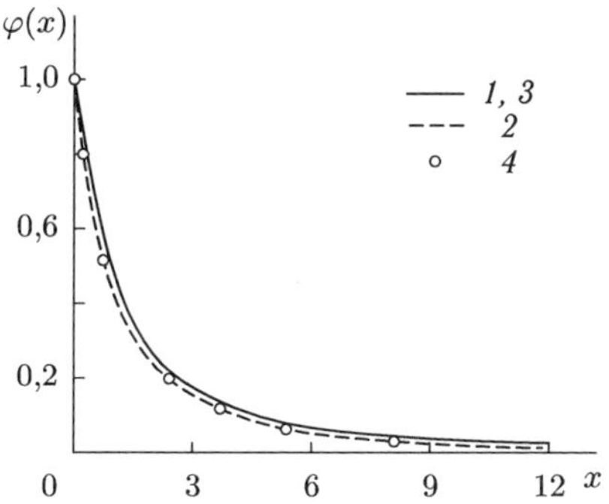

Fig. 6.10   The atomic potential ($\varphi(x) = (r/Z)p_{\mathrm{F}}^2$, $r = 0.885Z^{-1/3}x$). *1* from the Hartree–Fock model, *2* from the TFM, *3* from the TFM with a shell correction, *4* from the QSM [334].

it was found that, in comparison with the TFM, the theoretical results come appreciably closer to the experimental data. Thus, according to the author of [243], the above quantities for xenon have (in certain units) the experimental values 4.10 and 43.9 respectively, the values 24.8 and 113.0 in the TFM, and the values 4.06 and 45.7 when quantum effects are taken into account. The behavior of $p_{\mathrm{F}}^2$ with shell effects taken into account has recently been obtained [333]; it is shown in Fig. 6.10.

We may state that, on the whole, the generalized TFM, incorporating quantum, exchange and shell effects, gives a considerably better description of the properties of a heavy atom than does the standard TFM. The contrary statement in [890] is based on the factual errors pointed out in Ref. [243]. The most important and widely-used applications of the TFM to the physics of extremal states of matter pertain to the thermodynamics.

## 6.6   Thomas–Fermi equation of state

The application of the TFM and its generalizations to the derivation of the equation of state of matter is realized, as a rule, in the framework of the Wigner–Seitz cell model. Without dwelling on the details (see Refs. [361, 544,1080]) we point out that the most complete data on the equation of state in the TFM are contained in the [605] (cold matter) and [604] (hot matter) by Latter. A considerable part of these data, which refer to a wide range of pressures and temperatures, is incorrect because of the inapplicability of the TFM itself; this conclusion is also confirmed by direct comparison with the available experimental data [27].

Refinement of the TFM by the inclusion of the effects considered above makes it possible to distinguish the reliable part of these data and obtain an improved equation of state of matter, valid in a wider range of pressures and temperatures. Above were given for the corresponding corrections to the equation of state, together with their explicit form in the uniformity regime.

In the region of not very high temperatures (see Fig. 6.6), the principal contribution to the equation of state is made by the exchange and quantum effects. The first have been investigated in most complete form in Ref. [194]. However, the results are invalidated to a considerable extent by the fact that the quantum effects were not taken into account at all. As was explained in Ref. [547], numerically the latter makes the same contribution (to within a coefficient $\approx 1/3$).

A quantitative theory of the quantum effects in the equation of state of matter was constructed in Ref. [539], and a numerical solution of the problem for cold matter is contained in Ref. [498], from which we borrow Fig. 6.11. It can be seen from this figure that the quantum and exchange corrections do indeed make the TFM result come closer to the experimental

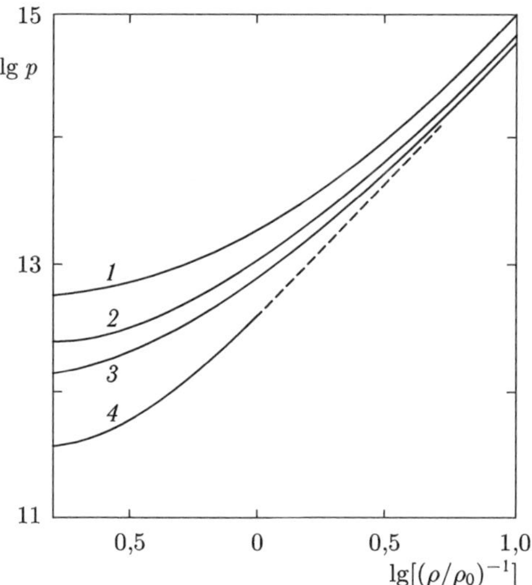



Fig. 6.11   Equation of state of iron ($\rho_0$ is the normal density). *1* from the TFM, *2* from the TFD model, *3* from the TFM with exchange and quantum corrections, *4* experiment [27].

data.[4] The results of Ref. [60], which, like Ref. [498], refer to the cold compression of iron, lead to an analogous conclusion. In Ref. [672] the equation of state with quantum effects taken into account was compared with experiment in the high-temperature region and convergence of the data was again noted.

Although the TFM, even when refined by allowance for quantum and other effects, is not directly applicable in the low-pressure region, it leads to reasonable results for the atomic volumes (or densities) of uncompressed matter [498, 500]. In this connection, we recall that at zero pressure the TFM itself leads to an infinite radius for the atom [253, 930]. In the framework of the TFD model, or when quantum effects are taken into account, the situation is changed: uncompressed matter acquires a finite density. Although these models are not capable of describing the shell oscillations

---

[4]The papers [498, 499] contain a complete thermodynamic description in the framework of the TFM with quantum and exchange effects taken into account. The paper [505] contains detailed tables of thermodynamic quantities calculated from the TFM with quantum and exchange corrections taken into account.

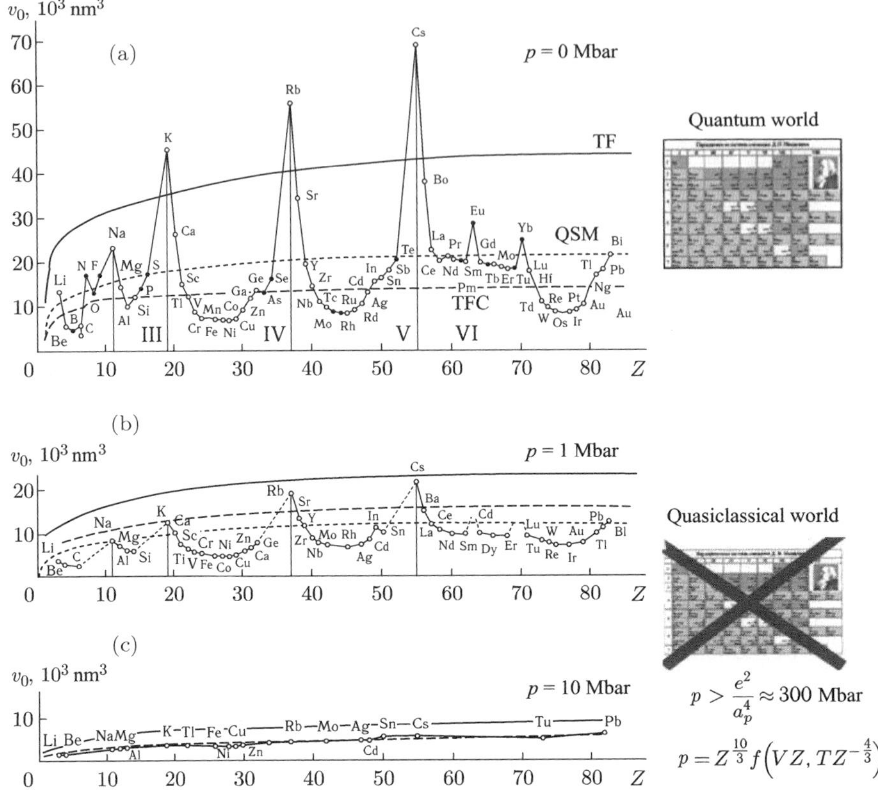


Fig. 6.12　Variation of elemental atomic volumes $v_0(Z)$ with increasing pressure [27, 40, 500].

of the atomic volume, they convey the behavior of this quantity averaged over the periodic table. The best results are given by the quantum-statistical model.

The above is illustrated by Fig. 6.12(a), borrowed from [500]. The empirical curve in this figure reflects the atomic-volume oscillations associated with shell effects. electron with a given value of the principal quantum number and $l = 0$ first appears in the atom. The descending parts of the curve correspond to close-packed structures; a description of these parts can be found in Ref. [254]. As $Z$ increases the curves pass through a minimum and the ascending parts correspond to structures that are not close-packed. This is the reason why the atomic volume of a Group-0 element, for which we might expect the lowest value, in fact differs little from the atomic volume of the alkali metal that follows it. This is connected with the fact that a crystal of a Group-0 element belongs to the class of loosely-packed van der Waals structures. Therefore, the curve under consideration is characterized by a discontinuity not of the function itself, but of its derivative.

The question of how the shell effects behave with increasing pressure is of considerable interest. The conventional point of view is that they decrease with increasing pressure, becoming negligibly small for $p > 1$ (in the usual units this corresponds to 300 Mbar). At higher pressures the properties of matter become universal, and the dependence of its characteristics on the atomic number becomes smooth and monotonic. At first sight, the empirical curves [27, 500]) shown in Figs. 6.12(b) and (c) justify this point of view.

However, the results of Ref. [547] show that, in fact, the situation with the shell effects is more complicated. Oscillations associated with discontinuity of the derivative of the function $v(Z)$ (see Fig. 6.12) are indeed "washed out" at a pressure of order 1, but at higher pressures sharper oscillations, associated with discontinuity of the function $v(Z)$ itself, appear in their place. Physically, this corresponds to the fact that even the structures corresponding to the ascending parts of Fig. 6.12(a); become close-packed with increasing pressure; because of this, the minima of the curve $v(Z)$ disappear and a discontinuity in the atomic volume arises between a Group-0 element and the neighboring alkali metal. The discontinuity points themselves remain the same as at zero pressure, so long as we are far from the limits of the uniformity regime ($n \ll Z^2$). However, at higher pressures the "magic" values of $Z$ begin to move in the direction of higher $Z$, as can be seen from formula (6.52).

This happens because, on account of the influence of the neighboring cells, the character of the occupation of the levels becomes less and less hydrogenlike. Only when $n > Z^4$ do the shell oscillations completely disappear, because all the levels have crossed into the continuous spectrum. In the language of the function $v(p)$ (the equation of state) the shell effects are manifested in the appearance of a series of first-order phase transitions, corresponding to the "squeezing out" of deep levels of the atoms into the continuous spectrum. Of course, such transitions will also occur outside the uniformity regime. This is illustrated by Fig. 6.13, borrowed from Ref. [908] (see also Ref. [549]. Such phase transitions will occur, e.g., inside "white-dwarf" stars, and are manifested in the form of discontinuities in the density distribution, and in the appearance of singularities on the mass–radius curve. Moreover, as was noted by A.S. Kompaneets, the latent heat

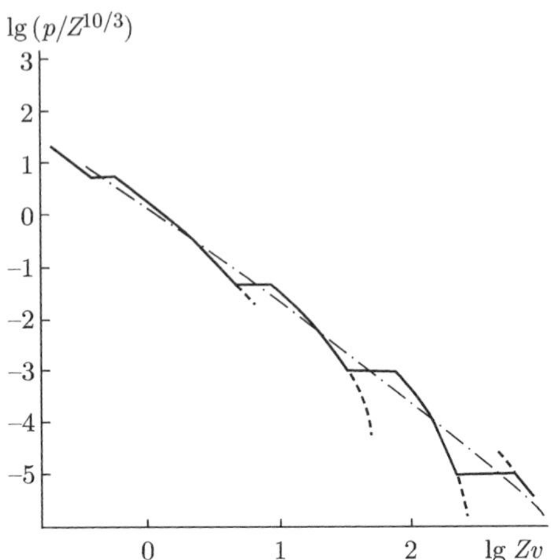



Fig. 6.13   Equation of state with allowance for shell effects ($Z = 100$).

of the phase transition could turn out to be an appreciable factor in the evolution of stars [547].

To conclude this section we shall touch briefly upon one further extremal factora high-intensity external magnetic field, confining ourselves to treating its influence on the structure of a many-electron atom (see Refs. [187, 191, 493, 494, 595, 721, 868]). In a strong field,[5] $H \gg Z^{4/3}$, all the electrons occupy the lowest Landau level and have spin opposite in direction to the external field. So long as $Z^3 \gg H \gg Z^{4/3}$, the atom remains spherically symmetric, but its radius decreases like $Z^{1/5}H^{-2/5}$ with increase of the field. At still higher fields $H \gg Z^3$, the atom is elongated in the direction of the field, and it becomes favorable for the system of atoms to form a distinctive polymeric structure.

Being interested in this problem from the point of view of the TFM, we should consider lower field values $1 \ll H \ll Z^{4/3}$, 3, for which the semiclassicality condition is fulfilled with respect to the field: $r_H p_{\mathrm{F}} \gg 1$, where $r_H = p_{\mathrm{F}}/H$ is the Larmor radius. Under these conditions the action of the field reduces to flipping the electron spins, and we are dealing with the atomic analog of Pauli paramagnetism. The electron-density distribution is now characterized by two functions: $n_+ = [-2(U + H)]^{3/2}/6\pi^2$, for $r < r_+$, and $n_+ = 0$, for $r > r_+$ ($U(r_+) + H = 0$) for a spin parallel to the field, and

---

[5]The field is measured in units $m^2 e^3 c/\hbar^3 = 10^9$ gauss.

$n = (-2U)^{3/2}/6\pi^2$ for a spin antiparallel to the field. The actual potential $U$ is determined by the Thomas–Fermi equation:

$$\Delta U = -4\pi(n_+ + n_-).$$

Referring to [595] for the details, we point out that the total spin $S$ of the atom increases with the field in accordance with the law $S = Zf(HZ^{-4/3})$ where $f$ is a certain universal function. The effects considered are important in the conditions obtaining on the surface of a pulsar.

## 6.7  Limits of TFM applicability. Experiment

A drawback of the TFM is an incorrect description of the electron density at the cell periphery and near the nucleus due to violation of the quasi-classicality condition there. Kirzhnits *et al.* [314] came up with a method for removing this drawback by employing the method of successive approximations in the solution of the Thomas–Fermi equations with quantum corrections without expanding in a series in a small parameter. This approach forms the basis of the quantum-statistical model, in which the solution rapidly converges near the nucleus and exhibits a qualitatively correct quantum-mechanical behavior away from the nucleus. However, the resultant formulae are no longer self-similar with respect to $Z$. Numerical calculations have to be carried out anew for each element, and the equations themselves are much more arduous for numerical calculations. Several other modifications [547, 710, 781, 782] to the quasi-classical model in the vicinity of the nucleus were also proposed, which differ in the way the corrections are introduced. However, in the thermodynamic description the difference between these models and the TFM with corrections is noticeable only outside of the domain of their formal applicability, which underlies the preference for the simpler TFM to carry out specific calculations.

When use is made of the quasi-classical method of description, it is well to bear in mind the specific inaccuracies introduced by the cell model itself. All electron correlations in this model are automatically limited to the dimensions of the atomic cell and therefore cannot exceed the average internuclear distance, and nucleus–nucleus correlation effects are absent. These circumstances obviously limit the applicability of the TFM for the description of plasma in conditions typical for this state of matter, when the screening sphere contains a substantial number of nuclei, whose correlations make the main contribution to the Debye correction [471]. That is why this model does not encompass the Debye limit in the plasma

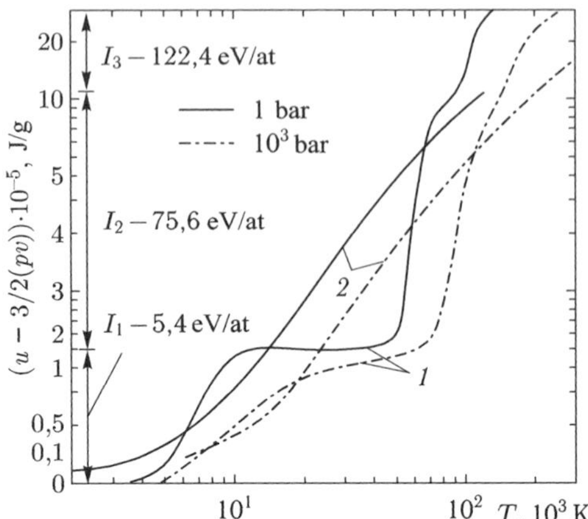


Fig. 6.14   Equation of state of lithium plasma [471]. Calculation: *1* chemical model, *2* TFM with quantum and exchange corrections. $I_1$, $I_2$, $I_3$ sequential ionization potentials.

domain and, furthermore, its extrapolation properties become worse as the plasma becomes more tenuous (see Ref. [471]), because the model does not describe the stepwise nature of thermodynamic functions in gas plasmas owing to ionization effects (Fig. 6.14). Inadequacy of TFM calculations and quantum-mechanical Hartee–Fock–Slater calculations [739] in the characteristic plasma range is also confirmed by Fig. 6.15.

Nuclear correlation effects may also be significant in the condensed phase, since the cell model ignores the deviation of the real cell volume from the average one due to nuclear motion, which is valid only for ordered systems. For nonideality parameter values $\Gamma \approx 100$ typical for the condensed state and dense plasmas, in the equation of state it is necessary to take into account the motion of nuclei, which also leads to atomic volume fluctuations. For the TFM an estimate of this effect using the example of $SiO_2$ showed [711] that the inclusion of nuclear motion in the range $1 < \Gamma < 100$ increases the pressure by $\approx 15\%$.

Insomuch as the quasi-classical model is widespread, great efforts were made to experimentally verify its applicability conditions [294, 295].

The physical conditions for the validity of the quasi-classical model, as noted above, correspond to extremely high pressures $p \gg 300\,$Mbar and temperatures $T \gg 10^5$K, which are realized in different astrophysical objects but are hard to attain by experimental techniques in terrestrial conditions. The presently attainable highest-pressure and highest-temperature

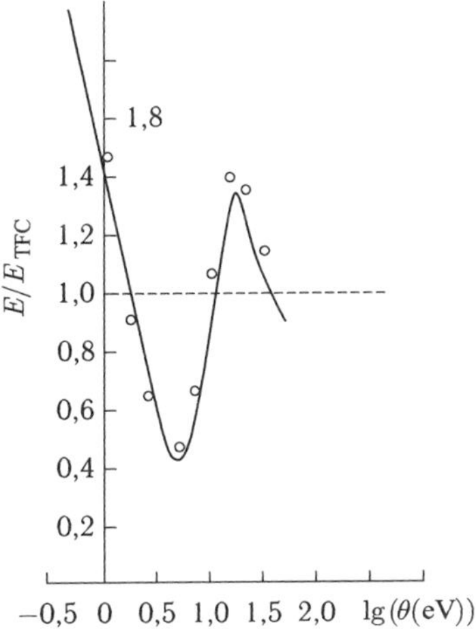

Fig. 6.15 Ratio of the lithium plasma energy by the Hartee–Fock–Slater model [739] to the energy by the TFC model [500, 505] at a pressure $p = 1\,\text{kbar}$. Points show the calculation by the chemical model.

states are realized, as is evident from the foregoing, with the help of dynamic methods that make use of the techniques of intense shock waves [64, 820–822, 985, 1023, 1098]. Although the majority of shock-wave experimental data do not correspond exactly to quantumstatistical conditions, they permit the properties of the quasi-classical models to be extrapolated beyond the limits of their formal applicability defined [35, 985] by the smallness of the corresponding literal criteria. These constructions suggest that the introduction of quantum, exchange, and correlation corrections (oscillation corrections were not included) improves extrapolation, which becomes possible in this case, in the view of Al'tshuler *et al.* [35], to pressures $p \gtrsim 300\,\text{Mbar}$ for zero temperature and to $p \approx 50\,\text{Mbar}$ for $T \gtrsim 10^5\,\text{K}$. At the same time, the data of comparative measurements made in underground nuclear explosions allow ambiguous interpretations and do not provide an unequivocal answer to the question about the superiority of one or other version of the quasi-classical model, and are at variance with the data of absolute measurements [820–822, 1023]. It is also noteworthy that experiments on the shock compression of highporosity copper performed for pressures of 10–20 Mbar and temperatures up to $2 \cdot 10^5\,\text{K}$ [1098] and of solid materials for pressures up to 160 Mbar [64] (Fig. 6.16, see also Fig. 6.17) are indicative of significant shell effects in the domain that was previously described by tradition by the standard TFM.

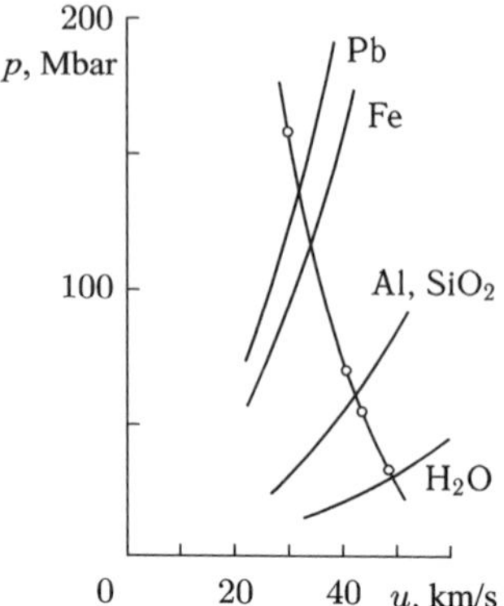



Fig. 6.16   Results of measurements of relative compressibility of substances at superhigh pressures [64]. Points stand for the experiment and solid lines the TFC calculations. The nonmonotonic difference of the experimental data from the calculations testifies to notable shell effects.

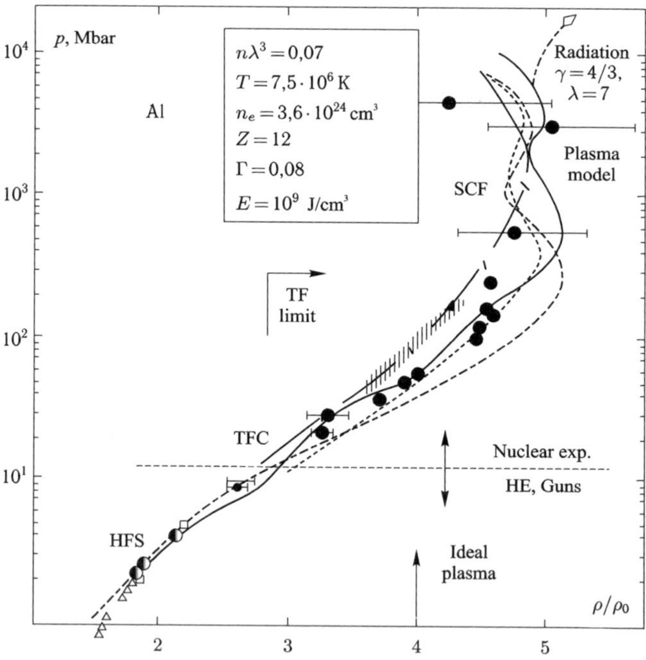



Fig. 6.17   Shock-wave compression of aluminum to gigabar pressures [62, 63, 1019].

The shock adiabat of Al [62, 1019] presented in Fig. 6.17 demonstrates the quality of the description by theoretical models of the data on shock compression of plasma in a wide range of parameters up to today's record pressures $p \approx 4\,\text{Gbar}$ obtained in the near zone of an underground nuclear explosion. It is of interest that at these superhigh pressures the specific energy of plasma reaches giant values of the order of $1\,\text{GJ/cm}^3$ close to the energy density of nuclear matter. In this compression regime, the pressure and energy of thermal radiation is comparable with the pressure and temperature of the electrons and ions.

It is obviously senseless to raise the pressure of shock compression above the indicated values because dominating in this case will be the thermodynamics of the photon gas and not of the substance itself.

On the strength of initial simplifications, the quasi-classical model with quantum and exchange corrections is inapplicable at low pressures, but it provides [500, 505] a reasonable averaging of elemental atomic volumes over the periodic system. The qualitative improvement of the description in the low-pressure region reflecting the individual features of substances was attained through a model unification of elements of the quasi-classical and band theories [616].

The large amount of thermodynamic information available at the present time demonstrates [27] (Fig. 6.12) the pressure-induced smoothing of specific atomic volumes of chemical elements. Indeed, the sharp non-monotonic dependence $v_0(Z)$ observed under normal conditions $p = 0$ (the lower curve in Fig. 6.12) is a manifestation of the quantum character of substance structure. This is in full measure reflected in the structure of Periodic system and in the chemical reactivity of elements. The substance compression leads to the "crush" and "stir" of electron shells, and so the elements lose their chemical identity and their behavior becomes increasingly universal [545, 547]. As is shown by the experiment [27], with rising pressure the observed oscillations become less and less pronounced (the curves $p = 1$ and $10\,\text{Mbar}$) thus confirming the main idea of the quasi-classical model (disregarding shells) of the defects of "simplification" of the structure and properties of matter with its increasing compression [294, 295].

It should be emphasized that in view of disregard of cohesive forces the TFM itself leads at zero pressure to an infinite radius of atomic cell, and it is only the introduction of corrections that gives a finite density of matter. It was noticed in Refs. [500, 505] that with rising pressure the oscillations of atomic volume decrease and approach the calculations by the modified TFM, although it is partly explained by the fact that at high pressures comparison was performed with the results of extrapolation [502] of experimental data to the quasi-classical calculations themselves.

In reality the tendency for substance properties to become simpler is violated at higher pressures due to the existence of inner atomic shells in atoms. As determined in Refs. [547, 548], shell effects may be qualitatively described in the framework of the quasi-classical approximation by including the correction irregular in $\hbar^2$ that corresponds to the oscillatory part of the electron density, which was previously disregarded by mistake. In this case it is significant that the shell effects are described even in the lowest quasi-classical approximation for the wave function and should therefore be taken into account along with the regular corrections considered above. The shell effects reflect the irregularities of substance characteristics arising from the discrete energy spectrum and emerge in the quasi-classical model as a result of the interference of de Broglie waves [192]. Therefore, the quasi-classical model in its most up-to-date version is much more substantive than considered before. It turns out that this model not only describes the averaged behavior of electrons in heavy and strongly compressed atoms, but it also qualitatively reproduces the atomic inner-shell structure [547] and in many cases yields results that are close to the data calculated by more accurate quantum-mechanical models, being profitably different from them in simplicity and obviousness.

The inclusion of shell effects appreciably changes the equation of state of substances, giving rise to discontinuities in the atomic volume curve $v(Z)$ in the high-pressure domain (where this approximation is justified) [547]. In this case, in the equation of state there emerge characteristic nonmonotonic features caused by electron phase transitions, when the energy levels are expelled from the discrete spectrum into the continuum. One might expect these features to smooth out with increasing temperature under the influence of resonance electrons, however, the data calculated in the central-field approximation [608] testify (Fig. 6.18) to a significant influence of the shell structure even for a strongly heated substance. The significance of the contribution from the shell effects at hyper-megabar pressures also follows from the data of quantum-mechanical calculations in the more exact Hartree–Fock–Slater approximation [547] and by the augmented plane wave method [676] (see Figs. 6.19 and 6.20). These effects are predicted for a broad parameter range [547] and should vanish, which we discuss in ch , for $n \gg Z^4$ in the homogeneity domain, when all atomic energy levels pass into the continuum. In this connection we note that the question of the asymptotic behavior of the quasi-classical model is not trivial, for it has

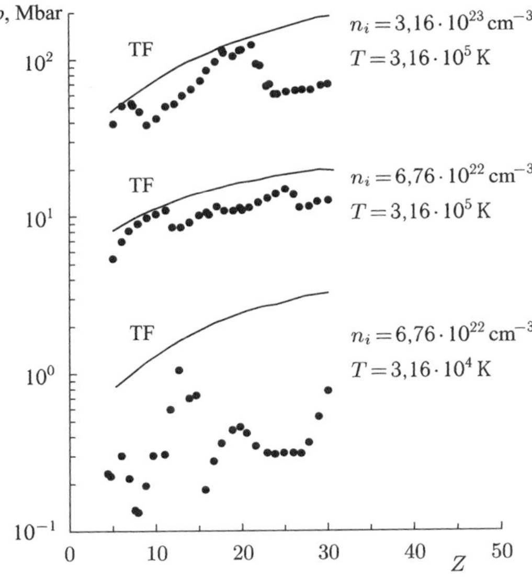

Fig. 6.18 Dependence of pressure on the atomic number at a fixed density and temperature [608]. Solid lines stand for calculations by the TFM, points demonstrate allowance for the shell effects in the central field approximation.

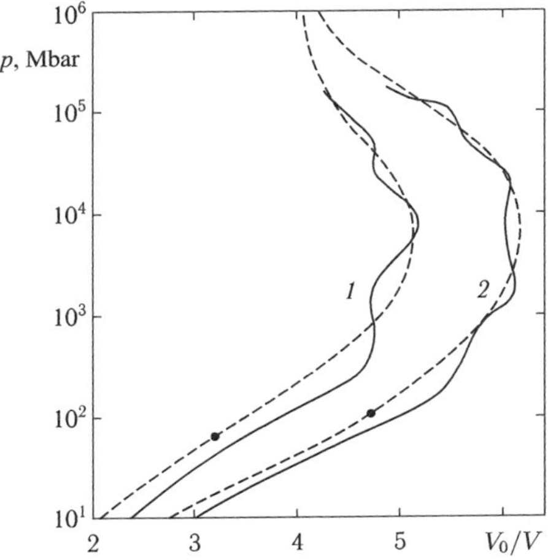

Fig. 6.19 Shock adiabats of copper *1* and lead *2* [915, 916]. Solid curves stand for calculations by the Hartree–Fock–Slater model and dashed lines show TFC calculations and interpolation [35].

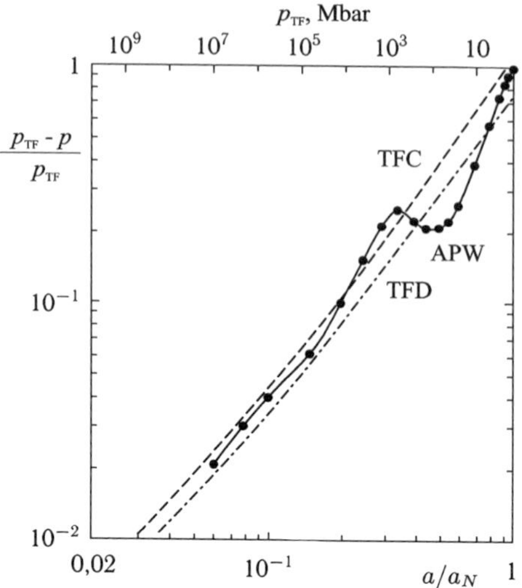


Fig. 6.20  Equation of state of aluminum ($T = 0\,$K) [676]. Calculation: TFD is Thomas–Fermi–Dirac model, TFC is the Thomas–Fermi model with quantum and exchange corrections, APW is the model of attached plane waves, $a$ is the lattice parameter.

been shown [618] that this model corresponds to the exact solution of the Schrödinger equation only for $Z \to \infty$, but not in the high-density limit.

Therefore, the question of the limits of validity of the quasi-classical model is still, to a large extent, open, and the behavior of substances in the $p > 300\,$Mbar domain turns out to be more varied than assumed on the basis of earlier simplified ideas [500, 505, 604]. The experimental verification of predictions of the quasi-classical shell model, the same as the search for the corresponding phase transitions is now the most interesting problem in superhigh pressure physics. Solution of this problem will obviously need special experimental technics using high-power lasers [979, 1015], electron beams [153] or underground nuclear explosions [820–822]. In this connection, of considerable interest is the reproduction of shell effects in the equation of state by direct quantum-mechanical methods to be discussed in Chapters 3 and 7

# Chapter 7

# Density Functional Method

## 7.1  Density functional method

The expression for the energy in Thomas–Fermi model (TFM) at $T = 0$ can be represented as an explicit density functional [547]:

$$E\{n\} = T[n] + E_{\mathrm{e}}[n] + E_{\mathrm{i}}[n], \tag{7.1}$$

$$T[n] \backsim \int d\mathbf{x}\, n^{5/3}(\mathbf{x}). \tag{7.2}$$

The minimum of the functional $E - \mu \int d\mathbf{x} n$ in $n$ leads to the Thomas–Fermi equation (Chapter 6) and determines the energy of the ground state of the system and the corresponding density distribution. It is readily seen that when going beyond TFM and allowing for the corresponding corrections, we are still dealing with the functional (7.1), but the functional (7.2) has a more complicated form. In particular, allowance for the quantum effects leads to the fact that instead of the quasi-homogeneous functional we obtain a functional depending on the derivatives of $n$. It is therefore not surprising that the following assertion holds true: in an exact formulation of the many-body problem (the multiparticle Schrödinger equation) the energy of the system is expressed in the form of the functional (7.1) with a single-valued universal functional $E_k$ of the general form where the minimum of the functional $E - \mu \int d\mathbf{x} n$ gives the energy and the density distribution in the ground state of the system. This assertion was supported by March and Murray [654] using perturbation theory and was then proved as a rigorous theorem by Hohenberg and Kohn [440].

The density functional method (DFM) is an exact quantum-mechanical theory for a system of interacting quantum numbers in an external potential $V_{\mathrm{ext}}(\mathbf{r})$. The method itself is based on two strictly proved theorems [440].

301

(1)	For any system of interacting particles in the external potential $V_{\text{ext}}(\mathbf{r})$, the potential $V_{\text{ext}}(\mathbf{r})$ is determined up to an arbitrary derivative by the electron density $n_0(\mathbf{r})$ in the ground state.

(2)	The energy of nondegenerate ground state of the system for any external potential $V_{\text{ext}}(\mathbf{r})$ is the electron density functional $V_{\text{ext}}(\mathbf{r})$. The ground state of the system is the minimum of this functional which is reached for a density $n_0(\mathbf{r})$ corresponding to the ground state of the system.

The first theorem implies that the Hamiltonian of the system is determined up to a constant value by the electron density $n_0(\mathbf{r})$ of the ground state; from this it follows that the multiparticle wave functions are defined for all (ground and excited) states. Thus, all the properties of the system are completely defined if the electron density $n_0(\mathbf{r})$ of the ground state is known. The second theorem implies that for a known functional $E[n]$ one can find the density and energy of the ground state.

In line with [440] the energy functional can be written as:

$$E_{\text{HK}}[n] = T[n] + E_{\text{int}}[n] + \int d^3 r V_{\text{ext}}(\mathbf{r})n(\mathbf{r}) + E_{\text{II}}$$

$$\equiv F_{\text{HK}}[n] + \int d^3 r V_{\text{ext}}(\mathbf{r})n(\mathbf{r}) + E_{\text{II}}, \tag{7.3}$$

where $E_{\text{II}}$ is the nuclear interaction energy. The functional $F_{\text{HK}}[n]$ includes the kinetic and potential energy of the system of interacting electrons:

$$F_{\text{HK}}[n] = T[n] + E_{\text{int}}[n]. \tag{7.4}$$

The minimum of the functional (7.3) determines the energy of the system in the ground state and its electron density. It should be noted that (7.3) carries no information on the excited states of the system.

The direct minimization of the functional $E_{\text{HK}}[n]$ encounters certain difficulties. An alternative way of finding the minimum of the functional $E[n]$ was proposed in papers [613, 617], where the energy is expressed in terms of the multiparticle wave function $\Psi$:

$$E = \langle \Psi | \bar{T} | \Psi \rangle + \langle \Psi | \bar{V}_{\text{int}} | \Psi \rangle + \int d^3 r V_{\text{ext}}(\mathbf{r})n(\mathbf{r}) + E_{\text{II}}. \tag{7.5}$$

The minimization goes in two stages. First, the energy is minimized on the class of multiparticle wave functions that yield one and the same electron density $n(\mathbf{r})$:

$$E_{\mathrm{LL}}[n] = \min_{\Psi \to n(\mathbf{r})} \left[ \langle \Psi | \bar{T} | \Psi \rangle + \langle \Psi | \bar{V}_{\mathrm{int}} | \Psi \rangle \right] + \int d^3 r V_{\mathrm{ext}}(\mathbf{r}) n(\mathbf{r}) + E_{\mathrm{II}}$$

$$\equiv F_{\mathrm{LL}}[n] + \int d^3 r V_{\mathrm{ext}}(\mathbf{r}) n(\mathbf{r}) + E_{\mathrm{II}}. \tag{7.6}$$

Here

$$F_{\mathrm{LL}}[n] = \min_{\Psi \to n(\mathbf{r})} \langle \Psi | \bar{T} + V_{\mathrm{int}} | \Psi \rangle \tag{7.7}$$

is the electron density functional which does not depend explicitly on $V_{\mathrm{ext}}$. At the second stage one seeks the minimum of the functional $E_{\mathrm{LL}}[n]$ in the electron density $n(\mathbf{r})$:

$$E = \min_{n(\mathbf{r})} E_{\mathrm{LL}}[n]. \tag{7.8}$$

Such a two-stage procedure can also be applied to a degenerate ground state. In this case the minimum is attained for any electron density corresponding to the ground state.

The density functional theory admits extension to the case of finite temperatures [683]. The energy functional of the electron density will then be replaced by the functional of grand thermodynamic potential $\Omega$ of the density operator $\hat{\rho}$:

$$\Omega[\hat{\rho}] = \mathrm{Sp}\left[ \hat{\rho}(\hat{H} - \mu \hat{N}) + \frac{1}{\beta} \ln \hat{\rho} \right]. \tag{7.9}$$

The minimum of this functional coincides with the thermodynamically equilibrium expression for the grand thermodynamic potential:

$$\Omega = \Omega[\hat{\rho}_0] = -\ln \mathrm{Sp}\, e^{-\beta(\hat{H} - \mu \hat{N})}, \tag{7.10}$$

where $\hat{\rho}_0$ is the density operator in the grand canonical ensemble:

$$\hat{\rho}_0 = \frac{e^{-\beta(\hat{H} - \mu \hat{N})}}{\mathrm{Sp}\, e^{-\beta(\hat{H} - \mu \hat{N})}}. \tag{7.11}$$

The Mermin theorem [683] states that it is not only the energy, but also all the thermodynamic functions, in particular, the entropy and heat capacity that are equilibrium density functionals.

The DFM has a rigorous theoretical substantiation, but establishing relation between the electron density of the ground state and the properties of the substance is a nontrivial task. The main problem on this way is the absence of direct relation between the kinetic energy and the electron density function in the general case. For the solution of this problem it was supposed in [563] that the ground state of the interacting system of particles coincides with the ground state of an equivalent system of noninteracting particles, and the interaction can be taken into account using the so-called exchange-correlation functional dependent on the electron density. In this formulation, the problem can be solved by numerical methods to an accuracy determined by the expression for the exchange-correlation functional. This technique appeared to be rather successful, and all the existing computations based on the DFM utilize this approximation. The exchange-correlation functionals developed to date describe well the II–V group semiconductors, simple and transition metals, insulators, for example, diamond, NaCl, and molecules with covalent or ionic bonds. Thus, the Kohn–Sham approach [563] is a very important step on the way of constructing the methods for calculation of different properties of many-electron strongly interacting systems.

The energy functional $E[n]$ in the Kohn–Sham approximation is written as follows:

$$E_{\mathrm{KS}} = T_s[n] + \int d\mathbf{r} V_{\mathrm{ext}}(\mathbf{r})n(\mathbf{r}) + E_{\mathrm{Hartree}}[n] + E_{\mathrm{II}} + E_{\mathrm{xc}}[n], \qquad (7.12)$$

where the electron density is determined by the expression

$$n(\mathbf{r}) = \sum_{\sigma} n(\mathbf{r}, \sigma) = \sum_{\sigma} \sum_{i=1}^{N^{\sigma}} |\psi_i^{\sigma}(\mathbf{r})|^2, \qquad (7.13)$$

the kinetic energy of a system of noninteracting particles looks like

$$T_s = -\frac{1}{2} \sum_{\sigma} \sum_{i=1}^{N^{\sigma}} \langle \psi_i^{\sigma} | \nabla^2 | \psi_i^{\sigma} \rangle = \frac{1}{2} \sum_{\sigma} \sum_{i=1}^{N^{\sigma}} |\nabla \psi_i^{\sigma}|^2, \qquad (7.14)$$

and the Coulomb energy of interacting electrons is

$$E_{\mathrm{Hartree}}[n] = \frac{1}{2} \int d^3r \, d^3r' \frac{n(\mathbf{r})n(\mathbf{r}')}{|\mathbf{r} - \mathbf{r}'|}. \qquad (7.15)$$

Here $\sigma$ is the summary projection of system's spin, $N^{\sigma}$ is the number of states for a given spin projection $\sigma$ characterized by the wave functions $\psi_i^{\sigma}(\mathbf{r})$ and the energy eigenvalues $\epsilon_i^{\sigma}$.

The exchange-correlation functional contains multiparticle exchange and correlation effects; in addition, it involves a part of the kinetic energy functional. In other words, the exchange-correlation functional is the difference between the kinetic energy and the electron interaction energy in a real interacting multiparticle system and a fictitious system of noninteracting particles in which the electron–electron interaction is replaced by the Hartree energy:

$$E_{\mathrm{xc}}[n] = \langle \hat{T} \rangle - T_s[n] + \langle \hat{V}_{\mathrm{int}} \rangle - E_{\mathrm{Hartree}}[n]. \tag{7.16}$$

The equations for determination of the wave functions $\psi_i^\sigma(\mathbf{r})$ and the energy eigenvalues $\epsilon_i^\sigma$ can be obtained from the condition of $E_{\mathrm{KS}}[n]$ functional minimum under condition of orthonormality of the system of functions $\langle \psi_i^\sigma | \psi_j^{\sigma'} \rangle = \delta_{i,j}\delta_{\sigma,\sigma'}$ [563]:

$$(H_{\mathrm{KS}}^\sigma - \epsilon_i^\sigma)\psi_i^\sigma(\mathbf{r}) = 0, \tag{7.17}$$

$$H_{\mathrm{KS}}^\sigma(\mathbf{r}) = -\frac{1}{2}\nabla^2 + V_{\mathrm{KS}}^\sigma(\mathbf{r}), \tag{7.18}$$

$$V_{\mathrm{KS}}^\sigma(\mathbf{r}) = V_{\mathrm{ext}}(\mathbf{r}) + \frac{\delta E_{\mathrm{Hartree}}}{\delta n(\mathbf{r},\sigma)} + \frac{\delta E_{\mathrm{xc}}}{\delta n(\mathbf{r},\sigma)}$$

$$= V_{\mathrm{ext}}(\mathbf{r}) + V_{\mathrm{Hartree}}(\mathbf{r}) + V_{\mathrm{xc}}^\sigma(\mathbf{r}). \tag{7.19}$$

Equations (7.17)–(7.19) are called Kohn–Sham equations. Formally this is a system of equations for noninteracting particles in the self-consistent potential $V_{\mathrm{xc}}^\sigma(\mathbf{r})$. For a known exact expression for the exchange-correlation functional $E_{\mathrm{xc}}[n]$ the solution of the system (7.17)–(7.19) gives exact values of the energy and the electron density of the ground state.

The exchange-correlation functional $E_{\mathrm{xc}}[n]$ can to a good accuracy be approximated by a local or almost local density functional (LDA):

$$E_{\mathrm{xc}}[n] = \int d\mathbf{r}\, n(\mathbf{r})\epsilon_{\mathrm{xc}}([n],\mathbf{r}), \tag{7.20}$$

where $\epsilon_{\mathrm{xc}}([n],\mathbf{r})$ is energy per electron at the point $\mathbf{r}$ only depending on the density $n(\mathbf{r},\sigma)$ in a certain neighborhood of the point $\mathbf{r}$. The expressions for $E_{\mathrm{xc}}[n]$ also exist in which $\epsilon_{\mathrm{xc}}$ depends not only on the electron density $n$, but also on its gradients,

$$E_{\mathrm{xc}}^{\mathrm{GGA}} = \int d\mathbf{r}\, n(\mathbf{r})\epsilon_{\mathrm{xc}}([n], |\nabla n(r)|\mathbf{r}), \tag{7.21}$$

This is the so-called generalized gradient approximation [777]. It should be emphasized that refinement of the approximation LDA does not always improve the results of simulation [778]. Moreover, the description of exchange-correlation effects with the help of exchange-correlation functionals in the approximations LDA and GGA is invalid for systems in which the electron density $n(\mathbf{r})$ is not a slowly varying function. In particular, such systems are the electron Wigner crystal, low-overlapping systems with Van der Waals interaction between them, the surface of a substance adjacent to a vacuum [562]. Theoretically, however, such exchange-correlation functionals can be constructed.

The Kohn–Sham equations (7.17) for finding the ground state of the system are solved in different ways. The most widespread of them are listed below.

The most natural basis for the expansion of wave functions in the Kohn–Sham equations is the plane wave basis. Almost free electrons are well described in this basis. Furthermore, plane waves are very convenient for calculation of the band structure of matter. Different pseudopotentials, both empirical and theoretical are used to describe the electrons of inner shells [586, 1008]. The plane wave method found wide application in quantum molecular-dynamic calculations [157].

Another approach consists in the construction of wave functions of a many-electron system by combining the wave functions of separate atoms, the so-called localized orbital or strong coupling method. This approach has many versions, the most well-known of which is the method of linear combination of atomic orbitals [929]. The strong coupling method is the simplest and fastest method to calculate the band structure of matter and it underlies the numerical approaches for which the simulation time depends linearly on the number of particles. This method is also the basis of more complicated methods, for instance, the linear combination of muffin-tin orbitals [46].

Finally, an approach exists that possesses advantages of the two previous ones. It is the so-called associated function method [926]. For the inner atomic shells the localized orbital method is used, and the spherically symmetric Kohn–Sham, Schrödinger, or Dirac equations are normally solved. For the space between the atoms the solution is found through expansion in basis functions (plane waves). Such a computational method is most often employed for so-called all-electron calculations where all the electrons of the atom are taken into account; on the other hand, this method is the most cumbersome.

The calculation of thermodynamic functions by the DFM proceeds as follows. The calculation of the total energy of the electron system is the most simple, at $T = 0$ one can also calculate the pressure, $p = -dE/dV$. At a nonzero temperature a self-consistent calculation utilizes the occupation numbers proportional to the Fermi–Dirac functions:

$$f(\epsilon, \rho, T) = \frac{1}{1 + \exp[(\epsilon - \mu(\rho, T))/T]}, \qquad (7.22)$$

where $\mu$ is the chemical potential of the system determined from the electroneutrality condition. If the occupation numbers are known, one can calculate the configurational entropy [741] and the free energy of the system and determine all the necessary thermodynamic parameters.

The striking efficiency of the electron DFM opened up wide prospect of its utilization for the solution of chemical, physical, biological, and nucleophysical problems.

Their detailed description can be found in monographs [222, 663, 773] and also in the collection of lectures [778].

Following paper [562], we shall only list some of the directions of development of this method:

- Spin-polarized systems — DFM with allowance for spin.
- Systems with degenerate ground states.
- Multicomponent systems (electron–hole drops of the nucleus).
- Statistic ensembles for degenerate ground states.
- Free energy at finite temperatures.
- Quasi-equilibrium ensembles for excited states.
- Relativistic electrons, astrophysics.
- Current-dependent functionals.
- Time-dependent phenomena, excited states.
- Bosons (instead of fermions).
- Combination of TFC with the molecular dynamic or Monte Carlo method (especially useful for determination of geometric structures). (The Car–Parrinello method).
- Combination of LDA and the Hubbard parameter.
- Properties of compressed nuclear matter and many other scientific directions.

We shall only give several examples of DFM close to the subject of our monograph [914].

## 7.2  Atomic and molecular structures

Table 7.1 [914] presents the experimental data on the total energy of light atoms and the deviations from the experiments of the total energies

Table 7.1   Deviations of the theoretical total energies of light atoms from the sum of experimental ionization potentials corrected for the effects disregarded in the calculations (relativism, etc.) (for the notation see paper [914]).

| $Z$ | Atom | $E_{\text{exp}}$ | $\Delta E_{\text{HF}}$ | $\Delta E_{\text{LSD}}$ | $\Delta E^{(1)}_{\text{LSD}-\text{SIC}}$ | $\Delta E^{(2)}_{\text{LSD}-\text{SIC}}$ |
|---|---|---|---|---|---|---|
| 2 | He | −79.0 | +1.1 | +1.9 | 0.4 | −0.4 |
| 3 | Li | −203.5 | +1.3 | +3.7 | −0.7 | −0.7 |
| 4 | Be | −399.1 | +2.6 | +6.1 | −0.7 | −0.7 |
| 5 | B | −670.8 | +3.4 | +8.3 | −1.2 | −0.7 |
| 6 | C | −1029.7 | +4.2 | — | −2.3 | −0.9 |
| 7 | N | −1485.3 | +5.0 | +12.6 | −3.6 | −1.1 |
| 8 | O | −2042.5 | +6.9 | — | −5.1 | −1.2 |
| 9 | F | −2713.5 | +8.6 | +16.9 | −7.2 | −1.7 |
| 10 | Ne | −3508.1 | +10.3 | +19.2 | −9.5 | −2.1 |
| 11 | Na | −4414.7 | +10.5 | +22.1 | −11.4 | −2.7 |
| 12 | Mg | −5443.1 | +11.6 | +24.8 | −13.3 | −3.3 |
| 13 | Al | −6594.0 | +12.5 | +28.0 | −14.8 | −3.2 |
| 14 | Si | −7873.2 | +13.5 | — | −16.3 | −3.2 |
| 15 | P | −9285.1 | +14.1 | +34.0 | −18.6 | −3.3 |
| 16 | S | −10832.3 | +16.2 | — | −19.7 | −3.0 |
| 17 | Cl | −12520.7 | +18.2 | — | −21.5 | −2.6 |
| 18 | Ar | −14354.6 | +19.9 | +44.1 | −23.7 | −2.3 |

*Note*: Energies are expressed in electron-volts. 1 at. un. $= 27.21$ eV.

calculated in the Hartree–Fock (HF) approximation ($\Delta E_{\text{HF}}$) and within the functional of the local spin density ($\Delta E_{\text{LSD}}$) disregarding correction and ($\Delta E_{\text{LSD}-\text{SIC}}$) allowing for the correction for the residual self-action. The data are presented for two versions of the correction: only depending on the orbital densities ($\Delta E^{(1)}_{\text{LSD}-\text{SIC}}$) and depending also on the angular momentum ($\Delta E^{(2)}_{\text{LSD}-\text{SIC}}$).

Table 7.1 [914] shows that the values of the total atomic energy in the HF approximation exceed the experimental values. In the LSD approximation they are even higher, with an error about twice as large as that of the HF approximation. Allowance for the correction for the residual self-action leads to energies lying below the experimental ones, but in most cases closer to them than the energies in the HF approximation. Allowance for an explicit dependence of the correction for residual self-action on the orbital momentum leads to remarkable agreement with the experiment for the total energies of the considered atoms.

The deviation from experiment of the calculated values of the first ionization potentials of light atoms as a function of $Z$ are given in Fig. 7.1 [914]. The data obtained in the HF approximation are joined by lines: the dashed

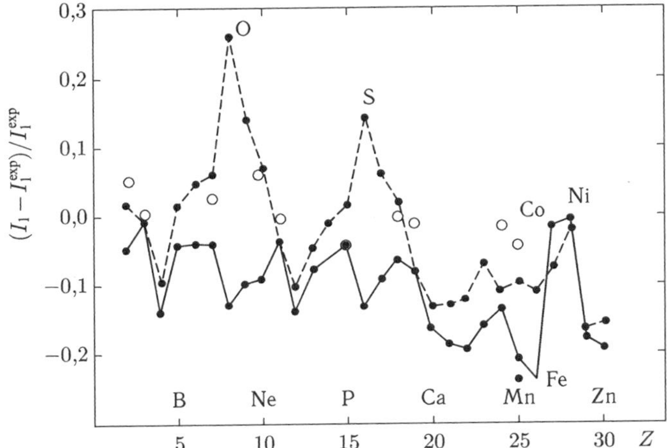


Fig. 7.1 Deviation of theoretical light-atom first ionization potentials from experiment. The dashed line is the HF approximation by the Koopmans' theorem; the solid line is the HF approximation with relaxation; the circles demonstrate the LSD method.

line for the case when the ionization potentials were estimated by the Koopmans' theorem as eigenvalues taken with the inverse sign, the solid line stands for the use of the difference of total energies of a neutral atom and an ion. The circles correspond to the errors in first ionization potentials obtained by the LSD method with correction for the residual self-action. The presented results show that the LSD method with correction for the residual self-action give first ionization potentials that agree better with the experiment. The data employed in Fig. 7.1 are demonstrated in Table 7.2.

Table 7.3 contains data on the electron coupling energy in negative ions.

The quantum-chemical methods for calculating the characteristics of molecules based on Slater determinants lead to a sharp increase in the volume of calculations with increasing number of atoms in a molecule and (or) and the charge of the nuclei of atoms constituting the molecule. In spite of the great difficulty of the calculation employing the Slater determinant, the results thus obtained agree with the experiment not so well as the results of a much less cumbersome calculation by the LSD functional method.

Table 7.4 [914] contains the dissociation energies, the equilibrium inter-atomic distances, and the oscillation frequencies of biatomic molecules consisting of some light atoms obtained in calculations by the LSD functional method. The corresponding experimental data are also given.

Table 7.2    Experimental light-atom ionization potentials and the results of their calculation by different methods [914].

| $Z$ | Atom | Experiment | HF Method | | LSD–S1C |
| --- | --- | --- | --- | --- | --- |
| | | | $-\varepsilon_{max}$ | $E_{HF}^{(+)} - E_{HF}^{(0)}$ | $-\varepsilon_{max}$ |
| 2 | He | 24.587 | 24.98 | 23.45 | 25.8 |
| 3 | Li | 5.392 | 5.341 | 5.341 | 5.4 |
| 4 | Be | 9.322 | 8.416 | 8.043 | — |
| 5 | B | 8.298 | 8.432 | 7.932 | — |
| 6 | C | 11.260 | 11.79 | 10.79 | — |
| 7 | N | 14.543 | 15.44 | 13.96 | 14.9 |
| 8 | O | 13.618 | 17.19 | 11.89 | — |
| 9 | F | 14.422 | 19.86 | 15.72 | — |
| 10 | Ne | 21.564 | 23.14 | 19.84 | 22.9 |
| 11 | Na | 5.139 | 4.955 | 4.952 | 5.1 |
| 12 | Mg | 7.646 | 6.884 | 6.615 | — |
| 13 | Al | 5.986 | 5.714 | 5.507 | — |
| 14 | Si | 8.151 | 8.081 | 7.660 | — |
| 15 | P | 10.486 | 10.66 | 10.04 | 10.0 |
| 16 | S | 10.360 | 11.90 | 9.031 | — |
| 17 | Cl | 12.967 | 13.78 | 11.80 | — |
| 18 | Ar | 15.759 | 16.08 | 14.78 | 15.8 |
| 19 | K | 4.341 | 4.011 | 4.005 | 4.3– |
| 20 | Ca | 6.113 | 5.320 | 5.118 | — |
| 21 | Sc | 6.54 | 5.717 | 5.347 | — |
| 22 | Ti | 6.82 | 6.008 | 5.513 | — |
| 23 | V | 6.74 | 6.275 | 5.668 | — |
| 24 | Cr | 6.766 | 6.038 | 5.88 | 6.7 |
| 25 | Mn | 7.435 | 6.743 | 5.90 | 7.1 |
| 26 | Fe | 7.870 | 7.026 | 6.05 | — |
| 27 | Co | 7.86 | 7.279 | 7.78 | — |
| 28 | Ni | 7.635 | 7.515 | 7.62 | — |
| 29 | Cu | 7.726 | 6.476 | 6.40 | — |
| 30 | Zn | 9.394 | 7.959 | 7.65 | — |

*Note*: The ionization potentials are expressed in electron-volts $E_{HF}^{0}$, $E_{HF}^{+}$ are energies of a neutral atom and an ion in the HF approximation, $\varepsilon_{max}$ is the maximum energy eigenvalue of occupied electron states.

Agreement of equilibrium distances and vibrational frequencies obtained in the framework of the LSD method with the experimental data is rather good: for the molecules included in Table 7.4 [914] the mean deviation from experiment makes up 0.05 at. un. for equilibrium states and $80\,\mathrm{cm}^{-1}$ for vibrational frequencies. The dissociation energies are somewhat overestimated, the mean deviation from the experiment, according to Table 7.4, is equal to 1.2 eV. As is shown by the calculation for the nitrogen molecule $_2$,

Table 7.3   Electron binding energy in negative ions [914].

| Ion | HF | LSD–SIC | Experiment | Ion | HF | LSD–SIC | Experiment |
|-----|------|---------|------------|--------|------|---------|------------|
| $H^-$ | −0.33 | 0.7 | 0.75 | $Cl^-$ | 2.58 | 3.8 | 3.61 |
| $O^-$ | −0.54 | 1.6 | 1.46 | $Br^-$ | 2.58 | — | 3.36 |
| $F^-$ | 1.36 | 3.6 | 3.45 | $I^-$ | 2.47 | — | 3.06 |

Table 7.4   Some characteristics of biatomic molecules obtained by the LSD functional method and their comparison with experiment [914].

| Molecule | $D$, eV | | $R$, at. un. | | $\omega$, cm$^{-1}$ | |
|----------|------------|------|------------|------|------------|------|
| | Experiment | LSD | Experiment | LSD | Experiment | LSD |
| $H_1$ | 4.8 | 4.9 | 1.40 | 1.45 | 4400 | 4190 |
| $Li_2$ | 1.1 | 1.0 | 5.05 | 5.12 | 350 | 330 |
| $B_2$ | 3.0 | 3.9 | 3.00 | 3.03 | 1050 | 1030 |
| $C_2$ | 6.3 | 7.3 | 2.35 | 2.35 | 1860 | 1880 |
| $N_2$ | 9.9 | 11.6 (9.94) | 2.07 | 2.07 (2.07) | 2360 | 2380 |
| $O_2$ | 5.2 | 7.6 | 2.28 | 2.27 | 1580 | 1620 |
| $F_2$ | 1.7 | 3.4 | 2.68 | 2.61 | 890 | 1060 |
| $Na_2$ | 0.8 | 0.9 | 5.82 | 5.67 | 160 | 160 |
| $Al_2$ | 1.8 | 2.0 | 4.66 | 4.64 | 350 | 350 |
| $Si_2$ | 3.1 | 4.0 | 4.24 | 4.29 | 510 | 490 |
| $P_2$ | 5.1 | 6.2 | 3.58 | 3.57 | 780 | 780 |
| $S_2$ | 4.4 | 5.9 | 3.57 | 3.57 | 730 | 720 |
| $Cl_2$ | 2.5 | 3.6 | 3.76 | 3.74 | 560 | 570 |

the results of which are given in brackets in the corresponding row of the table, the allowance for the correction for residual self-action to LSD must improve substantially the agreement between the calculated dissociation energies with experiment.

## 7.3   Condensed media

The deviations from experiment of the calculated atomic binding energies in a crystal and the densities corresponding to zero temperature and pressure as functions of $Z$ are shown for a number of metals in Figs. 7.2 and 7.3 [914].

The presented results were obtained, as has already been mentioned [914], by far not the most perfect method of band structure calculations and should, therefore, be regarded as only the upper estimate of precision now attainable in calculations of crystal properties on the basis of the DFM.

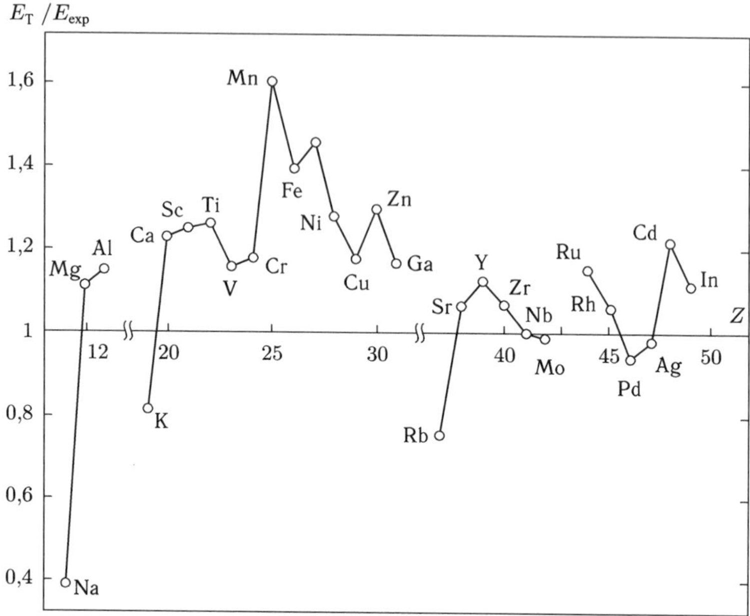



Fig. 7.2   Deviation of the theoretical values of atomic binding energy in some metals from experiment [914].

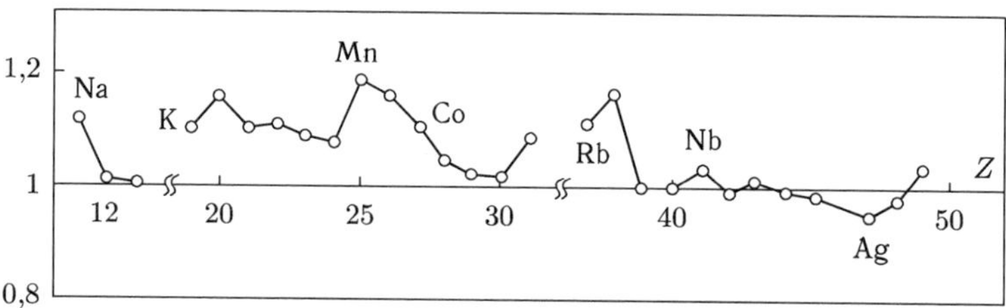



Fig. 7.3   Deviation from experiment [914] of the theoretical values of crystalline density of some metals at zero pressure and temperature.

The calculations showed that except for lanthanides and light actinides the theory predicts the stable structure for the most part correctly. It is only for sodium and gold that the BCC structure turns out to be more advantageous than respectively the hexagonal close-packed structure and the face-centered structure observed in experiment. Less successful was the prediction of stable structures of lanthanides and light actinides, although the number of coincidences with experiment is also considerable. This is

Table 7.5    Energy gap in noble metal crystals.

| Crystal | $\Delta E_{\exp}$ | $\Delta E_{\mathrm{HF}} - \Delta E_{\exp}$ | | $\Delta E_{\mathrm{LSD}} - \Delta E_{\exp}$ | $\Delta_{\mathrm{SIC}}$ |
| | | without correlation | with correlation | | |
|---|---|---|---|---|---|
| Ne | 21.4 | +3.8 | +0.9 | −10.2 | +9.9 |
| Ar | 14.2 | +4.3 | +1.0 | −5.9 (−0.7) | +5.8 |
| Kr | 11.6 | +4.8 | +1.8 | −4.9 | +4.9 |

obviously due to inadequacy of the computational method for metals with unoccupied $f$-shell.

The known problem in the band structure calculations is the discrepancy between the experimental values of the energy gap width between the filled and empty bands in insulators and the results of computations in the HF and the LSD functional approximations. In the HF approximation the gap width is overestimated because of a disregard of correlations, while in the LSD approximation the underestimation of this quantity may be due to the fact that because of the different degree of orbital localization in the valence and conduction bands the residual self-action affects differently the corresponding eigenvalues. For example, in noble gas crystals, $p$-like valence bands are narrow and lie as far from the other bands as atomic $p$-states. It can therefore be expected that with allowance for the correction for residual self-action the same shift of $p$-states as is observed in atomic calculations will take place in crystals. Table 7.5 [914] presents the experimental values of the gap width ($\Delta E_{\mathrm{Exp}}$) in noble gas crystals and the deviations from them of the HF calculations disregarding and approximately allowing for the correlations and the LSD functional. The table also gives the differences ($\Delta_{\mathrm{SIC}} = E_{\mathrm{app}}^{\mathrm{LSD}} - E_{\mathrm{app}}^{\mathrm{LSD-SIC}}$) between the external $p$-orbital energies calculated without and with allowance for the correction for residual self-action. All the data are given in electron-volts.

The data presented show that the shift of the eigenvalues of external p-orbitals in noble gas atoms with account of the correction for the residual self-action virtually coincides with the error of theoretical gap value in the LSD approximation. This suggests that the difficulties of the local density approximation should in this case be associated with residual self-action. The result obtained in the calculation for an argon crystal with allowance for the correction for residual self-action and given in brackets in the corresponding row of the table confirms this assumption thus testifying to the fact that one should not *a priori* ignore the residual self-action in the crystal.

The DFM are successfully applied to calculate the thermodynamic properties of substances and, most often, to calculate the cold curves, isotherms $T = 300\,\mathrm{K}$ and shock adiabats of the crystal phase [791, 918–920, 1032]. In paper [1032] the authors present room-temperature isotherms and shock adiabats of Al, Cu, Ta, Mo, and W, in paper [919] the cold curve of Zn, in [919] the cold curve of Al and the room-temperature isotherm of Be, in [920] the isotherm $T = 300\,\mathrm{K}$ for Mg, in [791] the cold curve of Al up to superhigh pressures of nearly 100 Mbar. As an example, Fig. 7.4 demonstrates the cold curve of aluminum [791] as compared to the existing equations of state, Fig. 7.5 presents the isotherm $T = 300\,\mathrm{K}$ for Mg [920] compared to the experimental data and Fig. 7.6 shows the shock adiabat of copper [1032]. The Fermi–Zel'dovich method described in Chapter 9 makes it possible to calculate the isentropy of aluminum release in good agreement with experiment (Fig. 7.7).

We can see that the DFM, although equivalent to the exact solution of the many-body problem, differs advantageously from direct quantum-mechanical methods by a much greater simplicity and is therefore considered to be a powerful tool of description of the wide spectrum of physical

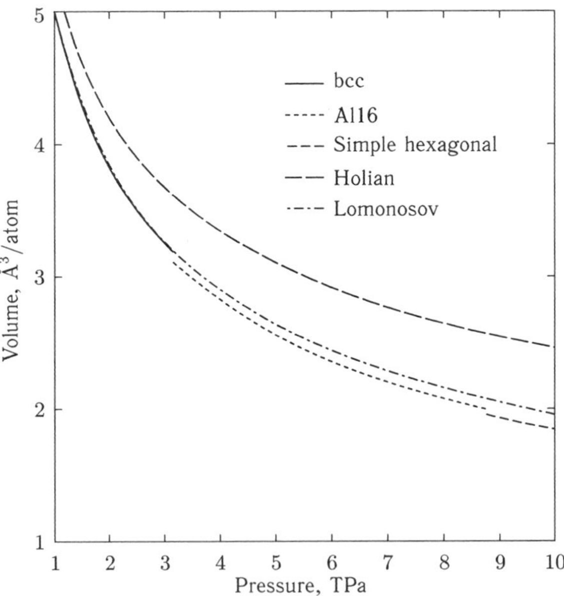


Fig. 7.4   Aluminum cold curve. The calculations were performed by the equations of state [634] and [441].

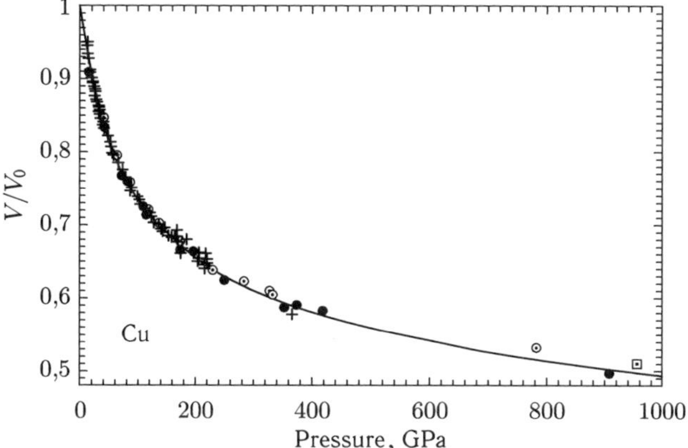

Fig. 7.5  Dependence of pressure on the degree of compression for a hexagonal close-packed (solid line) and body-centered cubic (dashed line) lattices on the isotherm $T =$ 300 K for Mg. Experimental data: $\triangledown$ is the point of going over from a HCP to a BCC lattice [757], $\bullet$ HCP [755], $*$ BCC [755], $\square$ [183], o [780], $\triangle$ [246].

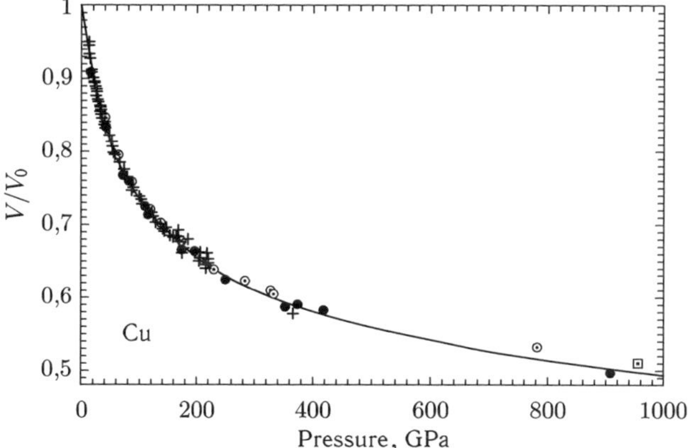

Fig. 7.6  Shock adiabat of copper. Solid line stands for the calculation by the DFM. Experiment $+$ [658], o [699, 700], $\bullet$ [32, 36, 37], $\square$ [576].

properties of complicated electron systems. In condensed state physics it is sometimes called the "standard model" thus emphasizing the high quality of its results.

This method does not substitute for, but supplements the earlier approaches and opens another way of constructing approximate solutions of the many-body problem. If the approximate solutions were as

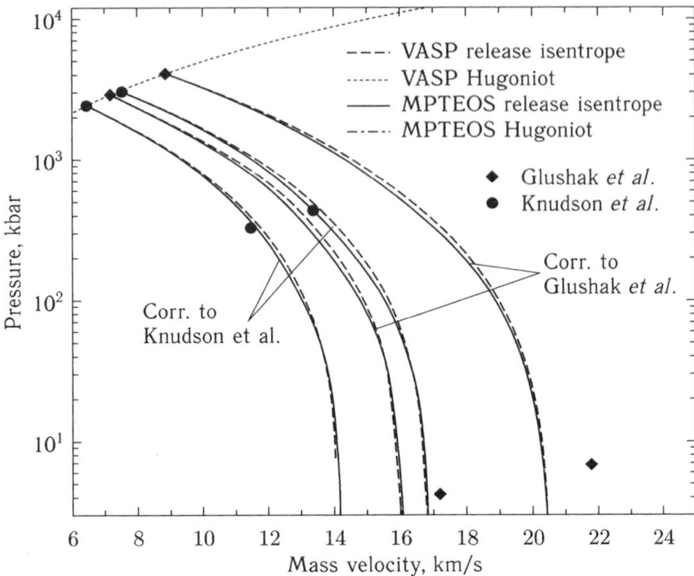


Fig. 7.7 Aluminum release isentropies in the pressure–mass velocity coordinates. Curves show the calculation by the semi-empirical equation of state [528] and by the DFM, points are experimental data [355, 553].

a rule found earlier by a cutoff or a partial summation of series, the DFM allows construction of approximations proceeding from analogies with other, simpler systems whose solution is found by independent methods. The resources of this highly physical approach are far from being exhausted, and it promises many interesting results in future.

Chapter 8

# Phase Transitions

The consideration of various models of the thermodynamic description of matter physical properties makes it clear that the conventional classification of states in the high-pressure and high-temperature region often loses its definiteness and becomes conditional. The phase boundaries either vanish or become smeared and actually correspond to continuous mutual transformation of the close states. In this chapter, we will consider the relationships between different phases of matter. This will help to present in a more definite way the general form of the phase diagram taking into account the real and hypothetical phase transitions.

## 8.1  Melting

Melting is one of the most universal and well studied phase transitions. Being a transition from the ordered to the disordered structure, it is associated with vanishing of the long-range order in the system. In this case there is an important problem of the existence of the critical point of the second kind in which the first-order phase transition is replaced by the second-order phase transition because, according to the results of Landau [597], a melting curve cannot end at a critical point. The necessary condition of that is the simultaneous vanishing of the volume $\Delta V_\mathrm{m}$ and pressure $\Delta S_\mathrm{m}$ jumps on melting. Therefore the experimental verification of the theoretical predictions reduces to the simultaneous measurement of these quantities at high pressures [454, 575, 697, 949, 950].

The available experiments on temperature measurements at high pressures of 0.5–3 Mbar [575] and volume measurements under static conditions [454, 697] testify to the hypothesis on the absence of the critical points of the second kind because the volume and entropy jumps on melting and

317

do not disappear with pressure increase. Interestingly, the limiting measured values of $\Delta S_\mathrm{m}$ turn out to exceed $R\ln 2$, which is the minimal value associated with vanishing of the long-range order on the transition from the ordered to the disordered structure [949, 950].

A detailed comparison of the available experimental data with the theoretical results obtained for some simple systems (noble gases, one-component plasma, charged spheres, alkali metals) indicates the leading role played in melting by the structure dependence of the potential energy [949, 950] which, in its turn, weakly depends on the concrete nature of the intermolecular forces and for model systems can be realized making use of the maximally simplified hard-sphere and soft-sphere potentials. It is important that the results of these models give also evidence for the absence of the critical points which is in agreement with the general conclusion [597].

At present, for the quantitative description of the melting curve a wide use is made of the Lindemann criterion according to which a solid body melts when the amplitude of thermal vibrations of the lattice atoms $q$ is a certain fraction $A$ of the atomic spacing in a crystal $R$:

$$A^2 = \frac{q^2}{R^2} = \frac{\langle E\omega^{-2}\rangle}{MR^2}, \tag{8.1}$$

The averaging is carried out over all modes of the thermal vibrations $\omega_i$, the energy of the $i$ mode in the quasiharmonic approximation is given by the formula

$$E_i = \frac{\hbar\omega_i}{e^{\hbar\omega_i/kT} - 1}. \tag{8.2}$$

The real spectrum of thermal vibrations and its volume dependence are in most cases extremely complicated; this makes it necessary to use model considerations. For high-temperature melting $E_i = kT$ and in the Debye approximation the traditional form of the Lindemann condition

$$\frac{T_\mathrm{m}}{\omega_\mathrm{D}^2 V_\mathrm{m}^{2/3}} = \text{constant},$$

follows from (8.1). Here $\omega_\mathrm{D}$ is the Debye frequency. In the general case of an arbitrary spectrum [130]

$$\frac{d\ln T_\mathrm{m}}{d\ln V_\mathrm{m}} = 2\left(\frac{\langle\gamma\omega^{-2}\rangle}{\langle\omega^{-2}\rangle} - \frac{1}{3}\right), \quad \gamma_i = -\frac{\partial\ln\omega_i}{\partial\ln V}. \tag{8.3}$$

The conventional assumptions which simplify equation (8.3) are the condition

$$\langle \gamma \omega^{-2} \rangle = \langle \gamma \rangle \langle \omega^{-2} \rangle, \tag{8.4}$$

and the replacement of the average value $\langle \gamma \rangle$ by the thermodynamic quantity — the Grüneisen parameter $\gamma = V(\partial p / \partial E)_V$. This replacement is valid at high $(T > 0.1 \hbar \omega_{\mathrm{D}}/k)$ temperatures. It follows from equation (8.3) that the melting temperature increases monotonically on compression because the experimental values of $\gamma$ exceed $1/3$.

In spite of its purely empiric origin, the Lindemann relationship is widely used in geophysical and astrophysical applications, in the investigation of Wigner crystallization of electrons, "cold" melting of strongly compressed matter, etc. Naturally, a problem of the theoretical justification of the Lindemann rule and its applicability limits arises.

It can be shown that the Lindemann relationship is a rigorous consequence of self-similarity of the nonideal part of the statistical sum of a system of particles interacting according to the law $U(r) \sim 1/r^n$.

Let us emphasize in this connection that the melting curves of the type $c\frac{dT}{dp} < 0$ and $c\frac{dT}{dp} = 0$ cannot be obtained in the class of systems with interaction of the form $U(r) \sim 1/r^n$. Actually, it is easy to show [949, 950] that for the melting of systems with a power-law interaction the following relations are valid:

$$p \sim T^{1+3/n}, \quad \frac{\Delta V}{V_S} = \text{constant}, \quad \Delta S = \text{constant},$$

Here, $p$ and $T$ are the coordinates of the melting curve, $\Delta V/V_S$ is the relative volume jump on melting, $\Delta S$ is the entropy jump on melting.

When attraction is turned on, these formulae acquire the meaning of high-temperature asymptotic relations which naturally excludes the equalities $dT/dp = 0$ and $\Delta V = 0$ at arbitrary finite values of temperature and pressure.

The concrete characteristics of melting can be calculated taking into account conditions (8.3) and (8.4), provided the volume dependence of the Grüneisen parameter is known. The generalized form of this dependence [1010] is as follows:

$$\gamma(V) = \frac{t-2}{3} - \frac{V}{2} \frac{\left(p_x V^{2t/3}\right)''_{V^2}}{\left(p_x V^{2t/3}\right)'_V}, \tag{8.5}$$

where $t = 0$ corresponds to the Slater–Landau approximation, $t = 1$ is related to the Dugdale–MacDonald one, and $t = 2$ — corresponds to the Vashchenko–Zubarev free volume theory [1081], $p_x = p(V, T = 0\,\mathrm{K})$ — being the pressure on zero isotherm. The methods of the determination of parameter $t$ in equation (8.5) and the variety of ways to find the dependence $\gamma(V)$ making use of the static and dynamic experimental data are considered in Chapter 9. We will only point out here that the optimal value of $t$ may be also noninteger. For instance, the magnesium melting curve experimentally registered at high pressures [995] was described at $t = 0.55$ (Fig. 8.1).

Note that the Simon and Kraut–Kennedy melting laws used in practice [949, 950] follow under certain assumptions from the Lindemann law. Generalization and reformulation of criterion (8.1) for an arbitrary interparticle interaction potential by the methods of the statistical mechanics [857] are an extra confirmation of its validity.

Monte Carlo computer calculations of the system of hard spheres [834] showed also the applicability of the Lindemann criterion up to a

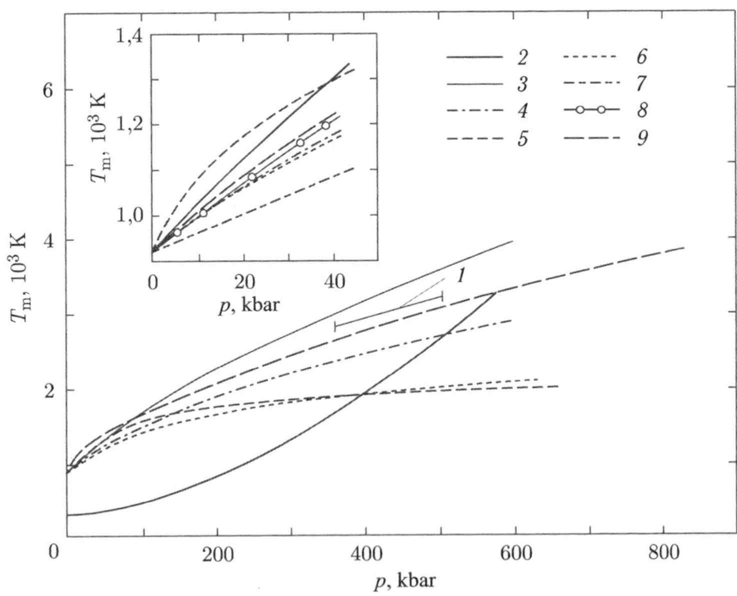

Fig. 8.1   Melting of magnesium at shock loading [857]. *1* experiment, *2* shock adiabat, *3*, *4*, *7*, *9* calculation using equation (8.5) at $t = 0$, 1, 2, 0.55, *5* Simon law, *6* Kraut–Kennedy law, *8* static data.

hundred-fold increase in temperature and four-fold increase in pressure, melting arising at 10% linear deviation from the close packing of the spheres. Being based on simple physical reasoning, this criterion became very popular in the thermal and cold melting problems and in the studies of plasma crystallization.

In the latter case direct observation of particle oscillations in the crystal phase (Chapter 2) makes it possible to find the mean-square deviation of the crystal particles from the equilibrium state $(q\sqrt{\langle r \rangle^2})$ and to show that on melting it is close to $(0.15 \div 0.25)$ of the interparticle distance $A$ according to criterion (8.1).

Interestingly, a study of melting at the kinetic level (by direct observation) shows (Fig. 2.7) that melting of a crystal occurs in a heterogeneous way rather than homogeneously over the overall volume. A homogeneous crystal breaks into moving clusters which have intrinsic crystalline structure and disintegrate into smaller particles as the temperature increases. As a result, the crystal transforms into a homogeneous liquid. These conclusions agree with the theory of heterogeneous melting performed by the molecular dynamics techniques: heating of the crystal lattice leads to the growth of dislocations which increase in number with the temperature increase and destroy the crystal state in the melting point in correspondence with the Kraft–Macher dislocation model.

When the real spectrum of the crystal oscillations from equation (8.3) is taken into account, it is as well possible to obtain a negative slope of the melting curve [764, 765] which follows from the anomalous volume dependence of one of the long-wave oscillation modes. This enables one to describe the nonmonotonic behavior of the melting curves. For instance, the experimentally found softening of the transverse acoustic modes in the phonon spectra of the barium body-centered lattice exposed to pressure made it possible to explain the maximum on the melting curve due to the decrease of the stability of this lattice [118]. The change in the character of the interparticle interaction induced by the rearrangement of the electrons may be also one of possible reasons of the nonmonotonic temperature variation on the melting curve, as it is exhibited by the cesium and cerium data.

Figure 8.2 [369] presents the Na melting curve with indication of the bcc and fcc solid state phases. It can be seen that the peculiarities of the melting curve correlate with the structure of the crystal in the pre-melting state and in the liquid phase the curve presumably keeps the features of the corresponding structure. Similar correlations can be seen also in case of uranium dioxide (Fig. 8.3) [704] and calcium (Fig. 8.4) [312].

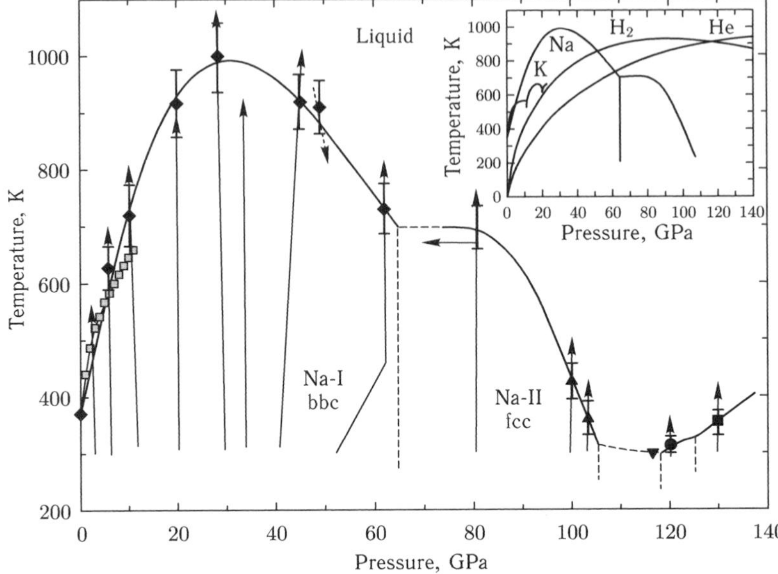


Fig. 8.2    Na melting curve up to 130 GPa. The filled symbols are the results of Ref. [369], open squares are the data from Ref. [1089]. The solid line is the melting curve (for details see Ref. [369]). The dashed lines show the tentative phase boundaries. The arrows show the $p$–$T$ variation in experiments [369] (a small pressure shift on heating is not presented). The inset shows the Na melting curve [369] in comparison with the melting curves of $H_2$ [371], He [201], and K [1089].

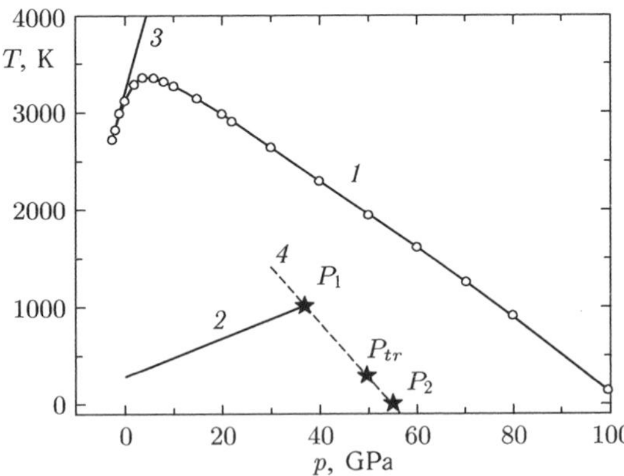


Fig. 8.3    Phase diagram of polymorphic modifications of $UO_2$ and their melts [704].

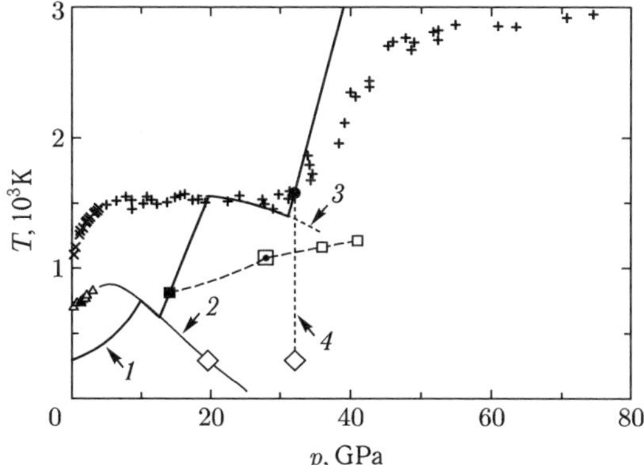

Fig. 8.4 Phase diagram and thermodynamic states of calcium on multiple shock compression. △ experimental data [485] for the fcc–bcc equilibrium line, x experimental data [485] for the bcc-phase melting curve, + – experimental data [247] for Ca melting curve, ◇ pressures of calcium phase transitions at room temperature [756, 1044]. Curves: *1* our calculations of shock adiabats for single compression of Ca monolith, *2* calculation of the region of the fcc–bcc equilibrium line, *3* calculation of the region of the bcc-melt equilibrium line, dashed lines calculated adiabats of step-wise shock loading, *4* (dashes) bcc–sc equilibrium line [247], ■ TD state of bcc calcium in the first shock wave, ⊡ TD state in the vicinity of the bcc–sc equilibrium line whose crystal phase is not defined unambiguously (see Ref. [312]), □ estimate of TD states in sc phase, ● TD state of Ca melt in the first shock wave [312].

In connection with the planetary applications, the melting curves were recently studied making use of the powerful laser-driven shock waves [555]. These investigations indicated a negative slope of the melting curve in the region of high pressures (Figs. 8.5, 8.6 and 8.7).

The nonsimple behavior of the high temperature melting curve for nitrogen [108] is related (Fig. 8.8) to the formation of the nitrogen polymer phase and its amorphization, while the maximum found in Ref. [245] on the hydrogen melting curve (Fig. 8.9) is explained due to dissociation and ionization effects at high pressures [123].

In some cases the negative value of $dT_{\rm m}/dp_{\rm m}$ can be also explained in the framework of the two-species liquid model [826, 827]. According to this model, a liquid is a mixture of two kinds of particles which are either the structures having different coordination numbers or atoms in which an electron transition has or has not occurred. The first option is realized for substances which have low-density phases of low pressure. In this case the slope of the melting curve is negative because there exist structures near the phase boundary whose coordination numbers correspond to the

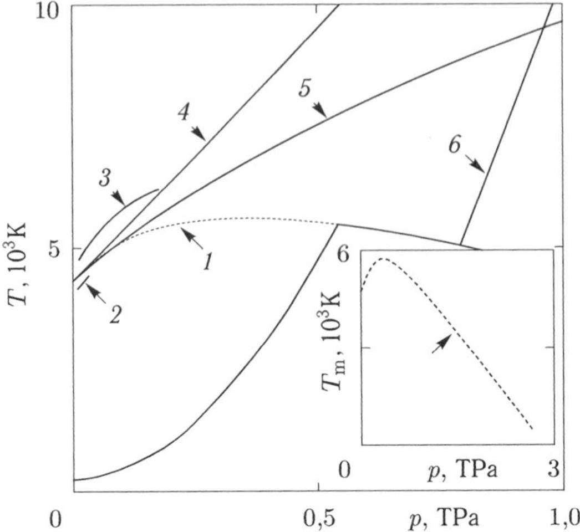

Fig. 8.5   Melting curve and temperature along the diamond shock adiabat. *1* – melting curve [705], *2* [61], *3* [143], *4* a tangent to the melting curve 10.2 K/GPa in the triple point (•) [977], *5* melting curve [705], *6* the temperature on the shock adiabat for monolith diamond [705].

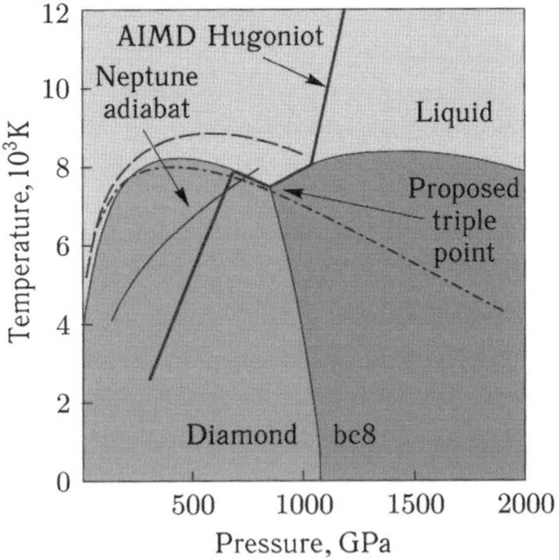

Fig. 8.6   Phase diagram of carbon at high energy densities. Full curve phase boundaries from Ref. [193]; dashed-dotted grey curve diamond melting curve [385], dashed grey curve diamond melting curve [1030], full grey curve adiabat predicted for Neptune [450], full black curve shock adiabat [555].

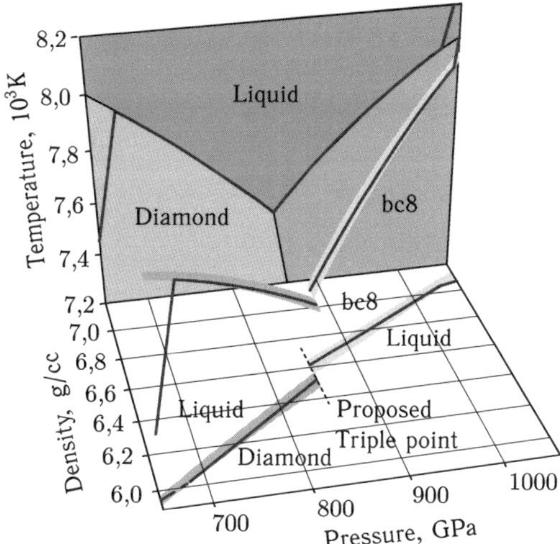


Fig. 8.7 Carbon phase diagram (with shock adiabat) in the $p$–$\rho$–$T$ space. The shock adiabat starts from the pressure $p \approx 680\,\mathrm{GPa}$ ($6{,}02\,\mathrm{g/cm}^3$ ), reaches the triple point at $p \approx 850\,\mathrm{GPa}$ and leaves the diagram at $p \approx 1040\,\mathrm{GPa}$ ($7{,}04\,\mathrm{g/cm}^3$). The shock adiabat has two solutions on the proposed triple point: (1) a mixture of diamond and liquid ($6{,}52\,\mathrm{g/cm}^3$) and (2) a mixture of the bc8 phase and liquid ($6{,}62\,\mathrm{g/cm}^3$) [555].

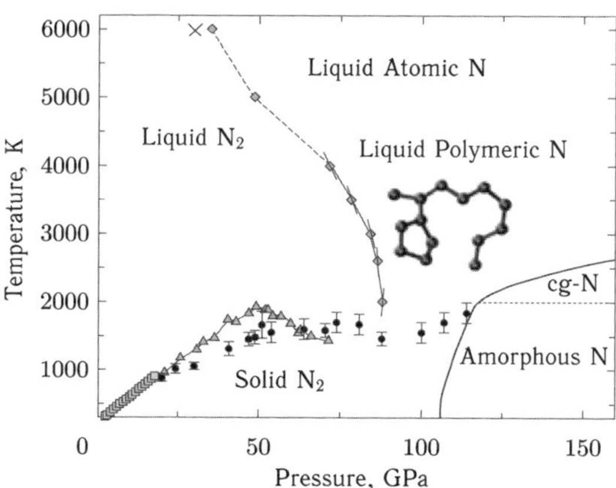


Fig. 8.8 Phase diagram of nitrogen (with the domain of metastable amorphous N). Experimental data: squares [1068], triangles [70], and circles [362]. The known strongly compressed solid polymer phase is separated by a full line (dashes denote the absence of accurate data) [244, 363, 370, 627]. The structural formula corresponding to polymeric liquid is depicted in the corresponding domain. The figure is adopted from Ref. [108].

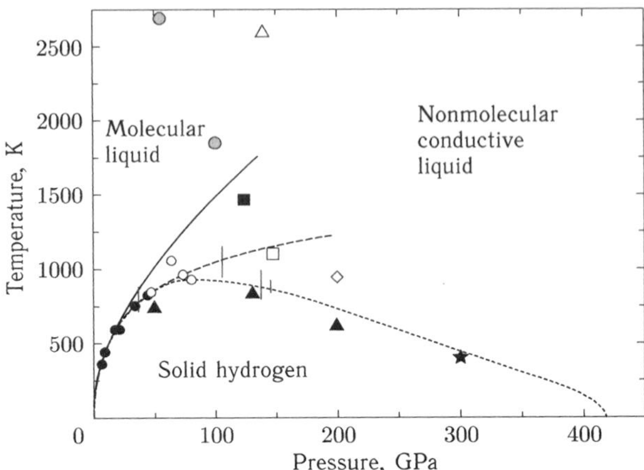


Fig. 8.9   Phase diagram of carbon at high energy densities. Full curve phase boundaries from Ref. [193], dashed-dotted grey curve diamond melting curve [385], dashed grey curve diamond melting curve [1030], full grey curve adiabat predicted for Neptune [450], full black curve shock adiabat [555].

high-pressure phase. This fact can be confirmed by the analysis of the structural data which indicates the existence (Hg, Sn) or even dominance (Bi, Sb, Ga) of the high-pressure phase structures in a liquid near the melting curve [826, 827].

A different situation is realized in case of melting of metals which allow, like e.g., lanthanides, a possible rearrangement of the electronic structure. The experimentally discovered kinks and significant decrease of compressibility on the shock adiabats of the rare-earth elements [30, 161, 407] were explained due to the termination of the electronic transitions [28, 30] or high rigidity of the internal closed $5p^6$ shell of xenon [407]. In paper [384] it was noted for the first time that the experimental kinks of the shock adiabats of the lanthanides are mostly located in the range which contains the melting curves extrapolated from the low-pressure region. The significant size of the kinks testifies that the quantity $dT_m/dp_m$ has large negative values when crossing the two-phase region.

The thorough measurements of the shock adiabats of lanthanides [161] made it possible, on the basis of the shock-wave and static data, to obtain melting curves up to high ($p \lesssim 1\,\text{Mbar}$) pressures and to locate the position of their intersection with the adiabats. For all metals this position matched fairly well the experimentally found kink (Fig. 8.10). Since the high pressure phases of the rare-earth elements have closest packed structures, the

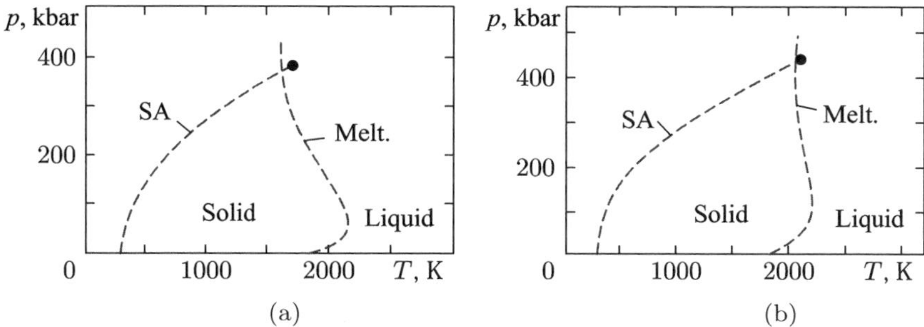



Fig. 8.10   Phase diagrams of thulium (a) and erbium (b) [161]. Dashed lines: the calculated temperatures on the shock adiabat (SA) and on the melting curve (Melt.). Symbols: melting on the shock adiabat from the experiments.

discovered anomalous behavior of the melting curves and, as a result, the increased density of the liquid cannot be ascribed to the presence of higher coordination numbers. On the basis of the two-species liquid model, we can make a conclusion that when the pressure grows, the atoms of the liquid undergo rearrangement of electrons which increases the liquid density. In the solid phase rearrangement of the electronic structure takes place either in a similar continuous manner (cerium type) but at a different speed, or beginning from the triple point (cesium type), which leads to the anomalies in the melting curves of the lanthanides.

The concrete mechanism of the rearrangement leading to the increase of the density and decrease of the compressibility may be different and can be explained either due to the termination of s–d-electronic transitions and formation of low-compressible shells (lanthanum [677]) or due to a sharp decrease of the ion metallic radius on delocalization of f-electrons (cerium [351], praseodymium [652]). Emphasize that it is not mandatory that the kink locations on the shock adiabats of lanthanides correspond to their intersections with the melting curves. A kink (or a sharp knee [677]) due to electronic rearrangement can arise as well in the solid phase. For lanthanides, however, the mentioned correspondence is quite convincing because of the anomalous behavior and low temperatures characteristic of the melting curves and as well due to significant values of pressures associated with the electronic rearrangement. A different situation is presumably observed for refractory metals (Ti, Zr, Hf) in which the termination of the s–d transitions on the shock adiabat occurs in a solid phase and is accompanied by a structural phase transition.

The maximum on the melting curve appears as well at extremely high pressures of matter because of the quantum effects [545]. As the crystal is compressed, the spatial localization of nuclei leads to the momentum uncertainty and, consequently, to the increase of the amplitude of zero-point oscillations whose energy $\hbar\omega_{\mathrm{D}} \sim \hbar\omega_0 \sim n^{1/2}$ ($\omega_{\mathrm{D}}$ and $\omega_0$ are Debye and plasma frequencies of ions) grows with the density increase faster than the lattice stabilizing Coulomb energy which is $\sim n^{1/3}$. Eventually, this leads to the crystal melting at $T = 0\,\mathrm{K}$, i.e., the existence region of the crystalline state turns out to be limited. The estimates of the crystal limiting densities are fairly uncertain, but in any case correspond to the extreme values of $\rho$ of the order of $10^3$–$10^8\,\mathrm{g/cm^3}$ [966, 996].

## 8.2  Polymorphic and electronic transformations

At pressures of hundreds of kilobars solid bodies, depending on thermodynamic conditions, may have different crystalline structures. Transitions between these structures induce the appearance of new phase boundaries. Classification and description of polymorphic phase transitions is a cumbersome and complicated job which requires precise calculation of the energy and photon spectra of the competing modifications. Therefore direct static and dynamic experiments [978] are the main source of information about the polymorphic transitions. Phase transformations share a common tendency that under pressure a crystalline substance undergoes a transition to more closely-packed structures with maximum coordination numbers. According to the data of the dynamic measurements [31, 679], the polymorphic phase transitions terminate at pressures of 0.5–1 Mbar.

The behavior of "simple" metals turned out to be much richer and more interesting than it followed from the simplified models [648].

The shock-wave compression [322] of "simple" metals demonstrates a surprising and rather nontrivial behavior of degenerate strongly nonideal plasma at megabar pressures [322, 648, 730]. According to the concepts [339] which were considered until recently, the electronic properties of alkali metals can be described by the simplest model of the homogeneous electron Fermi gas which contains point ions. However, more modern and sophisticated quantum-mechanical models [648, 730] predict the formation at elevated pressures of complex crystalline structures with large coordination numbers. Then "pairing" of conduction electrons occurs in these structures and, as a result, the electric conductivity decreases over the pressure range of 0.3–1.0 Mbar.

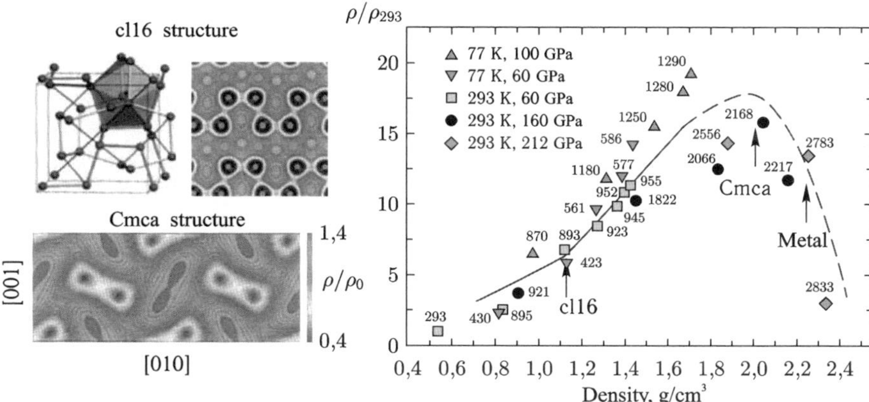



Fig. 8.11 Pressure-induced "dielectrization" of degenerate plasma [323]. The left panel illustrates the structure and the distribution of the electronic density in the high-pressure phase.

The experiments on Li, Na, and Ca quasiadiabatic compression carried out over the above-mentioned range of the dynamic pressures [322, 323, 648] demonstrated unambiguously this unusual effect, namely, pressure-induced "dielectrization" of "simple" metals [322, 323, 648] (Fig. 8.11). It can be seen that the compression of these metals first decreases their electric conductivity ("dielectrization"), but beginning from 1.2–2 Mbar they revert again to the "metallic" state which should presumably remain completely unchanged on further compression (Fig. 8.11).

The qualitative peculiarities of the phase diagram of substances with account of the polymorphic transitions between the close-packed structures of different types were successfully reproduced by means of the molecular dynamics calculations using the square-well interaction potential [1063, 1064] which corresponds to an effective attraction between the lattice atoms (Fig. 8.12). Calculations with the step potential [1063, 1064] which simulates the ion radius variation allowed to explain the details of isostructural transformations in metals and to obtain anomalous phase diagrams characteristic of cesium and cerium.

The electronic phase transitions (predicted already by Fermi) also occur in compressed matter. They are not directly associated with the change in the lattice symmetry and are caused by the rearrangement of the electronic structure of elements. In this respect the most illustrative are the observed transformations of metals belonging to the long periods whose energy levels are filled in an inverted order. As pressure increases, the

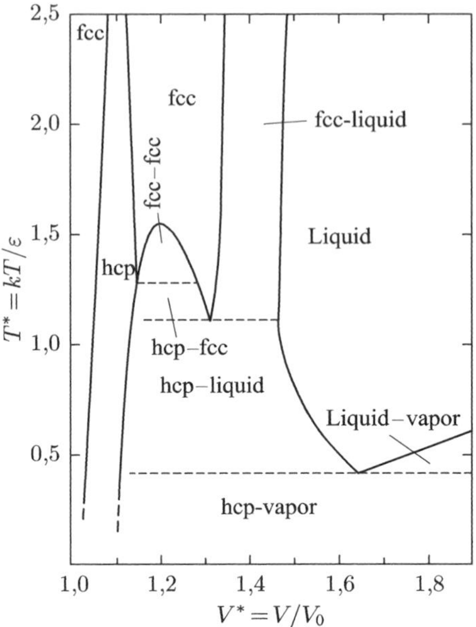


Fig. 8.12   Phase $p$–$V$ plane for a system of particles with the square-well interaction potential [1063, 1064]. The polymorphic modifications of the solid phase have face-centered cubic (fcc) and hexagonal close-packed (hcp) lattices. Of interest is the isostructural fcc-hcp transition with a critical point observed for cerium.

electrons redistribute among the shells and inversion gradually disappears which causes the change in the crystal binding energy and leads to the structural phase transitions.

The main factor determining the type of the metal lattice is the number of binding d-electrons. It can be seen from the calculations [713,789] that the s–d-transition and the increase of the d-band occupancy on compression of transitional and even simple metals are the principal reasons of the sequence of phases observed in these elements. Similar binding energy calculations for various structures of trivalent lanthanides [227] indicated that the obtained optimal type of lattice corresponds to the experimentally observed one, with the number of d-electrons increasing on compression from 1.5 to 2.7 and without any participation of f-electrons in the rearrangement process.

An additional confirmation of the fact that on moderate change of volume trivalent lanthanides behave like transition metals is the experimental discovery [1022] of the fact that the sequence of structural phases of yttrium, which belongs to the same Group 3B but contains no f-electron, is similar to that characteristic of the members of the series.

The existing picture was broken by the unexpected information that at high pressures scandium, which is the first fourth-period transition metal of Group 3B, has the lattice of the rare neptunium $\beta$-phase type [1020]. This fact points out in the first place the insufficiency of the approximate band-theory calculations [227, 713, 789], which do not take into account s–d-hybridization leading to such a lattice, and also the small contribution given by 5f-electrons to the binding energy of actinides since the neptunium $\beta$-phase-type structures are typical of them. The experimental paper [960] which studies the structure of lanthanides at high static pressures ($p \lesssim 500\,\mathrm{kbar}$) is of substantial interest. It was discovered that on the further growth of pressure the sequence of the first-order structural phase transitions leading to the closest-packed fcc lattice is supplemented with a second-order phase transition. This phase transition manifests itself as a progressive distortion of the fcc structure and arises due to a complicated hybridization of bands at high compressions. This phase transition is also an evidence of the fact that the single-particle model [227, 713, 789] has limited capability under such conditions.

There are peculiarities inherent in the electronic transitions in lanthanides associated with squeezing out narrow essentially localized f-bands to the conduction band. Such an electronic transition from a localized level to a collectivized state is accompanied by the intermediate valence phenomenon [533] which consists in that each atom possesses a noninteger number of conduction electrons. Ytterbium is an element in which this transition can be seen. When ytterbium is compressed from 40 to $300\,\mathrm{kbar}$, one can observe [380], without any change in the lattice structure, the gradual increase of the number of valence electrons from 2 to 3 accompanied by a corresponding rearrangement of the electronic structure from the state $4\mathrm{f}^{14}\,(5\mathrm{d}6\mathrm{s})^2$ to the state $4\mathrm{f}^{13}\,(5\mathrm{d}6\mathrm{s})^3$. It cannot be excluded however that this transition is due to the delocalization process considered below.

The electronic phase transitions are not at all always associated with the change of the lattice symmetry. Moreover, they can be realized in both solid and liquid phases. The most typical example is provided by the Mott transition which corresponds to delocalization of the electronic shells on compression. In this case the population of levels varies weakly, but when compression reaches its critical value, the electronic states change in character and the electronic shell turns into a conduction band. The Mott transition can proceed without structural rearrangement and end at a critical point which fact gives grounds to consider it as a "gas–electron liquid" phase transition [488].

Cerium is the best experimentally studied metal with such a transition [978]. At low temperatures the 4f-electron delocalization in cerium leads on compression to a considerable (15%) decrease of the ion size which induces an isostructural phase transition with a critical point. Praseodymium, which is the next after cerium element in the lanthanide series, reveals at $p \approx 200$ kbar pressure a structural phase transition between close tightly-packed structures [652] associated with delocalization of 4f-electrons and as well accompanied by a sharp ($\approx 19\%$) decrease of the metallic radius of the ion. Both examples confirm the validity of the two-species liquid model (see above) for the explanation of the anomalous behavior of the lanthanides' melting curves and show that f-electrons undergo stronger collectivization in a liquid phase. The latter results in the density increase and compressibility decrease on melting.

In metals the electron rearrangement or shell delocalization effects are currently discovered in the dynamic experiment on the decrease of compressibility of the new phase [30, 161, 407] as well as in the static experiment studying jumps of the electrical resistance [939] and variation of the structural characteristics of a crystal [978]. The electron phase transitions are also a subject of numerous theoretical studies [41, 351, 352, 483, 677, 783, 811, 922–924, 1027] employing the band theory methods (see Chapter 5). The calculation results enabled the correct explanation of the experimentally observed effects and confirmed that under compression the electronic structure rearrangement reveals a general tendency to reach the hydrogen-like level scheme.

For instance, the application of the linear-muffin-tin-orbitals (LMTO) method to the lanthanum band-structure calculation [677] showed the presence of the 6s–5d-electronic transition and the quantitative matching of its termination location and the experimentally detected decrease of the shock compressibility (see Fig. 3.6). It turned out that as temperature increases the transition effects become smoother because of broadening of the energy bands. Note that in paper [619] the peculiarities of the electronic characteristics of metals are as well studied in the vicinity of the singular point of the electronic transition induced by the change of the Fermi surface topology under pressure.

The description of the delocalization of f-electrons in lanthanides and actinides on compression is an interesting theoretical result. The calculation carried out for cerium [351] showed that the 4f-band narrows with the density decrease and splits into two states corresponding to different spin polarizations. This effect is associated with the 4f-electron transition from metallic to localized state and corresponds to the experimentally well studied [978] $\gamma$–$\alpha$-transition accompanied by arising of the $\gamma$-phase anti-ferromagnetism. Emphasize that at low temperatures this is a pure Mott transition rather than an f–d-one because calculations [353] showed that in the $\alpha$-phase the electronic configuration $4f^1$ $(5d6s)^3$ is energetically more beneficial than $(5d6s).^4$ Besides, it provides an essentially better agreement of the experimentally measured values of the transition energy and of the bulk modulus with the calculated ones. At over-critical temperatures, however, the difference between f-electron delocalization and f–d-electronic transition becomes presumably fairly conditional. The LMTO results gave also a qualitative description of the sharp change of the magnetic properties of transition metals at the temperatures exceeding the Curie point. The reason of this change is that due to the energy decrease the nonpolarized-spin states are preferable when the transition occurs [783].

f-electron delocalization on compression depends on the atomic number. The actinide series is an example of this dependence. Figure 8.13 presents the experimental and calculated values of the equilibrium atomic radii for actinides [922,923]. It can be seen that for the first elements of the series the experimental data are in total consistence with the calculations,

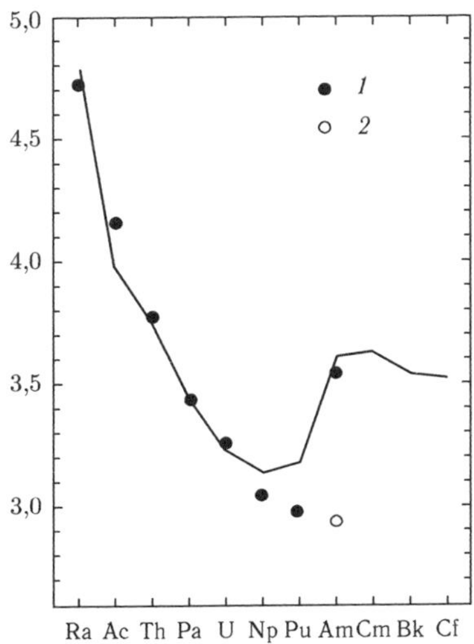


Fig. 8.13 Comparison of the experimental and theoretical *1* values of the atomic radii of actinides. *2* calculation made for americium without taking into account the localization of the 5f-electrons.

but the agreement breaks down beginning with neptunium and plutonium. The correlation effects become more important for these elements and just at small negative pressure the polarized spin solution becomes beneficial, which is optimal for americium also at normal conditions. Thus, in the actinide series americium is the first element with localized 5f-electrons which exhibits properties characteristic of the rare-earth elements. Calculations [922, 923] showed that on compression in americium, similarly to the cerium and praseodymium cases, the 5f-electron delocalization process must occur accompanied by the decrease of the metallic radius. Interestingly, this fact was really confirmed experimentally [854] and, although the volume jump on transition turned out to be smaller than the theoretical prediction [922, 923], the high-pressure phase has a low-symmetrical uranium $\alpha$-phase structure observed as well in cerium $\alpha'$-phase studies. It can be also expected that on compression other actinides, following americium, must transform into exotic structures with low coordination numbers characteristic of the first elements of the series.

Temperature, like pressure, can significantly influence the character of the f-electron states. Under ambient conditions, plutonium and neptunium possess delocalized 5f-electrons. However, already at a small volume increase they undergo a transition to the localized state [922, 923]. Taking into account the large value of the neptunium and plutonium bulk modulus and also the leading role played by the s–d-band hybridization in the neptunium $\beta$-phase formation [1020], the authors of Ref. [1021] come to the conclusion that, in high-temperature solid phases of these metals, the 5f-electron delocalization process occurs which just explains the observed anomalies in their properties.

The electronic phase transitions also arise when the internal electronic shells are squeezed out from the discrete to the continuous spectrum on compression, which according to Ref. [548] corresponds to a first-order phase transition. Such an electronic phase transition is naturally a series of phase transformations which are reduced to a sequence of pressure-induced electronic shells. The quasiclassical estimate of parameters of such phase transitions (see Chapter 6) and calculations carried out by the direct quantum-mechanical methods (see Chapter 3) yield pressures essentially exceeding 300 Mbar, which complicates the experimental verification of these predictions. These transitions may be responsible for the appearance of the anomalies in the "mass–radius" dependence of the white dwarfs [548].

## 8.3  High-temperature boiling

Interesting peculiarities are typical of the phase transformations in unordered structures where the competition between attraction and repulsion forces in particle interaction leads to the liquid–vapor transition which ends at a critical point. The characteristics of the boiling points and the parameters of the critical points have been measured for a considerable number of chemical compounds and can be found in a bulk of reference literature. Of special scrutiny are the experiments carried out in the near-critical region where the data on critical indices, spatial-temporal fluctuations, kinetic coefficients, etc. serve as a basis for the development of modern theories of critical phenomena [966]. At present, the information on the boiling curves for the most part of the chemical elements is however fairly limited. For instance, the critical point parameters are determined only for the three lowest-boiling metals [297] of all metals which constitute about 80% of The Periodic Table of Elements, not to mention any more detailed phase boundary investigations or near-critical and especially supercritical states.

Parameters of the critical points can be estimated by the tabular heat, temperature, and density values on the liquid–vapor equilibrium line making use of the thermodynamic analogy in their properties (Fig. 8.14). The detailed consideration of the thermodynamic similarity principles and the

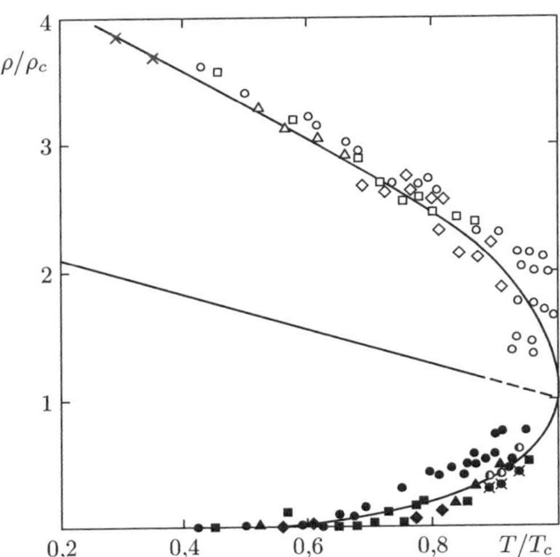

Fig. 8.14  Liquid–vapor coexistence curve for alkali metals [297].

values of the critical point parameters estimated for the most part of the elements by various methods are presented in papers [297,753]. Generally, different estimation techniques yield similar results, but the qualitative difference in the values of the critical parameters can reach sometimes 30–50%. Close values can be also obtained in the framework of the model equations of state (EOS) (see Chapter 9). They are independently confirmed by the experimental data on adiabatic expansion of shock-compressed metals [34] and on slow electric explosion of metallic wires [34,340,341,1061]. In Chapter 9 the experimental results on adiabatic expansion of shock-compressed uranium are given (Fig. 9.5).

The kinks on the release adiabats mark the location where the isentropes $S$ enter the two-phase boiling domain, and the dashes correspond to the metastable states on the isentropes $S$. The use of these experimental data makes it possible to specify the position of the high-temperature boiling curve for metals and to make a conclusion, in the framework of the semi-empiric EOS (Chapter 9), about the location of the critical points for metals.

The corresponding data on the parameters of the critical points [287,305] are presented in Table 8.1.

The essential peculiarity of the evaporation of metals is the fact that, due to the appreciable values of the critical temperatures comparable with the corresponding ionization potentials, the high-temperature part of the boiling curve is located in the plasma phase which, due to the large values of the critical densities (see Table 8.1), is strongly nonideal with respect to the wide spectrum of interparticle interactions. Thus, at high pressures metals evaporate directly to the plasma phase skipping, as distinct from low-boiling substances, the gas region. This fact may distort the existing estimates of the boiling phase boundary and lead to the appearance of new phase boundaries.

## 8.4   Plasma phase transitions

As we have seen (Chapter 4), for a qualitative description of the physical properties of nonideal plasma, the heuristic models are widely used based on the extrapolation of the notions of collective and quantum effects in Coulomb interaction developed for weakly nonideal plasma. It is characteristic that some of these models lose their thermodynamic stability at high nonidealities. This fact is explained [234,745,995] due to a possible first-order phase transition and separation of a strongly compressed Coulomb system into phases of different densities [347].

Table 8.1 Critical parameters of metals.

| Element | $T_c$, K | $p_c$, $10^8$ Pa | $\rho_c$, g/cm$^3$ | $S_c$, cal$\cdot$mol$^{-1}$K$^{-1}$ |
|---|---|---|---|---|
| Zr | 16250 | 7.52 | 1.79 | 40.45 |
| Zn | 3196 | 2.63 | 2.29 | 31.27 |
| Y | 10800 | 3.74 | 1.30 | 38.92 |
| Yb | 4280 | 1.38 | 2.36 | 36.92 |
| V | 12500 | 10.78 | 1.86 | 37.26 |
| U | 11630 | 6.11 | 5.30 | 43.07 |
| W | 21010 | 15.83 | 5.87 | 43.34 |
| Ti | 11790 | 7.63 | 1.31 | 37.92 |
| Sn | 8200 | 3.35 | 2.05 | 39.44 |
| Th | 14950 | 4.88 | 3.21 | 43.67 |
| Tu | 5910 | 2.65 | 3.22 | 38.52 |
| Tl | 4470 | 1.63 | 3.16 | 38.67 |
| Tb | 8060 | 3.08 | 2.57 | 42.34 |
| Te | 1850 | 0.75 | 2.21 | 35.22 |
| Ta | 20570 | 13.5 | 5.04 | 43.07 |
| Sr | 3860 | 0.90 | 0.86 | 35.23 |
| Ag | 7010 | 4.50 | 2.93 | 36.86 |
| Se | 8350 | 4.08 | 0.93 | 36.82 |
| Sm | 5340 | 2.10 | 2.51 | 40.62 |
| Ru | 15500 | 13.74 | 3.79 | 37.14 |
| Rh | 13510 | 11.23 | 3.62 | 39.66 |
| Re | 19600 | 15.7 | 6.32 | 43.42 |
| Ra | 3830 | 0.77 | 1.93 | 38.32 |
| Pr | 9160 | 2.85 | 1.86 | 41.39 |
| Tc | 15930 | 11.81 | 3.09 | 41.00 |
| Pa | 12650 | 4.82 | 3.72 | 43.34 |
| Po | 2050 | 0.62 | 2.67 | 37.52 |
| Pt | 14330 | 8.70 | 5.02 | 41.8 |
| Pd | 10760 | 7.64 | 3.20 | 36.64 |
| Os | 17110 | 14.49 | 6.83 | 42.60 |
| Nb | 19040 | 12.52 | 2.59 | 41.65 |
| Ni | 10330 | 9.12 | 2.19 | 36.50 |
| Nd | 7920 | 2.65 | 2.05 | 41.33 |
| Mo | 16140 | 12.63 | 3.18 | 41.21 |
| Lu | 7060 | 2.84 | 2.97 | 39.41 |
| Fr | 1810 | 0.12 | 0.65 | 39.50 |
| Bi | 4200 | 1.26 | 2.66 | 40.2 |
| Hg | $1763 \pm 15$ | $1.51 \pm 0.025$ | $4.2 \pm 0.4$ | 33.790 |
| Hg | $1753 \pm 10$ | $1.52 \pm 0.01$ | $5.7 \pm 0.2$ | 33.241 |
| Na | 2573 | 0.275 | 0.206 | 31.683 |
| K | 2223 | 0.152 | 0.194 | 33.931 |
| Rb | 2093 | 0.159 | 0.346 | 36.485 |
| Cs | $2057 \pm 40$ | 0.144 | $0.428 \pm 0.012$ | 38.111 |
| Li | 3223 | 0.689 | 0.105 | 27.666 |

*(Continued)*

Table 8.1  (*Continued*)

| Element | $T_c$, K | $p_c$, $10^8$ Pa | $\rho_c$, g/cm$^3$ | $S_c$, cal $\cdot$ mol$^{-1}$K$^{-1}$ |
|---|---|---|---|---|
| Mn | 5940 | 6.28 | 2.46 | 35.68 |
| Mg | 3590 | 1.98 | 0.56 | 29.48 |
| Pb | 4980 | 1.84 | 3.25 | 39.81 |
| La | 11060 | 3.35 | 1.78 | 40.58 |
| Fe | 9600 | 8.25 | 2.03 | 37.20 |
| Ir | 15380 | 12.78 | 6.77 | 41.91 |
| In | 6120 | 2.43 | 1.84 | 37.95 |
| Ho | 7240 | 2.94 | 2.84 | 40.63 |
| Hf | 18270 | 9.38 | 3.88 | 42.59 |
| Au | 8970 | 6.10 | 5.68 | 39.40 |
| Ge | 9170 | 4.90 | 1.64 | 37.77 |
| Ga | 7210 | 4.31 | 1.77 | 35.91 |
| Gd | 8670 | 3.25 | 2.50 | 42.79 |
| Er | 8250 | 3.34 | 2.86 | 40.51 |
| Eu | 4680 | 1.21 | 1.67 | 41.36 |
| Dy | 7240 | 2.91 | 2.76 | 41.05 |
| Cu | 8390 | 7.46 | 2.39 | 35.30 |
| Co | 10460 | 9.23 | 2.20 | 37.12 |
| Cr | 9620 | 9.68 | 2.22 | 37.35 |
| Ce | 8860 | 3.03 | 2.03 | 40.51 |
| Ca | 4180 | 1.21 | 0.49 | 32.70 |
| Cd | 2790 | 1.60 | 2.74 | 33.20 |
| Be | 8080 | 11.70 | 0.55 | 27.02 |
| Ba | 4100 | 0.81 | 1.15 | 37.06 |
| Al | 8000 | 4.47 | 0.64 | 33.52 |
| Sb | 2570 | 1.30 | 2.61 | 36.1 |
| B | 8200 | 9.57 | 0.69 | 28.9 |
| Se | 1010 | 0.37 | 1.60 | 31.8 |
| S | 730 | 0.30 | 0.72 | 27.7 |

The model analysis of strongly compressed Coulomb systems leads to
the possible formation of an electronic Wigner crystal. There are a considerable number of theoretical works devoted to its investigation [1040]
and experiments [192, 407] on its search (see Chapter 4). The possible existence of anomalies in the thermodynamic functions upon metallization of
the dense vapors of metals were discussed in paper [1078] which predicted
that, along with the regular boiling point, additional metallization-induced
phase transitions may arise in the gaseous and liquid phases.

It is probably worth to note here that strongly compressed systems have
a general tendency to spatial ordering [294, 295], that is to a phase transition
with production of plasma liquids and crystals. Up to date such Coulomb
crystals were observed in some exotic experimental situations in nonideal

"dust" plasma [302] and in colloid [487] plasma, in laser-cooled ion bunches in electrostatic traps [223] and cyclotron accelerators [894], and also in a two-dimensional electron gas on the liquid helium surface [339, 902].

Let us consider a series of phase transitions [142, 229].

The plasma phase transition, first studied by Norman and Starostin [744, 745], describes the plasma transition from a weakly-ionized state to a highly-ionized state.

The authors of paper [229] developed two approaches to the calculation of the EOS for a dense hydrogen plasma. In both approaches the Pade approximants and the chemical model (PACH) are used, but the interactions of charged and neutral particles are taken into account in two different ways. The first model is called PACH-DE ("density-dependent energy levels"). The second model is called PACH-EV ("excluded volume"). In brief, the two approaches differ in that the PACH-DE model includes the interaction of charged and neutral particles by consideration of density-dependent energy levels and PACH-EV accounts for the corresponding effects by means of the virial expansion. Both EOS provide the realistic description for low (neutral fluid) and high (fully ionized plasma) temperatures. The thermodynamic instabilities arise in the intermediate region associated with a first-order phase transition or a soft-mode first-order transition.

Compared to the previous approaches [238, 426, 456, 457, 882, 883], the PACH and HACH-DE approaches provide the more systematic account of correlations in the dense neutral component as well as of the interaction of neutral and charged particles by going beyond the scope of the approximation methods, such as the hard-sphere model or the perturbation theory.

Figures 8.15 and 8.16 were obtained making use of the PACH-EV model. They illustrate the transition domain in which two phases co-exist; the co-existence area is fairly narrow. The experimental point in which metal-like conductivity was detected [1037] lies above the co-existence line. This corresponds to the obtained result with a considerable ionization in this area.

Another EOS, which is built making use of the excluded-volume method [229], proceeds from the formula for the free energy of a two-component system consisting of neutral and charged particles. The free energy of such a system consists of the contributions coming from neutral ($F_{\mathrm{nl}}$) and charged ($F_{\mathrm{pl}}$) particles and each contribution splits into an ideal component and an interaction component.

The energy levels are considered to be density independent and the interaction between charges and neutral particles is taken into account making

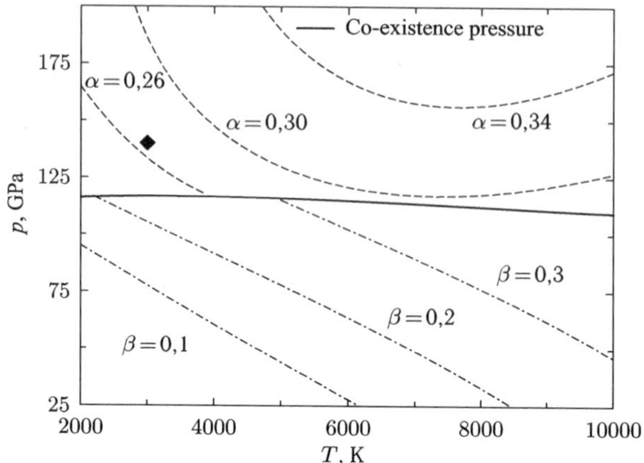


Fig. 8.15  Pressure on the coexistence line and on the lines of constant degree of dissociation $\beta$ and ionization $\alpha$, correspondingly, versus temperature. The calculations were
carried out on the basis of the PACH-EV model and the computer code BERO-H0.
The conditions corresponding to the metal-like conductivity observation are denoted by
diamond [1037].

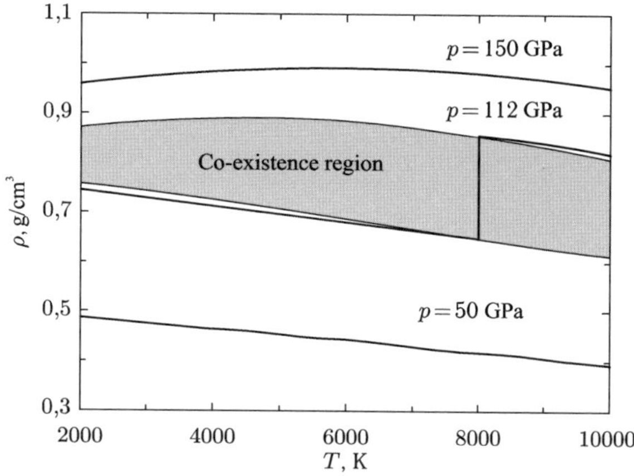


Fig. 8.16  The co-existence region and three isobars versus temperature (calculated on
the basis of PACH-EV and BERO-H0) [1037].

use of the reduced volume concept [229]. This approach is called PACH-EV;
the computations are carried out employing the BERO-HO code.

This model involves also the plasma phase transition (PPT) (Fig. 8.17).
In the co-existence region, where the temperature exceeds $10000\,\mathrm{K}$, plasma
consists of two phases I and II whose mass densities differ by approximately

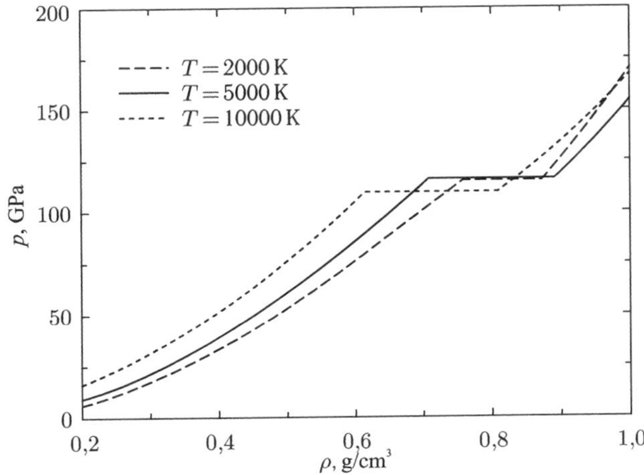

Fig. 8.17 Pressure dependence on density for three temperature values. In the instability region we made use of Maxwell construction and as a result obtained constant pressure in the co-existence region. The calculations were carried out employing the BERO-EV and BERO-H0 codes.

$0.2\,\mathrm{g/cm^3}$. The phase co-existence densities $\rho_\mathrm{I}$ and $\rho_\mathrm{II}$ decrease with the temperature increase. PPT is characterized by the jumps of the mass density, of the ionization and dissociation degrees. For the temperature range under consideration phase I corresponds to almost nonionized atomic-molecular fluid, while phase II corresponds to partially ionized plasma. The transition pressure decreases with the increase of temperature and varies within the limits from $120\,\mathrm{GPa}$ through $110\,\mathrm{GPa}$ in the temperature range $T = (2\text{--}10) \cdot 10^3\,\mathrm{K}$. Figure 8.15 presents the line of two-phase co-existence $p^\mathrm{PPT}(T)$ and the lines of constant dissociation and ionization degrees.

Model [238], which takes into account the Coulomb interaction by means of the Pade approximation and the interaction of neutral particles, leads to the plasma transition (Fig. 8.18) with a critical point $c_2$: $T_\mathrm{c} = 16500\,\mathrm{K}$, $p_\mathrm{c} = 22.8\,\mathrm{GPa}$, $\rho_\mathrm{c} = 0.13\,\mathrm{g/cm^3}$, $\alpha_\mathrm{c} = 0.32$, $\beta_\mathrm{c} = 1$.

Figure 8.18 [238] presents the critical point of the PPT, co-existence line, and other characteristic lines. The straight lines $\alpha = \mathrm{constant}$ and $\beta = \mathrm{constant}$ connect the states which are characterized by equal degrees of ionization and dissociation, respectively. The plasma composition sharply changes in the vicinity of the critical point $c_2$. We can see also that the two co-existing phases near the critical point are only partially ionized. At low temperatures the molecular gas and metallic liquid can co-exist close to each other. The region of solid molecular hydrogen is adopted from the paper by

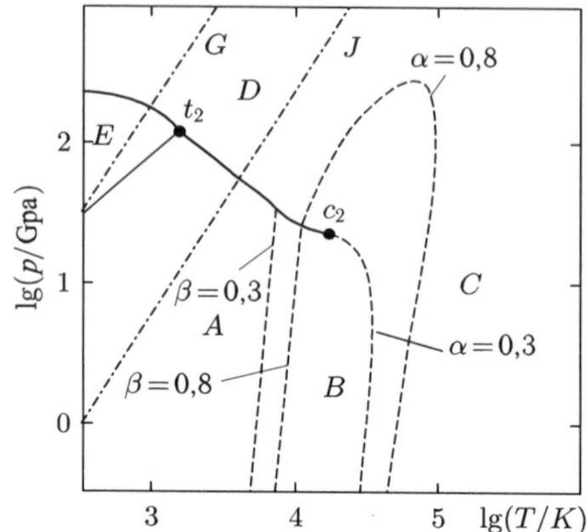


Fig. 8.18   Phase diagram of hydrogen with the second critical point $c_2$ and the co-existence line $c_2$–$t_2$ [238]. *A*: region of liquid molecular hydrogen. *B*: region of weakly-ionized plasma. *C*: region of strongly-ionized plasma. *D*: region of metallic hydrogen. *E*: estimate for the solid molecular hydrogen region. *G*: estimate for shock adiabat [373]. *J*: estimate for the Jupiter adiabat [844].

Kerley [519]. Three phases (metallic liquid, solid molecular hydrogen, and liquid molecular hydrogen) can co-exist in the triple point characterized by

$$T_{tr} \approx 1500\,\mathrm{K}, \quad p_{tr} \approx 100\ \mathrm{GPa},$$

Let us compare the obtained results with the critical data calculated by Frank [328] or by Robnik and Kundt [844]:

$$T_\mathrm{c} = 19000\,K, \quad p_\mathrm{c} = 24\ \mathrm{GPa}, \quad \alpha = 0.5, \quad \beta_c = 1,$$

and also with the earlier estimates by Ebeling and Sandig [240]:

$$T_\mathrm{c} = 12600\,\mathrm{K}, \quad p_\mathrm{c} = 94\ \mathrm{GPa}, \quad \rho_\mathrm{c} = 1\ \mathrm{g/cm}^3, \quad \alpha = 0.008, \quad \beta_\mathrm{c} = 1.$$

It can be seen that the estimates match in the order of magnitude. Note, however, that the critical data substantially depend on the chosen approximation. Therefore further improvements of the theory are necessary.

The other investigators [745] used a different method to calculate the co-existence line and the critical point. They employed the improved model of the EOS based on the realistic values of the neutral particle interaction potentials and obtained the following results:

$$T_\mathrm{c} = 15000\,\mathrm{K}, \quad p_\mathrm{c} = 65\ \mathrm{GPa}, \quad \rho_\mathrm{c} = 0.36\ \mathrm{g/cm}^3.$$

Moreover, in the framework of this model the phase beyond the limits of the co-existence line is characterized by moderate ionization rather than being fully ionized ("metallic").

In various models the slope of the co-existence line is either positive [426, 661, 844] or negative [238, 745]. In the general case the slope of the transition line obeys the Clausius–Clapeyron law:

$$\frac{dp}{dT} = \frac{\Delta s}{\Delta v},\tag{8.6}$$

where $\Delta s$ and $\Delta v$ are the difference of the specific entropy and specific volume values for the two phases. From the phenomenological viewpoint, the negative slope of the co-existence line in a PPT is more logical because, when the co-existence line intersects the compression line, the entropy increases [231].

But the most important issue which deserves attention is the sharp pressure decrease upon the transition to the metallic hydrogen phase compared to the corresponding pressure upon the transition to the solid state (expected to exceed 200 GPa) [374, 375, 519, 653, 706]. It is assumed that on the transition from the liquid molecular hydrogen phase to the liquid metallic hydrogen phase the pressure will be by a factor of 10 less than on the transition from the solid molecular hydrogen phase to the solid metallic hydrogen phase.

The phase diagram (Fig. 8.19) illustrates numerous predictions of the PPT for that moment. We will discuss the experimental evidence for this effect in the end of this chapter.

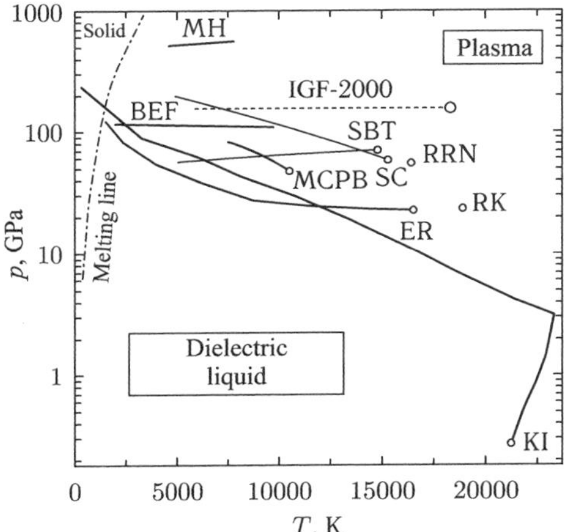


Fig. 8.19  Hydrogen phase $p$–$T$ diagram with different versions of the theoretically predicted PPT. Notation: RK [844], ER [238], MH [661], SC [745], SBT [892], RRN [835], MCPB [646], KI [550], BEF [101], IGF [396].

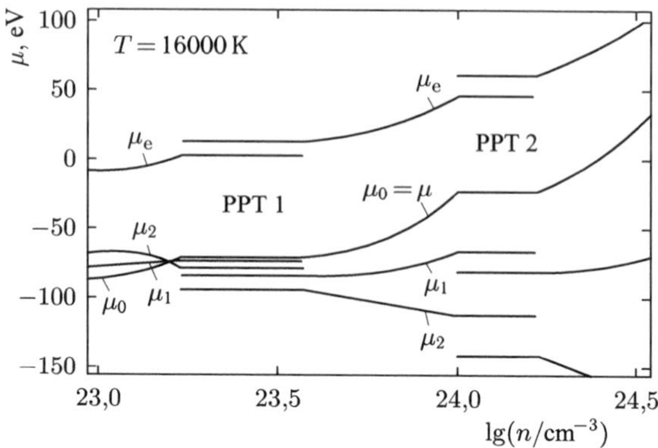

Fig. 8.20   Chemical potentials of atoms (subscript 0), singly charged (subscript 1) and doubly charged (subscript 2) ions, electrons (subscript e) of helium plasma. $\mu$ is the total chemical potential. Temperature equals 16000 K.

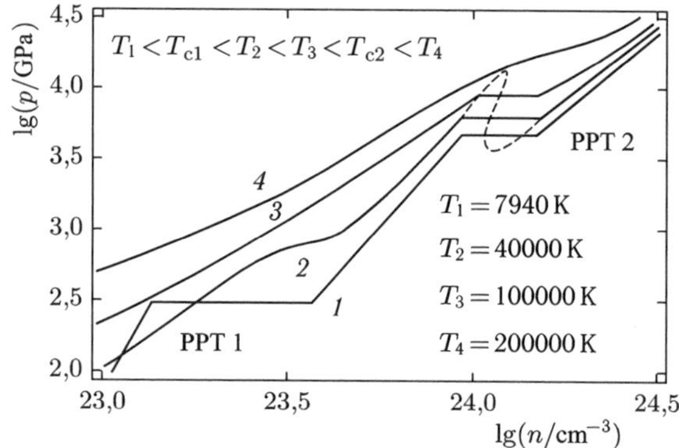

Fig. 8.21   Pressure isotherms for subcritical and supercritical temperatures for two phase transitions of helium plasma. For one of the temperature values the phase transition is denoted by the van der waals loop (dashed curve).

On the basis of papers [283, 284, 495] we depict the sequence of the phase transitions on the phase diagrams (Figs. 8.20, 8.21, 8.22, and 8.23). In helium plasma they correspond to the ionization steps $He^0 \rightarrow He^+$ and $He^+ \rightarrow He^{++}$.

Table 8.2 contains the data on the system at the critical points as well as the estimates for the melting curve.

The helium metallization phenomenon was intensively studied for the case of solid helium at low temperatures. A special attention was paid to the metallization pressure. On the basis of zone structure calculations the

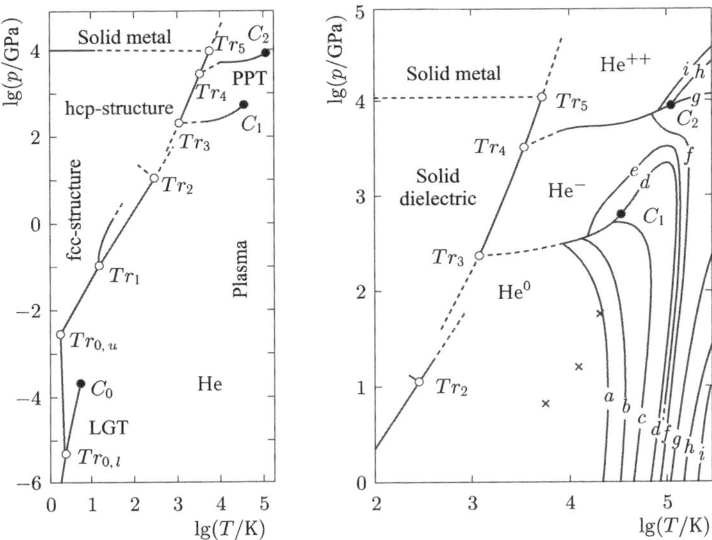

Fig. 8.22  Phase diagrams for helium-4 in the pressure vs temperature plane. $Tr$ – triple points: subscripts $0, l$ and $0, u$ lower and upper $\lambda$-points, subscripts: 1 hcp structure/fcc structure/gas, 2 fcc structure/hcp structure(?)/gas, 3 solid helium/He$^+$/He$^\circ$(PPT1), 4 solid helium/He$^{++}$/He$^+$(PPT 2), 5 solid dielectric/solid metal/He$^{++}$. $C$ critical points, subscripts: 0 liquid–gas transition (LGT), 1 PPT 1, 2 PPT 2. Solid curves phase transitions (dashes denote extrapolation). Lines marked by letter symbols: lines of constant mean charge of heavy particles $\bar{z}$: a 0.01, b 0.1, c 0.5, d 0.9; e 1.0, f 1.1, g 1.5, h 1.9, i 1.99. Shock-wave experiment: × direct wave [230].

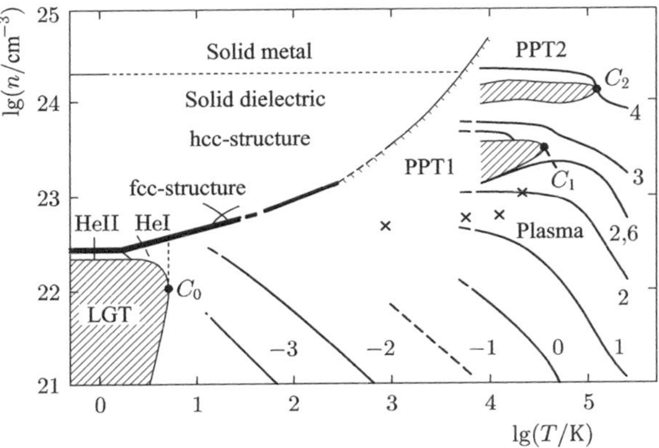

Fig. 8.23  Phase diagram of helium-4 in the density-temperature plane. The domains of the first-order phase transitions are hatched: LGT liquid–gas transition, PPT 1 and PPT 2 two PPT corresponding to the ionization steps He$^0 \to$ He$^+$ and He$^+ \to$ He$^{++}$. Thin lines denote isobars. Asterisks stand for shock-wave points (see Fig. 8.22).

Table 8.2   Critical and triple points corresponding to the phase transitions of helium plasma [230].

| | $T/10^3$ K | $n/10^{23}$ cm$^{-3}$ | $p/$TPa | $\bar{z}$ |
|---|---|---|---|---|
| $C_1$ | 35 | 3.3 | 0.66 | 0.8 |
| $C_2$ | 120 | 13 | 10 | 1.8 |
| $Tr_3$ | 1 | — | 0.2 | $0 \to 1$ |
| $Tr_4$ | 4 | — | 3 | $1 \to 2$ |

following values of the metallization pressure were obtained: 220 TPa [897], 3 TPa [980], 11.2 TPa [1065], and 11.0 TPa [685]. Similarly to the case of hydrogen plasma, helium plasma is characterized by a substantial decrease of pressure on the transition to metallic plasma.

The calculations show that in the narrow density region (of the order of $10^{23}$ particles per cm$^3$) at $T = 30000$ K two free energy minima are observed, i.e., two values of the degree of ionization exist. This is an evidence for the thermodynamic instability

$$\left[\frac{\partial \ln p}{\partial \ln \rho}\right]_T < 0, \tag{8.7}$$

where $\rho$ is mass density. The thermodynamic instability leads to a phase transition in hydrogen–helium plasma when the critical temperature $T \approx 25000$ K is reached. Remind that for pure hydrogen plasma the critical temperature was $T_c = 16500$ K [238, 239].

We see again that double ionization induces instability and phase transitions [241]. The ionization energy of a helium ion decreases drastically due to the relatively high value of the electronic density. As a result, double ionization turns out to be more preferable than it could be expected on the basis of the theory of ideal plasma.

Remind that in the framework of the stellar dynamics [294, 295], besides the breakdown of equilibrium involved in equation (8.7), there are as well other instabilities of interest, for instance,

$$\left[\frac{\partial \ln p}{\partial \ln \rho}\right]_S < \frac{4}{3}. \tag{8.8}$$

We will see later whether our EOS is valid under the conditions of such instabilities.

The matter in the solar interior can be treated as a prototype of the hydrogen–helium plasma (see Sec. 4.5). From the viewpoint of the plasma physics, the Sun is a large plasma sphere which can be considered as a gigantic laboratory of plasma studies in a wide range of the thermodynamic parameters. For instance, in the center of the Sun where the temperature exceeds $10^7$ K the plasma density is very high (more than $10^{25}$ particles per cm$^3$), while in the vicinity of the Sun periphery where the temperature falls down below $10^4$ K plasma is very rarefied. For an approximate physical consideration we will model the Sun in the following way: the Sun is a plasma sphere which (in the units of molar fractions) is composed of 90–93% of hydrogen, 6–8% of helium, and less than 2% of heavier particles. It is assumed that the plasma sphere is in the mechanical and thermodynamic equilibrium and the temperature and density distributions along its radius are known. We will neglect the hydrodynamic flows and the transfer phenomena different from radiation, but we will take into account the non-equilibrium induced by nuclear reactions, ionization reactions, and radiative transitions. From the physical viewpoint the solar radiation accepted and re-emitted by the Earth serves as a driving force of all self-organization processes which take place on our planet. The so-called "photon factory" provides the Earth with the entropy which is necessary for the support and development of life, mankind, science, and culture. The entropy flux, generated by the Sun–Earth system, in the background radiation can be easily estimated under the assumption that the surface of the Earth absorbs and re-irradiates on an average $230\,\mathrm{W/m^2}$. Taking the temperature of the incident photons to be about $5800\,\mathrm{K}$ and assuming that the Earth irradiates heat in the same way as a black body whose temperature equals $260\,\mathrm{K}$, we obtain that the entropy flux is $1\,\mathrm{W/(m^2\,K)}$. Any essential change of these parameters will obviously lead to dramatic consequences which will affect the life conditions on our planet.

Since the Sun plays the principal role in our life, this circumstance becomes the strongest motivation for the growth of the number of professionals who provide astrophysicists with the instrumentation for studying the processes in the solar interior. The question of the solar plasma thermodynamics was considered in Sec. 4.5.

Continuing the astrophysical topic it should be noted that for many years the existence or absence of phase transitions (PT) in the interiors of Jupiter (J), Saturn (S), and brown dwarfs (BD) has been one of the central problems in the evolution theory of these objects (see reviews [342], [145, 170, 171, 285, 405, 448, 1004], etc.). Over the last decade this collection was joined by the numerous team of the so-called extrasolar giant planets (EGP), e.g., [145]. Each of these objects is mainly a helium–hydrogen mixture ($\approx 25\%$ by weight) with small admixtures of the heavy components, which goes through various states from the hot superdense plasma in its center ($p \approx 1000\,\mathrm{GPa}$, $T \approx 10-100\,\mathrm{kK}$) to the relatively cold helium–hydrogen fluid on its surface ($p = 1\,\mathrm{bar}$, $T \approx 100$–$1000\,\mathrm{K}$). The phase transitions in such a mixture split into two kinds. (A) Separation into phases of different chemical proportion due to the limitations on the mutual solubility of dense hydrogen and helium. A phase transition of this kind in principle does not exist in pure hydrogen or helium by very definition. (B) The above-mentioned phase transitions which already exist in hydrogen or helium and transform into the mixture conditions. For many years the so-called "plasma phase transition" (PPT) has been the main transition of the (B)-type. Since the pioneering work [744], there have been predicted many PPT versions obtained as a rule within the framework of the far extrapolation of some model description of nonideality (see Refs. [101, 230, 832, 884, 892, 1043], etc.). As we have seen above, the principal peculiarity of PPT is the jump of the equilibrium degree of ionization and some models [230] predict for helium two PPTs rather than one, i.e., one PPT per each degree of ionization. On the contrary, other models [1043] predict the unique PPT, but with double ionization $\Leftrightarrow \mathrm{He}^{(++)} + 2e$. Over the last decade the predictions of the "plasma"-type phase transitions have been essentially supported by the categorical recognition of the very fact of the PPT existence within the framework of the so-called rigorous first-principles (RFP) approaches by the path integral Monte Carlo technique (PIMC) [102, 264, 268].

Other versions of the RFP approaches (the so-called density functional method for electrons and the molecular dynamics method for nuclei — (DFT + MD)) were also employed to predict [122, 123, 887] besides PPT the existence of one more, dissociation-driven, phase transition in hydrogen (DPT). The specific feature of these DPT is the jump of the equilibrium degree of dissociation of the $H_2$ molecule. Just recently these predictions were supported by one more version of the first-principles approaches (the so-called "wave-packet method" for electrons and the molecular dynamics method for nuclei — (WPMD)) which also predicted the phase-transition-type anomaly with a jump-wise hydrogen dissociation [481, 482]. It should be noted that the idea of such a PT was already discussed in Refs. [572, 574] and developed using models in Ref. [722].

Figures 8.24 and 8.25 present the phase diagrams of xenon plasma together with the data from the dynamic experiments.

In many respects plasma of inert gases manifests itself as hydrogen plasma if only one electron is ionized. In particular, it is associated with

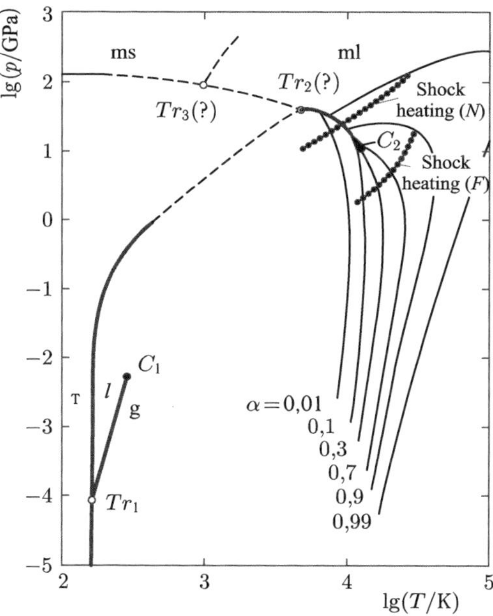



Fig. 8.24  Xenon phases in the pressure-temperature plane in the framework of the hydrogen-like model [231, 282]. The thick lines represent the first-order phase transitions. $Tr_1$ triple point between solid s, liquid (l), and gaseous (g) phases. $C_1$ critical point for the LGT. $C_2$ critical point for the PPT. The dashed lines illustrate possible supplements to the phase diagram: $Tr_2$ triple point between the phases of solid dielectric, liquid dielectric, and metallic liquid (ml); $Tr_3$ the triple point between the metallic solid (ms) and metallic liquid (ml) phases and the solid dielectric phase. The thin lines show the regions of constant degree of ionization $\alpha_1 = \bar{z}$. The metallization process at low temperatures is described according to Ref. [862]. The dot lines depict shock adiabats drawn using the experimental results.

the relatively high ionization energy of the inert gases. It directly follows from this fact that two different critical points can exist. In the first critical point which corresponds to the gas–liquid transition the information on the system is well known (within the limits $10^1 \div 10^2$ K and $10^5 \div 10^7$ Pa). The second critical point is assumed to be located within the strong interaction region (about $10^4 \div 10^5$ K and $10^9 \div 10^{11}$ Pa). The phase transition corresponding to this point is a first-order transition. It is accompanied by the sharp change of the mass density and degree of ionization. This PPT is, so to say, a high-temperature modification of the well-known dielectric–metal transition.

Among the noble gases xenon is the main candidate for PPT. The results of the experiments on static compression at low temperatures prove that

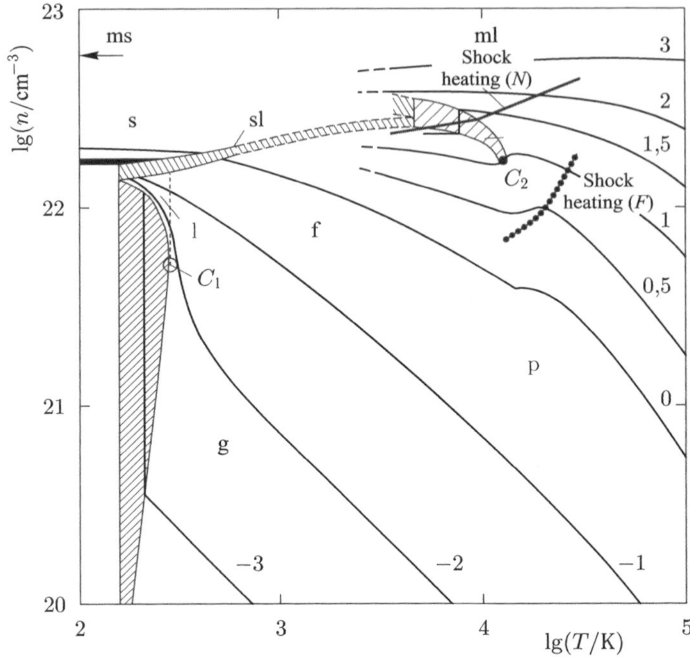


Fig. 8.25　Xenon phases in the density-temperature plane obtained using the model of hydrogen-like plasma. The following different regions are considered: s–g solid substance evaporation region, l–g liquid and gas co-existence region, s–l melting region, s solid, l liquid, g gas, f fluid, p plasma, PPT co-existence region on PPT, ml metallic liquid, ms metallic solid. The thin lines represent the isobars intersecting the co-existence region. $\lg(p/\text{GPa})$ is taken as a pressure unit. The arrow shows the density on transition to metallic state at $T = 0\,\text{K}$ (according to Ref. [862]. The dot line corresponds to the shock-wave experiments.

the dielectric–metal transition occurs at pressures exceeding $100\,\text{GPa}$. Ross and McMahan [862] carried out the necessary calculations and came to the conclusion that the transition takes place at the pressure value about $130\,\text{GPa}$. The precise phase diagram for xenon at high temperature values has not been developed so far.

Figures 8.24 and 8.25 illustrate the possible shape of the phase diagram for xenon which was built on the basis of the one-electron theory [231, 282]. The shock-wave data are obtained in different experiments on xenon compression from liquid (N) and gaseous (F) states. Data (N) [514, 735] intersect the PPT instability region and data (F) [695, 1072] lie beyond it which corresponds to the initial low density.

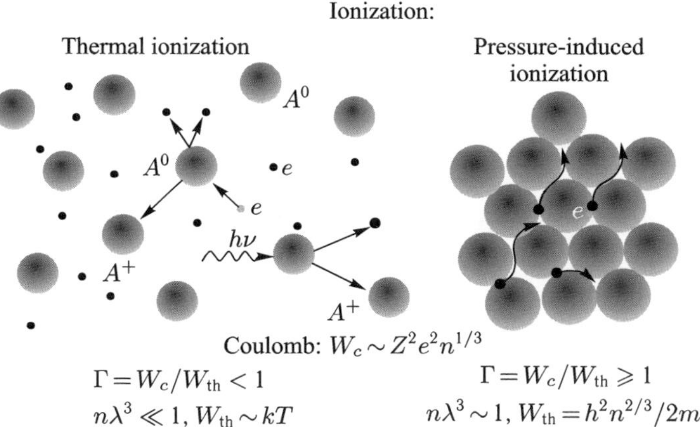



Fig. 8.26　Thermal and pressure-induced ionization of a substance.

The thermodynamic properties of xenon described so far in the literature do not provide any clear evidence for a phase transition. Whether it is possible to speak in this case about a PPT will be clear from the further careful measurements of the electric conductivity or reflection coefficient of xenon.

Let us consider the experiments on quasi-adiabatic compression of hydrogen and deuterium stimulated by the study of the pressure-induced metallization effect and by the search for PPT [294, 295].

It is well known (Fig. 8.26) that a substance can be transformed into a conductive state (ionized) either by heating or by compression [291, 305]. At present, thermal ionization is the main and best investigated ionization mechanism in the plasma physics [291]. It is associated with heating of rarefied plasma to temperatures comparable with the ionization potential of the substance $T \sim J$. This mechanism was discussed in Chapter 4. The pressure-induced ionization mechanism is an alternative to thermal ionization and is associated with strong compression of "cold" substance to the densities $n \sim a_0^{-3}$ sufficient to induce the overlapping of the atomic orbitals which have a characteristic size of the Bohr radius $a_0 \approx a_{\rm B} = \hbar^2/me^2$. The realization of this criterion requires advancing to the region of the condensed plasma densities and, as a consequence, generation of Mbar pressures. In order to separate these two ionization mechanisms it is necessary to realize the "cold" ($T \ll J$) compression of the substance in the experiment by decreasing the thermal heating effects. From this viewpoint hydrogen

Table 8.3   Chronology of hydrogen studies.

| | |
|---|---|
| 1766 | Cavendish — discovery of "burning gas" — hydrogen |
| 1898 | Dewar — liquid and solid ($H_2$); is it a hydrogen-alkali metal? |
| 1925 | Fowler — $H_2$ in the stars — plasma |
| 1925 | Herzfeld, Clausius-Mossotti — dielectric dielectric catastrophe at $0.6\,g/cm^3$ |
| 1935 | 1935 Wigner and Huntington — metallization at $2.5\,GPa$, $T = 0\,K$ |
| 1968 | Ashcroft — (BCS theory) high-temperature superconductivity of atomic hydrogen |
| 1972 | Kormer *et al.* multiple explosive spherical compression up to $4\,Mbar$ |
| 1978 | Hawke *et al.* — explosive magnetic compression up to $2\,Mbar$, $4000\,K$ |
| 1980 | Mao, Hemley, and Silvera — static compression up to $1\,Mbar$ |
| 1987 | Pavlovski *et al.* — explosive magnetic compression up to $1\,GPa$, $3000\,K$ |
| 1990 | Ashcroft — dissociation and metallization at $3\,Mbar$ |
| 1993 | Nellis *et al.* — reverberation of shock waves — "metallization" |
| 1997 | Fortov and Ternovoi — explosive quasi-adiabatic compression — nonideal plasma |
| 1997 | DaSilva, Cauble *et al.* — laser-induced shock waves — nonideal plasma |
| 2001 | Trunin, Fortov *et al.* — explosion-induced spherical shock waves — nonideal plasma |
| 2001 | Asay, Knudson *et al.* — electrodynamical shock-wave compression up to $1\,Mbar$ |
| 2005 | Zhernokletov, Mochalov, Fortov *et al.* — explosive cylindrical compression, plasma phase transition |

turned out to be the most appropriate object because its small molar mass leads to the lowest temperatures of shock compression.

Hydrogen is the most abundant (90%) and, at the same time, the simplest chemical element in nature. Hydrogen has been attracting the attention of the researches of different disciplines during almost 250 years. A brief chronology of the main stages of hydrogen studies is presented in Table 8.3.

Active investigations of the EOS and electric conductivity are stimulated by the importance of the hydrogen plasma for astrophysics and physics of giant planets as well as by the search for the high-temperature superconductivity of hydrogen metallic phase. This has been a permanent pragmatic incentive to the studies over the last 50 years [649] (Fig. 8.26). The experiments on pressure-induced ionization of hydrogen plasma and noble gases were carried out employing the multiple shock compression technique which enables the quasi-adiabatic compression to be performed, thereby essentially increasing the degree of compression and decreasing (up to $4$–$5 \cdot 10^3\,K$) the temperature of the substance [377, 701, 734, 967, 1037]. The experiments were done in the planar and cylindrical geometry using light–gas guns and explosion launching devices and also employing the cylindrical explosive

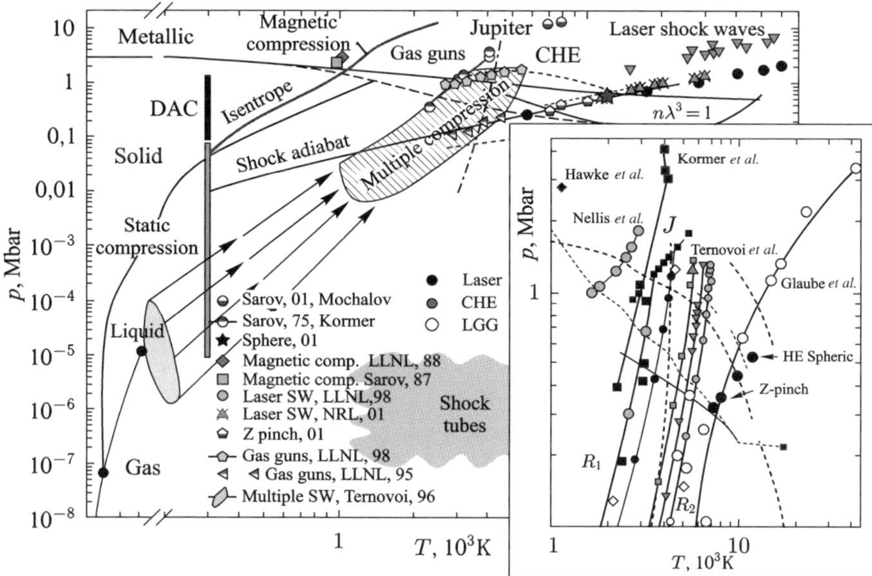



Fig. 8.27   Phase diagram of hydrogen. The boundaries of PPT are denoted by dashed lines [319].

magnetic cumulative generators [431, 774] in which "isentropization" of the compression process is achieved by a powerful magnetic field.

The numerous experiments showed a sharp (up to 5–6 orders of magnitude) growth of the static electric conductivity of hydrogen in a narrow range of the condensed densities at megabar pressures (Figs. 8.27 and 8.28). In this case the electric conductivity maximal values achieved under these conditions are of several hundreds of $ohm^{-1}cm^{-1}$. These values are close to the conductivity of alkali metals and not far from the Ioffe–Regel "minimal metal" conductivity [339]. For this reason the effect under consideration is often called "metallization" which is not completely correct because according to Refs. [339, 1078] the notions of metal and dielectric can be separated only at $T = 0$. We believe that the case in point is "pressure-induced ionization" [230, 319] caused by the overlapping of the wave functions of neighboring atoms which simplifies their ionization in dense medium. Figure 8.29 presents the geometrical characteristics of a hydrogen molecule in an isolated state compared to the volume of accessible space per molecule (Wigner–Seitz cell radius) at the chosen density $\rho$ value. It can be seen that at density values $\rho > 0.3\,g/cm^3$ the hydrogen molecules

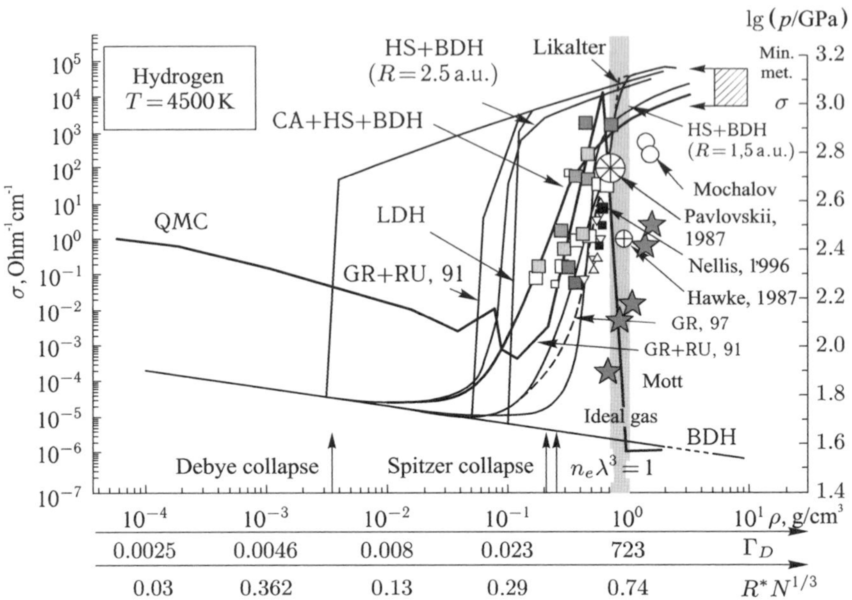



Fig. 8.28    Pressure-induced ionization of non-ideal hydrogen plasma [377, 431, 701, 734, 774, 967]. The area of the thermodynamic phase transition is marked in grey, asterisks indicate the density measurements by pulsed X-ray radiography [701]. QMS stands for the calculations by the quantum Monte Carlo method [102, 264, 267, 270].

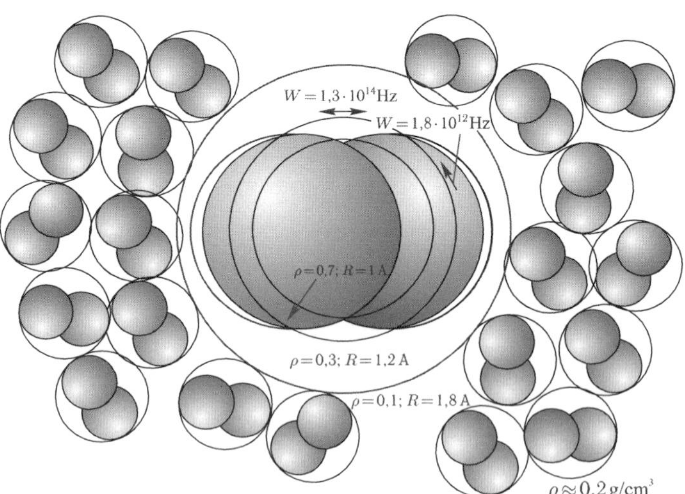



Fig. 8.29    Hydrogen in the megabar pressure range [377, 701, 734, 967, 1037]. Dark spheres represent a molecule. Circles denote the Wigner-Seitz cells at density $\rho$.

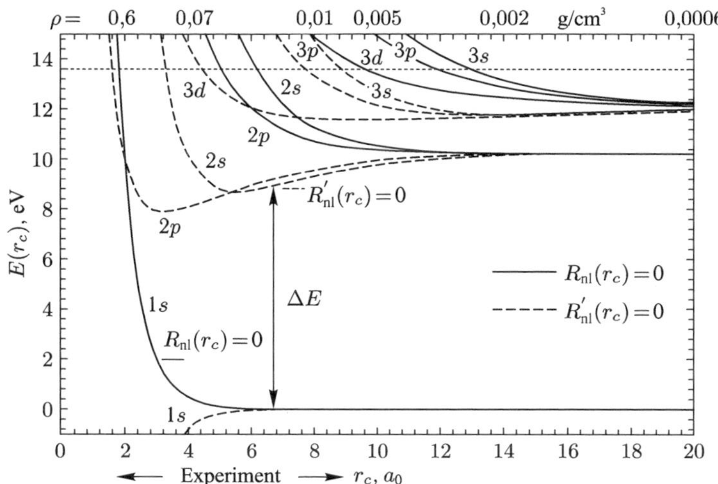



Fig. 8.30 Energy spectrum of compressed hydrogen as a function of the radius of the WZ sphere $r_c$.

first become comparable in size with the Wigner–Seitz (WZ) sphere and later their size decreases below the WZ sphere.

Physically it corresponds to strong overlapping of the wave functions of the electrons in neighboring atoms even in the ground energy state. This overlapping provides conditions for the delocalization [339] of the electrons and, consequently, for their quasi-free motion in plasma. The energy spectrum and the effective potential $\Delta E$ of hydrogen are presented in Fig. 8.30 as a function of the size of the WZ sphere. Solid lines mark the upper region of the energy band calculated using the condition that the radial part of the wave function $R_{nl}(r_c)$ equals zero on the cell boundary and the lower boundary of the band (dashed lines) was determined from the similar condition imposed on its derivative $R'_{nl}(r) = 0$ (for more details see Ref. [319]). It can be seen that in the process of compression, on $r_c$ decrease under the experiment conditions, the energy levels are broadening and transforming into energy bands with their subsequent overlap. As a result, the effective ionization potential of the substance $\Delta E$ decreases. The obtained value of $\Delta E$ is in reasonable agreement with the corresponding value provided by the experimental measurements of the conductivity dependence on temperature.

The similar data on the measurement of the conductivity of quasi-adiabatically compressed plasma were obtained for a number of other

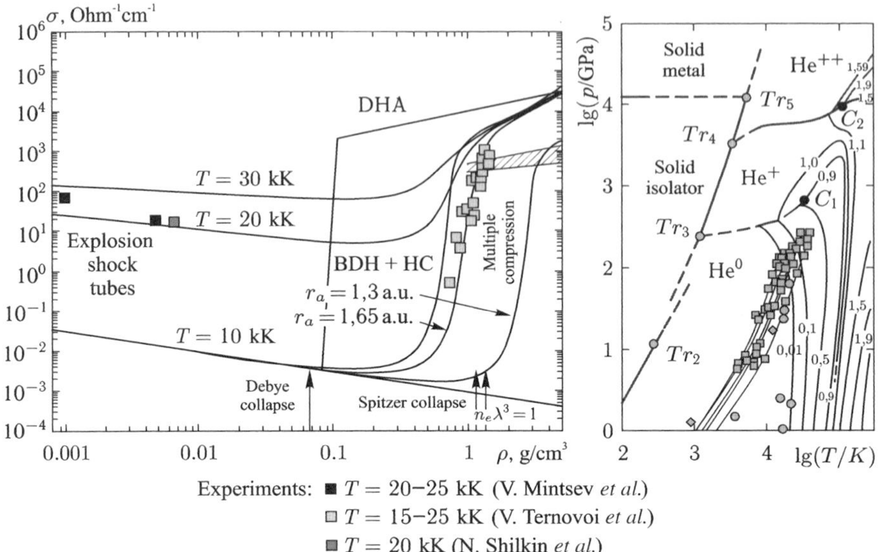


Experiments:   ■ $T = 20{-}25$ kK (V. Mintsev et al.)<br>
□ $T = 15{-}25$ kK (V. Ternovoi et al.)<br>
■ $T = 20$ kK (N. Shilkin et al.)

Fig. 8.31  Phase diagram and pressure-induced ionization of helium plasma.

elements, namely, He (Fig. 8.31), $D_2$, Ar, Xe, and for the hydrogen–helium mixture which forms the plasma of the Jupiter atmosphere [319].

For the additional study of the electronic shell overlapping effect the experiments were performed [760] on quasi-adiabatic compression of the fullerene $C_{60}$ whose molecule has a typical size which significantly exceeds the size of the hydrogen atom (7Å against 1Å). According to the expectations, the "metallization" pressure of the fullerene $C_{60}$ turned out (Fig. 8.32) to be smaller than that for hydrogen by approximately an order of magnitude.

Note that the models of substance transition to a conductive state suggested by Mott, Anderson, Lifshits, Herzfeld, Likalter (for details see Refs. [319, 339]) also predict transitions in the range of parameters close to the experimental one.

The typical feature of most of the physical models of nonideal plasma is their thermodynamic instability in the range of high nonidealities $\Gamma > 1$ [83, 102, 125, 230, 237, 264, 267, 270, 745, 881, 883, 1039, 1078]. It is the range where the experiments on the dynamical compression of plasma were planned and performed [26, 39, 88, 299, 311, 377, 695, 701, 734, 967, 1037, 1070]. This instability of strongly compressed Coulomb systems corresponds to the "plasma" phase transition which was predicted using simplified models by

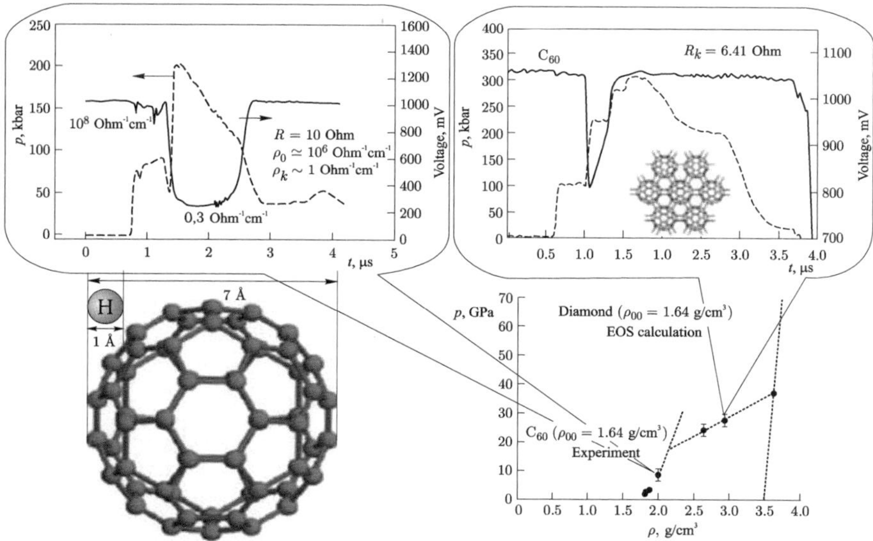

Fig. 8.32 "Metallization" of the $C_{60}$ fullerene at high dynamic pressures [760].

Wigner [1039], Landau and Zeldovich [1078], Norman and Starostin [745], Ebeling *et al.* [230, 339], Saumon and Chabrier [881, 883] and reproduced by the parameter-free methods of the molecular dynamics [83] and by the QMS [66, 267]. The corresponding region of plasma instability ("Debye collapse") predicted by the ring Debye approximation is denoted in Fig. 8.28 by the vertical arrow on the left. Fig. 8.33 represents the results for the hydrogen nonideal plasma structure obtained by the Monte Carlo method which takes into account the quantum effects by means of the path integral technique [102, 264, 267, 270]. It clearly illustrates plasma separation into phases followed by the formation of an ordered plasma structure in the nonideal plasma.

Aiming at the search for the phase transition in a real electron–ion plasma, the experiments [701] were performed on the explosive quasientropic compression of the deuterium plasma in the cylindrical geometry making use of the pulsed X-ray radiography for the measurement of plasma density (Figs. 8.33 and 8.35). The results of the experiments revealed the abrupt ($\approx 25\%$) plasma density jump at pressure $p \approx 1.2\,\mathrm{Mbar}$ just in the density range where the electrophysical measurements [377, 431, 701, 734, 774, 967, 1037] undoubtedly demonstrate the sharp (of 5–6 orders of magnitude) growth of the electric conductivity (see Fig. 8.28 and

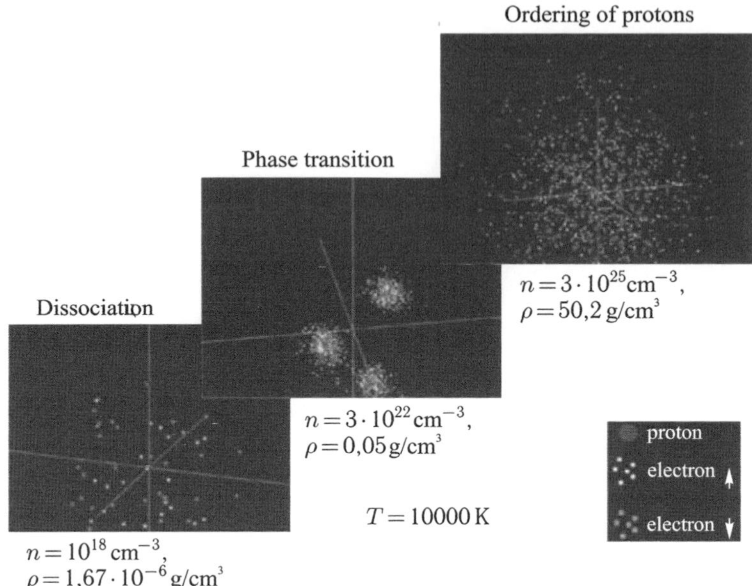



Fig. 8.33   Ordered structures of nonideal hydrogen plasma in the range of plasma phase transition. Calculation was made by the QMS [66, 102, 264, 270, 649].

the conductivity data in Fig. 8.34) and where the quantum Monte Carlo calculations [102, 264, 267, 270] lose their stability. The nonideality parameter estimates obtained under these conditions give $\Gamma \approx 150$–$200$ for partial plasma degeneration $n\lambda^3 \approx 1$.

Figure 8.36 illustrates the comparison of the data taken directly from the experiments [701] with the theoretical results [301] in $p$–$T$ coordinates where the disruption temperature on the experimental isentrope is calculated using the initial experimental $((p, V))$ data with the aid of the SAHA-H thermodynamic code [386].

The thermodynamic and electro-physical measurements are considered to be an evidence for the experimental detection of the phase transition in nonideal plasma on its quasientropic compression [301].

## 8.5   Noncongruent phase transitions

The above-mentioned phase transformations were predicted by the models developed for the description of the one-component systems, namely, of hydrogen, helium, or inert gases. The transition to the hydrodynamics of

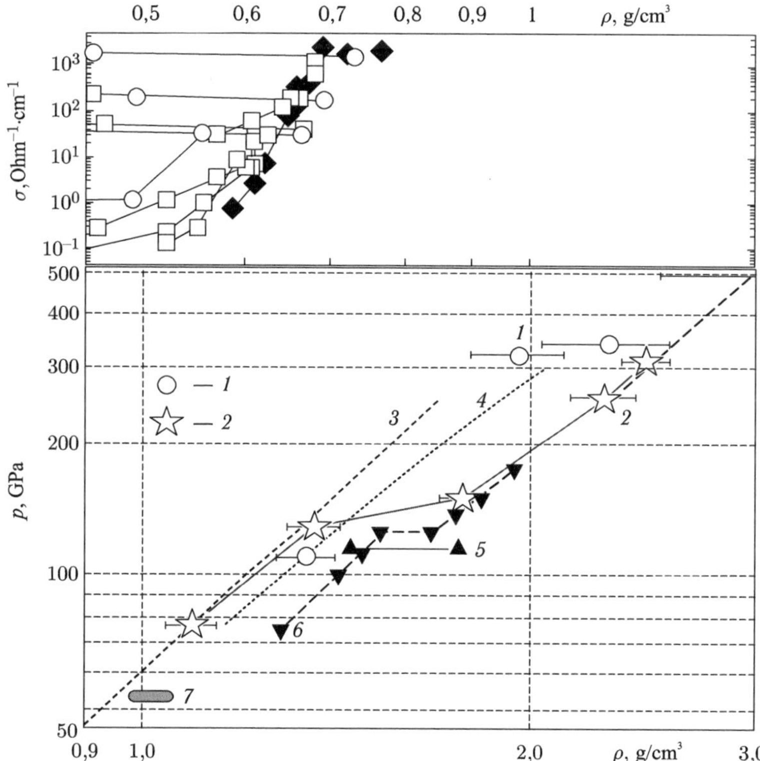

Fig. 8.34 Isentropic compression of deuterium. Comparison of the experimental data and theoretical predictions. Top: The results of the electric conductivity measurements. Notation: *1, 2* experiments (*1* 1972 [374], *2* 2005 [299], *3–4* isentrope calculated for deuterium (*3* chemical model of plasma (code SAHA-D [299], *4* "compressible covolume" model [572, 574]), *5–7* "plasma" phase transition predictions: *5* plasma chemical model ($T \div 2000$–$10000$ K) [230], *6* wave packet method (WP/MD) (isotherm $T = 1500$ K) [481, 482], *7* density functional method (DFT/MD) (EOS anomalies at $T \approx 3000$ K) [122, 123, 887].

the $H_2$ + mixture is very often made in the framework of the simplified "additivity approximation" (e.g., Refs. [884, 892], etc.). In this approximation the specific volume and the enthalpy of the mixture are replaced by the sum of the specific volumes and enthalpies of the components:

$$V_{H + He}(p, T) = (1 - Y)V_H(p, T) + Y V_{He}(p, T),$$
$$(Y \equiv He/H \text{ by weight}).$$

$$(8.9)$$

Approximation (8.9) has an important peculiarity just from the viewpoint of the transformation of phase boundaries [386]. The structure of

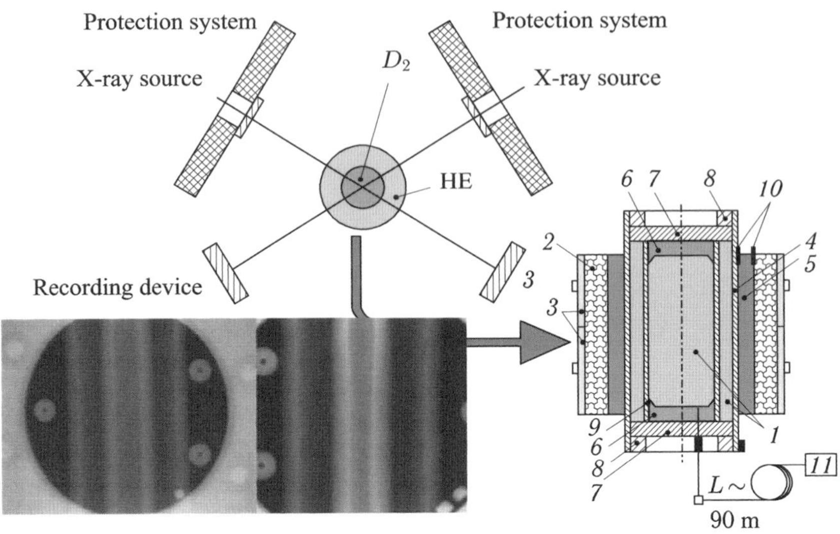


Fig. 8.35 Explosion cylindrical devices for quasiisentropic compression of plasma [299, 301, 311, 701]: *1* cylindrical sample, *2* HE charge, *3*, *4* inner and outer metallic liners, *5* X-ray source, *6* X-ray detectors.

Equation (8.9) is such that the $p$–$T$ coordinates of the spinodals, binodals, and the critical point of the phase transition can be transferred without changes from pure substances to the mixture, i.e., the final phase diagram of the mixture in the $p$–$T$ (as well as $\mu$–$T$ and $\mu$–$p$) coordinates is the direct superposition of the phase diagrams of the components [476]. This means, in particular, that all phase diagrams in the $H_2 + He$ mixture keep being, as in a simple substance, one-dimensional $p_s$–$T_s$ curves which end at the critical points of the traditional kind $(\partial p/\partial V)_T = 0$; $(\partial^2 p/\partial V^2)_T = 0$ [170, 171, 636]. In the general case just this result is wrong and may be valid only as an exception. Figure 8.37 shows as an example the hypothetical phase diagram of the $H_2 + He$ mixture calculated using approximation (8.9) and presented as a superposition of four separate phase transitions: two hydrogen-like PT, namely, plasma PT (according to Ref. [101]) and DPT (according to [122, 887]), and two PPT in helium (one PT per each ionization degree) [230].

It should be noted that such a picture is an exaggeration because at least for the "plasma" PT it is known [892] that in reality the presence of helium decreases the hydrogen tendency to PPT (such data on DPT are not available) and on the contrary the presence of hydrogen

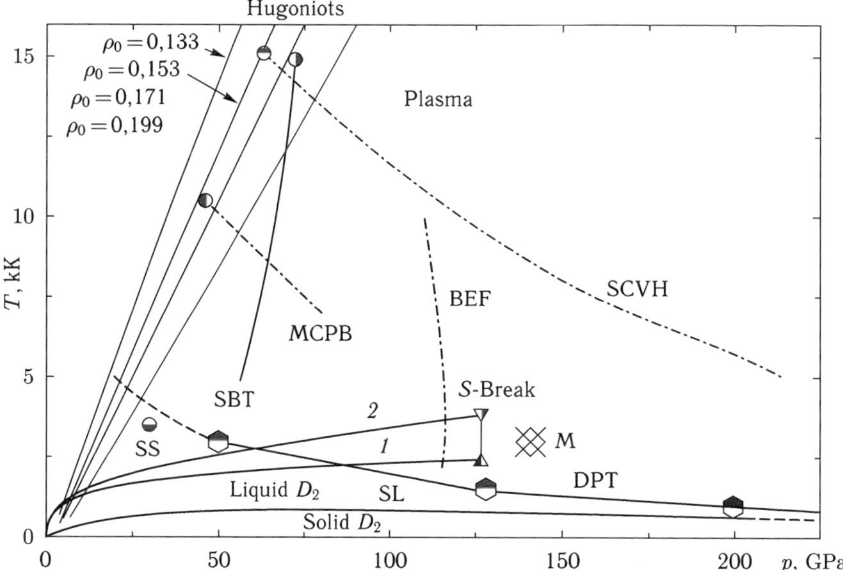



Fig. 8.36   See Ref. [386]. Isentropic compression of deuterium. Comparison of the experimental and theoretical results: *1*, *2* isentrope calculation by the SAHA (1) [386] and QMC (2) [572, 574] models, S-break -the isentrope break (calculation), PPT hypothetical plasma phase transitions: SS [948], SCVH [884], SBT [892], MCPB [646], BEF [101], DPT "dissociative" PP [887], [122], solid line first-order PT, dashed line "smeared" PT, SL melting curve (calculation [123]), M metallization of hydrogen [1037], Hugoniots shock adiabats (calculation, SAHA code [386]) at different densities of gaseous ($\rho_0 = 0.133/0.153\,\text{g/cm}^3$), liquid ($\rho_0 = 0.171\,\text{g/cm}^3$), and solid ($\rho_0 = 0.199\,\text{g/cm}^3$) deuterium realized in the experiments performed by the All-Russian Research Insitute of Experimental Physics (Sarov) [377] and Sandia Lab. [558].

decreases the helium tendency to PPT. It should be emphasized, however, that even in the model calculations of the $H_2 + He$ mixture [892] the mixture phase boundaries are traditionally found by the flattening of the "van der Waals loops" making use of the Maxwell "equal area" rule or of the "double tangency" rule. Then for various $He/H_2$ proportions the phase boundaries are drawn as the one-dimensional $p$–$T$ curves which describe a regular phase transition with a standard critical point (points) [892].

In the general case the properties of the phase transitions, which occur in the systems formed by two and more chemical elements, essentially differ from the PT properties in simple elements (see Ref. [175]). The major difference of such PT is that they have the noncongruent character, i.e., they

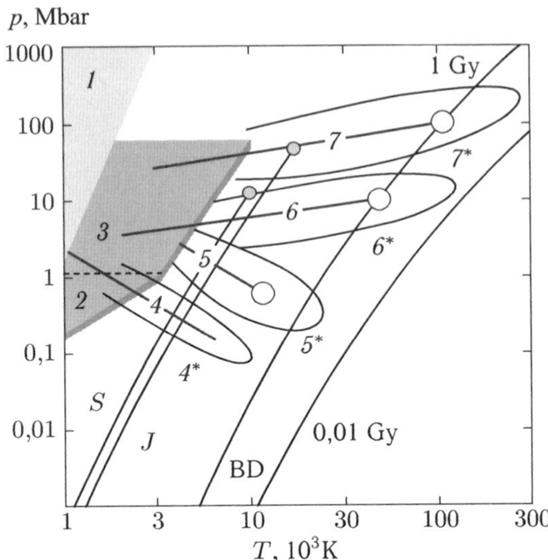


Fig. 8.37   Phase diagram of the H₂ + He mixture under the conditions of the interiors of the astrophysical objects: J Jupiter, S Saturn, BD brown dwarfs (1 and 0.01 million years); *1* solid hydrogen, *2*, *3* hydrogen–helium immiscibility zones: *2* neutral liquid [636], *3* plasma [790], *4*, *5*, *6*, *7* boundaries of phase transitions in the mixture in the additivity approximation (1): *4*, *5* hydrogen-like PTs: dissociative PT according to Refs. [122, 481, 482, 887] and plasma PT according to Ref. [884]; *6*, *7* helium-like PTs according to Ref. [230]: *6* 1st ionization, *7* 2nd ionization; ∘ critical points; *4**, *5**, *6**, *7** hypothetical smearing of interphase boundaries into zones due to noncongruence of PT in the H₂ + He mixture [476].

separate into phases with different stoichiometry. In this case the presence of elements of different kind is not important for the description of the dissociation and ionization process. The mentioned peculiarities will be considered below in detail using as an example the noncongruent evaporation in the high-temperature uranium-oxygen system which is a product of the extreme (emergency) heating of the nuclear reactor fuel, namely, of uranium dioxide (UO₂) [473, 476, 853]. Now let us consider the issues most important in the present context.

The authors of paper [476] emphasize that, along with the evaporation in the uranium–oxygen mixture, the indicated peculiarities of noncongruent PT are inherent in principle in any transitions in the mixture of chemical elements and in particular in the plasma or dissociative PT in the helium–hydrogen mixture in the interior of the giant planets. Study of the scale of this phenomenon and of its influence on the evolution of planets requires full-size calculations of this effect to be carried out making use of

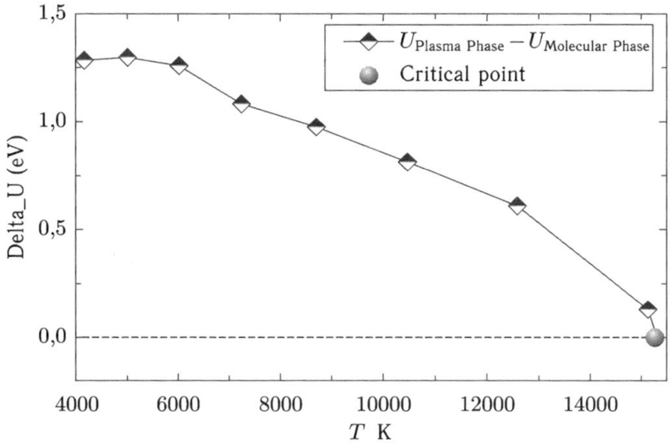



Fig. 8.38   [990]. The potential of the interphase boundary for the hypothetical "plasma" phase transition (PPT) (version of Ref. [884]).

the techniques described in Refs. [473, 476, 853]. The expedience of such calculations is justified by the estimate of the extent of noncongruence for one of the versions of plasma PT [884] in the $H_2 + He$ mixture, as it was reported in Ref. [472]. The deviation of the stoichiometry of the co-existing phases obtained in Ref. [472] turned out to be of the same sign as and comparable in value with the observed value of the helium depletion in the Jupiter and Saturn atmospheres [285]. The obtained result justifies the study of the noncongruency calculations besides [884] for all other hypothetical phase transitions predicted for hydrogen and helium and transferred to the conditions of the $H_2 + He$ mixture (see Figs. 8.20–8.24 and 8.38). The qualitative effect of the discussed noncongruency of the hypothetical phase transitions in the interior of the giant planets is shown in Fig. 8.37. The main predicted change is the separation of the one-dimensional (in approximation (8.9)) hydrogen-like and helium-like interphase boundaries in the two -dimensional zones 4*–7*. It should be emphasized however that, besides the non-congruency, the transition from pure $H_2$ and He to their mixture suppresses the initial phase transitions, i.e., the presence of helium decreases the hydrogen tendency to PPT (such data on DPT are not available) and, on the contrary, the presence of hydrogen decreases the helium tendency to PP [892]. This fact allows us to assume that the final phase diagram of the mixture has a fairly complicated structure compared to the simple schemes presented in Fig. 8.37.

The noncongruency effect in the H–He mixture under the conditions of Jupiter and Saturn was considered in Ref. [990] in the framework of the plasma phase transition model [884]. Using the tabular data of Ref. [884], some thermodynamic quantities of the hydrogen–helium plasma in the interiors of Jupiter and Saturn were reconstructed in the range of parameters corresponding to the conditions in the vicinity of the PPT co-existence boundaries [884]. Making use of the general thermodynamic relationships, the characteristics of the Coulomb and density corrections (the so-called "non-ideality corrections") were reconstructed. This made it possible to estimate for this PPT version two previously unknown quantities:

(A) The jump of the electrostatic potential at the PPT inter-phase boundary which is generally inherent in any phase transition in equilibrium Coulomb systems [465], and

(B) The typical PPT noncongruence scale (i.e., the differences in the stoichiometry of the co-existing phases) in the hydrogen–helium plasma of the interiors of Jupiter and Saturn.

While the first quantity, namely, the potential at the PPT inter-phase boundary, turned out to be within the typical scale of contact electrochemical potentials [465], i.e., $\Delta\varphi \approx 1\text{--}2\,\text{eV}$ (Fig. 8.38), the second reconstructed quantity, PPT noncongruency [884], turned out to have a significant value under the conditions of Jupiter and Saturn (Fig. 8.39). It is important that it coincides in sign and is comparable in magnitude with the helium deficiency experimentally observed in the atmosphere of the giant planets. According to the data of review [285], $Y(\text{He})_{\text{Jupiter}} \approx 0.231$, $Y(\text{He})_{\text{Saturn}} \approx 0.215$. It is interesting to emphasize that one of the goals (among others) of the Cassini–Huygens terrestrial apparatus, which is now on the orbit of Saturn, is to verify and improve the value of the above-mentioned helium "deficiency" in the atmosphere of Saturn.

Let us consider the phase transitions for high-temperature evaporation of multi-component mixtures taking as an example the uranium dioxide which is the main fuel of the present nuclear power industry [472] (see also Refs. [399, 466, 473, 476, 853]).

The knowledge of the EOS of the $UO_2$ heating products is necessary for the analysis of the possible consequences of a large nuclear accident on the existing and future advanced types of nuclear reactors. According to the available scenarios, in the process of such an accident the temperature can really reach several thousands of kelvins which corresponds to the estimated vapor pressures of many hundreds of atmospheres. The principal peculiarity of the uranium dioxide evaporation is its non-congruent

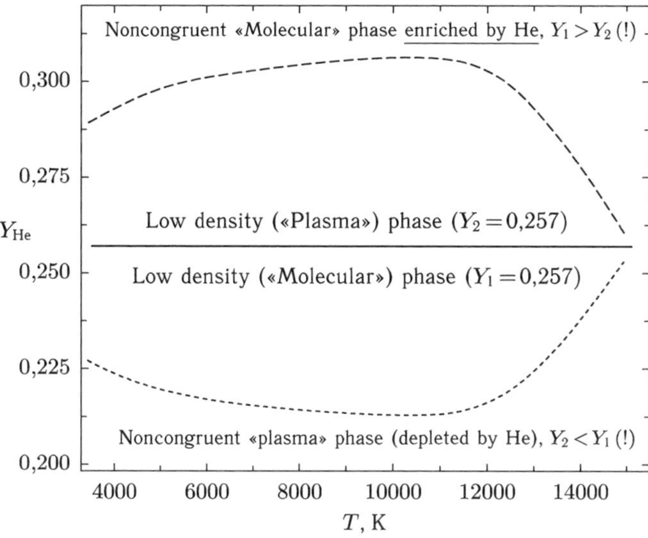


Fig. 8.39   Ref. [990]. Noncongruence of the hypothetical PPT (version of Ref. [884]) in hydrogen–helium mixture under the conditions of the interiors of Jupiter and Saturn ($Y_{\mathrm{He}} \approx 0.257$). The temperature dependence of the helium abundance for the co-existing "plasma" and "molecular" phases.

character which manifests itself as possible co-existence of phases with different stoichiometry. As a result, an abrupt oxygen enrichment of the uranium dioxide evaporation products may occur which creates an additional thread in the series of the possible hypothetical consequences of the nuclear accident.

The noncongruent character of evaporation is a distinctive feature of the phase equilibrium in high-temperature chemically active nonideal plasma [473, 476, 853]. The noncongruence dramatically complicates the structure and properties of the high-temperature part of the phase diagram of a wide class of substances which are called chemical compounds. The problems which stem from this fact are of specific importance for thermal physics of phase equilibrium for many objects used in nuclear power industry, such as the uranium dioxide $UO_2$ for active reactors, especially for fast breeders [473]. It is also important for many future nuclear devices, first of all, for the family of gas-phase reactors (so-called "B" scheme [207, 398, 467, 474, 480]). In different models of such reactor uranium mixtures with various metals, hydrogen, and other substances [398, 474] are widely used as working media. The above-mentioned noncongruency is of principle importance also for a gas-phase reactor using uranium hexafluoride $UF_6$ [207, 467]. According to the presently accepted practice, the exploitation and modernization of active reactors as well as the development of future reactor designs must be accompanied not only by the theoretical and experimental investigation of the design mode, but also by theoretical (and, whenever possible, experimental) investigation of possible consequences of the hypothetical emergency situations, including the extreme cases of the catastrophic nuclear accidents [473].

The EOS developed in Refs. [399, 465, 472, 473, 476, 853, 990] is intended for the description of evaporation in uranium–oxygen mixture which may have hypo- or hyper-stoichiometry. As distinct from models [274–276, 752], this model for the first time takes consistently into account the above-mentioned noncongruency of evaporation over the whole temperature range and, as a consequence, for the first time consistently reproduces the general structure of the total phase boundary, including the regions of the so-called retrograde condensation, points of extreme pressure and temperature on the phase boundary, and finally the true critical point of the noncongruent evaporation. The new EOS of the uranium–oxygen system describes the known properties of liquid uranium dioxide near its melting temperature and is used further for the extrapolation of the thermodynamic description of the $UO_{2\pm X}$ properties to the high-temperature region, including the parameters of the gas–liquid transition in the vicinity of the transition critical point.

Both co-existing phases (liquid- and gas-plasma) are described in a unified representation as a multi-component strongly interacting ("nonideal") mixture of chemically reacting atoms, molecules, atomic and molecular ions, and electrons.

The following set of components was used in the direct calculations of the thermodynamically equilibrium uranium–oxygen system:

$$U, U^+, U^{++}, UO, UO_2, UO_3, \{U_2O_n\}(n = 0, 1, 2, 3, 4, 5, 6),$$
$$O, O^-, O_2, UO^+, UO_2^+, UO_2^-, UO_3^-. \tag{8.10}$$

The principle point of the developed approach was that the same ideal-gas thermodynamic functions of the components of the uranium–oxygen system were extrapolated also for the calculation of the chemical and ionization equilibrium in the strongly nonideal uranium–oxygen system corresponding to the liquid uranium dioxide. In the real phase equilibrium calculations the above-mentioned ideal-gas characteristics were supplemented with a whole complex of the so-called "nonideality" corrections which took self-consistently into account the effects produced by the whole spectrum of intense interparticle interactions in the system [398]. The numerous parameters (virtually unknown in advance) of the interparticle interactions of all charged and neutral components of the system were first estimated theoretically and after that were corrected ("calibrated") with the goal to reproduce the known thermodynamic properties of liquid uranium dioxide (density, vapor pressure, etc.) at the melting point ($T = 3120\,\mathrm{K}$).

The most interesting result of approach [399, 465, 472, 473, 476, 853, 990] is the unusual structure of the predicted phase boundary of the uranium dioxide evaporation process (Fig. 8.40). The calculation results demonstrate the principle difference of this structure from the known analogs of the high-temperature boundary in "simple" substances, e.g., in metals (see 8.3). This difference is the direct consequence of the above-mentioned noncongruent nature of evaporation in nonideal chemically reacting plasma, in particular in the uranium dioxide plasma. The major distinctive feature of this evaporation is the sharp dependence of its parameters (vapor pressure and the extent of the vapor oxygen enrichment) on the evaporation rate [109].

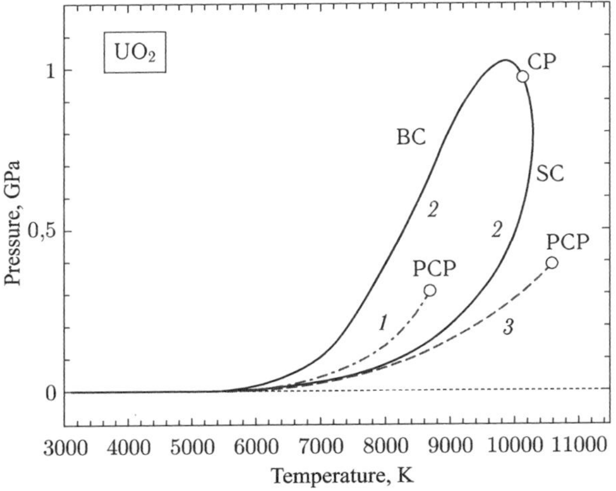


Fig. 8.40  Phase $p$–$T$ diagram of the uranium dioxide ($UO_{2.0}$) evaporation. *1* – forced-congruent equilibrium curve built making use of the standard "double-tangency" rule (BC–SC) with the pseudo-critical point (PCP); *2* – true two-phase boundary of totally equilibrium noncongruent evaporation with the boiling curve (BC), saturation curve (SC), and true critical point (CP) drawn using the results of the present work; *3* – total vapor pressure on the boiling curve and the pseudo-critical point (PCP) according to data [274–276]. (The calculations were performed within the framework of the "significant structure" theory.)

In terms of thermodynamics it is equivalent to the separation of the being previously unique temperature dependence of the saturated vapor pressure $p_{sat}(T)$ into two different boundaries — boiling curve (BC) and saturation curve (SC) (Fig. 8.40).

The first boundary (BC) corresponds to the chemical, ionization, and phase equilibrium of the stoichiometric liquid uranium dioxide $UO_{2.0}$ with non-stoichiometric (oxygen-enriched) vapor phase, $UO_{2+X}(X \geq 0)$. Dynamically this case corresponds to the regime of slow, totally thermodynamically equilibrium $UO_{2.0}$ evaporation. Such a regime is the closest one to the regime which must really set up in the process of a hypothetical nuclear reactor accident. On the contrary, the second boundary (SC) corresponds to the equilibrium between the vapor phase with stoichiometric composition $UO_{2.0}$ and non-stoichiometric (oxygen depleted) liquid $UO_{2+X}(X \leq 0)$. Dynamically it corresponds to the superfast forced-congruent mode (FCM) of only partially equilibrium evaporation which is too short for the stoichiometry of the evaporating substance to change [109]. The direct consequence of the noncongruence of evaporation in the uranium dioxide is the significantly high level of the maximal vapor pressure which was theoretically predicted on the melting curve ($p_{max} \approx 1\,GPa$) (Fig. 8.40).

Another important consequence predicted by the present theory is the extremely high extent of maximal oxygen enrichment of the vapor phase which is in equilibrium with the boiling uranium dioxide with stoichiometry $UO_{2.0}$ {max $(O/U)_{BC} \approx 7$ at $T \approx 8000\,K$) (Fig. 8.41).

One more consequence predicted by the present theory is the unusual shape of the phase diagram of non-congruent evaporation in the enthalpy-temperature coordinates. This phase diagram is essentially different from the similar $H$–$T$ diagram describing evaporation of "common" substances (Fig. 8.42). Its most prominent feature is that the analog of the evaporation heat $Q_{vap} \equiv H(T)_{vap} - H(T)_{liquid}$ displays the nonmonotonic temperature dependence. All the above-quoted peculiarities are extremely important for the applications associated with the nuclear safety problem.

The merit of the theoretical approach used in papers [473, 476, 853] is the possibility of the unified and self-consistent description of EOS of the strongly interacting uranium–oxygen system in the vicinity of the true critical point of the noncongruent phase transition. The properties of this critical point as well as the whole phase transition are in principle different from the properties of the critical point of a "regular" gas–liquid phase transition which corresponds to the fulfillment of the standard relations

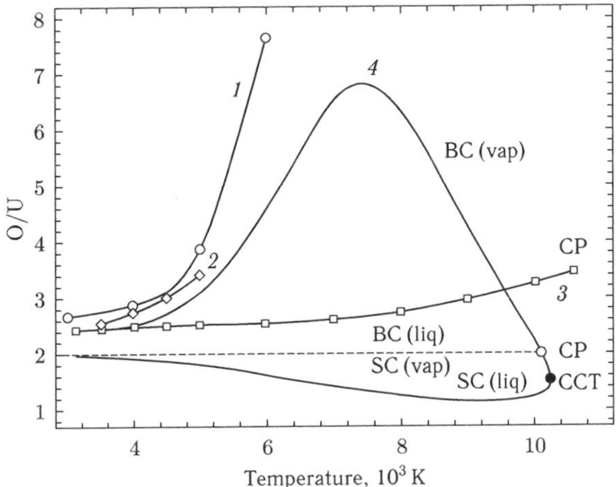



Fig. 8.41   See Ref. [473]. The ratio O/U in the coexisting phases in uranium dioxide ($UO_{2.0}$) noncongruent evaporation. *1* stoichiometry of the gaseous phase by model [368], *2* same as 1 using the results of the equilibrium composition calculation in Ref. [752], *3* same as 1 according to the calculations within the framework of the "significant structure" theory (SST) [274–276], *4* the boundary of the two-phase region of noncongruent evaporation calculated in Refs. [399, 472, 473, 476, 636, 790, 853]. Marks: stoichiometry of the co-existing vapor and liquid in the boiling mode (BC) and saturation mode (SC), as well as the true critical point (CP) and the maximal temperature point on the saturation curve (CCT).

$[(\partial p/\partial V)_T = (\partial^2 p/\partial V^2)_T = 0; (\partial^3 p/\partial V^3)_T < 0]$. In particular, the isothermal compressibility of a system at the critical point of the noncongruent phase transition does not tend to infinity, but turns out to be close to the ideal-gas value $(\partial \ln p/\partial \ln V)_T \approx 1 \neq 0$. The determining feature of the critical point of the noncongruent phase transition is that the matrix $||\partial \mu_i/\partial n_j||$ loses its positive definiteness ($\mu_i$ and $n_i$ are the chemical potential and the concentration of particles of the $i$th kind).

The present model predicts the following parameters for the critical point under consideration:

$$T_{\mathrm{cr}} \approx 10120\,K, \quad p_{\mathrm{cr}} \approx 965\ \mathrm{MPa}, \quad \rho_{\mathrm{cr}} \approx 2.61\ \mathrm{g/cm}^3,$$

$$S_{\mathrm{cr}} \approx 1.64\ \mathrm{kJ/kg \cdot K}, \quad \Gamma_{\mathrm{D}} \equiv e^2/kTR_{\mathrm{D}} \approx 1.2,$$

$$C_p \approx 1.8\ \mathrm{kJ/kg \cdot K}, \quad \beta_T \equiv \rho^{-1}\left(\frac{\partial \rho}{\partial p}\right)_T \approx 1.03 \cdot 10^{-4}\ \mathrm{1/bar}, \tag{8.11}$$

$$\alpha_p \equiv \rho^{-1}\left(\frac{\partial \rho}{\partial p}\right)_T \approx 5.15 \cdot 10^{-4}\ \mathrm{1/K}.$$

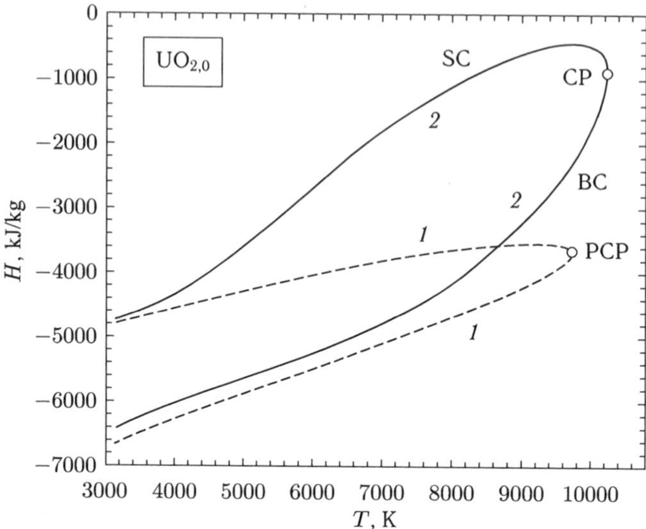


Fig. 8.42   Enthalpy-temperature phase diagram for uranium dioxide (UO$_{2.0}$) noncongruent evaporation. *1* the boundary of the two-phase region of uranium dioxide described in the monomolecular model for the system of interacting UO$_2$ molecules (phase diagram of the standard kind with the critical point (PCP), *2* the true boundary of the two-phase region of noncongruent evaporation with the boiling curve (BC), saturation curve (SC), and true critical point (CP) drawn using the results of Refs. [472, 473, 476, 853].

In recent years interesting results were obtained by using the pulse heating method to measure the vapor pressure for noncongruently evaporating uranium dioxide [135]. The details of the experimental technique and special procedures, which were used to carry out the correct measurements and to obtain relevant experimental results, are extensively discussed in Ref. [80]. The theoretical analysis [990] led to the conclusion that due to the essential noncongruency of the uranium dioxide evaporation, the vapor pressure measured in experiment [135] does not correspond to the conditions of fully equilibrium (quasistationary) boiling mode. On the contrary, it corresponds to the fast regime of "forced-congruent" evaporation, which in the language of thermodynamic modes of noncongruent evaporation corresponds to the saturation regime, i.e., to the equilibrium of hydrogen-depleted liquid and uranium dioxide vapor with strictly stoichiometric composition (O/U = 2.00) (see Fig. 8.41). The comparison of the experimental

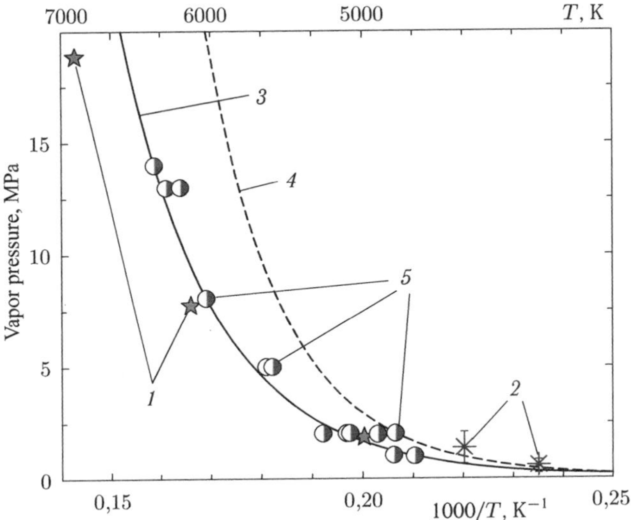

Fig. 8.43  See Ref. [476]. The pressure-temperature phase diagram of uranium dioxide ($UO_{2.0}$) noncongruent evaporation. Comparison of the theoretical calculations with the experimental results: *1* boiling boundary in "ampoule" experiment [133–135], *2* boiling boundary in the laser heating experiment [110], *3* calculated boundary of noncongruent boiling of the $UO_{2.0}$ liquid phase according to the results of Refs. [473, 476, 853], *4* calculated boundary of noncongruent saturation of the gaseous phase with $O/U = 2,0$ according to the results of Refs. [473, 476, 853], *5* the gaseous phase saturation boundary in IVTAN laser heating experiment [80, 560].

results [80] with the calculations performed by the above-described theoretical model [473] confirms this conclusion and provides an additional argument in favor of the constructed theoretical EOS for the high-temperature uranium–oxygen system (uranium dioxide). This comparison is presented in Fig. 8.43.

# Chapter 9

# Semi-Empirical Equations of State

It follows from the previous chapters that the description of the thermodynamic properties of nonideal media requires the application of complicated models which, due to their narrow specialization, can be used in a limited range of matter state parameters. As a rule, these models incorporate cumbersome quantum-mechanical numerical calculations and provide information in the form of diagrams or big tables that makes it difficult to use these data in practical calculations. Such equations of state (EOS) are sometimes called "global" [1002]. They are generally based on using modern powerful theoretical models for the calculation of the separate domains of the phase diagram. The obtained tables of the thermodynamic functions should be then smoothly sewed by means of the interpolation procedure.

An alternative way to describe the properties of nonideal media in a wide range of parameters is to build semi-empirical models [27, 1081, 1090] in which the general form of functional dependences is derived on the basis of the theoretical grounds and the experimental data are used for the determination of the numerical values of the coefficients entering these dependences.

The description of the dynamic adiabats and the dimensionless EOS are presented in Refs. [49, 173, 812–815].

It can be said that the modern semi-empirical EOS models in a wide phase diagram range employ analytic and interpolation formulae adopted from rigorous theories, including the "first-principles" ones, and are essentially based on the experimental data. In these models, the functional dependences of the thermodynamic potential are chosen proceeding from the theoretical considerations and the concrete constants are given by the best fit to the theoretical or experimental data. The numerical coefficients in these formulae are divided into two groups: the coefficients in the first group are constants inherent to the specific substance, including the fundamental constants like the charge and mass of a nucleus, normal density, bulk modulus under normal conditions, etc.; the coefficients in the second group are fitting parameters chosen from the best description of a bulk of the experimental data and theoretical calculations at high pressure and temperature values.

## 9.1 Quasi-harmonic approximation

The most traditional approach to semi-empirical models is the separation of the thermodynamic potential (e.g., of the free energy) into a cold component $E_c(V)$ and thermal terms determined by thermal excitation. The thermal components, in their turn, are presented as a sum of the contributions given by the thermal motion of the lattice atoms or molecules $F_a(V, T)$ and by the thermally excited conduction electrons $F_e(V, T)$. This leads to the general expression

$$F(V, T) = E_c(V) + F_a(V, T) + F_e(V, T). \tag{9.1}$$

Here, the specific form and expressions used for separate terms depend on how general the corresponding semi-empirical models are.

For the calculation of $F_a(V, T)$ the quasiharmonic approximation is widely used. A crystal is treated as a collection of $3N$ harmonic oscillators whose normal oscillation frequencies $\omega_i$ depend only on the volume $V$. The thermal energy and pressure of such a crystal have the form:

$$\varepsilon_a(V, T) = \sum_{i=1}^{3N} \frac{\hbar\omega_i}{\exp\left(\frac{\hbar\omega_i}{kT}\right) - 1}, \tag{9.2}$$

$$p_a(V, T) = \sum_{i=1}^{3N} \frac{\gamma_i}{V} \frac{\hbar\omega_i}{\exp\left(\frac{\hbar\omega_i}{kT}\right) - 1}, \tag{9.3}$$

where $\gamma_i = -d\ln\omega_i/d\ln V$. The free energy is given by the expression

$$F_a(V, T) = kT \sum_{i=1}^{3N} \ln\left[1 - \exp\left(-\frac{\hbar\omega_i}{kT}\right)\right], \tag{9.4}$$

if the summation in Equation (9.4) is replaced by the integration.

$$F_a(V, T) = kT \int Z(\omega) \ln\left[1 - \exp\left(-\frac{\hbar\omega}{kT}\right)\right] d\omega, \tag{9.5}$$

In Equation (9.5) the weight function $Z(\omega)$ determines the vibration spectrum of the lattice atoms. The real form of this spectrum and its volume dependence are extremely complicated, therefore the concrete calculation of $F_a$ becomes possible only if additional model simplifications are introduced. The Debye model defines the phonon spectrum in the acoustic approximation which is valid for weak sound waves:

$$Z(\omega) = \begin{cases} \dfrac{\omega^2}{2\pi}\left(\dfrac{1}{c_l^3} + \dfrac{2}{c_t^3}\right) & \text{at } \omega \leq \omega_D, \\ 0 & \text{at } \omega > \omega_D, \end{cases} \tag{9.6}$$

where $c_l$ and $c_t$ are the longitudinal and transverse components of the sound velocity and the Debye frequency $\omega_D$ is defined by the normalization to the entire number of the vibration modes:

$$V \int_0^{\omega_D} Z(\omega) d\omega = 3N. \tag{9.7}$$

In this case after introducing the new notation $R = kN$ and $\theta = \hbar\omega_D/k$ formula (9.5) for the free energy takes the form

$$F_a(V, T) = RT\left(3\ln\left\{1 - \exp\left[-\frac{\theta(V)}{T}\right]\right\} - D\left[\frac{\theta(V)}{T}\right]\right), \tag{9.8}$$

where

$$D(x) = \frac{3}{x^3} \int_0^x \frac{t^3 dt}{\exp t - 1}. \tag{9.9}$$

The thermal components of the energy and pressure are, correspondingly, equal to

$$\varepsilon_a(V, T) = 3RTD\left[\frac{\theta_D(V)}{T}\right], \tag{9.10}$$

$$p_a(V, T) = \frac{\gamma(V)}{V} 3RTD\left[\frac{\theta_D(V)}{T}\right], \tag{9.11}$$

the Debye function $D(\theta_D/T)$ ensuring the low-temperature $C_V \sim T^3$ and high-temperature $C_V \sim 3R$ asymptotics for the lattice heat capacity. $\gamma = -d\ln\theta_D/d\ln V$ is the Grüneisen parameter which is only volume dependent. In the Debye model a solid is assumed to be uniform and isotropic and thermal excitations of a crystal are reduced to sound waves with density $Z(\omega)$. In this case the external pressure just affects the boundary frequency $\omega_D(V)$, whereas the whole spectrum stays harmonic and has the form (9.6).

Due to the relatively low temperature arising in the shock compression of matter to the pressure of tens of megapascals, the contribution given by the thermal excitation of electrons to the EOS (9.1) can be neglected. Then the thermodynamics of the solid in this domain of the phase diagram can be successfully decribed by the quasiharmonic model (9.2)–(9.11) in the form of the Mie–Grüneisen caloric EOS:

$$p(V,\varepsilon) = p_c(V) + \frac{\gamma(V)}{V}[\varepsilon - \varepsilon_c(V)], \tag{9.12}$$

where $p_c = -d\varepsilon_c(V)/dV$, and for the Debye model $\varepsilon_c(V) = \varepsilon_0(V) + (9/8)\theta_D(V)$ is a sum of the crystal elastic energy $\varepsilon_0$ at $T = 0$ and the energy of zero vibrations of the atom.

If the experimental shock adiabat of matter,

$$p = p_H(V), \tag{9.13}$$

is known, then the energy conservation law at the shock jump (assuming $\varepsilon_0 = 0$, $p_H \gg p_0$) written as

$$\varepsilon_H = \frac{1}{2}p_H(V_0 - V), \tag{9.14}$$

additional model assumptions on the (9.12) dependence allow to determine the characteristics of the cold curve $\varepsilon_c$. Then it is possible to obtain the thermodynamically complete description of matter in the framework of the quasi-harmonic model with a given form of the electron component contribution (at low temperatures it is usually assumed $F_e = 0$).

The most obvious way to obtain the cold curve parameters is to employ the shock-wave data under the assumption $\gamma/V = \text{constant}$ which is widely used by the American researchers (see, for instance, Ref. [679]). This simplification reduces the system (9.12)–(9.14) to the first-order differential equation with respect to the unknown function $\varepsilon_c(V)$:

$$p_H(V) = -\varepsilon_c' + \frac{\gamma_0}{V_0}[\varepsilon_H(V) - \varepsilon_c].$$

The temperature closing the thermodynamic description can be found directly from Equation (9.10) because the volume dependence of the Debye temperature takes the simple form:

$$\theta_{\mathrm{D}}(V) = \theta_{\mathrm{D}0} \exp\left[\gamma_0\left(1 - \frac{V}{V_0}\right)\right],$$

where $\theta_{\mathrm{D}0}$ and $\gamma_0$ correspond to the normal conditions.

The condition $\gamma/V =$ constant which describes well the experimental data at moderate compression $V_0/V \leq 1.5$ leads to a very rapid decrease of $\gamma$ with the further decrease of the volume. This contradicts the theoretical concepts and the experimental data. The more appropriate relationship is $\gamma/\sqrt{V} =$ constant, but it neither provides the required asymptotic behavior $\gamma \to 2/3$ at $V \to 0$. Finally, direct experimental check of the applicability of the approximation $\gamma/V^q =$ constant made under static conditions at low matter compression $V_0/V \leq 1.2$ [115] gives diverse values of the power $q$ ranging from 0.6 (Fe) to 1.8 (In).

A variety of interpolation formulae were suggested to describe the volume dependence of the Grüneisen parameter. They have the required asymptotic behavior $\gamma(V \to 0) = 2/3$, the value of the derivative $d\gamma/dV$ at $V = V_0$ being either small,

$$\gamma(V) = \gamma_0\frac{V}{V_0} + \frac{2}{3}\left(1 - \frac{V}{V_0}\right),$$

or large,

$$\gamma(V) = \gamma_0\frac{V}{V_0} + \frac{2}{3}\left(1 - \frac{V}{V_0}\right)^2,$$

Other relationships have been also proposed which contain the fitting parameter for more accurate description of the experimental data. Using approximation [146]

$$\gamma(V) = \frac{2}{3} + \left(\gamma_0 - \frac{2}{3}\right)\frac{1 + \sigma_{\mathrm{m}}^2}{\sigma^2 + \sigma_{\mathrm{m}}^2}\sigma,$$

where $\sigma = V_0/V$ is the degree of compression, one can carry out hydrodynamic calculations in a wide range of the phase diagram. This formula satisfies the asymptotics $\gamma = 2/3$ in the limit of high compression and low density and, also, gives the correct value of the Grüneisen parameter at normal conditions. On the basis of the data on shock compression of porous samples, it takes also into account the peculiarities of the behavior of the Grüneisen parameter, the values of $\sigma_m$ for various substances being within the interval 0.5–0.9. The most popular way to present the results of the dynamic experiment at moderate compression is to draw the linear dependence between the wave-front velocity and the velocity of the substance motion [27], $D = c_0 + \lambda u$. This gives the simple form of the shock adiabat: $p_{\mathrm{H}}(V) = c_0^2(V - V_0)/[(V_0 - \lambda(V_0 - V)]^2$. The existence of the second-order tangency of the shock adiabat and the normal isentrope [1081] makes it possible, neglecting the shock load strength effects, to relate the parameters of the linear relationship $D(u)$ to the thermodynamic characteristics at normal conditions:

$$c_0 = \sqrt{V_0 B_{s0}}, \quad \lambda = \frac{1}{4}(B'_{s0} + 1), \tag{9.15}$$

where $B_{s0}$ and $B'_{s0}$ are the values of the adiabatic bulk modulus $B_s = -V(\partial p/\partial V)_s$ and its isentropic derivative with respect to pressure $B'_s = (\partial B_s/\partial p)_s$ at normal conditions. It follows from the comparison of the static and dynamic experiments that for closely packed materials relationships (9.15) are satisfied within a good accuracy. However, if there are structural phase transitions, these relationships are violated on compression because $c_0$ and $\lambda$ found from the results of the dynamic experiments correspond to the high pressure phase.

Another way to determine the cold curve parameters from the shock wave data by means of the Mie–Grüneisen state equation is based on using the different versions of the small oscillation theory. According to this theory, the volume dependence of the Grüneisen parameter is related to the characteristics of the cold curve and in the generalized form is given as

$$\gamma(V) = \frac{t - 2}{3} - \frac{V}{2} \frac{(p_c V^{2t/3})''_{VV}}{(p_c V^{2t/3})'_V}. \tag{9.16}$$

Here, index c is associated with the elastic component ($T = 0$); $t = 0$ corresponds to the Slater–Landau approximation, $t = 1$ is related to the Dugdale–MacDonald one, and $t = 2$ to the Vashchenko–Zubarev free volume theory [1081]. At $p_c = 0$ different versions give close values of the Grüneisen parameter, $\gamma^{\mathrm{SL}} = \gamma^{\mathrm{DM}} + 1/3 = \gamma^{\mathrm{VZ}} + 2/3$, the difference decreasing on compression and, if the pressure varies as $p_c \sim \sigma^n$ at $\sigma \to \infty$, all calculated values of $\gamma$ tend to the same limit $\gamma_\infty = (n - 1/3)/2$. In practical calculations the Slater–Landau or Dugdale–MacDonald formulae are commonly used because they provide values $\gamma_0 = \gamma(V_0)$ which are the closest to the experiment. However, the comparison of various calculation curves $\gamma(V)$ with the experimental data for aluminum and beryllium shows that none of the modifications of Equation (9.16) has a crucial advantage in the description of the results of the dynamic experiments. In the general case the optimal value of the parameter $t$ may be also non-integer.

It is experimentally known that the wave and mass velocities are related by the linear dependence as $D = C_0 + \lambda u$ [27, 153]. This fact leads to the shock-adiabat equation which has the form

$$p(V) = \frac{C_0^2(V_0 - V)}{(V_0 - \lambda(V_0 - V))^2}.$$

Then, making use of the known technique [153, 1081] which takes into account the existence of the third-order tangency of the shock adiabat and zero isentrope, it is possible to write an approximate formula relating the parameter $t$ and the experimentally measured quantities — Grüneisen parameter $\gamma_0 = \gamma(V_0)$ and the adiabat slope $\lambda$:

$$\gamma_0 = 2\lambda - \frac{2 + t}{3}.$$

There are also many other ways to determine $t$ from the results of static and dynamic measurements [851,852]. The comparison of the various methods of $\gamma(V)$ calculation with the experimental data for aluminum [729] shows, however, that none of the quasiharmonic models has a crucial advantage in the description of the results of the dynamic experiment [851,852]. The simplest approximation $\gamma/V = $ constant often used in calculations gives equally bad results when compared to the data [729]. Let us also note that application of formula (9.16) to the calculation of $\gamma(V)$ is, strictly speaking, possible only for isotropic structures or for the structures which have a cubic symmetry, while in the general case the tensor structure of the Grüneisen parameter should be taken into account. The authors of Ref. [205] considered the determination of the Grüneisen parameter for some metals, characterized by cubic and hexagonal symmetry, making use of the measured elastic constants and their derivatives with respect to pressure. Emphasize once more that the $\gamma(V)$ behavior problem can be entirely solved only taking into account the real frequency spectrum of crystal vibrations (see Chapter 3).

The Mie–Grüneisen EOS (9.12), being supplemented with the known matter shock adiabat and the Grüneisen parameter definition according to the generalized formula (9.16), leads to the differential equation of the third order with respect to the elastic compression energy $\varepsilon_c(V)$. This equation can be solved numerically, with the given initial conditions

$$\varepsilon_c(V_{0c}) = 0, \tag{9.17}$$

$$p_c(V_{0c}) = 0, \tag{9.18}$$

$$B_c(V_{0c}) = B_{0c}, \tag{9.19}$$

where $V_{0c}$ is the value of the specific volume at $p_c = 0$ and $T = 0$, $B_{0c}$ is the value of the bulk modulus $B_c = -V\,dp_c/dV$ at $V = V_{0c}$. To solve the system one must reduce the thermodynamic parameters of the original state on the shock adiabat to the initial state on the zero isotherm, i.e., determine the values of $V_{0c}$ and $B_{0c}$.

However, in practice the suggested method of calculation of the $\varepsilon_c(V)$ value does not allow to find unambiguously the volume dependence of the Grüneisen parameter because the values of the $p_c$ derivatives required for this procedure can be found using the experimentally found shock adiabat which does not provide by itself any reliable information on the derivatives. It is shown in Ref. [578] that in this case various analytic formulae for the shock adiabat, which describe equally well the initial experimental data, give the qualitatively different dependence $\gamma(V)$.

An alternative way to build the EOS for different substances and to describe the shock-wave measurement results reduces to the semi-empirical form of the determination of the cold curve as the analytic dependence of pressure on the degree of compression $\sigma_c = V_{0c}/V$. The fitting parameters in such dependences should ensure validity of conditions (9.18) and (9.19) and also correspond to the given value of the derivative of the isothermal bulk modulus with respect to pressure at $T = 0$:

$$\left.\frac{dB_c}{dp}\right|_{V=V_{0c}} = B'_{0c}, \tag{9.20}$$

whose value can be experimentally determined making use of the measured data on the isothermal compressibility or from the ultrasound measurements.

The simplest Murnaghan potential satisfies all three conditions (9.18)–(9.20) if $B_p$ is replaced by $B'_{0c}$. The simplicity of this potential enables one to write the Mie–Grüneisen equation in the economic algebraic form [593] (here, $n = B'_{0c}$):

$$p(V,\varepsilon) = \frac{B_{0c}}{n}\left[\sigma_c^n\left(1 - \frac{\gamma_0}{n-1}\right) - \gamma_0 - 1\right] + \gamma_0\left[\frac{\varepsilon}{V} + \frac{B_{0c}\sigma_c}{n-1}\right],$$

which ensures time-efficient computer calculations and provides a satisfactory accuracy at not too high pressures. Then the value of the parameter $n$ is chosen from the best fit to the experimental shock adiabat.

When describing the static and dynamic experimental data, more sophisticated potentials are as well used. Their form is obtained by means of various physical models. Besides the Birch and Murnaghan potentials, there should be mentioned the widely used Morse potentials

$$p_{\mathrm{c}}(V) = A\sigma_{\mathrm{c}}^{2/3} \left\{ \exp\left[2\alpha\left(1 - \sigma_{\mathrm{c}}^{-1/3}\right)\right] - \exp\left[\alpha\left(1 - \sigma_{\mathrm{c}}^{-1/3}\right)\right] \right\}, \quad (9.21)$$

the Born–Mayer potential:

$$p_{\mathrm{c}}(V) = Q\left\{ \sigma_{\mathrm{c}}^{2/3} \exp\left[q\left(1 - \sigma_{\mathrm{c}}^{-1/3}\right)\right] - \sigma_{\mathrm{c}}^{4/3} \right\}, \quad (9.22)$$

the Rose potential:

$$p_{\mathrm{c}}(V) = 3B_{0\mathrm{c}}\left(1 - \sigma_{\mathrm{c}}^{-1/3}\right)\sigma_{\mathrm{c}}^{2/3} \exp\left\{ \frac{3}{2}(B_{0\mathrm{c}}' - 1)\left[1 - \sigma_{\mathrm{c}}^{-1/3}\right] \right\}, \quad (9.23)$$

and the Parsafar–Mason potential:

$$p_{\mathrm{c}}(V) = \frac{1}{2}B_{0\mathrm{c}}\left[(B_{0\mathrm{c}}' - 7) - 2(B_{0\mathrm{c}}' - 6)\sigma_{\mathrm{c}} + 5(B_{0\mathrm{c}}' - 5)\sigma_{\mathrm{c}}^2\right]\sigma_{\mathrm{c}}^2. \quad (9.24)$$

All of them satisfy condition (9.18) and contain the coefficients $A$, $\alpha$, $Q$ and $q$ defined according to Equations. (9.19) and (9.20). The Born–Mayer potential, following from the ion crystal theory, gives the best results for the description of shock compressibility of various substances. This potential was used to obtain the EOS of some metals and minerals. The Birch–Murnaghan potential obtained on the basis of the finite deformation theory is usually applied for the analysis of the isothermal compressibility experimental data, but at the same time it shows good extrapolation properties with respect to the Thomas–Fermi model calculations at strong compression. The Rose potential, when integrated over $V$, gives the sublimation energy values within 10–30% of the tabular one and does not contain any fitting coefficients. However, this is its only merit because on compression the exponential dependence turns out to be too "stiff". Although the authors declare that formula (9.23) is applicable for the inertial thermonuclear fusion calculations, this potential does not ensure the correct asymptotics $\sigma_{\mathrm{c}}^{5/3}$ in the strong compression limit and it makes obviously not much sense to use it for the construction of wide-range EOS. The same holds true for the most recent Parsafar–Mason potential (9.24).

In some cases, instead of satisfying the additional condition (9.20), the fitting parameter $t$ in formula (9.16) defining the Grüneisen parameter was chosen using the joint condition that the shock wave data and the equality $\gamma(V_{0\mathrm{c}}) = \gamma_0$ be simultaneously best described. Actually, the use of this condition is equivalent to replacing the $B$ value by the $B_{0\mathrm{s}} = 4\lambda - 1$ value in Equation (9.20) and satisfying the relationship

$$t = 3(2\lambda - \gamma_0) - 2.$$

which follows then from Equation (9.16). There are also other ways to determine the parameter $t$, namely, using the experimental pressure dependence of the Poisson coefficient obtained from ultrasound measurements or applying the Lindemann criterion which relates the melting curve slope to the Grüneisen parameter.

Making use of potentials (9.21) and (9.22) for the representation of the cold melting curves has the advantage of their regular behavior which ensures the monotonic volume dependence of the Grüneisen parameter according to Equation (9.16). However, in practice the calculation results appreciably contradict the shock compression experimental data already at $\sigma_c = 1.5$–$2$ and do not agree with the theoretical concepts on the asymptotic behavior at $\sigma \to \infty$. This fact restrains the applicability range of the obtained potential curves and leads to the necessity of sewing both curves with the expressions corresponding to the quantum-statistical asymptotics.

It is convenient to represent the cold compression curve by an approximate series in powers of the inverse atomic spacing $r_{\mathrm{a}}^{-1} \sim \sigma_{\mathrm{c}}^{1/3}$ [576]:

$$p_{\mathrm{c}}(V) = \sum_{i=1}^{N} a_i \sigma_{\mathrm{c}}^{1+i/3}, \tag{9.25}$$

where for the determination of the coefficients $a_i$ one uses conditions (9.18) and (9.19) and some additional requirements whose number is determined by the maximal number $N$ of the expansion terms. As these requirements, the following conditions are imposed on the tabular values of the sublimation energy [576, 578],

$$\varepsilon_{\mathrm{c}}(V \to \infty) - \varepsilon_{\mathrm{c}}(V_{0\mathrm{c}}) = \varepsilon_{\mathrm{sub}}, \tag{9.26}$$

and of the Grüneisen parameter making use of some modification of Equation (9.16), commonly [576] in the Salter–Landau form:

$$\gamma^{\mathrm{SL}}(V_{0\mathrm{c}}) = \gamma_0. \tag{9.27}$$

There is also a requirement that the cold curve and its slope must be in correspondence with the quantum-statistical model calculations at $\sigma_{\mathrm{c}} = \sigma^*$:

$$p_{\mathrm{c}}(\sigma^*) = p^*, \tag{9.28}$$

$$\left.\frac{dp_{\mathrm{c}}}{d\sigma}\right|_{\sigma_{\mathrm{c}}=\sigma^*} = c^*, \tag{9.29}$$

where the value of $\sigma^*$ was chosen from the condition $p_{\rm c}(\sigma^*) \approx 300\,{\rm TPa}$ which is the characteristic pressure for the applicability of these models. The system of six equations (9.18), (9.19), and (9.26)–(9.29) defines six coefficients $a_i$ of the series (9.25). To improve the accuracy of the experimental data fitting the series can be supplemented with a seventh term and an additional condition can be imposed on the parameters of the reference experimental point on the shock adiabat [576, 578].

Thus obtained cold curves enable one, according to Equation (9.16), to calculate the Grüneisen lattice parameter whose values describe the shock-wave data in the entire experimentally investigated parameter range. The cold curves presented in the form (9.25) were successfully applied to the construction of the EOS for a wide variety of substances: metals, ion crystals, hydrogen, and argon. However, due to the formal character of the approximate expansion (9.25), its application region is limited within the interval from $\sigma_{\rm c} = 1$ to $\sigma^*$ leading to the nonphysical behavior of the Grüneisen parameter already at compression $\sigma_{\rm c} \approx \sigma^*$ and, at still stronger compression, to progressive deviation of the calculation curve $p_{\rm c}(V)$ from the quantum-statistical calculation results (see the detailed analysis of the results of various theoretical calculations in monograph [150]).

The inapplicability of the approximate series in the superhigh pressure region led to further modification of the form of the cold curve and of the method of defining its coefficients. To ensure the correct quantum-statistical asymptotics in the high compression region the authors of [502] suggested to use the power series of the type of (9.25) (but lower order), such that the higher-order terms ($\sigma_{\rm c}^{4/3}$ and $\sigma_{\rm c}^{5/3}$) have a precise quantum-statistical value. The lower-order terms of the series must ensure the fulfillment of conditions (9.18), (9.19) and the adequate description of the cold curves in the experimentally studied range of parameters. The obtained interpolation cold curves have the correct behavior at low and superhigh pressures. However, in the calculation of the lattice Grüneisen parameter using formula (9.25) they give unrealistic values which do not provide the correct description of shock-wave experiments.

## 9.2 Anharmonicity. Melting of the lattice

As the temperature increases, the anharmonic effects produced by thermal vibrations of the lattice atoms and the contribution given by the thermally excited conduction electrons become essential. In the limit of high temperature the anharmonicity of vibrations results in that the behavior of the lattice component of the condensed phase potential is characterized by the properties of an ideal gas of nuclei. This effect is taken into account in Ref. [576] by introduction of an additional term into Equation (9.1):

$$F_{\mathrm{a}}(V,T) = \frac{3}{2}RT\ln(1+z),\qquad(9.30)$$

where the quantity $z = lRT/c_{\mathrm{c}}^2$ is proportional to the ratio of thermal to elastic parts of pressure, $l$ is an empirical coefficient determined from the experimental shock compression of porous specimens. Under the assumption that the parameters of the cold curve and the Grüneisen parameter are related by Equation (9.16), the Slater–Landau approximation can be written as

$$\gamma(\rho) = \frac{1}{3} + \frac{d\ln c_{\mathrm{c}}(\rho)}{2d\ln\rho}{}',$$

and the expression for the free energy takes the form

$$F_{\mathrm{a}}(V,T) = \frac{3}{2}RT\ln\frac{\theta^2(V)(1+z)}{T^2},$$

which shows that the model involves a transition from the condensed $((z\approx 0))$ to the ideal gas $((z\to\infty))$ state. In the quasiharmonic model [150] the same approach is applied for taking into account the anharmonicity of the atomic thermal vibrations, the fitting parameter $T_{\mathrm{a}}$ in the free energy formula

$$F_{\mathrm{a}}(V,T) = \frac{3}{2}RT\ln\frac{\theta^2(V) + \sigma^{2/3}\frac{T}{T_{\mathrm{a}}}}{T^2},$$

having the meaning of the characteristic temperature when going from the lattice free energy (9.8) to that of an ideal gas

$$F_{\mathrm{a}}(V,T) = \frac{3}{2}RT\ln\frac{\sigma^{2/3}}{T}.$$

In order to ensure the ideal gas asymptotics, in a series of papers [381–383] the temperature dependence of the heat capacity was taken into account in the simplest analytic form. An example of such dependence is given by the expression

$$C_V = \frac{3}{2}RT\left(1 + \frac{1}{1 + 0.1\frac{T}{T_{\mathrm{m}}}}\right),\qquad(9.31)$$

suggested in Ref. [383]. This expression is obtained making use of the data on the simplest liquid models, namely, a system of soft spheres and the Coulomb plasma. The anharmonic effects were also taken into account in the framework of the cell model in the EOS [728] with a specific interparticle interaction potential

$$\varphi(R) = \frac{2ma^2\omega^2}{\pi^2} \tan^2\left(\frac{\pi r}{2a}\right),$$

which allows one to find the exact solution to the Schrödinger equation. This EOS has the correct asymptotics for a gas at $T \to \infty$ and $V \to \infty$, but provides only a schematic description of the high pressure dynamic experiments.

The expression for the free energy suggested in Ref. [576] which includes the term (9.30), taking into account the anharmonic effects of atomic vibrations, has been written in a more general form for an arbitrary density dependence of the Grüneisen coefficient of the thermodynamic potential of the liquid phase [992]. A generalized expression for the thermal component of the free energy of the liquid phase, which has the ideal gas asymptotics at high temperature as well as at low density, is presented in monograph [146]. The anharmonic effects can be also taken into account by other, purely approximation, methods [878].

Calculations show that melting weakly affects the kinematic characteristics of the shock adiabat, while the temperature varies more appreciably on melting (Fig. 9.1). In the earlier models of the EOS the melting effects are disregarded and the experimentally observed caloric $p(V, \varepsilon)$ properties of medium are described by these models within the accuracy of the shock-wave experiments. However, the discovered anomalies, specifically in the kinematic $D(u)$- diagram for lanthanides [161], can be

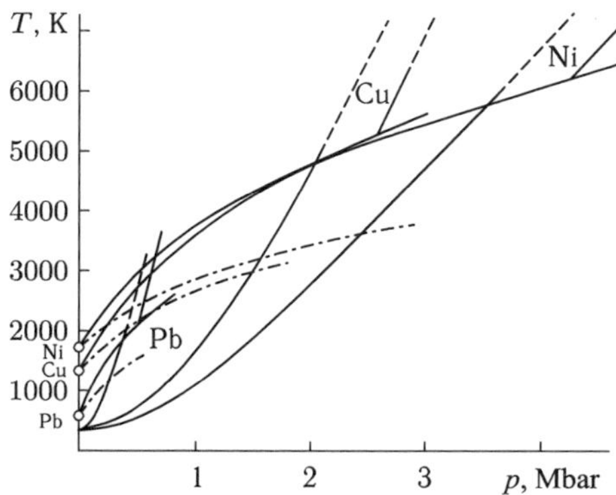


Fig. 9.1 Melting and temperature curves for shock compressed lead, copper, and nickel [993]. Dashed curves: the overheated solid phase, dashed-dotted curves: the Lindemann melting law.

explained just due to melting, more precisely, due to its anomalous behavior with a negative derivative $dT/dp$ on a part of the curve. The direct measurements of the temperature of shock-compressed metal [1058] indicate a strong dependence of this quantity on the melting effect. These facts, along with the necessity of the temperature calculation for solving the gas-dynamic problems which involve heat conductivity, need to be correctly taken into account in the EOS of the melting effect.

The melting contribution can be effectively included into the EOS making use of the Lindemann criterion $T_m/(\omega_D^2 V_m^{2/3}) = \text{constant}$. Then the melting entropy value is assumed to be the average for all elements and equals $\Delta S_m = 1.15R$ [381] and the internal energy of the liquid phase $\varepsilon_l$ differs from the energy of the solid phase $\varepsilon_s$ by the additional thermal component which corresponds to melting:

$$\varepsilon_l(V,T) = \varepsilon_s(V,T) + RT_m \left\{ \frac{\Delta S_m}{R} - \frac{3\alpha}{4} + 1.5\tau \left[ \frac{\ln(1+\alpha\tau)}{\alpha\tau} - 1 \right] \right\}.$$
$$(9.32)$$

Here, $\tau = T/T_m$, $T_m$ is the melting temperature at the density value corresponding to the given temperature and pressure, $\alpha = 0.1 = \text{constant}$. Equation (9.32) can be derived by integrating the heat capacity $C_V$ in the two-phase crystal–liquid region with the entropy jump $\Delta S_m$. Actually, the physical meaning of the model [381] is incorporated in the relatively simple formula for the heat capacity:

$$C_{VL} = C_{VS} - 1.5R \frac{\alpha\tau}{\alpha+\tau}.$$
$$(9.33)$$

In Russian literature [146, 992] melting is taken into account in a more consistent way with an explicit separation of the liquid and crystal phases [146]. In this case, the lattice term in the formula for the liquid phase potential is supplemented with a correcting function whose free parameters are chosen from the best fit to the experimental values of the density and entropy jumps on melting. This approach ensures the correct theoretical asymptotics (blurring of distinction between a crystal and a liquid) in the limit of high pressures and temperatures.

In the van der Waals EOS models, i.e., those which contain the thermodynamic potential terms with a repulsive part and attraction in the rarefaction region $V_0/V < 1$, the liquid–vapor two-phase area arises in a natural way. As an example of such a potential for a liquid we can consider the simplest two-term formula for the elastic energy [150]:

$$\varepsilon_{\mathrm{c}}(V) = \frac{B_{0\mathrm{c}}V_{0\mathrm{c}}}{m-n}\left(\frac{\sigma_{\mathrm{c}}^{m}}{m} - \frac{\sigma_{\mathrm{c}}^{n}}{n}\right) + \varepsilon_{\mathrm{sub}}, \tag{9.34}$$

where the sublimation energy $\varepsilon_{\mathrm{sub}} = \varepsilon_{\mathrm{c}}(V \to \infty)$. Models of the EOS of metals incorporate more sophisticated functional dependences of the potential [146, 529], including those in the form of series (9.25) with seven coefficients [1025] as well as the potentials of the system of soft or hard spheres with weak attraction [612, 1062]. The "compressible covolume" model with the explicit contribution of attraction which depends in a power-law form only on density [680] is an extension of the Van der Waals model.

## 9.3 Thermal excitation of electrons

In addition to the anharmonicity of the lattice vibrations at high temperatures, the contribution introduced by thermally excited electrons should be also taken into account in the EOS. In general, this contribution is unique for each substance and is determined by the scheme of filling the electronic energy bands. The simplest way of accounting for the electronic subsystem contribution is to use the nearly-free electron model [27, 1027, 1090] in which the number of electrons is defined by the element valency and considered to be constant. At not too high temperature the conduction electrons are degenerate and their properties are described by the ideal Fermi-gas model:

$$\varepsilon_{\mathrm{e}}(V,T) = \frac{\pi^2}{6}(kT)^2\nu(\varepsilon_{\mathrm{F}}), \tag{9.35}$$

$$p_{\mathrm{e}}(V,T) = \frac{\gamma_{\mathrm{e}}\varepsilon_{\mathrm{e}}}{V}, \tag{9.36}$$

where $\nu(\varepsilon_{\mathrm{F}})$ is the density of the electron states on the Fermi surface, $\gamma_{\mathrm{e}} = -d\ln\nu/d\ln V$, which has the value $\gamma_e = 2/3$ for the degenerate electron gas in the nearly-free electron model. The nearly-free electron model, due to its certain conditionality, does not describe the experimental data on the Fermi surface obtained in low-temperature experiments. The quantum-mechanical calculations of $\nu(\varepsilon_{\mathrm{F}})$ [1027] show that for transition metals, in which the conduction electrons are localized in a narrow d-band, the d-band undergoes broadening and transformation on compression. Obviously, in this case the density of the electron states strongly changes. This fact induces the appearance of anomalies in the coefficient $\gamma_{\mathrm{e}}$ which can even become negative.

The fine effects of the variation of $\nu(\varepsilon_{\mathrm{F}})$ and, moreover, of the Fermi surface topology are completely disregarded in the semi-empirical EOS aimed at the description of the thermodynamic properties of medium at the macroscopic level. Traditionally, the contribution introduced by the electron component is treated in an *a priori* simplified manner: the value of the electronic heat capacity coefficient $\beta_0 \sim \nu(\varepsilon_{\mathrm{F}})$ is extracted from the heat capacity measurements at low temperatures (where, according to the Debye model for the lattice, the relation $C_V \sim T^3$ is valid, whereas for the electron subsystem $C_V \sim T$) and the Grüneisen parameter of an electron is equal to $1/2$ in agreement with the Thomas–Fermi theory. Under this assumption, the contribution to the free energy has the simple form

$$F_e(V,T) = \frac{1}{2}\beta_0 T^2 \sigma^{-1/2},$$

which is widely used [27] in semi-empirical models for the description of the conduction electron contribution. This assumption is true at strong compression, but it appreciably worsens the description of the real situation at normal conditions. For metals at $T = 0$ the experimental values are $\gamma_{\mathrm{e}} = 1$ and the Thomas–Fermi asymptotic value $1/2$ is approached according to the calculations carried out for aluminum [356] and thorium [357] at the temperature of the order of $3 \cdot 10^4$ K. As a result of the further increase of the temperature up to $T \sim \varepsilon_{\mathrm{F}}$ (about $10^5$ K for metals) the degeneracy vanishes and, instead of expressions (9.35) and (9.36), one should use the exact formulae for the ideal electron gas [598].

In semi-empirical EOS the contribution of the electron subsystem is usually introduced making use of various approximations which contain, as a rule, several fitting parameters and have the correct asymptotics at low and high temperatures. As an example of such dependence one can consider the expression suggested in [576]

$$\varepsilon_{\mathrm{e}}(V,T) = \frac{b^2}{\beta}\ln\cosh\left(\frac{\beta T}{b}\right),$$

where $\beta = \beta_0 \sigma^{-\gamma_e}$. At low temperature $((T \ll b/\beta))$ this expression leads to the formula for a degenerate electron gas and at finite temperature, owing to the fitting parameter $b$, agrees with the results of the Thomas–Fermi model calculation. The approximation which has a similar meaning

$$\varepsilon_{\mathrm{e}}(V,T) = \frac{\beta T^2}{2}\frac{T_{\mathrm{F}}}{T_{\mathrm{F}} + T\sigma^{-\gamma_{\mathrm{e}}}},$$

is used for the description of the electron component of the free energy in the condensed and gaseous phase [878]. Modern models of the semi-empirical EOS use also more refined methods to account for the electron subsystem. In the "compressed covolume" model [680] particles are divided into charged and neutral ones, the equilibrium composition being calculated proceeding from the condition of the equality of the chemical potentials of the components. For the case of metals, the attraction effects being taken into account, at fixed degree of ionization $\alpha = 4$ one can see a certain "softening" of the shock adiabats of porous lead and nickel (the adiabat of a crystalline metal being virtually unchanged) [680].

The iron semi-empirical EOS [517] provides for the description of the electron subsystem smoothly sewed results of the calculations made in the framework of the self-consistent cell model [615] and plasma model of the ionization equilibrium [523]. The quotidian equation of state (QEOS) model [712] and its further development with a large number of fitting parameters in the liquid phase [1059] describe the electron subsystem making use of the calculation results obtained in the framework of the Thomas–Fermi automodel theory. Realizing that the quasiclassical description can be applied in a limited range, the authors of Ref. [712] make a phenomenological correction of the free energy (and, correspondingly, of its derivatives — pressure, entropy, etc.) by means of the fitting function $f(T)$ such that $F_e^* = f(T)F_e$. In the "average-ion" plasma model [81] the contribution introduced by the electron component to the pressure is assumed to be equal to the pressure of the ideal Fermi gas of free electrons whose concentration is determined from the ionization equilibrium equation extended in order to describe the ionization of compressed cold matter.

Building the EOS of dielectrics, for instance, of ion crystals [577] or of inert gases [856], one must take into account thermal excitation of electrons into the conduction band. In this case the number of free electrons depends on the density and temperature $N_e \sim T^{3/2} \exp(-\Delta E/2kT)$, where $\Delta E(V)$ is the energy gap between the valence and conduction band which is either given by a model [531, 577] or is calculated by means of direct quantum-mechanical methods ([673–675] (see Chapter 3). Besides the term responsible for the free electron contribution, in the formula for the free energy an expression should be added which takes into account thermal excitation of electrons [526, 530, 531, 577, 856] (Fig. 9.2). Since the gap width decreases on compression, the additional term introduces a negative contribution whose value strongly depends on the speed of the gap collapse [862].

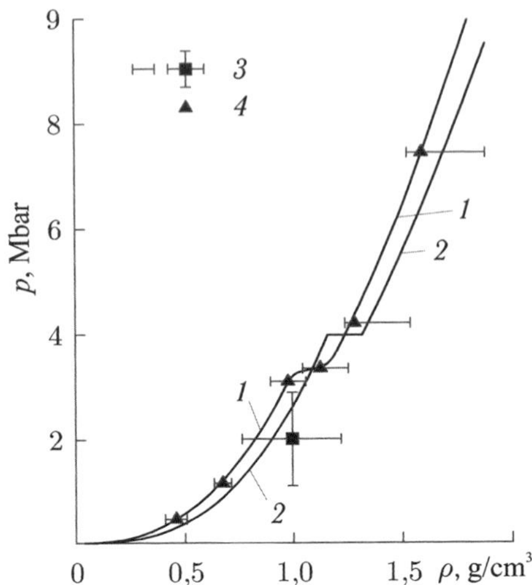

Fig. 9.2   Zero isotherms *1* and isentropes *2* of molecular and metallic hydrogen [373]. *3* experimental data, *4* gas-dynamic calculation.

## 9.4   Evaporation. Wide-range EOS

Description of the processes occurring under the conditions of high-temperature expanding of matter led to the necessity to construct the wide-range semi-empirical models which provide a consistent account of the evaporation effects. One of the methods [119, 1060] used to describe the transition of a system from a condensed state to a gaseous state is to sew the quasiharmonic [381] model with the modified van der Waals liquid state model [445, 446] which is a purely fitting procedure and does not ensure the correct quantitative agreement. At the same time, the available experimental data on isentropic expanding of shock-compressed metals [29, 34] give necessary information for the construction of the semi-empirical EOS which correctly describe the liquid–vapor transition (plasma).

Straight-through gas-dynamic calculations carried out in a broad range of physical conditions requested the creation of the wide-range EOS providing an adequate thermodynamically complete description of matter properties in a large domain of the phase diagram and taking consistently into account melting, evaporation, dissociation, ionization effects, etc. As a result, the extension of the applicability region of the semi-empirical EOS with explicit separation of the physical asymptotics, led to the appreciable complication of the terms in formula (9.1) for the thermodynamic potential and to the increase of the number of the free parameters associated with the need to describe the corresponding limiting cases. On the basis of various EOS construction principles, wide-range EOS were developed for many substances, namely, for metals [34, 148, 151, 303, 310, 527, 634, 877], water [111, 506, 580, 737, 738], sodium chloride [112, 506].

Figure 9.3 shows an example of the consistent description, in the framework of the unique EOS [34], of different-type experimental data for Pb: shock adiabats, sound velocity at shock loading, temperature in the unloading wave. It can be seen as well from Fig. 9.3 that the calculation [34] is adequate to the experimental data. Figure 9.4 presents the calculation and experimental results on the thermodynamic properties of aluminum in the near-critical region and in the liquid phase. The semi-empirical EOS of the

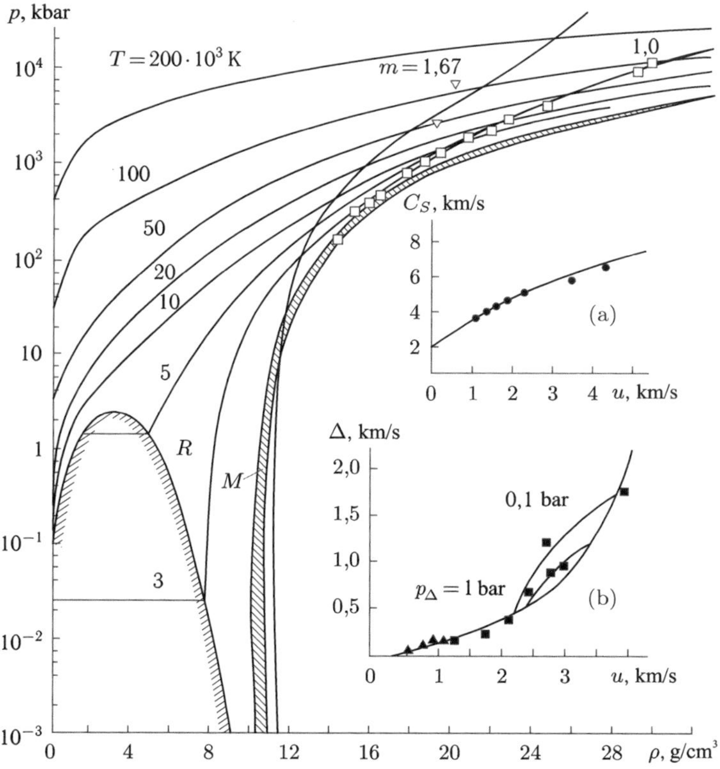


Fig. 9.3  Phase diagram of lead [34]. Symbols are experimental points. $M$ melting domain, $R$ boiling curve, $m$ shock adiabats, $T$ isotherms. Inserts: (a) sound velocity on shock adiabate, (b) $\Delta = W - 2u$ in isentropic unloading wave.

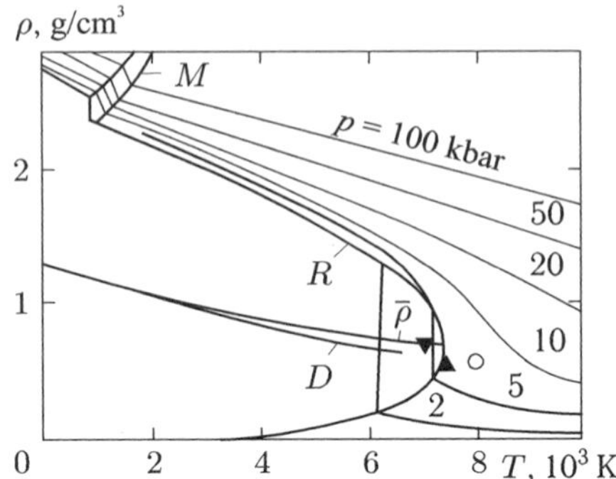



Fig. 9.4   EOS of aluminum in near-critical region [34]. $M$ melting domain, $R$ boiling curve, $D$ straight-line diameter, $\bar{\rho}$ half-sum of the densities of the liquid and gaseous phase. Symbols: critical point parameters estimated basing on the thermodynamic similarity principles [753].

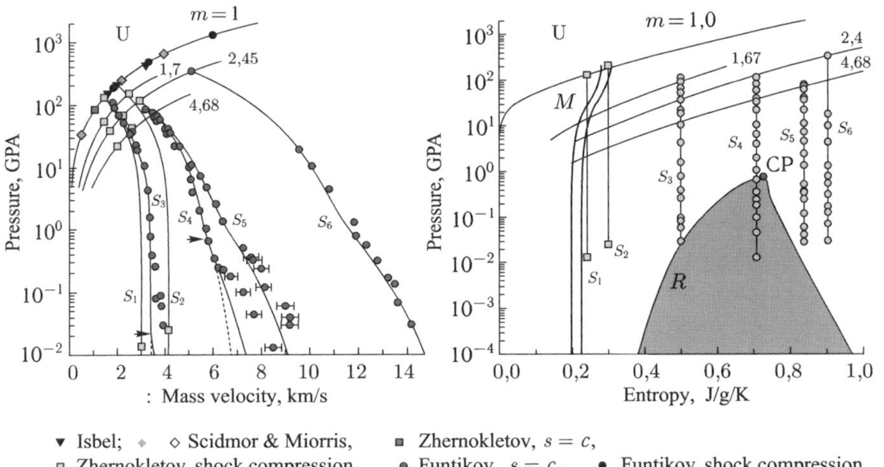



▼ Isbel;   ◇ Scidmor & Miorris,    ▪ Zhernokletov, $s = c$,
▫ Zhernokletov, shock compression,    ● Funtikov, $s = c$,    ● Funtikov, shock compression.

Fig. 9.5   Adiabatic expansion of shock-compressed uranium plasma [292].

adiabatically expanded uranium plasma are shown in Fig. 9.5. The broad spectrum of the statistical and dynamical experimental results listed in Table 9.1 was used as the initial data.

At high pressures and/or temperatures these EOS have as asymptotics the Thomas–Fermi model of compressed matter (Chapter 6) and the Debye–Huckel plasma model (Chapter 4).

Table 9.1    Initial data for building semi-empirical EOS.

---

Static experiments
— Isotherms $T = 293\,\mathrm{K}$
— Pressure on melting curve
— Jumps of volumes and enthalpy on melting
— Boiling temperature at $p = 1$ atm
— Binding energy
— Electron heat capacity
Dynamic experiments
— Shock adiabats of solid samples
— Shock adiabats of porous samples
— Release adiabats
— Electric explosion data
— Temperature of shock compression
— Sound velocity in shock-compressed state
Theoretical asymptotics
— Thomas–Fermi theory
— Ionization model of plasma
— Electron spectrum at low temperatures

---

An example of the practical application of the wide-range semi-empirical EOS is shown in Fig. 9.6, where the entropy diagram of states (built using a semi-empirical EOS [146]) is presented which makes it possible to determine the physical conditions arising when high-power relativistic heavy-ion beams interact with a substance (project FAIR [300]).

The presented curves (Figs. 9.3 and 9.4) show that the calculated characteristics are in precise (within the experimental errors) agreement with the experimental results obtained in a broad range of the variation of thermodynamic parameters — from the normal conditions to the pressures of hundreds of megabars. Along with the adequate description of the experimental data, including those obtained under the conditions of melting and evaporation, the aluminum wide-range EOS [151] take as well into account the ionization effects and have the correct asymptotics in the limit of high temperatures and large volumes. It should be emphasized that nowadays semi-empirical wide-range EOS are the most acceptable pragmatic technique for the consistent and precise description of thermodynamic properties of substances in the broad domain of the phase diagram, which generalizes experimental results and model calculations.

The concrete methods and results of developing the wide-range equations of state for some construction materials are contained in Refs. [303, 310, 526, 527, 529–532, 612, 1002].

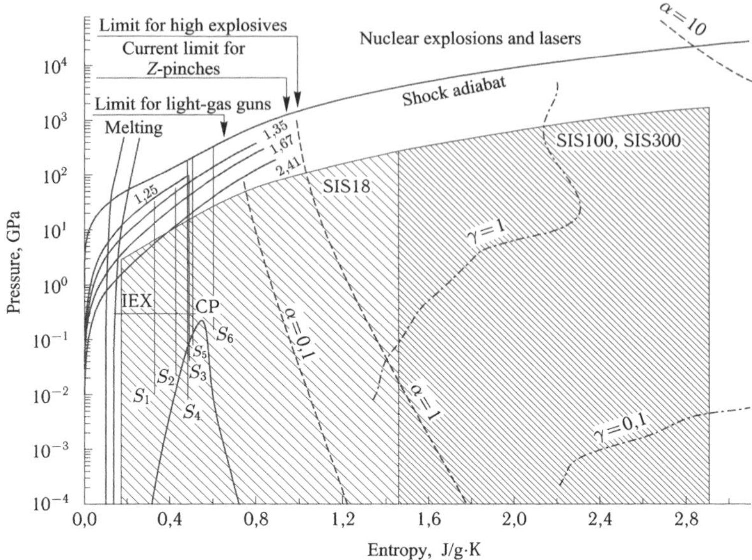


Fig. 9.6    Capabilities of heavy-ion accelerators SIS18, SIS100, SIS300 for the generation of high energy densities in lead [300, 437, 962, 963].

Some useful information on the thermodynamics of elements and compounds along with the parameters of the semi-empirical EOS can be found in Ref. [634] and in the electronic database www.ihed.ras.ru/rusbank.

## 9.5   The Fermi–Zeldovich problem. Construction of the thermodynamically complete EOS using the results of dynamic measurements

A peculiarity of the dynamic experiments, which are the basis for the development of the EOS semi-empirical models, is that the experimental data make it possible to build the EOS of a substance only in the thermodynamically incomplete form $E = E(p, V)$, because the internal energy is not a complete thermodynamic potential with respect to the variables $p$ and $V$. Therefore building of the closed thermodynamics needs additional information on the temperature $T = T(p, V)$ [1081] or entropy $S = S(p, V)$ which is of the utmost importance for the development of the adequate semi-empirical EOS.

The point is that the dynamic methods of investigation [1081] are based on recording the mechanical parameters of hydrodynamic motion which do not provide information about thermal or entropy characteristics of shock-compressed matter.

Upon propagation of a stationary shock-wave discontinuity through matter, the mass, momentum, and energy conservation laws are satisfied on its front [1081]. The conservation laws relate the kinematic parameters (the shock wave velocity $D$ and the matter mass velocity $u$ behind the wave front) with the thermodynamic quantities — specific internal energy $E$, pressure $p$, and specific volume $V$:

$$\frac{V}{V_0} = \frac{(D-u)}{D}, \quad p = p_0 + \frac{Du}{V_0},$$

$$E = E_0 + \left(\frac{1}{2}\right)(p + p_0)(V_0 - V), \qquad (9.37)$$

Here the subscript 0 marks the parameters of matter at rest before the shock wave front. These equations make it possible to determine hydrodynamic and thermodynamic properties of matter by recording any two out of the five parameters $E$, $p$, $V$, $D$, $u$, which characterize the shock-wave discontinuity. The basis methods ensure the easiest and most accurate way to measure the shock wave velocity $D$. The choice of the second measured parameter depends on particular conditions of the experiment.

Analysis of the errors of relationships (9.37) indicates [315, 731] that in case of strongly compressed ("gaseous") media it is reasonable to perform the recording of the concentration $p = V^{-1}$ of shock-compressed matter. At present, the method developed for these measurements is based on the detection of the soft X-ray absorption by cesium, argon, and air plasma. At lower compressibility of the system (condensed matter) the acceptable accuracies are achieved [27, 40] by recording the mass velocity of motion $u$. The described techniques were applied to study the degenerate plasma states of metals [287, 294, 295, 305] and the states of dense Boltzmann plasma of argon and xenon.

In the experiments which involve the recording of the isentropic expansion curves for shock-compressed matter [287,305], the states in the centered unloading wave are described by the Riemann integrals [1081]:

$$V = V_{\mathrm{H}} + \int_{p}^{p_{\mathrm{H}}} \left(\frac{du}{dp}\right)^2 dp, \quad E = E_{\mathrm{H}} - \int_{p}^{p_{\mathrm{H}}} p\left(\frac{du}{dp}\right)^2 dp, \qquad (9.38)$$

which are calculated along the measured isentrope $p_s = p_s(u)$. By recording at different initial conditions and shock-wave intensities one can determine the caloric EOS $E = E(p, V)$ in the $p$–$V$-diagram domain overlapped by the Hugoniot and (or) Poisson adiabats. In the experiments on plasma dynamic stimulation performed by now the shock-wave intensity was varied by varying the power of the excitation sources, such as the pressure of the propelling gas, types of explosives, projectile launchers, and targets. Furthermore, different ways were used to vary the parameters of the initial states, such as the variation of initial temperatures and pressures (plasma of inert gases, cesium, liquid), as well as application of fine- dispersed targets intended to increase the irreversibility effects [287, 305, 1081].

Thus, the dynamic methods of diagnostics, being based on the general conservation laws, make it possible to reduce the problem of determination of the caloric EOS $E = E(p, V)$ to the measurement of the kinematic parameters of motion of shock waves and contact surfaces, i.e., to the registration of distances and times which can be done with high accuracy. As we have already mentioned, the internal energy, however, is not the thermodynamic potential with respect to the variables $p$, $V$ and the construction of closed thermodynamics requires an additional dependence of the temperature on the system parameters $T(p, V)$.

In optically transparent and isotropic media (gases, ion crystals, etc.), the temperature is measured together with other shock-compression parameters. Condensed media and, primarily, metals are, as a rule, non-transparent. Therefore, the light radiation of shock-compressed medium is inaccessible to detection.

According to the idea put forward by E. Fermi [256] and Ya. B. Zeldovich [1077], a thermodynamically complete EOS can be constructed directly from the results of dynamic measurements without introducing *a priori* considerations on the properties and nature of the substance under investigation [287, 305], but basing on the first law of thermodynamics and the experimentally known dependence $E = E(p, V)$ (or more convenient function $\gamma(p, V) = pV/E(p, V)$). This leads to the linear inhomogeneous equation for $T(p, V)$:

$$\left[ p + \left( \frac{\partial E}{\partial V} \right)_p \right] \frac{\partial T}{\partial p} - \left( \frac{\partial E}{\partial p} \right)_V \frac{\partial T}{\partial V} = T, \qquad (9.39)$$

whose solution can be obtained by the method of characteristics:

$$\frac{\partial p}{\partial V} = \frac{p + \left( \frac{\partial E}{\partial V} \right)_p}{\left( \frac{\partial E}{\partial p} \right)_V}; \quad \frac{\partial T}{\partial V} = - \frac{T}{\left( \frac{\partial E}{\partial p} \right)_V}, \qquad (9.40)$$

or

$$E = E_0 \exp\left\{ -\int_{V_0}^{V} \gamma(V, E) d\ln V \right\},$$

$$T = T_0 \frac{pV}{p_0 V_0} \exp\left\{ -\int_{V_0}^{V} \left( \frac{\partial \ln \gamma(E, V)}{\partial \ln V} \right) d\ln V \right\}. \qquad (9.41)$$

Equations (9.39)–(9.41) are complemented by the initially conditions, namely, the temperature is given in the low-density region where its reliable theoretical calculation is possible (cesium plasma) or the temperature is known from a static experiment [287, 305].

The relationships $E(p, V)$ required in the calculation of the right-hand parts of Equations (9.40), (9.41) were determined from the experimental data presented as power polynomials

$$E(p, V) + \sum_i \sum_j a_{ij} p^i N^j$$

and for $\gamma(E, V)$ in the form of fractional rational functions.

The accuracy of the obtained solution as a function of the experimental errors and inaccuracy of the initial data was determined using the Monte Carlo method by numerical simulation of the probability structure of the measurement procedure [287, 305].

This method is of general nature because it is free of restricting assumptions on the properties, nature, and phase composition of the medium under

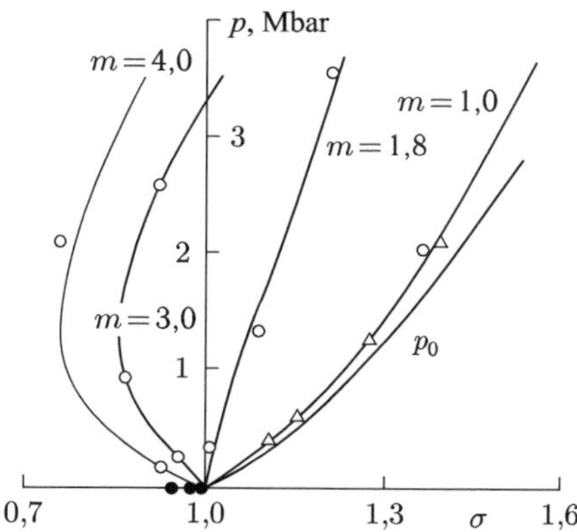


Fig. 9.7   Shock adiabats $m$ and the cold curve $p_0$ for tungsten [289].

investigation. This is so because the method makes use of the first principles of the continuum mechanics, namely, conservation laws (9.37), (9.38) and the fundamental thermodynamic identity (9.39). This thermodynamic universality enabled one to construct, using a unified procedure, the EOS for the wide spectrum of condensed media and to use it for the description of phase transformations [287, 305]. The method proved to be especially effective in the study of nonideal cesium plasma on the basis of the experiments on the shock and adiabatic compression of saturated vapors of the main-group and transition metals, ion crystals, silicon oxide. Figure 9.7 illustrates the calculation results obtained for tungsten making use of this method.

Chapter 10

# Relativistic Plasma.
# Wide–Range Description

In this chapter, following Ref. [414] we will consider the thermodynamic description of the electromagnetic (Coulomb) plasma arising in various astrophysical objects — in white, brown, and yellow dwarfs, red giants, and neutron and quark–gluon stars.

Evolving under the action of two "driving" forces — gravitational compression and thermonuclear energy release — matter in the stellar objects goes through an extremely wide spectrum of thermodynamic states ranging from the rarefied intergalactic gas, $n_e \approx 10^{-31}\,\mathrm{cm}^{-3}$, $T \approx 10^3\,\mathrm{K}$ to supercompressed neutron matter and quark–gluon plasma in the cores of neutron and "strange" stars $\rho \approx 10^{14}$–$10^{17}\,\mathrm{g/cm}^3$, $T \approx 10^8$–$10^{11}\,\mathrm{K}$. While at the lower values of these parameters only the applicability of the thermodynamic equilibrium conditions is discussed, in the upper range of supercompression one must provide the inclusive description of the relativistic effects, degeneration of the electrons and ions, thermal and pressure-induced ionization, thermodynamic effects of radiation, as well as of complicated transformations of the nuclear component of matter.

The construction of adequate thermodynamic models (for the description of this spectrum of states) turned out to be a serious challenge for the contemporary physics because it required the use of the most recent information provided by nuclear and high-energy physics [294, 295]. Differently from the previous chapter which was to a certain extent based on experiments, in this case our experimental capabilities are naturally more than restrained and are reduced to the observational data on distant astrophysical objects located at many parsecs from us and high-energy heavy-ion collisions in terrestrial accelerators. Unfortunately, the experiment cannot provide any reliable support in this case.

In the further presentation we will follow the wonderful book [414] where the main attention is given to the equations of state of neutron stars being the final stage of the evolution of the stars whose mass is $M \gtrsim 8M_\odot$, which are a result of a powerful explosion of the supernovae after they exhaust thermonuclear fuel in their cores. These are the most interesting and "extreme" astrophysical objects possessing record — high magnitudes of the density, pressure, temperature, and magnetic induction which reach, correspondingly, $\rho \approx 10^{15}\,\mathrm{g/cm^3}$, $p \approx 10^{34}\,\mathrm{dyn/cm^2}$, $T \approx 10^{11}\,\mathrm{K}$, $B \approx 10^{15}\,\mathrm{Gs}$. In what follows we will consider these values as limiting [414].

Let us consider the electromagnetic plasma with Coulomb interparticle interaction, the particles charges being $Z$ and mass numbers $A_j$. In such a plasma the particle density is $n_N = \sum n_j$. The electric neutrality condition has the form $n_\mathrm{e} = \langle Z \rangle n_j$, where the brackets $\langle \ldots \rangle$ mean averaging over the components $j$ with statistic weights proportional to $n_j$:

$$\langle f \rangle = \frac{1}{n_N} \sum_j n_j f_j. \tag{10.1}$$

Taking into account nuclear transformations and possible nucleon transitions from the bound states in nuclei to the free states, we can express the baryon concentration as $n_\mathrm{b} \approx A' n_N = [\langle A \rangle + A''] n_N$, where $\langle A \rangle$ is the mean concentration of baryons bound in nuclei, $A''$ is the concentration of free neutrons per nucleus. We have $A'' n_N = n_n(1 - \omega)$, where $n_n$ is the concentration of free neutrons and $\omega$ is the fraction of volume occupied by the atomic nuclei.

The mass density in the relativistic case can be expressed in terms of the obtained energy density $\rho = \varepsilon/c^2$ which in the classical case has the conventional form $\rho \approx m_\mathrm{u} n_\mathrm{b}$.

## 10.1 Description of electrons

The relativistic free electrons are described by the Salpeter parameter [874]:

$$x_\mathrm{r} = \frac{p_\mathrm{F}}{m_\mathrm{e}c} \sim 1.00884 \left( \frac{\rho_6 \langle Z \rangle}{A'} \right)^{1/3}, \tag{10.2}$$

where $\rho_6 = \rho/10^6\,\mathrm{g/cm^3}$,

$$P_\mathrm{F} = \hbar k_\mathrm{F} = \hbar(3\pi^2 n_\mathrm{e})^{1/3}, \tag{10.3}$$

$$\epsilon_\mathrm{F} = c\sqrt{(m_\mathrm{e}c)^2 + p_\mathrm{F}^2}. \tag{10.4}$$

The rest energy $m_e c^2$ is usually included in the electron energy. The characteristic Fermi temperature is

$$T_F = T_r(\gamma_r - 1), \qquad (10.5)$$

where

$$T_r = \frac{m_e c^2}{k_B} \approx 5.930 \cdot 10^9 \text{ K}, \qquad (10.6)$$

is the characteristic relativistic temperature and $\gamma_r = \sqrt{1 + x_r^2}$. The electron gas is nonrelativistic at $T \ll T_r$ and $x_r \ll 1$ and it is ultrarelativistic at $x_r \gg 1$ or $T \gg T_r$. It is nondegenerate at $T \gg T_F$ and strongly degenerate at $T \ll T_F$.

Figure 10.1 presents as an example the characteristic parameters of plasma in the pulsar outer envelope [414]. The relativistic and plasma degeneracy domains are shown.

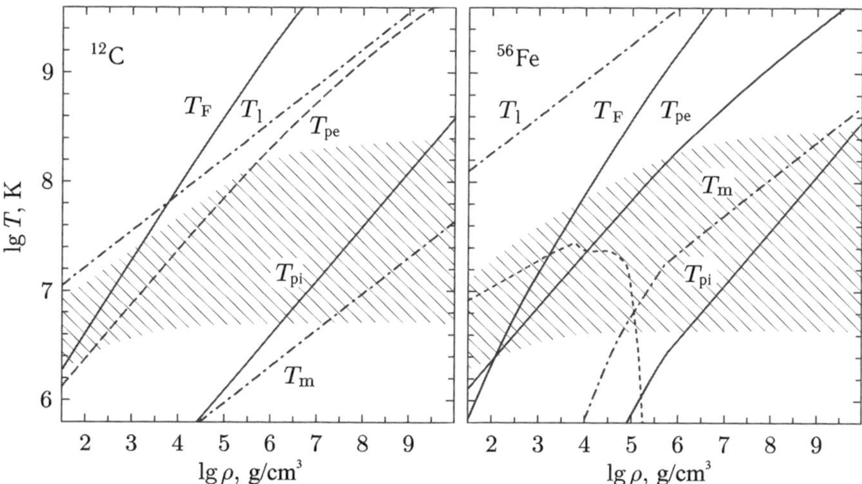



Fig. 10.1   Density-temperature diagram for the outer envelope composed of carbon (left) and iron (right). The Figure presents the electron Fermi temperature ($T_F$), the electron and ion plasma temperature ($T_{pe}$ and $T_{pi}$), the temperature of the gradual gas–liquid transition ($T_l$), and the temperature of the sharp liquid–solid phase transition ($T_m$). The shaded areas are the regions of typical temperatures in the outer envelopes of the middle-aged cooling neutron stars (of the order of $10^4$–$10^6$ years old). The lower left domain on the right panel separated by the dashed line is characterized by strong electron response or bound state formation (it corresponds to the effective charge $Z_{ef} < 20$). The figure is reproduced from Ref. [414] with kind permission of the authors.

The frequently used quantities are the electron Fermi velocity $v_\mathrm{F} = \frac{\partial \epsilon_\mathrm{F}}{\partial p_\mathrm{F}}$ and the corresponding dimensionless parameter

$$\beta_\mathrm{r} = \frac{v_\mathrm{F}}{c} = \frac{x_\mathrm{r}}{\gamma_\mathrm{r}}, \tag{10.7}$$

as well as the density parameter $r_\mathrm{s}$:

$$r_\mathrm{s} = \frac{a_\mathrm{e}}{a_0} = \left(\frac{9\pi}{4}\right)^{1/3} \frac{\alpha_\mathrm{f}}{x_\mathrm{r}} = \frac{0.01400}{x_\mathrm{r}}. \tag{10.8}$$

Here $a_\mathrm{e} = \left[\frac{3}{(4\pi n_\mathrm{e})}\right]^{1/3}$, $a_0 = \frac{\hbar^2}{m_\mathrm{e} e^2}$ is the Bohr radius, $\alpha_\mathrm{f} = \frac{e^2}{\hbar c} = \frac{1}{137.036}$ is the fine-structure constant.

The Coulomb interaction of nondegenerate electrons is characterized by the nonideality parameter

$$\Gamma_\mathrm{e} = \frac{e^2}{a_\mathrm{e} kT} \approx \frac{22.75}{T_6} \left(\rho_6 \frac{\langle Z \rangle}{A'}\right)^{1/3}. \tag{10.9}$$

For degenerate electrons the degeneracy parameter $\theta = T/T_\mathrm{F}$ is often used. In the nonrelativistic $x_\mathrm{r} \ll 1$ and ultrarelativistic $x_\mathrm{r} \gg 1$ case we have $\theta = 0.543 r_\mathrm{s}/\Gamma_\mathrm{e}$ and $\theta \approx (263\Gamma_\mathrm{e})^{-1}$, correspondingly.

As we have already noted in Chapter , at high densities $x_\mathrm{r} > 1$ the electrons become relativistic. In this regime the energy $\epsilon_\mathrm{F}$ is quite high, so that the electron gas is almost ideal.

## 10.2 Ions

Ions have large masses, therefore the boundary of their relativism shifts to the range of ultrahigh densities. So at most values of parameters the properties of ions are determined by Coulomb and nuclear interactions:

$$\Gamma_j = \Gamma_\mathrm{e} Z_j^{5/3} = \frac{(Z_j e)^2}{(a_\mathrm{i}^{(j)} k_\mathrm{B} T)}, \tag{10.10}$$

where

$$a_\mathrm{i}^{(j)} = a_\mathrm{e} Z_j^{1/3}, \tag{10.11}$$

is the ion-sphere radius. For a multicomponent plasma, $\Gamma = \Gamma_\mathrm{e} \langle Z \rangle^{2/3}$.

At sufficiently high temperatures ions are described by the model of a classical Boltzmann gas. When the temperature decreases, the nonideality effects become stronger and the plasma gradually transforms into a strongly nonideal Coulomb liquid which in its turn transforms via a phase transition into a Coulomb crystal [1012].

The gaseous regime corresponds to $T \gg T_\mathrm{i}$, where

$$T_\mathrm{i} = \frac{Z^2 e^2}{a_\mathrm{i} k_\mathrm{B}} = 2.27 \cdot 10^7 \times Z^2 \left(\frac{\rho_6}{A'}\right)^{1/3} \ \mathrm{K}.$$

For a plasma of light chemical elements $T_\mathrm{i} < T_\mathrm{F}$, whereas for heavy elements $T_\mathrm{i} > T_\mathrm{F}$. In the ion-sphere model for strongly compressed Coulomb systems $(T < T_\mathrm{e})$ plasma breaks up in ion spheres whose radius equals $a_\mathrm{i}$. Each sphere has an ion in its center surrounded by a quasi-uniform electron background [901]. The sphere radius is chosen employing the electric neutrality condition and the total charge of the sphere is zero.

In the low-density regime, a characteristic length of the Coulomb screening is the Debye length

$$r_\mathrm{D} = q_\mathrm{D}^{-1} = \left[\frac{4\pi}{k_\mathrm{B} T} \sum_j n_j (Z_j e)^2\right]^{-1/2}. \tag{10.12}$$

For a one-component plasma (OCP) $r_\mathrm{D} = a_\mathrm{i}/\sqrt{3\Gamma}$. For a gaseous plasma $T \gg T_\mathrm{e}$ the screening wave number

$$q_\mathrm{D}' = (q_\mathrm{D}^2 + k_\mathrm{TF}^2)^{1/2}, \tag{10.13}$$

where $k_\mathrm{TF} = r_\mathrm{e}^{-1} = \left(4\pi e^2 \frac{\partial n_\mathrm{e}}{\partial \mu_\mathrm{e}}\right)^{1/2}$.

If the electrons are nondegenerate and nonrelativistic $(T \gg T_\mathrm{F}, x_\mathrm{r} \ll 1)$, then $q_\mathrm{D}'$ equals $q_\mathrm{D}$ from Equation (10.12), with the summation extended over all ion and electron types.

It follows from [3] that ions crystallize at the temperature

$$T_\mathrm{m} = \frac{Z^2 e^2}{a_\mathrm{i} k_\mathrm{B} \Gamma_\mathrm{m}} \approx 1.3 \cdot 10^5 \times Z^2 \left(\frac{\rho_6}{A'}\right)^{1/3} \ \mathrm{K}, \tag{10.14}$$

where $T_\mathrm{m}$ is the melting temperature.

In the low-temperature regime the behavior of electrons and ions is strongly different. Since the electrons are degenerate, they behave like an ideal gas, whereas ions stay being a strongly interacting Coulomb system. In this limit, $T \ll T_\mathrm{pi}$ the quantum effects of the ion motion turn out to be important. Here,

$$T_{\mathrm{pi}} \equiv \frac{\hbar\omega_{\mathrm{pi}}}{k_{\mathrm{B}}} \approx 7,832 \cdot 10^6 \left( \frac{\rho_6}{A'} \left\langle \frac{Z^2}{A} \right\rangle \right)^{1/2} \mathrm{K}, \tag{10.15}$$

is the ion temperature related to the plasma frequency,

$$\omega_{\mathrm{pi}} = (4\pi e^2 n_N \langle Z^2/m_{\mathrm{i}} \rangle)^{1/2}. \tag{10.16}$$

The corresponding dimensionless parameter is

$$t_{\mathrm{p}} = \frac{T}{T_{\mathrm{pi}}} \approx \frac{2,92 \langle Z^{1/2} \rangle \left\langle \frac{Z^2}{A} \right\rangle^{-1/2}}{(\Gamma_{\mathrm{e}} \sqrt{x_{\mathrm{r}}})}. \tag{10.17}$$

For the OCP $t_{\mathrm{p}} = \sqrt{R_{\mathrm{S}}/3}/\Gamma$, where

$$R_{\mathrm{S}} = \frac{a_{\mathrm{i}} m_{\mathrm{i}} (Ze)^2}{\hbar^2} = r_{\mathrm{s}} \left( \frac{m_{\mathrm{i}}}{m_{\mathrm{e}}} \right) Z^{7/3}, \tag{10.18}$$

is an ion parameter similar to the electron parameter $r_{\mathrm{s}}$ (10.8).

At $T_{\mathrm{pi}} \gg T_{\mathrm{m}}$ ion vibrations reduce the melting temperature (10.14). This decrease continues with the increase of the density, and the melting temperature drops down to zero at a certain density (pressure) value (see Fig. 10.1).

## 10.3   Fully ionized plasma

Fully ionized dense plasma does not contain neutral particles and satisfies the requirement that the mean spacing between ions $a_{\mathrm{i}}$ is small compared to the radius of the ion core $r_{\mathrm{a}} \sim \frac{a_0}{Z^{1/3}}$ (the Thomas–Fermi radius) [414]. This corresponds to the condition $\rho \gg \rho_{\mathrm{eip}} = \left( \frac{m_{\mathrm{u}}}{a_0^3} \right) AZ \approx 11 AZ \, \mathrm{g/cm^{-3}}$ [786] when we can use an electron–ion plasma (eip) model which is quite popular in plasma physics. In this model the electron–ion plasma is composed of point-like ions with the charge $Z_{nucl}$ immersed in a structureless charge-compensating background formed by electrons which provide electric neutrality (and, hence, stability) of such a plasma model.

The free energy in this model has the form

$$F = F_{\mathrm{id}} + F_{\mathrm{id}}^{e} + F_{\mathrm{ex}} + F_{\mathrm{ii}} + F_{\mathrm{ei}}. \tag{10.19}$$

Besides the obvious terms, $F_{\mathrm{ei}}$ takes into account the interaction of ions with the electron background (the background polarization). $F_{\mathrm{ex}}$ — stands for the correlations in the electron system. In case when the spacing between electrons is considerably smaller than the $K$-shell radius of an atom, the charge neutralizing electron background may be treated as "rigid" and the electron-ion correlations may be neglected. This leads to the condition $r_{\rm s} \lesssim 1/2Z$ or $\rho > \rho_{\rm rigid} \approx 22Z^2 A\,{\rm g/cm}^3$. The same criterion results from the condition for the electron–ion Coulomb potential $\sim Ze^2/a_{\rm i}$ to be small compared to the electron Fermi energy $\sim \hbar/m_a a_{\rm e}^2$. These conditions correspond to the model of plasma of ions immersed in the charge neutralizing structureless background formed by an ideal electron gas [414].

The thermodynamics of an ideal electron gas is described by the Thomas–Fermi distribution

$$f^{(0)}(\epsilon - \mu_{\rm e}, T) \equiv \frac{1}{\exp[(\epsilon - \mu_{\rm e})k_{\rm B}T] + 1}, \tag{10.20}$$

where

$$\epsilon = \sqrt{m_{\rm e}^2 c^4 + P^2 c^2}, \tag{10.21}$$

is the energy of electrons and $P$ is their momentum. It is also convenient to introduce the quantity:

$$\chi = \frac{(\mu_{\rm e} - m_{\rm e}c^2)}{k_{\rm B}T}. \tag{10.22}$$

The free energy of the electron gas is:

$$F_{\rm id}^{(e)} = \mu_{\rm e}N_{\rm e} - p_{\rm id}^{(e)}V, \tag{10.23}$$

where $p_{\rm id}^{(e)}$ is the ideal-gas pressure. The pressure and the concentration are functions of $\mu_{\rm e}$ and $T$:

$$p_{\rm id}^{(e)} = 2k_{\rm B}T \int \ln\left[1 + \exp\left(\frac{\mu_{\rm e} - \epsilon}{k_{\rm B}T}\right)\right] \frac{d^3p}{(2\pi\hbar)^3}$$

$$= \frac{8}{3\sqrt{\pi}} \frac{k_{\rm B}T}{\lambda_{\rm e}^3} \left[I_{3/2}(\chi, t_{\rm r}) + \frac{t_{\rm r}}{2}I_{5/2}(\chi, t_{\rm r})\right], \tag{10.24}$$

$$n_{\rm e} = 2 \int f^{(0)}(\epsilon - \mu_{\rm e}, T) \frac{d^3p}{(2\pi\hbar)^3}$$

$$= \frac{4}{\sqrt{\pi}\lambda_{\rm e}^3}[I_{1/2}(\chi, t_{\rm r}) + t_{\rm r}I_{3/2}(\chi, t_{\rm r})], \tag{10.25}$$

where $\lambda_{\rm e}$ is the electron de Broglie wave length and

$$I_\nu(\chi, T) = \int_0^\infty \frac{x^\nu(1 + \tau x/2)^{1/2}}{\exp(x - \chi) + 1}dx, \tag{10.26}$$

is a Fermi–Dirac integral. The internal energy has the form

$$\frac{U_{\mathrm{id}}^{(e)}}{V} = \frac{4}{\sqrt{\pi}}\frac{k_{\mathrm{B}}T}{\chi_{\mathrm{e}}^3}\left[I_{3/2}(\chi,t_{\mathrm{r}}) + t_{\mathrm{r}}I_{5/2}(\chi,T_{\mathrm{r}})\right]. \qquad (10.27)$$

In the limit $t_{\mathrm{r}} \to 0$, integrals (10.26) turn into nonrelativistic expressions for which Pade approximations are presented in Ref. [51].

The detailed formulae for the Fermi–Dirac integrals and the corresponding thermodynamic expressions for the whole range of the degeneration and relativism parameters are contained in Ref. [414]. Fig. 10.2 presents as an example the plasma pressure and the corresponding relativistic asymptotes [414].

We see (Fig. 10.2) that when the plasma density rises up, $2\rho Z/A \sim N_{\mathrm{e}}$, as the nonrelativistic regime is replaced by the ultrarelativistic one ($x_{\mathrm{r}} \gg 1$), the plasma equation of state becomes "softer". This is especially important for the analysis of the Chandrasekhar limit on the mass of white dwarfs.

This model, notwithstanding strong simplifications, describes well the thermodynamics of the core of neutron stars (pulsars) and the interiors of the white dwarfs [294, 295] and serves as a basis for the calculation of their structure and evolution.

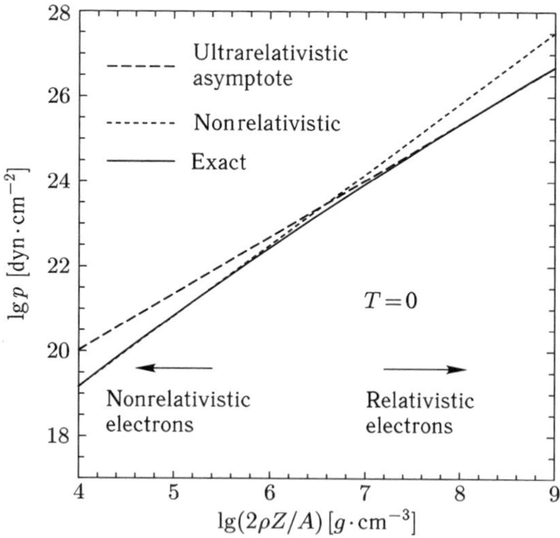


Fig. 10.2 The electron pressure $p = p_{\mathrm{id}}^{(e)}$ versus density for an ideal strongly degenerate electron gas (the solid line) together with nonrelativistic and ultra-relativistic asymptotes (the dotted and dashed lines). The figure is reproduced from Ref. [414] with kind permission of the authors.

## 10.4 Ion liquid. OCP

The free energy of ideal classic nonrelativistic ions has the form:

$$F_{\mathrm{id}}^{(i)} = Nk_{\mathrm{B}}T[\ln(n_N\lambda_{\mathrm{i}}^3/g_{\mathrm{i}}) - 1],\qquad(10.28)$$

where $g_{\mathrm{i}}$ is the spin degeneracy.

The electron–ion interaction (ei) can be described by the OCP model in which all thermodynamic quantities are functions of one dimensionless parameter $\Gamma$ [1053].

In the weak nonideality regime ($\Gamma \ll 1$) the Debye–Hückel model is applicable [203]:

$$\frac{F_{\mathrm{ex}}(\Gamma \to 0)}{V} = -\frac{2e^3}{3}\left(\frac{\pi}{k_{\mathrm{B}}T}\right)^{1/2}\left(\sum_j n_j Z_j^2\right)^{3/2}.\qquad(10.29)$$

For the OCP it gives

$$U_{\mathrm{ii,DH}} = -(\sqrt{3}/2)Nk_{\mathrm{B}}T\Gamma^{3/2},$$

$$F_{\mathrm{ii,DH}} = -(1/\sqrt{3})Nk_{\mathrm{B}}T\Gamma^{3/2}.\qquad(10.30)$$

Abe [2] used the perturbation theory to calculate corrections of the order of $\Gamma^2$ to Equation (10.30).

The next-order corrections were calculated in paper [185]. The internal energy in the low-$\Gamma$ limit has the form

$$\frac{U_{\mathrm{ii}}}{k_{\mathrm{B}}TN} = -\frac{\sqrt{3}}{2}\Gamma^{3/2} - 3\Gamma^3\left[\frac{3}{8}\ln(3\Gamma) + \frac{C_{\mathrm{E}}}{2} - \frac{1}{3}\right]$$

$$-\Gamma^{9/2}(1.6875\sqrt{3}\ln\Gamma - 0.23511) + \cdots,\qquad(10.31)$$

where $C_{\mathrm{E}} = 0.57721\ldots$ is the Euler constant. It follows from Equation (10.30) that

$$\frac{F_{\mathrm{ii}}}{k_{\mathrm{B}}TN} = -\frac{\Gamma^{3/2}}{\sqrt{3}} - \Gamma^3\left(\frac{3}{8}\ln\Gamma + 0.24225\right)$$

$$-\Gamma^{9/2}(0.64952\ln\Gamma - 0.19658) + \cdots.\qquad(10.32)$$

At large values of the nonideality parameters $\Gamma$ the OCP is described employing various numerical methods (see Chapter 5) [83, 458] and the integral equations methods (see Chapter 2).

In the latter case it is shown [414] (including direct experiments with dust plasma (Chapter 2)) that the hypernetted chain (HNC) equations have certain advantages for the case of nonideal plasma.

By now, there are comprehensive calculations of the OCP properties done by the Monte Carlo and integral equations (HNC) methods for a wide range of the nonideality parameters $1 < \Gamma < 160$ [211].

According to Ref. [421], the quantum correction to the Helmholtz free energy of the OCP has the form

$$F_q = \frac{\pi \hbar^2 N n_N (Ze)^2}{6 m_i k_\mathrm{B} T} = N \frac{k_\mathrm{B} T_\mathrm{pi}^2}{24 T}. \tag{10.33}$$

This correction is called the Wigner–Kirkwood term or the diffraction term. It is the same for the liquid and solid phases.

These corrections were calculated by the path-integral Monte Carlo technique in Refs. [491, 694].

For the wide-range calculation of the OCP thermodynamics the approximate formula was suggested in paper [809]:

$$\frac{U_{\mathrm{ii}}}{k_\mathrm{B} T N_N} = \Gamma^{3/2} \left( \frac{A_1}{\sqrt{A_2 + \Gamma}} + \frac{A_3}{1 + \Gamma} \right) + \frac{B_1 \Gamma^2}{B_2 + \Gamma} + \frac{B_3 \Gamma^2}{B_4 + \Gamma^2}, \tag{10.34}$$

where

$$A_1 = -0.9070, \quad A_2 = 0.62954, \quad A_3 = -\frac{\sqrt{3}}{2} - \frac{A_1}{\sqrt{A_2}},$$

$$B_1 = 0.00456, \quad B_2 = 211.6, \quad B_3 = -0.0001 \quad \text{and} \quad B_4 = 0.00462.$$

In Fig. 10.3 this fit is compared to the Monte Carlo calculations [209], the hypernetted approximation HNC [169], and the Debye–Hückel model (10.29).

The integration of equation (10.34) over $\ln \Gamma$ gives

$$\frac{F_{\mathrm{ii,liq}}}{N_N k_\mathrm{B} T} = A_1 \left[ \sqrt{\Gamma(A_2 + \Gamma)} - A_2 \ln \left( \sqrt{\Gamma/A_2} + \sqrt{1 + \Gamma/A_2} \right) \right]$$

$$+ 2A_3 \left[ \sqrt{\Gamma} - \arctan(\sqrt{\Gamma}) \right]$$

$$+ B_1 \left[ \Gamma - B_2 \ln \left( 1 + \frac{\Gamma}{B_2} \right) + \frac{B_3}{2} \ln \left( 1 + \frac{\Gamma^2}{B_4} \right) \right]. \tag{10.35}$$

This formula approximates $F_{\mathrm{ii}}$ within the accuracy of 0.02% for any $\Gamma$ in the gaseous and liquid phases.

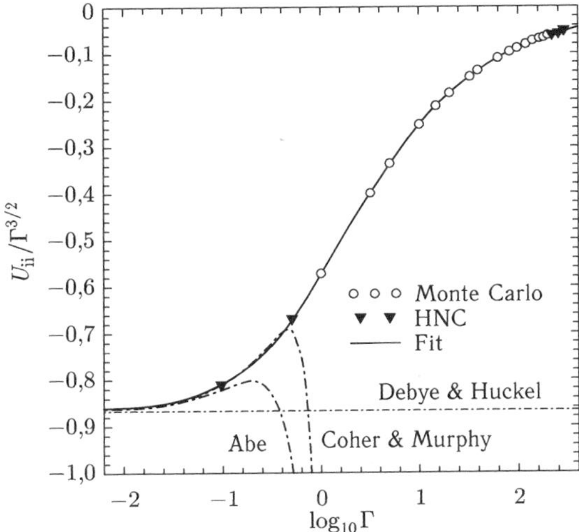


Fig. 10.3 Comparison of the fit (10.34) with the Debye–Hückel, Abe, and Cohen–Murphy approximations at small $\Gamma$ and with Monte Carlo and HNC calculations. The figure is reproduced from Ref. [414] with kind permission of the authors.

The corresponding contributions to the pressure and the heat capacity have the form

$$p_{\mathrm{ii}} = \frac{1}{3}\frac{U_{\mathrm{ii}}}{V}, \tag{10.36}$$

$$C_{V,\mathrm{ii}} = N_N k_{\mathrm{B}} \left[ \frac{\Gamma^{3/2}}{2}\left( -\frac{A_1 A_2}{(A_2 + \Gamma)^{3/2}} + \frac{A_3(\Gamma - 1)}{(1 + \Gamma)^2} \right) \right.$$
$$\left. - \frac{B_1 B_2 \Gamma^2}{(B_2 + \Gamma)^2} + \frac{B_3(\Gamma^4 - B_4\Gamma^2)}{(B_4 + \Gamma^2)^2} \right]. \tag{10.37}$$

The Coulomb pressure $p_{\mathrm{ii}}$ is exactly $1/3$ of the Coulomb density energy because $U_{\mathrm{ii}}/V$ depends in the classical case only on $\Gamma$. The Coulomb $p_{\mathrm{ii}}$ and $U_{\mathrm{ii}}$ are negative because the Coulomb interaction corresponds to an additional binding of ions.

This binding is small in the weak nonideality limit $|U_{\mathrm{ii}}| \ll kT$, therefore ions move as almost free particles. Actually, in the limit of strong nonideality ($\Gamma \gtrsim 1$) $|U_{\mathrm{ii}}| \gtrsim NkT$ the Coulomb liquid is realized. Ions in this liquid are oscillating in the Coulomb potential wells (the depth being $\sim \frac{Z^2 e^2}{a_{\mathrm{i}}}$) which in turn migrate within the liquid. In the limit of $\Gamma \gg 1$ one obtains from equation (10.34) the linear dependence on $\Gamma$, $\frac{U_{\mathrm{ii}}}{NkT} \approx A_1\Gamma$, and the quantities $U_{ii} \approx N A_1 \frac{Z^2 e^2}{a_i}$ and $P_{ii} \approx A_1 n_N e^2 / 3 a_i$ weakly depend on temperature. The parameter $A_1$ is close to the value $A_1^{IS} = -0.9$ in the ion-sphere model (Chapter 3 and 4). In this model the Coulomb energy of an ion sphere $\left( - \frac{0.9 Z^2 e^2}{a_i} \right)$ includes the electron–electron and electron–ion interactions $\frac{0.6 Z 62 e^2}{a_i}$ and $-1.5 \frac{Z^2 e^2}{a_i}$ [901].

## 10.5  Coulomb crystal

At $T < T_m$ the random ion motion is replaced by oscillations near equilibrium positions which means that a crystal state is formed. The crystal state becomes the ground state of the OCP ions in the high-density limit $r_s \ll 1$. In this limit the OCP forms a body-centered cubic (bcc) lattice, whereas at $r_s \gtrsim 1$ [561] face-centered cubic (fcc) and hexagonal close-packed (hcp) structures are formed. Note that simple cubic Coulomb lattices (like all cubic lattices with central forces) turn out to be unstable [129].

While studying strongly compressed Coulomb systems, the actual Wigner–Zeitz cell is often replaced by a sphere whose radius equals that of the ion sphere $a_i$ (10.11) with a spherical Brillouin zone having an equivalent radius $q_{BZ} = (6 \pi n_N)^{1/3}$.

For a classical Coulomb crystal we can obtain in the harmonic approximation [414]:

$$
F_{ii,\ \mathrm{harm}} = U_M + N_N k_B T \left( \frac{3}{2} \ln \Gamma + 1 + 3\bar{l} + \frac{1}{2} \ln \frac{6}{\pi} \right),
\qquad (10.38)
$$

where $U_M$ is the classical crystal lattice energy, $\bar{l} = \langle \ln(\omega_{ks}/\omega_{pi}) \rangle_{ph}$, $\omega_{ks}$ stands for the eigenfrequencies of ion oscillations in the lattice, $\omega_{pi}$ is their plasma frequency.

In the quantum limit, $T \ll T_{pi}$,

$$
F_i = U_0 - \frac{\zeta N_N k_B T}{12} \left( \frac{T}{T_{pi}} \right)^3 .
\qquad (10.39)
$$

where $U_0 = U_M + U_q$, $U_q = 3/2 N_N \hbar \omega_{pi} \cdot u_1$.

The interpolation formulae for the intermediate case are presented in paper [414].

Anharmonic oscillations of a Coulomb crystal are studied in detail by numerical methods [16, 252, 726, 797].

There are many approximate formulae suggested for this case, for instance [797]:

$$\frac{U_{\text{anharm,cl}}}{N_N k_{\text{B}} T} = \sum_{p \geq 1} \frac{a_p}{\Gamma^p}, \tag{10.40}$$

$$\frac{F_{\text{ii,anharm,cl}}}{N_N k_{\text{B}} T} = -\sum_{p \geq 1} \frac{a_p}{p\Gamma^p}. \tag{10.41}$$

In paper [252] calculations and the corresponding approximations are given in a wide range of the crystal states up to $\Gamma \lesssim 3000$.

The melting of the Coulomb crystal takes place when the chemical potentials of the liquid and the crystal equal to each other $\mu_e(T) = \mu_c(T)$ at the point $T_{\text{m}}$. The magnitude of $T_{\text{m}}$ can be determined within a poor accuracy because the slopes of the curves $\mu_e(T)$ and $\mu_c(T)$ are close to each other.

The authors of Ref. [1006] used the Lindemann criterion to calculate the value $\Gamma_{\text{m}} = 170 \pm 10$. After that this value $\Gamma_{\text{m}}$ was many times corrected employing analytic and numerical methods but did not undergo any substantial change. Note that the direct observations of the ion crystal melting in the experiments with dust plasma [294, 1012] as well give the values close to the above mentioned values of $\Gamma_{\text{m}}$. Being a first-order transition, this melting corresponds to a weak change of the thermodynamic functions. According to Ref. [414], the energy density jump associated with the crystallization to the BCC lattice is $\triangle U_{\text{i}}/V \approx 0.76 n_N k T_{\text{m}}$ with $\triangle U_{\text{i}}/U_{\text{i}} \approx 0.5\%$, and the jump of the pressure $\triangle p_{\text{i}}/p_{\text{i}}$ is three times smaller. The corresponding density jump is negligibly small. Nevertheless, even this small quantity plays an important role in the calculation of the structure and evolution of such astrophysical objects as the white dwarfs [168, 420].

At high plasma densities the amplitudes of the quantum oscillations (at $T = 0$) become comparable to the inter-ion spacing, which leads to "quantum" melting of ion crystals (see Chapter 8 and Refs. [164, 167, 702, 726]). It leads to the decrease of the melting temperature $T_{\text{m}}$ compared to the classical value and even to its vanishing $T_{\text{m}}(\rho_{\text{m}}) = 0$. At $\rho > \rho_{\text{m}}$ the liquid does not crystallize at all and becomes essentially quantum at $T \lesssim T_{\text{pi}}$. This quantum melting effect is particularly important for the oxygen and helium plasma and is hardly manifested for heavy elements like carbon, oxygen, and nitrogen even under the conditions of the neutron star envelopes [414].

A simplified EOS of a degenerate electron gas [414] is the EOS of the ideal electron gas corrected for the Coulomb interaction which is described within the framework of the ion-sphere model $p_{\mathrm{ii}} \approx -0.3 n_N Z^2 e^2 / a_{\mathrm{i}}$:

$$p = \frac{p_{\mathrm{r}}}{8\pi^2}\left[ x_{\mathrm{r}}\left(\frac{2}{3}x_{\mathrm{r}}^2 - 1\right)\gamma_{\mathrm{r}} + \ln(x_{\mathrm{r}} + \gamma_{\mathrm{r}})\right] - 0.3 n_N \frac{Z^2 e^2}{a_{\mathrm{i}}}. \qquad (10.42)$$

In the ultrarelativistic case ($x_{\mathrm{r}} \gg 1$) both terms $p_{\mathrm{id}}^e$ and $p_{\mathrm{ii}}$ are proportional to $x_{\mathrm{r}}^4$, so that their ratio is density independent

$$\frac{p_{\mathrm{ii}}}{p_{\mathrm{id}}^{(e)}} = -\frac{6}{5}\left(\frac{4}{9\pi}\right)^{1/3}\frac{Z^{2/3}e^2}{\hbar c} \approx -0.0046 Z^{2/3}. \qquad (10.43)$$

For the iron plasma this ratio equals $-0.04$. For the nonrelativistic electron gas $p_{\mathrm{id}}^e \sim x_{\mathrm{r}}^5$ and $p_{\mathrm{ii}} \sim x_{\mathrm{r}}^4$, which gives:

$$\frac{p_{\mathrm{ii}}}{p_{\mathrm{id}}^{(e)}} = -\frac{3}{2}\left(\frac{4}{9\pi}\right)^{1/3}\frac{Z^{2/3}e^2}{\hbar c \beta_{\mathrm{r}}} \approx -0.0057\frac{Z^{2/3}}{x_{\mathrm{r}}}. \qquad (10.44)$$

With decreasing $\rho$ the Coulomb interaction becomes stronger and the pressure, according to Equation (10.42), drops down to zero at $\rho \approx 0.18 Z A\,\mathrm{g/cm^3}$. However, Equation (10.42) itself becomes invalid still earlier because the approximation of the rigid incompressible electron background, being the basis of the ion-sphere model, breaks down.

Therefore the applicability of equation (10.42) is limited by the conditions $T \ll T_{\mathrm{F}}$, $\rho \gg \rho_{\mathrm{eip}}$ and $T \lesssim T_{\mathrm{e}}$. This corresponds to the strong compression regime. It is violated at high $T$ when the Coulomb interaction is small and the total pressure is determined by the electron pressure. Therefore equation (10.42) is often applied beyond its formal validity range.

The intermediate parameter range of moderate temperatures and densities, which lies between the fully ionized plasma and "thermally"– nondegenerate and quasi-ideal plasma, is usually described employing the interpolation method [414]. This method is often based on the "mean"- ion model which assumes that plasma consists of electrons and ions with the characteristic charge $Z_{\mathrm{ef}}$ which in the simplest case equals the ensemble-averaged ion charge. The interpolations [281] and the Opacity Library OPAL [810, 848] are constructed following precisely this way.

Another way to describe the "intermediate" parameter range under consideration is to employ the family of the quasiclassical Thomas–Fermi models (Chapter 6).

At extremely low $T$ and $\rho$ corresponding to large $\Gamma$ and $r_s$, the OCP of electrons behaves similarly to that of ions. This allows one to apply to the electron plasma the formalism developed earlier for ions. But in this case the quantum exchange effects begin to play a special role (Pauli principle).

A bulk of literature is devoted to the study of these exchange-correlation effects. In Ref. [796] a variational method was used. The local-density approximation of the density functional method with the effective screened potential enabled the authors of Ref. [784] to calculate the corresponding corrections at small $\Gamma_{\mathrm{e}}$ and arbitrary $\theta$. It is shown in Ref. [726] that the value of $\Gamma_{\mathrm{m}}$ increases with growing $r_{\mathrm{s}}$ and tends to infinity at $r_{\mathrm{s}} \leqslant 200$. Hence, the electron liquid does not freeze at any $T$ if $n_{\mathrm{e}} > 1.3 \cdot 10^{17}$ cm$^{-3}$. In Ref. [784] the exchange energy was calculated (Hartree–Fock method): $F_{\mathrm{NF}}(\theta) = \left(e^2/a_0\right) a(\theta)/r_{\mathrm{s}}$, where

$$a \equiv \left(\frac{9}{4\pi^2}\right)^{1/3} \frac{0.75 + 3.04363\theta^2 - 0.09227\theta^3 + 1.70350\theta^4}{1 + 8.31051\theta^2 + 5.11050\theta^4} \mathrm{th}\,\frac{1}{\theta}. \qquad (10.45)$$

In Ref. [964] the available analytic and numerical results are generalized in the form of the interpolation formula:

$$-\frac{F_{xc}^{\mathrm{nonrel}}}{k_{\mathrm{B}}TN_{\mathrm{e}}} = q\Gamma + \frac{2A}{f}\sqrt{\Gamma} + \left(\frac{B}{f} - \frac{dA}{f^2}\right) \ln|f\Gamma + d\sqrt{\Gamma} + 1|$$

$$-\frac{2(dB + CA)}{fD}\left[\arctan\left(\frac{2f\sqrt{\Gamma} + d}{D}\right) - \arctan\left(\frac{d}{D}\right)\right].$$

$$(10.46)$$

where $A = b - qd$, $B = a - q$, $C = 2 - d^2/f$, $D = \sqrt{4f - d^2}$, $a = a(\theta)$ are described by the relationships from Ref. [414] and $b$, $d$, $f$, and $q$ are the following functions [458]:

$$b = \sqrt{\theta}\,\mathrm{th}\left(\frac{1}{\sqrt{\theta}}\right) \frac{0.341308 + 12.07080\theta^2 + 1.1488890\theta^4}{1 + 10.4953460\theta^2 + 1.3266230\theta^4},$$

$$d = \sqrt{\theta}\,\mathrm{th}\left(\frac{1}{\sqrt{\theta}}\right) \frac{0.614925 + 16.9966055\theta^2 + 1.1489056\theta^4}{1 + 10.109350\theta^2 + 1.221840\theta^4},$$

$$f = \theta\,\mathrm{th}\left(\frac{1}{\theta}\right) \frac{0.539409 + 2.5222060\theta^2 + 0.1784840\theta^4}{1 + 2.5555010\theta^2 + 0.1463190\theta^4},$$

$$q = 0.872496 + 0.025248\exp\left(-\frac{1}{\theta}\right).$$

Although this result is nonrelativistic, it can be widely applied because the electron gas is almost ideal when the parameters are in the relativistic region. In this region (relativism) [903]

$$\frac{F_x^{(\mathrm{rel})}}{N_e k_B T} = \frac{3\alpha_f}{4\pi} \frac{1}{x_r^3 t_r} \left[ f_0(x_r) + f_2(x_r, t_r) t_r^2 + f_4(x_r) t_r^4 \right], \qquad (10.47)$$

where

$$f_0(x) = \frac{3}{3} \frac{L^2(x)}{1+x^2} - 3xL(x) + \frac{3}{2}x^2 + \frac{x^4}{2},$$

$$f_2(x,\tau) = \frac{\pi^2}{3} \left[ 2\ln\left(\frac{2x^2}{\tau}\right) + x^2 - \frac{3L(x)}{x} - 0.7046 \right],$$

$$f_4(x) = \frac{\pi^4}{18} \left( 1 - \frac{1.1}{x^2} - \frac{3.7}{x^4} - \frac{6.3}{x^5}L(x) \right),$$

$$L(x) = \sqrt{1+x^2}\,\ln(x + \sqrt{1+x^2}).$$

So far, we considered the model of the OCP of point-like charges on a rigid uniform background formed by the opposite-sign compensating charge. This background was assumed to be structureless and uniform. It was introduced in order to satisfy the electric neutrality (stability) condition of the Coulomb system.

In a real plasma of electrons and ions the ion interaction is screened by electrons. In case when the electron polarization is small (e.g., at $Ze^2/a_e \ll kT_F$), the polarization can be taken into account by the linear response model [336].

At $r_s \ll 1$ and $T \ll T_F$ the polarization corrections can be calculated for instance in the random-phase approximation [484]. In a more complicated case if $r_s \lesssim 1$ and $T \lesssim T_F$, the Monte Carlo methods [793] or pseudopotential models [458] can be employed.

Approximate formulae for the calculation of the polarization corrections are presented in Ref. [809]. The numerical results obtained using these formulae are given in Fig. 10.4 [414] for the relativistic and nonrelativistic case.

The account of the electron polarization in the crystal model of plasma is a particularly sophisticated problem [414]. To do that for classical ions one uses a model which consists in replacing the Coulomb potential by the Yukawa potential and subsequent application of the molecular dynamics numerical simulations [417]. These calculations showed that with decreasing the screening radius the melting parameter $\Gamma_m$ shifts to higher values. Moreover, when the Yukawa screening radius (in the Yukawa potential formalism this radius is an external parameter) becomes smaller than the

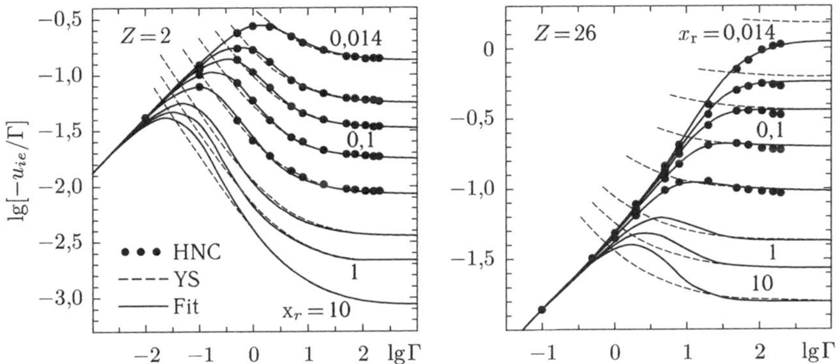

Fig. 10.4 Screening part of the free energy for an electron–ion plasma with $Z = 2$ (left) and $Z = 26$ (right) versus $\Gamma$. Calculation results for finite temperatures [169] (dots) and for a zero temperature [1053] (valid at small $r_\mathrm{s}$ and large $\Gamma$; dashed lines). Numbers at the curves indicate the values of $x_\mathrm{r} = 0.014/r_\mathrm{s}$ (from top to bottom, $x_\mathrm{r} = 0.014$, 0.034, 0.058, 0.1, 0.22, 0.55, 1, 10). The Figure is reproduced from [809] with kind permission of the authors

Debye radius, there appears a second phase transition at large $\Gamma$ from bcc to fcc lattice. The experiments with the dust plasma [304,1012] presumably confirm this conclusion.

In the classical domain the following approximations [809] are proposed for taking into account the polarization effects in the crystal plasma:

$$\frac{F_{\mathrm{ei}}}{NkT} = -f_\infty(x_\mathrm{r})\Gamma\left[\frac{1 + A(x_\mathrm{r})}{\Gamma^s}\right], \qquad (10.48)$$

where

$$f_\infty(x) = a_{\mathrm{TF}}Z^{2/3}b_1\sqrt{\frac{1+b_2}{x^2}}, \quad A(x) = \frac{b_3 + a_3 x^2}{1 + b_4 x^2},$$

$$s = \left[1 + 0.01(\ln Z)^{3/2} + 0.097Z^{-2}\right]^{-1},$$

$$b_1 = 1 - a_1 Z^{-0.267} + 0.27Z^{-1}, \quad b_2 = 1 + \frac{2.25}{z^{1/3}}\frac{1 + a_2 Z^5 + 0.222Z^6}{1 + 0.222Z^6},$$

$$b_3 = \frac{a_4}{(1 + \ln Z)}, \quad b_4 = 0.395\ln Z + 0.347Z^{-3/2}.$$

The parameter $a_{\mathrm{TF}}$ chosen in such a way as to reproduce the Thomas–Fermi asymptotic behavior at $Z \to \infty$:

$$a_{\mathrm{TF}} = \left(\frac{9\pi}{4}\right)^{1/3}\frac{a_\mathrm{f}}{Z^{7/3}}c_\infty = \frac{54}{175}\left(\frac{12}{\pi}\right)^{1/3}\cdot\alpha_\mathrm{f} \approx 0.00352.$$

The coefficients $a_1$–$a_4$ can be found in Ref. [809]. The modification of this approximation which is valid not only in the classical, but also in the quantum region is presented in Ref. [1050].

The results of the analysis presented above indicate [414] that the polarization effects in a Coulomb crystal are usually weak; they do not vanish even at large $\Gamma$ and $x_{\mathrm{r}}$ and manifest themselves as an increase of the crystal binding. In a strongly compressed Coulomb liquid this correction is approximately $\sim (k_{\mathrm{TF}}a)^2$ and tends to a finite value in the relativistic region. At relativistic densities the polarization contribution to the pressure $p_{\mathrm{ei}}$ has the same density dependence as in the nonrelativistic case $p_{\mathrm{ei}} \sim \rho^{4/3}$. In spite of its smallness, this correction can significantly affect the melting of the Coulomb crystal and can even change the type of the crystal symmetry on the melting curve (from bcc to fcc lattice).

Note that in the above analysis, differently from the well studied solids, the authors do not touch upon the complicated problem of the electron energy spectra. Consideration of this issue may appreciably change the situation.

In the low-temperature range the thermal and density dissociation and ionization in plasma are important. Therefore the plasma composition is a complex function of temperature and density. The methods of calculation of the thermodynamic properties and composition of such multicomponent plasma have been already considered in Chapter 4 taking into account the interparticle interaction and degeneracy effects.

In this chapter, we have considered the wide-range description of the relativistic plasma in the absence of magnetic field. A similar consideration of plasma in a quantizing magnetic field is presented in paper [1051].

Chapter 11

# Nuclear Transformations Under Strong Compression

Figure 11.1 shows the characteristic dimensions and times of the motion of molecular, atomic, and nuclear objects [525], and Fig. 11.2 shows the characteristic dimensions of different structural substance states. Because atomic and nuclear dimensions differ by 5–6 orders of magnitude away from nuclear densities $\rho \ll \rho_0 \approx 2.8 \cdot 10^{14}$ g/cm$^3$, these processes may be treated independently. Despite the fact that they are, of course, rigidly bound for any $\rho$, since the nuclear structure and charge directly define the electron structure of an atom or ion as well as its chemical properties and the aggregate substance state.

Proceeding from this fact, in the preceding chapters we focused our attention on the electron and ion substance components and the description of their thermodynamic properties in different parts of the phase diagram. In this case, the processes in nuclei were not considered owing to the insufficiently high ($\rho \lesssim 10^{10}$ g/cm$^3$) plasma density, so that the nuclear properties were assumed to be the same as under normal laboratory conditions.

Ascending the pressure and temperature scales and approaching the nuclear range of interparticle distances (the neutrino size is $\approx 0.16$ fm) and densities $\rho \sim \rho_0$, we ought to account for nuclear transformations, which correspond to the emergence of new forms of nuclear matter under ultra-extreme conditions [294, 414]. The possibilities that appear in this case are shown on the phase diagram of nuclear matter (Fig. 1.13), where the experimental capabilities of modern accelerators: RHIC, LHC (high temperatures and moderate densities of nuclear matter), as well as FAIR SIS 300 (a high baryon density). Also given in the diagram are the parameters of neutron stars [414, 808] emerging in the gravitational collapse of massive, $M > M_\odot$, stars at the final stages of their evolution.

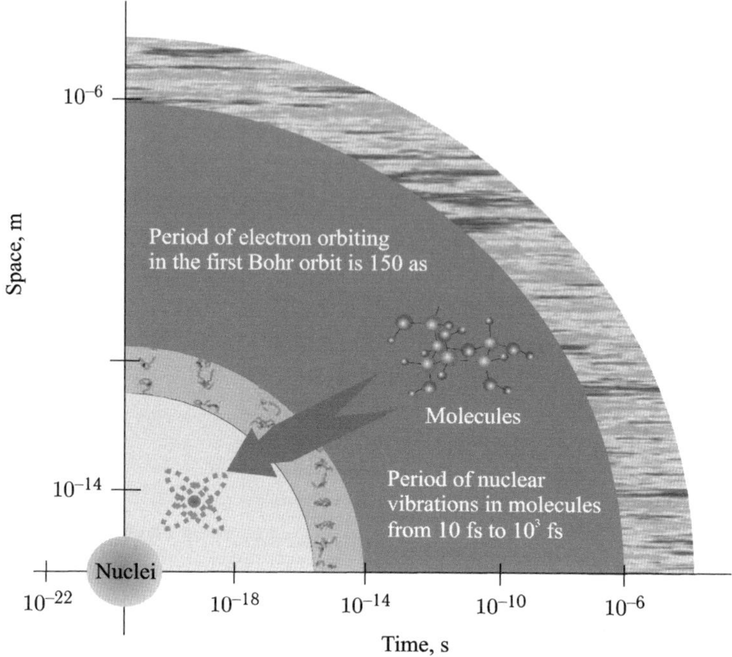


Fig. 11.1   Spatiotemporal characteristics of atomic and nuclear objects [525].

Leaving the consideration of the experiments on relativistic nuclear collisions to the Chapter 12, we shall briefly consider the physical conditions in neutron stars [294, 414].

## 11.1   Extreme states of neutron stars

Depending on the initial mass of a star of solar chemical composition, three types of compact remnants may emerge upon completion of the thermonuclear evolution in the stellar interior: white dwarfs, neutron stars, as well as black holes [545, 1073] and heretofore undiscovered quark stars [414, 808].

Neutron stars are perhaps the most exotic astronomical objects, in which there is a wide spectrum of superextreme states of matter [414, 1052] practically inaccessible to laboratory investigation. That is why neutron stars in a sense play the role of a "cosmic laboratory", when their observable manifestations permit judging about the behavior of substance under superextreme conditions — at supernuclear densities, superhigh magnetic fields, the

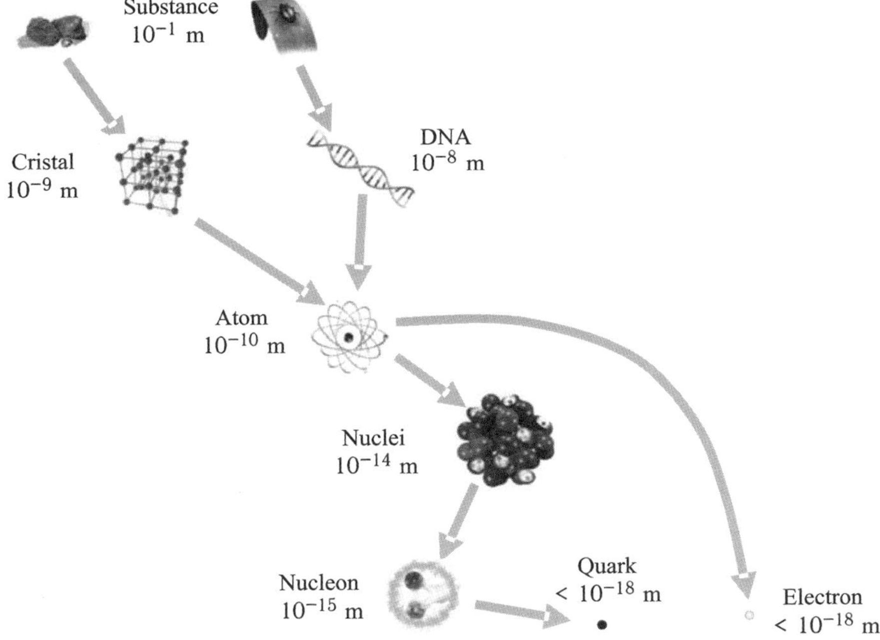


Fig. 11.2   Typical dimensions of the structural elements of substances.

superfluidity of the baryon component, and high-intensity nuclear transformations of ultracompressed substance. Naturally, the observable manifestations of these processes are also highly diversified: radio and X-ray pulsars, flaring X-ray sources, $\gamma$-ray sources, X-ray transients, and magnetars, etc. [901, 1052, 1073].

Neutron stars are the smallest observable stars in the Galaxy [477]. Their radii are of the order of $10\,\text{km}$, which is $10^5$ times smaller than the dimensions of ordinary stars. However, the neutron star masses $M$ are of the order of the solar mass $M_\odot$ and group about a value $1.4\,M_\odot$. The average substance density $\overline{\rho} = 3M/4\pi R^3 = 7 \cdot 10^{14}\,\text{g/cm}^3$ of neutron stars is several times higher than the ordinary nuclear density $\rho_0 = 2.8 \cdot 10^{14}\,\text{g/cm}^3$. Therefore a neutron star may be conventionally represented as a huge atomic nucleus of size $\approx 10\,\text{km}$. At the center of the star the density may range up to 10–20 times the nuclear density. Possible at these densities at the neutron star center is the condensation of pions, hyperons, and kaons. In this case we are dealing with strange or quark stars. The possibility of the formation of strange quarks is also under discussion.

A neutron star consists of a crust — inner and outer, — in which there occurs substance neutronization, and a core — also inner and outer. The number of protons and electrons in the inner crust and outer core amounts to several percent of the number of neutrons. The gravitational energy of a neutron star is equal to an appreciable fraction $(0.2Mc^2)$ of its rest energy.

The existence of neutron stars was predicted by Baade and Zwicky [67] in 1934, two years after the discovery of neutrons. Despite their small size, neutron stars are among the most active stars: they emit energy in the entire range of electromagnetic waves — from radiowaves to superhigh-energy photons above $1\,\mathrm{TeV}$. Rapidly revolving neutron stars experience high energy losses. In particular, the pulsar in Crab Nebula radiates an energy of $10^{38}$ erg/s, which is many orders of magnitude greater than the energy of solar radiation. The energy emanated in the radio frequency band accounts for only small fraction, $10^{-5}$–$10^{-6}$ of the energy loss. The highest-power radio pulsars also radiate in other ranges: optical, X-ray, and $\gamma$-ray ranges.

Neutron stars, which are associated with short-period radiation sources — pulsars, which were discovered in 1967–1969, — are the final stage of evolution of the ordinary stars with $M > 8\,M_\odot$, when gravitational forces compress substance to nuclear densities to give rise to neutron substance [545, 546, 1073], which was predicted by L.D. Landau back in 1932. Neutron stars exist owing to the mutual repulsion — between neutrons and protons rather than electrons, — which emerges due to the Pauli principle.

Apart from being radio pulsars, neutron stars are also sources [477, 808] of high-power X-ray radiation (X-ray pulsars), $\gamma$- and X-ray flares (magnetars), and permanent X-ray radiation emanating from the centers of supernova remnants, very weak optical stars. The radiation of neutron stars permits solving several basic physical problems. First and foremost, the case in point is the investigation of the equation of state (EOS) of superdense substance, $\rho > \rho_0$.

The presently accepted dynamics of the production of neutron stars is as follows [294, 295, 414]. The thermonuclear burn of silicon $^{32}\mathrm{Si}$ with the production of the $^{56}\mathrm{Fe}$, $^{58}\mathrm{Fe}$, $^{60}\mathrm{Fe}$, $^{62}\mathrm{Ni}$, etc. isotopes closes the chain of thermonuclear reactions in the nondegenerate core of a massive star, because the subsequent thermonuclear fusion is possible only with energy absorption. An important process which stimulates gravitational collapse is the photodissociation of iron nuclei into 13 alpha particles: $\gamma + {}^{56}_{23}\mathrm{Fe} \rightarrow 13\,{}^{2}_{4}\mathrm{He} + 4\mathrm{n}$, and substance neutronization.

Particles (protons and neutrons) which constitute the atomic nucleus can form (besides the nucleus) one more stable system which is neutron substance [545, 546]. Neutron substance mainly consists of neutrons with a small admixture (of the order of a few percent) of protons and also of electrons. Due to the Pauli principle, the neutron substance is stable against the neutron decay $n \to p + e^- + \bar{\nu}$, because the energy level of the electron, which could be emitted in the decay, has been already occupied by other electrons contained by the substance.

Neutron substance is produced under the superheavy compression of ordinary substance consisting of electrons and nuclei, which emerges in the gravitational collapse of a star that goes through the stage of supernova explosion. Under this compression, when the substance density ranges up to about $10^{11}$ g/cm$^3$, it and, hence, the electron energy become high enough to initiate the inverse $\beta$ decay: $p + e^- \to n + \bar{\nu}$ — the electron energy suffices to overcome the neutron–proton mass difference. As a result, with increasing pressure there occurs an accelerating process of electron capture by nuclei with the transformation of protons to neutrons. The neutron levels inside a nucleus turn out to be occupied and the next neutrons pass into the continuum to form a neutron Fermi liquid. And so, for a density which is not much lower than the nuclear one there emerges the neutron substance state, which fills the inner part of neutron stars.

A characteristic property of neutron stars is their superhigh (nuclear) density. However, unlike atomic nuclei in which nucleons are confined by the strong interaction, in a neutron star the nucleons are confined by gravitational forces and the neutron $\beta$ decay is suppressed by the strong electron degeneracy of compressed substance.

In the neutronization the elastic modulus of the degenerate substance becomes lower [1073], because the electron density decreases with retention of the baryon density and the substance becomes "softer". The growth of pressure with density moderates and the effective adiabatic exponent of the substance, $\gamma = d \log P / d \log \rho$, decreases from 5/3 to 4/3 and leads, in accordance with the virial theorem, to a violation of mechanical stability of the object [1073]. That is why the substance neutronization is one of the main physical processes which results in the collapse of massive stars at the late stages of their evolution.

The substance neutronization is extremely difficult to model in laboratories even with the use of superheavy compression in laser thermonuclear targets or in relativistic nuclear collisions; however, perhaps it will be possible to employ its equivalent process of the absorption of electron antineutrinos from, for instance, high-power sources with reactor strontium [460].

An additional cause for the loss of hydrostatic stellar stability is due to the effects of general relativity, when the substance pressure makes a contribution to attractive force and thereby increases the force which tends to compress the substance [545, 546]. In this case, the collapse of the core of a massive star is accompanied with a type II or Ib/c supernova outburst.

In the substance neutronization a star rapidly loses stability: the loss of elasticity results in compression and heating, but the negative heat capacity of ordinary stars no longer works in this case, because the degenerate gas pressure, which opposes compression, depends only slightly on the temperature. Furthermore, the bulk of energy released in the gravitational compression is carried away by neutrinos produced in the neutronization. And even if the temperature growth in the collapse removes degeneracy and increases the electron gas pressure, the energy removal persists due to antineutrinos in the course of beta decays of neutron-excess nuclei.

The collapse of a star terminates at densities of the order of atomic nuclear density, whereby neutron degeneracy effects become significant and the substance compressibility becomes capable of opposing the action of gravitational forces. The freely falling outer stellar layers strike against the solid surface of the compressed core and bounce off, which is believed to be the cause of shell shedding and the observable effect of a supernova outburst. But numerical simulations suggest that this mechanism is insufficient for the explosion of a supernova shell. Today it is commonly assumed that the cause of explosion is some nonone-dimensional effects: magnetic field, supersonic convection, stellar core rotation, etc. [221, 461, 463, 535].

The transformation of stars from dense white dwarfs to superdense neutron and quark stars shows (Fig. 11.3) that in the process there emerges an instability stage. Eventually there forms a compact star with $M \approx M_\odot$, a size of only about 10 km (Fig. 11.4), an initial temperature of the order of $10^{11}$ K, and a core density of $\approx 1.5$–$15\rho_0$, which possesses intense magnetic ($B \approx 10^{11}$–$10^{16}$ Gs) and gravitational (acceleration of gravity of $\approx (2$–$3) \cdot 10^{14}$ cm/s$^2$) fields, which calls for invoking general relativity for its description. Neutron stars (radio pulsars), which rotate with a period of 0.0016–1 s, are the only astrophysical objects in which the revolution slowing-down (and hence the evolution) is determined by electrodynamic forces. The neutron stars that are parts of binary systems manifest themselves as compact X-ray sources [1073].

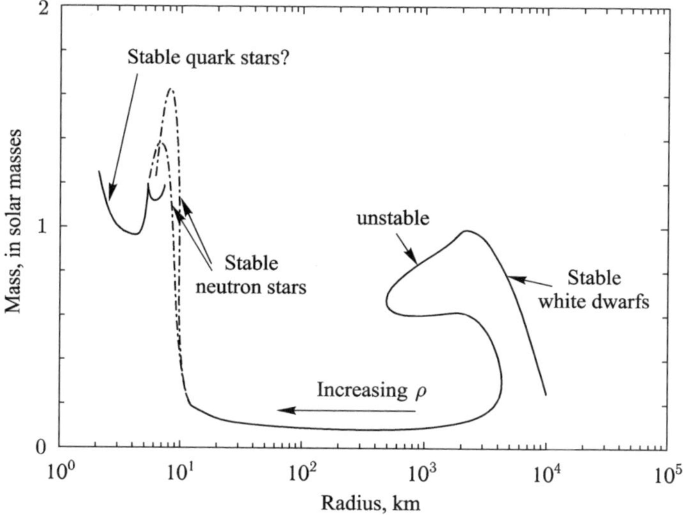



Fig. 11.3   Transition to neutron and quark stars [409].

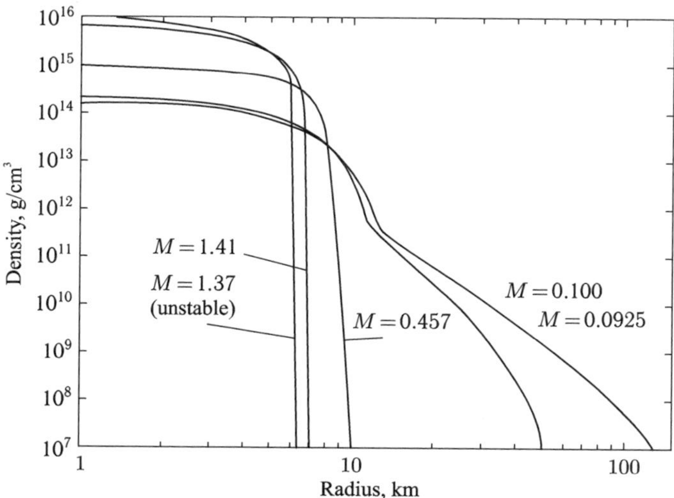



Fig. 11.4   Density distribution in neutron stars of different mass $M$ (in units of solar mass $M_\odot$) [87].

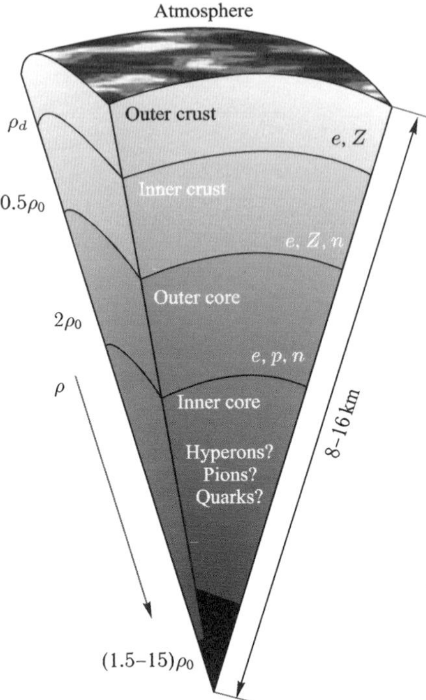


Fig. 11.5    Schematic sectional view of a neutron star of mass $1.4\,M_\odot$. The parameters of the star depend heavily on the EOS of its layers [1052, 1073]; $\rho_0 = 2.8 \cdot 10^{14}$ g/cm$^3$.

Despite the small size, the spectrum of substance states (Fig. 11.5) and the physical processes (Figs. 11.5 and 11.6) in a neutron star are extremely diversified. The atmosphere of a neutron star has a thickness ranging from tens of centimeters to several millimeters and a density of $0.1$–$100$ g/cm$^3$; it consists of a nonideal plasma with $T \leq 10^6$ K and possesses a huge magnetic field ($B \approx 10^{15}$ Gs).

The atmosphere of a neutron star may consist of the elements of the iron group formed at the stage of star production or of light elements like hydrogen and helium due to their accretion. When the stellar temperature is not too high, the crust may see the formation of light atoms, molecules, metal droplets and clusters. In the case of a high magnetic field the crust may be condensed, with a small amount of gas above it. Superhigh magnetic fields may result in quantum electrodynamic effects (like vacuum polarization) significant for radiative processes. All these problems are a strong hindrance to the calculation of the composition and optical properties of the

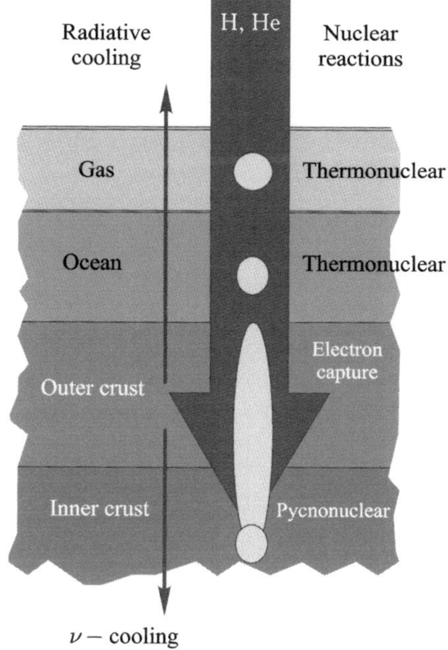


Fig. 11.6   Physical processes in a neutron star.

atmosphere and, consequently, of the radiative properties of the star itself and are therefore of major interest for theoretical astrophysics, nonideal plasma physics, and nuclear physics [901, 1052].

The outer crust several hundred meters in thickness [414, 414] consists of a dense plasma, in which the electron state undergoes a Boltzmann-to-degenerate state transition with depth and then (for $\rho \gg 10^6 \, \mathrm{g/cm^3}$) a transition to the degenerate relativistic gas. For $\rho \geq 10^4 \, \mathrm{g/cm^3}$ there occurs a pressure-induced complete plasma ionization. On further compression there occurs $\beta$ capture and substance neutronization.

At the boundary of the core of a neutron star atomic nuclei vanish, and neutrons in the inner core may be superfluidic, has an effect on the cooling dynamics and neutrino luminosity of the object. The superfluidity of neutron substance may be observed in the sudden rotation failure of some pulsars in the course of slowing down, which is caused by the interaction of normal and superfluid substance components in the inner core. In some cases the failure dynamics is attributable to the superfluid vortex separation from the stellar crust. The evolution of magnetic field "frozen" into the star permits drawing conclusions about the superconductivity of the stellar core. Since the inner stellar temperature depends heavily on the critical temperature of nucleons' transition to the superfluid state, it has been suggested [763] to employ a neutron star as a "thermometer" for measuring the critical temperatures of nucleons in asymmetric nuclear matter, which would yield indirect information about the EOS of nuclear substance.

The composition of the inner core of a neutron star is inexactly known, because the physics of strong interactions in superdense substance is insufficiently known [545, 1073]. The core perhaps consists of nucleon–hyperon matter, pion condensate, quark–gluon plasma, or some other exotic states. According to [1052], if the properties of the crust ($\rho < 0.5\rho_0$) of a neutron star are described by nonideal plasma models, the description of the properties of supernuclear density matter (for $\rho \geq \rho_0$) is extremely difficult because of the incompleteness of laboratory data and the absence of a closed theory of supernuclear density matter [754].

Many neutron stars that are parts of binary systems manifest themselves as X-ray pulsars in the flow of substance to the star with a high ($>10^{10}$ Gs) magnetic field. For a lower magnetic field the substance accumulates on the stellar surface and there occurs a thermonuclear explosion on the attainment of some critical value, which shows up as regular X-ray bursts. Unlike explosions on the surface of a white dwarf, the powerful gravitational field of a neutron star does not permit the explosion products to separate from the star and returns them back.

The substance structure under high magnetic fields ($B > 10^{12}$ Gs) inherent in neutron stars is quite unusual [477]. Under such fields the cyclotron radius of atomic electrons is shorter than the Bohr radius, an atom is strongly compressed in the direction perpendicular to the magnetic field and is needle-shaped. The properties of the substance formed by suchlike atoms may be judged by the interaction of the surface of a neutron star, at which the substance density amounts to $10^5$ g/cm$^3$, and its magnetosphere.

Finally, studying active neutron stars makes it possible to investigate the electrodynamic processes occurring in superhigh magnetic fields, $B > 10^{12}$ Gs, inherent in neutron stars.

When the magnetic field of a neutron star in a close binary system is extremely high ($10^{12} - 10^{14}$ Gs), a special kind of accretion on the neutron star may occur, whereby the matter of the normal star together with the magnetic field frozen in it are incident along the induction lines in the regions of the magnetic poles. In this case, the excess angular momentum is transferred to the star by magnetic field rather than hydrodynamic motion.

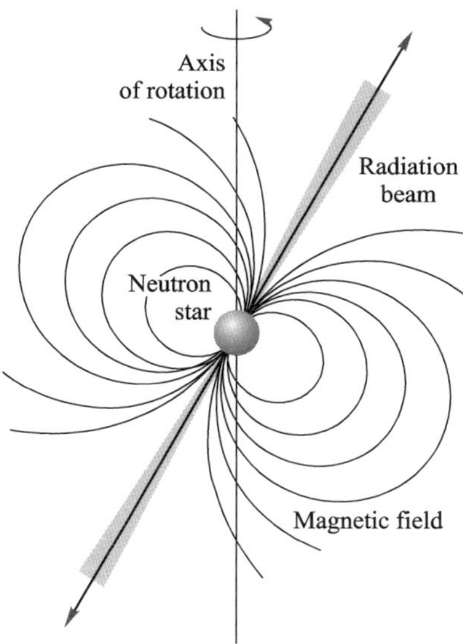


Fig. 11.7   Neutron star with a magnetosphere. Shown are the magnetic lines of force of the neutron star, with the magnetic dipole axis inclined to the axis of stellar rotation. The pulsar radiation in one of the models emanates from two cones (the conic beams in the drawing) coaxial with the magnetic dipole [806].

The velocities of falling on the surface of the neutron star can reach up to several hundred thousand kilometers per second, and small (hundreds of square meters) polar regions of the surface experience tremendous fluxes of matter and energy, which generate a plasma temperature of $\approx 10^9 - 10^{10}$ K. The released energy is radiated in the form of hard photons emanating from two hot "X-ray" spots (Fig. 11.7). The strong magnetic field makes the radiation of these spots anisotropic.

Since the magnetic axis in the general case does not coincide with the axis of mechanical rotation, an observer will record one or two X-ray radiation pulses during one revolution of the neutron star about its axis. Such sources are referred to as X-ray pulsars [1073] (Fig. 11.7).

A magnetic intensity on the order of $10^{18}$ Gs, where the energy of the magnetic field is comparable to the gravitational energy $E_s$, is believed to be the limiting magnitude of the magnetic field of a neutron star [113,477,596].

Observations suggest [477] that the energy lost by a gyrating neutron star (radio pulsar) goes primarily to form a relativistic particle stream, which is referred to as the pulsar wind. This particle flux amounts to $10^{40}$ particles per second.

As mentioned above, there are neutron stars which make up a nonnumerous (10% of all neutron stars) class of highly active stars; they produce bright gamma- and X-ray flares, and the magnetic field of some of them is much higher than that of pulsars.

Such stars — magnetars — rotate relatively slowly with a period $P \approx$ 5–10 s, but decelerate much faster than pulsars, $dP/dt \approx 10^{-10}$–$10^{-12}$ s/s. The X-ray radiation flux generated by such a star, $W_x \approx 10^{35}$–$10^{36}$ erg/s, is far greater than the rotation energy lost by the star. The energy stored in its magnetic field is also higher than the stellar rotation energy. This is indication that the source of stellar activity is the magnetic field rather than the stellar rotation, as with radio pulsars. Such stars are therefore referred to as magnetars [807].

The high magnetic field in the magnetosphere of a neutron star furnishes conditions for plasma production and the formation of wind — the flux of relativistic electrons and positrons emitted by active stars.

We will see in Chapter 12 that the quark–gluon plasma is an exotic super dense state of matter. Such plasma has been recently discovered in the laboratory. It consists of quarks, antiquarks, and gluons [350, 419, 746, 867]. The quark–gluon plasma has a maximal density and can emerge in the centers of neutron stars or in the collapse of ordinary stars (Fig. 11.8). In this case the quark stars and the hybrid stars are considered which consist

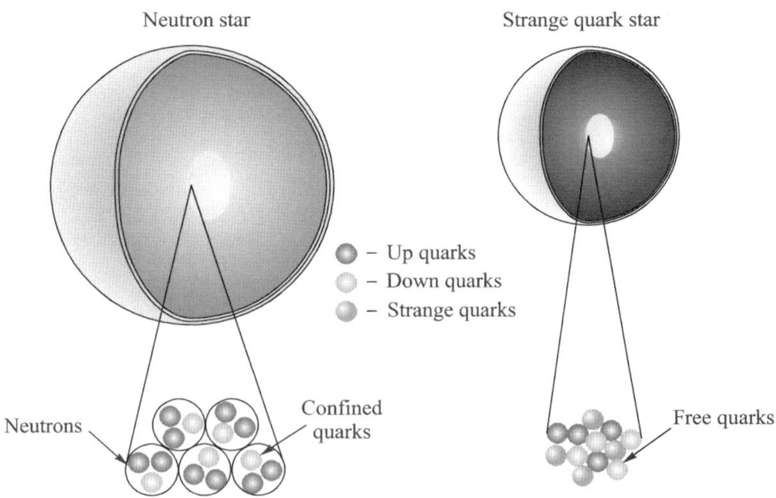


Fig. 11.8   Neutron and quark stars.

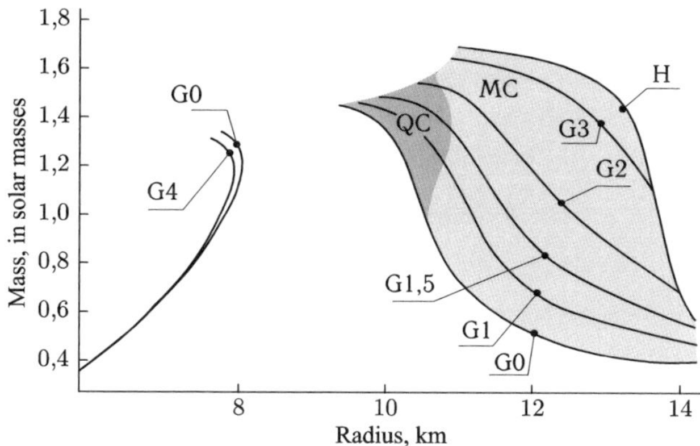


Fig. 11.9   Dimensions of quark and hybrid stars [889].

of the hadron shell and quark–gluon matter in the center [889]. The quark stars should be smaller than the neutron stars because the quark–gluon plasma is characterized by higher compressibility (Fig. 11.9).

An important feature inherent in relativistic astrophysical objects is the existence of gigantic magnetic fields, which determine to a large measure the dynamics of their motion and radiative characteristics.

The recorded radiation of a neutron star is primarily soft X-rays; this radiation becomes accessible $10^5 - 10^6$ years after the stellar birth. Recently recorded was a new class of neutron stars — magnetars — which possess an ultraintense magnetic field up to $10^{15}$ Gs; this field affects their gamma-ray radiation formed together with the thermal radiation of the surface [579, 639, 807, 1073]. An example of such an object is provided by the brightest gamma-ray burst recorded in our galaxy to date; its source is the neutron star SGR-1806-20, which belongs to the class of magnetars and is at a distance of $\approx$50000 light years from the Earth. The burst lasted for only $\approx$0.1 s, and the amount of energy released within several seconds after the burst was greater than the energy radiated by our Sun in 250000 years.

In magnetars there occur extremely strong magnetic bursts. In this case, the stellar surface cracks under the action of the Lorentz force and protons escape from these cracks, which interact with the magnetic field and radiate energy. The stellar magnetic field is determined from the amount of this energy [639, 766]. Zasov and Surdin [1074] came up with the following model of magnetic magnetar bursts.

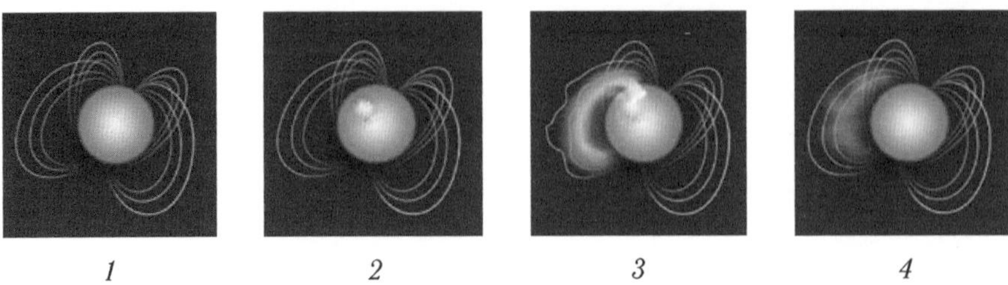



Fig. 11.10    Magnetar burst formation [579].

A magnetar is calm most of the time, but the stress induced by the magnetic field in its solid crust gradually increases (phase *1* in Fig. 11.10). At some point in time, the stress in the crust exceeds its ultimate strength, and the crust breaks into a multitude of small fragments (phase *2*). This "starquake" gives rise to a pulsating electric current, which rapidly decays to leave an incandescent plasma ball (phase *3*). The plasma ball radiates X-rays from its surface, cools down, and vaporizes within a few seconds (phase *4*).

The extremely short duration of the burst testifies [639] to the fact that the source of its energy is the dissipation of magnetic field stored in the magnetosphere and not in the core of the neutron star. The resultant expansion of the "magnetic cloud" with a small fraction of the baryon component (with $M < E/c^2$) is determined by the magnetic field. It is strongly relativistic and anisotropic, and retains these properties for several weeks after the burst.

## 11.2    Compression. Nuclear structures

As we move to the right along the nuclear density scale in Fig. 11.1, for $\rho \gtrsim 10^{10}\,\mathrm{g/cm^3}$ [414, 544, 545] nuclei become unstable relative to $\beta$ decay with a maximum neutron excess $\delta \sim \frac{N-Z}{A} \sim 0.3$, where $N$, $Z$, and $A$ are the respective numbers of neutrons, protons, and nucleons. With increasing compression the neutron excess $\delta$ increases and for $\delta > 3$ the neutron energy levels cease to be bound. The nuclei turn out to be immersed in the medium of free neutrons and cannot exist under these conditions, decaying with the emission of neutrons in a time of $\approx 10^{-20}\,\mathrm{s}$.

Our knowledge about this matter for $10^{11}\,\mathrm{g/cm^3} \lesssim \rho \lesssim 10^{14}\,\mathrm{g/cm^3}$ and $0.3 \lesssim \delta \lesssim 0.8$ relies only on theoretical constructions partly confirmed by astronomical observations of neutron stars — exotic and, as we saw in the foregoing, the most extreme astronomical objects emerging in the gravitational collapse of a variety of stars at the final stages of their evolution. However, the capabilities of the theory of multiparticle nuclear systems are quite limited in this area, like are the capabilities of experiments, especially in the $\rho > \rho_0 = 2.8 \cdot 10^{14}\,\text{g/cm}^3$ density domain characteristic of symmetric nuclear matter. These substance neutronization processes are the basis for the calculation of the structure and dynamics of neutron stars [414]; to describe their thermodynamics, highly sophisticated models have been developed, which are generalized in monograph [414]. In the subsequent discussion we follow this work.

In the density range under consideration the characteristic distance between the nuclei is comparable to their size, which corresponds to "nuclear" plasma densities of $\approx 10^{14}\,\text{g/cm}^3$. To describe these supercompressed situations, use is made of three models [414]. These are Hartree–Fock calculations with an effective nucleon interaction potential [121, 732], calculations by means of the Thomas–Fermi method at finite temperatures [659, 749], calculations by liquid drop model [606], etc.

The thermodynamics of nuclear matter under the conditions close to the onset on neutronization is described in [87]. This permitted calculating the properties of neutron-excess nuclei at different temperatures. In particular, the highest density whereby the $^{84}$Se nucleus (with $\frac{Z}{A} \approx 0.405$) exists was estimated at $\approx 8.2 \cdot 10^9\,\text{g/cm}^3$. Many neutron-excess nuclei and several new nuclear structures stable at a heavy compression have been discovered to date [414].

Using the theoretical methods developed to date it has been possible to calculate the composition and energy of several nuclear structures; those of which that are equilibrium are referred to as "cold catalytic matter".

Calculations show [414] that the minimal value of $E$ equal to 930.4 MeV is realized for a bcc lattice of $^{56}$Fe. This corresponds to $\rho \approx 7.86\,\text{g/cm}^3$ and $n_b = 4.73 \cdot 10^{24}\,\text{cm}^{-3}$. We note that $^{56}$Fe is not the weakest-bound free nucleus. The highest binding energy per nucleon

$$
b = \frac{[(A - Z)m_n c^2 + Z m_p c^2 - M(A, Z)c^2]}{A},
$$

is reached for $^{62}$Ni: $b(^{62}\text{Ni}) = 8.7945\,\text{MeV}$, for $^{56}$Fe this quantity $b(^{56}\text{Fe}) = 8.7902\,\text{MeV}$, and $b(^{58}\text{Fe}) = 8.7921\,\text{MeV}$.

The energy of the bcc lattice of $^{56}$Fe remains the main state of the cold substance up to $10^{30}\,\text{dyne/cm}^2$ and $\rho \approx 10^6\,\text{g/cm}^3$ [874]. Under these conditions the substance is a plasma, to which a one-component plasma model may be applied (see Chapters 4 and 10).

For $\rho \gg 10^9\,\text{g/cm}^3$ the electrons become ultrarelativistic and $\mu_e \sim P^{1/4}$. As $P$ increases, replacement of $((A, Z))$ by $((A', Z'))$ with a higher nuclear energy $W_N$ but lower $\frac{A'}{Z'}$ becomes energy-advantageous, because the increase in $\frac{W_N}{A}$ is greater than the decrease in $\frac{Z\mu_e}{A}$. This accounts for the effect of substance neutronization under its heavy compression [414, 544, 545].

The density jump in the neutronization [544, 545] obeys the relation [414]:

$$\frac{\Delta\rho}{\rho} \approx \frac{\Delta n_{\text{b}}}{n_{\text{b}}} \approx \frac{Z}{A}\frac{A'}{Z'} - 1. \tag{11.1}$$

The equilibrium parameters of nuclei in cold dense matter are collected in Table 11.1 borrowed from [413], which shows the highest densities at which these nuclei still persist; $\mu_e$ is the chemical potential at this density.

The transformation of a nucleus to the next nucleus brings about the phase transition of the Ist kind with a density jump $\frac{\Delta\rho}{\rho}$ given in the last column in Table 11.1 [413].

Table 11.1   Nuclei in the ground state in cold dense matter [413]. The upper part was obtained with experimentally measured nuclear masses and the lower part was obtained from the mass Moeller formula. The last line corresponds to neutron stability boundary. The table was borrowed from Ref. [414].

| Element | $Z$ | $N$ | $Z/A$ | $\rho_{\max}$, g/cm$^3$ | $\mu_e$, MeV | $\Delta\rho/\rho$ (%) |
|---|---|---|---|---|---|---|
| $^{56}$Fe | 26 | 30 | 0.4643 | $7.96 \cdot 10^6$ | 0.95 | 2.9 |
| $^{62}$Ni | 28 | 34 | 0.4516 | $2.71 \cdot 10^8$ | 2.61 | 3.1 |
| $^{64}$Ni | 28 | 36 | 0.4375 | $1.30 \cdot 10^9$ | 4.31 | 3.1 |
| $^{66}$Ni | 28 | 38 | 0.4242 | $1.48 \cdot 10^9$ | 4.45 | 2.0 |
| $^{86}$Kr | 36 | 50 | 0.4186 | $3.12 \cdot 10^9$ | 5.66 | 3.3 |
| $^{84}$Se | 34 | 50 | 0.4048 | $1.10 \cdot 10^{10}$ | 8.49 | 3.6 |
| $^{82}$Ge | 32 | 50 | 0.3902 | $2.80 \cdot 10^{10}$ | 11.4 | 3.9 |
| $^{80}$Zn | 30 | 50 | 0.3750 | $5.44 \cdot 10^{10}$ | 14.1 | 4.3 |
| $^{78}$Ni | 28 | 50 | 0.3590 | $9.64 \cdot 10^{10}$ | 16.8 | 4.0 |
| $^{126}$Ru | 44 | 82 | 0.3492 | $1.29 \cdot 10^{11}$ | 18.3 | 3.0 |
| $^{124}$Mo | 42 | 82 | 0.3387 | $1.88 \cdot 10^{11}$ | 20.6 | 3.2 |
| $^{122}$Zr | 40 | 82 | 0.3279 | $2.67 \cdot 10^{11}$ | 22.9 | 3.4 |
| $^{120}$Sr | 38 | 82 | 0.3167 | $3.79 \cdot 10^{11}$ | 25.4 | 3.6 |
| $^{118}$Kr | 36 | 82 | 0.3051 | $(4.32 \cdot 10^{11})$ | (26.2) | |

The line above the horizontal bar in Table 11.1 corresponds to the highest density whereby the ground state of the matter contains nuclei with laboratory-measured masses. The last line in Table 11.1 corresponds to the density at the boundary of neutron stability (neutron drip), which is estimated [414] at $\rho_{\mathrm{ND}} \approx 4.32 \cdot 10^{11} \, \mathrm{g/cm^3}$.

As already noted, to describe the neutron matter for $\rho \lesssim 10^{14} \, \mathrm{g/cm^3}$ use is made of the methods for calculating the properties of atomic nuclei under ordinary (laboratory) conditions. These are the Hartree–Fock method for a given neutron–neutron interaction potential, the quasiclassical Thomas–Fermi approximation, and the compressible liquid drop model.

According to these calculations, the number of neutrons outside of the atomic nucleus increases and the number of protons inside of it decreases with increasing compression.

The results of these calculations were generalized [414, 732] in the form of analytical formulas convenient for calculating the EOS of the inner crust of neutron holes.

## 11.3   Thomas–Fermi model

According to Ref. [414], in the vicinity of neutron stability boundary the number of nucleons in an elementary cell increases rapidly with compression and may range up to $A_{\mathrm{cell}} \approx 1000$ for $\rho \approx 10^{13} \, \mathrm{g/cm^3}$ [732]. This fact strongly complicates the application of the Hartry–Fock method because in the calculations of atomic structures (see Chapters 3 and 7) it significantly increases the scale of the necessary computations (see the Kohn exponential barrier problem [562]). On the other hand, the increase in the number of nucleons in an elementary cell justifies the application of a simpler quasiclassical approximation (Chapter 6).

Numerous calculations by this model (for more details, see Ref. [414]) yielded a neutron evaporation (neutron drip) density $\rho_{\mathrm{ND}} \approx 4 \cdot 10^{11} \, \mathrm{g/cm^3}$. According to [141], $Z$ remains practically invariable ($Z \approx 30$) for $10^{11} \, \mathrm{g/cm^3} \leq \rho \leq 5 \cdot 10^{13} \, \mathrm{g/cm^3}$ and then drops to $Z \approx 20$ for $\rho \approx 10^{17} \, \mathrm{g/cm^3}$.

The liquid-drop model [414] proceeds from separating out the bulk $E_{\mathrm{n,bulk}}$ and surface $E_{\mathrm{n,surf}}$ parts of the total energy:

$$E_{\mathrm{cell}} = E_{\mathrm{n,bulk}} + E_{\mathrm{n,surf}} + E_{\mathrm{coul}} + E_e, \qquad (11.2)$$

where the first three terms depend on the size and shape of a nucleus. The nucleons are then divided into three subsystems: the neutron bulk liquid "i", the lower-density liquid "o", and the surface liquid at the "i–o" interface denoted by "s". The entire system is in mechanical and "chemical" equilibrium. Inside the nucleus in phase "$i$" the nucleon densities $n_{p,i}$, $n_{n,i}$ are constant, while $n_{n,0}$ and $n_{p,0} = 0$ in phase "o". The nucleus size itself can be defined in terms of the proton radius $r_p$ so that $(4\pi r_p^3 n_p)/3 = Z$.

The equilibrium parameter values are found by minimizing $\varepsilon = \frac{E_{\text{cell}}}{V_c}$ for a given $n_b$. In doing this the conditions of the chemical and mechanical equilibrium (for more details, see Ref. [414]) are taken into account. These conditions give seven independent variables on which $E$ depends. Calculations by one of the most modern versions of the liquid-drop model [219] as applied to the conditions of the lower crust of a pulsar suggest that the number of nucleons in a nucleus $A$ increases with density and ranges up to $\approx 300$ for $\rho \approx 10^{14}$ g/cm$^3$ and $A_{\text{cell}} \approx 1000$. In this case, the number of protons varies quite slowly, $Z \approx 40$. For $\rho \approx 10^{14}$ g/cm$^3$ the nuclei turn out to be quite heavy and neutron-excessive, which brings up the question about their stability with respect to deformations and fission.

According to the calculations of Ref. [414], for $\rho < 10^{13.5}$ g/cm$^3$ the ground state of nuclear matter consists of spherical nuclei, which are stable with respect to decay into differently shaped structures, fragments, or *npe*-gas. The nuclear shape sphericity in this domain follows, for instance, from the liquid-drop model, which minimizes the spherical shape owing to the contribution $E_{\text{n,surf}} + E_{\text{coul}}$ for $r_p \ll r_c$. It is likely that the situation changes for $\rho > 10^{13.5}$ g/cm$^3$, where $r_p > r_c \gtrsim 0.5$.

Exotic nuclear shapes emerge [414] when the nuclear volume comes to exceed $\approx 50\%$ of the total volume of the system, and the nuclear matter with a spherical neutron "bubble" becomes energy-prominent.

The historically first works of this kind were performed in the calculation of the gravitational collapses of massive stars. As shown in Ref. [414], for $T \gtrsim 10^{10}$ K and an entropy of (1–2) K (and an entropy of $(1$–$2)k_{\text{B}}$) per nucleon, prior to transition to the homogeneous plasma state there occurs a series of phase transitions involving changes in nuclear shapes. Spherical nuclei in a nucleon gas ($3N$) as well as spherical "bubbles" ($3B$) in dense nuclear matter were considered. In the two-dimensional case a study was made of nuclear cylinders ($2N$) and hollow cylinders ($2B$) in a nuclear gas. In one case, parallel layers of nuclear matter separated by the nuclear gas $1N$ were analyzed. It was determined that with increasing density there occurs a $3N \to 2N \to 1N \to 2B \to 3B$ transformation with the subsequent transition to homogeneous matter. These transitions go in the direction of increasing fraction of the higher density (nuclear) phase.

In the analysis of the situation for $\rho \gtrsim 10^{14}\,\text{g/cm}^3$, in Ref. [635] it was shown that the very position and existence of exotic nuclear structures depends on the form of the N–N interaction potential. The $3N \rightarrow 2N \rightarrow 1N \rightarrow 2B \rightarrow 3B$ transitions begin for $n_{cc} \approx 0.064\,\text{fm}^{-3} \approx \frac{1}{3}n_0$ ($\rho \simeq 1.1 \cdot 10^{14}\,\text{g/cm}^3$) and terminate with the transition of $3B$ to a homogeneous $npe$-substance for $n_{cc} \approx 0.096\,\text{fm}^{-3}$; $\rho_{cc} \approx 1.6 \cdot 10^{14}\,\text{g/cm}^3$. The energy between the exotic phases is quite low, $\lesssim \text{keV} \cdot \text{fm}^{-3}$.

As shown in Ref. [762], in the relativistic Thomas–Fermi (RTF) model the existence domain of exotic phases is quite narrow: $(1.0–1.5) \cdot 10^{14}\,\text{g/cm}^3$. In Ref. [959], an even more narrow range, $0.050 \div 0.058\,\text{fm}^{-3}$, was obtained for the structural transitions.

Outside of the domain of experimentally studied nuclei, the EOS of the nuclear substance has been poorly studied. For $\rho > \rho_{\text{ND}}$ the nuclear properties depend progressively heavier on the ambient neutron gas, whose pressure increases with increasing total pressure. The EOS of nuclear matter consists of nucleon Hamiltonian, which describes the nucleon interaction for $\rho \leq 2\rho_0$. Nonnucleon degrees of freedom will be significant at higher densities. An effective nuclear Hamiltonian is introduced is introduced for the description in the $10^{11} < \rho \leq \rho_0\,\text{g/cm}^3$ domain. In this case, the effective interactions are defined in such a way as to describe the properties of the ordinary (laboratory) nuclei, because the main contribution to the pressure for $\rho \geq 10^{13}\,\text{g/cm}^3$ is made by neutrons and the effective nuclear forces provide an adequate description of the neutron gas of subnuclear density. Although we have no direct experimental information about the neutron substance, there are reasonably reliable numerical data about its properties for $n_n \lesssim n_0$ [786, 787].

A comparison of two EOS of the neutron substance is shown in Fig. 11.11 [414]. One can see that the transition to the $npe$-liquid appreciably increases the slope of the curves: there occurs an increase in substance "rigidity".

The slope of the curve is described by the adiabatic exponent

$$\gamma = \left(\frac{\rho}{p}\right)\left(\frac{\partial p}{\partial \rho}\right),$$

(see Fig. 11.12).

For $\rho < \rho_{\text{ND}}$ the adiabatic exponent $\gamma \sim \frac{\rho}{p}\left(\frac{\partial p}{\partial \rho}\right)$ is close to 4/3 [414], because in this domain the ultrarelativistic gas $P_e$ and lattice $P_i$ pressures are proportional to $\rho^{4/3}$. The pressure above the density jump is defined by the "soft" EOS of the liquid-drop model, which lowers $\gamma$. After the

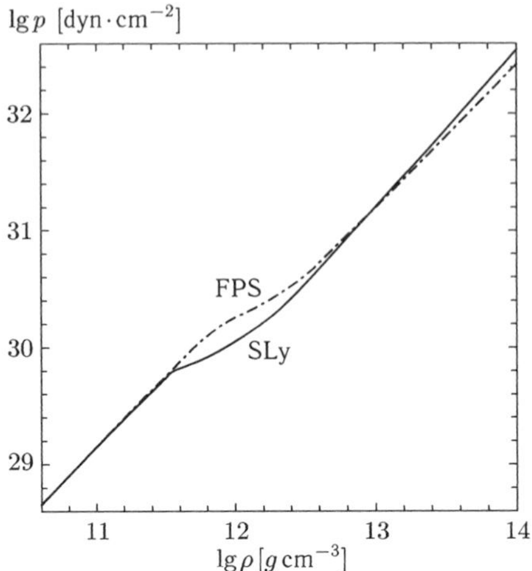



Fig. 11.11   Comparison of SLy and FPS EOS. Reproduced from Ref. [414] with kind permission of the authors.

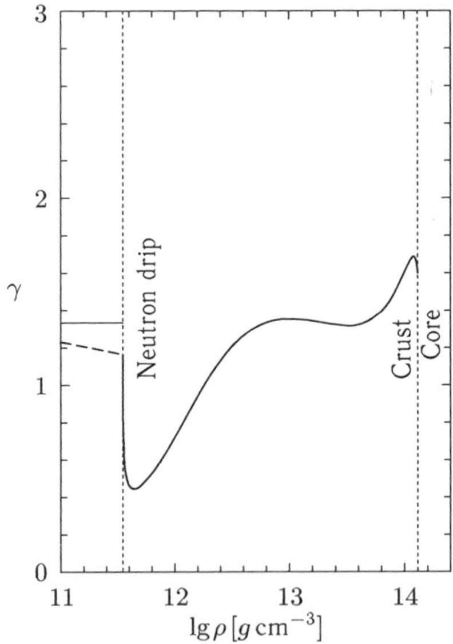



Fig. 11.12   Adiabatic exponent $\gamma$ for the ground state of the substance of a neutron star for the HP model up to the neutron evaporation point and for the CLDM model [219,220] for higher $\rho$. Reproduced from Ref. [414] with kind permission of the authors.

domain of "stability curve" the EOS becomes much more rigid, with an appreciable increase in $\gamma$ in the $\rho_{cc}$ domain. The subsequent growth of $\gamma$ is associated with the disappearance of nuclei and the emergence of the liquid of neutrons, protons, and electrons.

Here, we shall not describe the strength properties of a nuclear crystal for $\rho < \rho_{cc}$, which were analyzed for the relativistic case in Ref. [160] and for the nonrelativistic case in Ref. [414]. These properties are significant in the description of the crust of a neutron star for $\rho \lesssim 10^{14}$ g/cm$^3$. Considered in this case are bcc lattices and polycrystalline structures. Data on the shear moduli of the Coulomb one-component bcc lattice plasma with neglect of screening by degenerate electrons are given in Ref. [750]. The following expression was obtained for the shear modulus:

$$\mu = \frac{1}{5}(2b_{11} + 3c_{44}) = 0.1194\frac{n_N(Ze)^2}{r_{\rm c}}. \tag{11.3}$$

In an isotropic medium this expression is of the form:

$$\mu = 0.0159 \left(\frac{Z}{26}\right)^{2/3} P_{\rm e}, \tag{11.4}$$

where $P_{\rm e}$ is the pressure of ultrarelativistic electron gas. Therefore,

$$\frac{\mu}{K} = 0.016 \left(\frac{Z}{26}\right)^{2/3} \left(\frac{P_{\rm e}}{\gamma P}\right) \ll 1. \tag{11.5}$$

For this crystal the Poisson coefficient is $\sigma \simeq 1/2$ and the Young modulus is $E \simeq 3\mu$.

The plasma in magnetic field has been the object of numerous investigations in connection with research on controlled thermonuclear fusion [294, 295] and magnetohydrodynamic generators, as well as in connection with astrophysical applications [414].

In the latter case, as we saw above, emphasis is placed on neutron stars, where typical fields range up to $\approx 10^{12}$ Gs or $\approx 10^{15}$ Gs for magnetars [294, 295]. Under such fields the electron and ion motion transverse to the magnetic field is quantized into Landau orbitals of length $a_m \sim (\frac{\hbar c}{eB})^{1/2}$. The ratio between the Bohr radius $a_0$ and this length is the field intensity parameter $\gamma = (\frac{a_0}{a_m})$. For large $\gamma$ the Lorentzian force acting on the valence atomic electrons exceeds the Coulomb nuclear forces. Relativistic effects affect the Landau electron energy levels when the magnitude of the field

(in relativistic units)

$$b = \frac{\hbar\omega_\mathrm{c}}{(m_e c^2)} = \frac{B}{B_\mathrm{r}} \gtrsim 1. \tag{11.6}$$

Here, $\omega_\mathrm{c} = \frac{eB}{(m_e c^2)}$ is the electron cyclotron frequency and $B_\mathrm{r} = m_e^2 c^3/(e\hbar) \approx 4.4 \cdot 10^{13}$ Gs is the relativistic magnetic field.

In this notation the magnetic field is called strong for $\gamma \gg 1$ (radio pulsars) and superstrong for $b \gtrsim 1$ (magnetars).

The thermodynamic functions of ideal electron gas in magnetic field are given in Ref. [414].

The effect of magnetic fields on the thermodynamics and structure of substance is set forth at length in Ref. [414], which presents an exhaustive literature list and convenient approximation formulas.

When the density becomes higher than $\rho > \rho_0 \approx 2.8 \cdot 10^{14}\,\mathrm{g/cm^3}$, the nuclear substance transforms to new forms, which are different from the well-known nuclear structures at normal conditions $\rho_0$ and are therefore referred to as "exotic" [414]. The existence of these states is caused by the properties of strong (hadronic) interaction (pion and kaon condensations) as well as by the quark structure of baryons.

In the next Sec. 11.4 we will follow monograph [414].

## 11.4 Meson, pion, and kaon condensations

Pions are the lightest mesons; they may be present in a system as free particles and replace electrons in a condensate for $\rho > \rho_0$ [68]. The idea of pion condensate formation due to $\pi$N attraction was developed in Ref. [686, 687, 885, 886], where a change in symmetry was noted and an important role of nuclear correlations for the existence of such a condensate was emphasized. The formation of a pion condensate, according to Ref. [414], "softens" the EOS of nuclear substance.

Kaons are the lightest strange mesons and, according to [510], must form a condensate for $\rho > 3\rho_0$ on the strength of $K^- N$ attraction. Kaon condensation, according to Ref. [414], sharply "softens" the EOS as well.

Since quarks are the main building elements of hadrons, for sufficiently high densities they may "dissociate" to form quark substance, or quark–gluon plasma. The EOS of quark–gluon plasma was calculated in the framework of the lattice model [179] developed for quantum chromodynamics. The properties and models of the quark–gluon plasma were considered in Chapter 12.

Like in a gas (Chapter 2), the strong NN interaction may be represented as the sum of a short-range repulsion and a long-range attraction, and therefore crystallization is possible in the system. In this case, possible in the neutron matter is the formation of complex structures like alternating layers of liquid crystals, quasi one-dimensional filaments, etc. As this takes place, each of the coexisting phases may be electrically charged. It is evident that the formation of crystalline structures may lead to strength effects.

A large uncertainty in the EOS of nuclear matter permitted advancing [1047] a heuristic hypothesis that the exotic structures emerging under heavy compression which are in $\beta$ equilibrium will also be stable under normal conditions, forming hypothetical "self-bounded" states [610] or $Q$ matter [69]. Suchlike hypotheses underlie the construction of modern models of "strange" stars [294, 295, 414].

Exchange of virtual pions is responsible for the long-range component of N–N interaction. The energy $\omega_n$ and momentum $K$ of a pion do not satisfy the ordinary formulas $\omega_n^2 = m_\pi^2 c^2 + k^2 c^2$. The virtuality of pions consists in that they live for a time $\sim \frac{\hbar}{m_\pi c^2} \approx 5 \cdot 10^{-24}$ s and travel a characteristic length $\sim \frac{\hbar}{m_\pi c} \approx 1.4$ fm.

Owing to a strong pion–nucleon interaction, the ground state of dense nuclear matter must comprise the Bose–Einstein condensate of pions [686, 687, 885, 886]. This condensation may be described on the basis of the theory of a nuclear Fermi liquid [689] or in terms of a Lagrangian [885, 886].

The former approach involves a stability analysis of the spatially homogeneous state of $npe\mu$ substance in $\beta$ equilibrium. It turns out [414, 686, 687] that the system for $\rho > \rho_0$ loses stability with respect to periodic structures — density waves. In the consideration of stability for $\rho > \rho_0$ it was possible to determine different perturbation branches and their stability.

Pion condensation plays an important role in the calculation of neutrino emission, because it opens the URCA process with momentum conservation in the normal $npe\mu$ substance.

Calculations of the EOS of the $npe\mu$ substance under the conditions of $\beta$ equilibrium between hadrons and leptons proceed [414] from minimization of the total energy density:

$$\varepsilon(n_\mathrm{b}, n_\mathrm{q}^{(\mathrm{h})}, n_\mathrm{e}, n_\mu) = \varepsilon(n_\mathrm{b}, n_\mathrm{q}^{(\mathrm{h})}) + \varepsilon_\mathrm{e}(n_\mathrm{e}) + \varepsilon_\mu(n_\mu), \qquad (11.7)$$

for a fixed baryon density and under the electroneutrality condition:

$$n_\mathrm{n} + n_\mathrm{p} = n_\mathrm{b}, \qquad n_\mathrm{p} + n_{\pi_\mathrm{s}^+} = n_{\pi^-} + n_\mathrm{e} + n_\mu. \qquad (11.8)$$

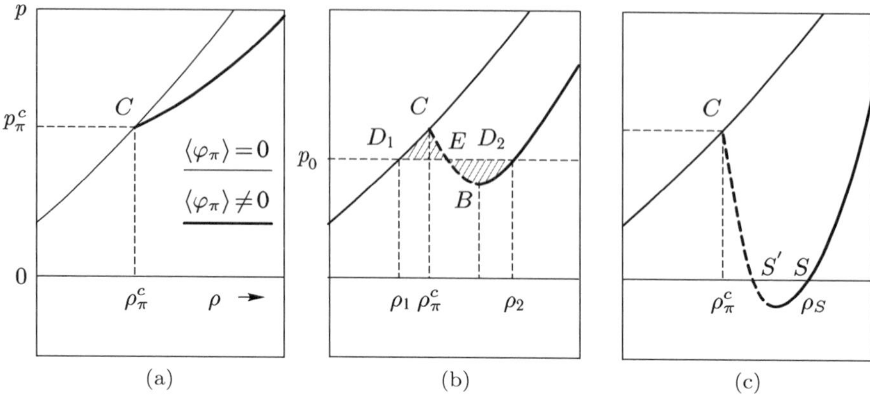


Fig. 11.13    Three qualitatively different EOS of pion-condensed media. Bold solid lines show pion-condensed phases, while thin solid lines refer to the normal phase, (a) phase transition of the second kind with a moderate softening, (b) a strong softening, which implies a phase transition of the first kind for $p = p_0$; (c) a very strong softening, which leads to stably "anomalous" superdense self-bounded states for $p = 0$ and $\rho = \rho_s$. Reproduced from Ref. [414] with kind permission of the authors.

The possibilities that open up in this case are depicted in Fig. 11.13 [414]. One can see a strong "softening" of the EOS due to pion condensation, which has the effect that $v_s^2 < 0$ (in the $CB$ portion in Fig. 11.13(b)).

Figure 11.13(c) deserves special mention, for it predicts [687,688] the formation of anomalous nuclei of significantly higher density than the ordinary ones. In astrophysics this leads to the possibility of existence of anomalous stars of size of about a golf ball [414, 428].

In the case of finite temperatures the peak of pion condensation remains invariable. This is a strong $\pi^- n$ attraction, which is opposed by an increase in kinetic energy due to the ordering of condensate state. An increase in $T$ shifts the condensation towards higher densities [592]. At the same time, for $k_B T \gtrsim 30\,\mathrm{MeV}$ the thermal pion gas also makes a contribution to the thermodynamics of the system. The critical temperature $T_{\mathrm{cr}}^\pi$ of pion condensation [564, 592] is equal to $k_\mathrm{B} T_{\mathrm{cr}}^\pi \gtrsim 60\,\mathrm{MeV}$.

Kaon condensation was long considered as a process of little significance due to the large kaon mass $m_\mathrm{K} c^2 = 493.6\,\mathrm{MeV}$. The stability condition for kaons $\mathrm{K}^-$ in the medium:

$$\mu_{\mathrm{K}^-} = m_\mathrm{K} c^2 = \mu_e. \tag{11.9}$$

A kaon may occupy the lower energy state only at high densities

$$n_{\rm b} > \frac{1}{3\pi^2} \left(\frac{m_{\rm K} c}{\hbar}\right)^3 \approx 33 n_0 \cdot \frac{x_{\rm e}}{0.1}, \tag{11.10}$$

where $x_{\rm e} = n_{\rm e}/n_{\rm b}$.

Due to a strong kaon–baryon interaction, kaons form a condensate at densities several times higher than the nuclear one $\rho_0$ [414, 825].

In the $npe\mu$ substance $K^-$ may be produced by way of the reactions:

$$e + N \to K^- + N + \nu_{\rm e}, \quad n + N \to p + K^- + N, \tag{11.11}$$

whereby its energy $\omega_{\rm K} < \mu_{\rm e}$. Nuclear experiments under normal conditions testify to a strong attraction of $K^-$ with an energy ranging from $-80\,{\rm MeV}$ to $-120\,{\rm MeV}$, and the potential depth is known for $n_{\rm b} \approx (2\text{–}3)n_0$. Like in the pion case, kaon condensation strongly increases the emission of neutrinos, resulting in enhanced stellar cooling.

Calculations of the effect of kaon condensate on the EOS of nuclear substance are performed in the framework of an average field model [414]. To construct the corresponding potentials and parameters of the average field model, advantage is taken of the experimental data on kaon-nucleon collisions as well as of the data of the quark model of hadrons and equilibrium nuclear properties [825]. Depending on the intensity of kaon-nucleon attraction, kaon condensation may be a phase transition of the first (for $U_{\rm K^-} \lesssim -90\,{\rm MeV}$) or second ($U_{\rm K^-} \gtrsim -80\,{\rm MeV}$) kind.

Estimates of the bounds of kaon condensation yield a value $n \gtrsim (3\text{–}5)n_0$ [565, 767] and predict a two-fold density jump in the transition and a strong "softening" of the EOS.

The effect of temperature on kaon condensation was investigated in Refs. [798, 799], where it was shown that the effect is small for $T \lesssim 60\,{\rm MeV}$ and consists in the shift of the condensation bound $\rho_c^{\rm K}(T) > \rho_c^{\rm K}(0)$. The data on the parameters of the corresponding critical points are supposedly absent [414].

For a collision energy $E_{e_{ab}} \sim 100\,{\rm MeV}/$nucleon of heavy nuclei the temperatures are not high enough for appreciable hadronization. Possible in this case is the effect of pressure decrease below the normal nuclear density, which is called the "liquid–gas" transition in the nuclear substance [954].

The critical point of this nuclear phase transition corresponds to $T \approx 18\,{\rm MeV}$, $\rho \sim 0.4\rho_e$. The occurrence of this transition may manifest itself in the features of fragment distributions in relativistic nuclear collisions. The details of this effect are described in Ref. [196].

## 11.5   Nucleons and hyperons under supercompression

As we saw above, nuclei cannot exist at a density $\rho > (1.5\text{–}2) \cdot 10^{14}\,\text{g/cm}^3$, and the system passes into a quasi-homogeneous plasma of neutrons, protons, and electrons. This state of matter is rather reliably described by the methods of many-body theory, which is validly employed for describing the nuclear structure. In this state, baryons are in $\beta$ equilibrium with electrons (and with muons if the electron Fermi energy is higher than $m_\mu c^2 \approx 105.7\,\text{MeV}$). Such states of neutron stars are referred to as "*npe$\mu$ matter*"; it is found in the outer core of neutron stars [414]. Under higher densities the substance passes into a so-called hyperon state.

Since we are dealing with densities higher than the ordinary nuclear density, $\rho \approx 2.8 \cdot 10^{14}\,\text{g/cm}^3$, our knowledge about such states are quite scarce, which opens the way to various qualitative assumptions. For instance, here the substance may lose homogeneity with the formation of pion and kaon condensates; the deconfinement of quarks with the production of quark–gluon plasma may also take place (see Chapter 12).

The calculation of the EOS for $\rho \gtrsim 10^{14}\text{–}10^{15}\,\text{g/cm}^3$ involves determination of the ground state of the system of hadrons and leptons and is made by minimizing the free energy for given electric and baryon charges.

This is an extremely intricate physical problem solved for an almost complete absence of experimental information — there are observational data only on neutron stars [414], supernova explosions [294, 295], and relativistic nuclear collisions.

Before setting out to discuss the EOS of nuclear matter, we emphasize once again that the case in point is the description of a homogeneous isotropic nuclear medium characteristic of astrophysical objects. In experiments on nuclear collisions (see Chapter 12) we are dealing with a finite system of nucleons, for which the conditions of homogeneity and thermodynamic equilibrium should be separately verified (see Refs. [294, 295]). Also at issue in this case is local thermodynamic equilibrium — an approximation which underlies transient gas dynamics.

For $\rho \sim \rho_0$, nuclei contain no more than 10% of protons and electrons, and so they may be neglected and the system may be treated as consisting purely of neutrons. A large number of calculations of the EOS have been made for this matter (see Refs. [414, 1045] and references therein), which suffer from large uncertainty for $\rho \sim 3\rho_0$, because about half of the nucleon energy is determined by the poorly known potentials of nucleon interactions.

The next step — calculation of the EOS for the $npe\mu$ substance in $\beta$ equilibrium is made proceeding from the data on asymmetric nuclear matter [414] — a homogeneous mixture of nucleons [11] — as well as the methods of Brueckner–Bethe–Goldstone (BBG) expansions [74].

To describe the contribution made by nucleons and hyperons to the EOS of matter requires knowledge of the corresponding interaction potentials [988] with the subsequent calculation of their thermodynamic potentials using, for instance, the Hartree–Fock technique. In the calculations [988] account was taken of nucleons, $\Sigma$, $\Lambda$, $\Xi$ hyperons and $\Delta$ resonances, which turned out to be present in the system for $\rho \gtrsim 1.2 \cdot 10^{15} \, \mathrm{g/cm^3}$. The results were found to depend heavily on the form of interparticle interaction.

Calculations of the EOS by the Green function [1035, 1036] and BBG techniques demonstrated a "softening" of the EOS of substance on including the hyperon subsystem.

The models of nuclear matter are developed for calculating the EOS of homogeneous and infinite nuclear substance as well as for describing heavy multinucleon nuclei. In the former case the Coulomb energy for an infinite system may be eliminated. In the EOS for the fluid nuclear model discussed above this is effected by the passage $E_{\mathrm{coul}} = 0$ and $A \to \infty$. In this limit the energy per nucleon $E$ depends only on the neutron and proton densities. The properties of this system are conveniently expressed in terms of the nucleon number density $n_b$ and the asymmetry parameter $\delta = (n_\mathrm{n} - n_\mathrm{p})/n_\mathrm{b}$. The case $\delta = 0$ corresponds to symmetric nuclear matter and $\delta = 1$ to the neutron matter. The former case is most simple: owing to the charge symmetry it is possible to consider only one type of particles — nucleons. The symmetric matter and the neutron matter are good approximations for describing the bulk properties of heavy nuclei and neutron stars, respectively. In the former case, for ordinary nuclei under terrestrial conditions $\delta^2 \lesssim 0.04$.

Calculations of nuclear matter properties serve as a material for comparison with multiparticle nuclear models (see Ref. [74]). $E$ and $n_e$ at the minima of the curves (Fig. 11.14) are denoted by $E_0$ and $n_0$, so that $B_0 = -E_0$ is the greatest binding energy per nucleon in nuclear matter. A comparison of short-range repulsion and long-range attraction determines the "saturation" in the nucleon–nucleon interaction. $B_0 = -E_0$ is the binding energy at saturation and $n_0$ is the saturation density.

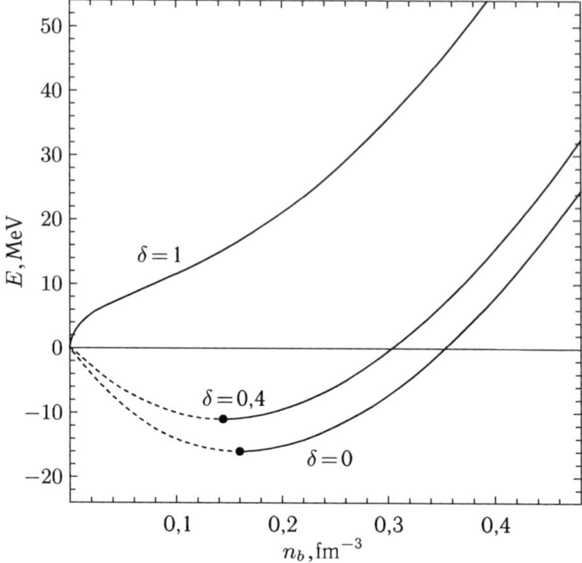

Fig. 11.14   Energy per nucleon as a function of baryon number density for symmetric nuclear matter ($\delta = 0$), for asymmetric nuclear matter with $\delta = 0.4$ (this asymmetry corresponds to the neutron evaporation point in crust of a neutron star and in the central core of a newly born protoneutron star), and purely neutron matter ($\delta = 1$). The minima of the $E(E(n_{\rm b}))$ curves are marked with points. The dashed segments correspond to negative pressure. These calculations were performed for the SLy4 model of the effective nuclear Hamiltonian, which was employed for calculating the SLy EOS [220]. This gives $n_0 = 0.16\,{\rm fm}^{-3}$ and $E_0 = -16.0\,{\rm MeV}$ [414].

Let us consider the case of small $\delta$ typical for nuclei at normal conditions:

$$E(n_{\rm b}, \delta) \simeq E_0 + S_0\delta^2 + \frac{K_0}{9}\left(\frac{n_{\rm b} - n_0}{n_0}\right)^2, \tag{11.12}$$

where $S_0$ and $K_0$ are the symmetry and compressibility energies:

$$S_0 = \frac{1}{2}\left(\frac{\partial^2 E}{\partial \delta^2}\right)_{n_{\rm b}=n_0,\,\delta=0}, \qquad K_0 = 9\left(n_{\rm b}^2\frac{\partial^2 E}{\partial n_{\rm b}^2}\right)_{n_{\rm b}=n_0,\,\delta=0} \tag{11.13}$$

$S_0$ defines the increase in energy due to a small asymmetry; $K_0$ defines the $E(n_{\rm b})$ curve at a point $n_{\rm b} = n_0$ and describes the increase in energy with increasing pressure. We will need these parameters in Chapter 13 for the development of the semi-empiric EOS of nuclear matter.

The numerical values of the quantities in equilibrium are determined from experimental data on the nuclear properties at normal conditions and are as follows (with different degrees of accuracy): $n_0 = 0.16 \pm 0.01\,\mathrm{fm}^{-3}$, $B_0 = 16.0 \pm 1.0\,\mathrm{MeV}$, $S_0 = 32 \pm 6\,\mathrm{MeV}$, $K_0 = 231 \pm 5\,\mathrm{MeV}$ [414].

The Hamiltonian of the system must comprise the contributions from particles or interactions. These contributions or usually determined by the methods of quantum chromodynamics (QCD). In this case, the Coulomb term in the Hamiltonian is negligible and the weak interaction is implicitly present in channels for the attainment of equilibrium.

Unfortunately, the practical realization of this formalism runs into serious difficulties, and therefore use is made of simplified models in which quarks are treated as being bound in hadrons (baryons and mesons) rather than treated separately. Next introduced into the Hamiltonian is a phenomenological model of strong (hadronic) interaction reliant on mesoscopic theories, where the strong hadronic interaction is described by a meson exchange model and the corresponding pseudopotentials are introduced. The parameters of these pseudopotentials are found on the basis of numerous experiments on nucleon scattering and experimental data on the properties of deuterium $^2$H. Experimental data on NH and HH interactions are confined to low-mass $\Lambda$ and $\Sigma$ hyperons. For other hyperons, experimental information is even more scarce.

Apart from pair interactions, for $\rho \geq 10^{15}\,\mathrm{g/cm}^3$ a significant contribution is made three- and multiparticle interactions, the data on them being quite scarce.

At first we consider a simplified minimalistic model [414] based on extension of the $npe\mu$ model to the domain $\rho \geq 2\rho_0$.

Historically, the main task of nuclear physics was to find the forces binding nucleons into a nucleus. According to Ref. [99], until the middle of the XX Century the mankind spent more time on this problem than on all other science problems altogether. Modern phenomenological potentials are constructed on the basis of a huge number of experiments on nucleon-nucleon scattering at energies of up to $350\,\mathrm{MeV}$, above which there set in inelastic pion production processes.

The effective nucleon–nucleon interaction potential is a complex and quite frequently cumbersome tensor function of spatial, spin, and isospin variables:

$$\hat{v}_{ij} = \sum_{u=1}^{18} v_u(r_{ij}) \hat{O}_{ij}^u, \qquad (11.14)$$

where the first 14 operators are charge-independent, i.e., are invariant with respect to time in the isotopic space:

$$\hat{O}_{ij}^{u=1,\ldots,14} = 1, \tau_i \cdot \tau_j, \sigma_i \cdot \sigma_j, (\sigma_i \cdot \sigma_j)(\tau_i \cdot \tau_j), \hat{S}_{ij}, \hat{S}_{ij}(\tau_i \cdot \tau_j),$$
$$\hat{\mathbf{L}} \cdot \hat{\mathbf{S}}, \hat{\mathbf{L}} \cdot \hat{\mathbf{S}}(\tau_i \cdot \tau_j), \hat{L}^2, \hat{L}^2(\tau_i \cdot \tau_j), \hat{L}^2(\sigma_i \cdot \sigma_j), \qquad (11.15)$$
$$\hat{L}^2(\sigma_i \cdot \sigma_j)(\tau_i \cdot \tau_j), (\hat{\mathbf{L}} \cdot \hat{\mathbf{S}})^2, (\hat{\mathbf{L}} \cdot \hat{\mathbf{S}})^2(\tau_i \cdot \tau_j).$$

Even this (11.14), (11.15) complex representation has a limitation: it is local and depends on $\vec{r}_{ij}$, and does not describe nonlocal terms — for instance, meson exchange.

Three-particle interactions begin to play a part with increasing $\rho$ and are required for describing the binding energy between $^3$H and $^4$He. One way to take them into account [816, 891] consists in modifying the medium-range potential and the single-pion exchange.

The relativistic corrections caused by the passage from the co-moving frame of reference to the laboratory one decrease the repulsive part of the potential and make a contribution to the EOS at high densities [11].

At the hadronic level, strong nucleon–nucleon interaction is the result of meson exchange between the nucleons. In field terms this is a result of the interaction between nucleon and meson fields.

The complete modern theory of strong interaction is the theory of QCD, where the main fields are those of quarks and gluons. Quarks and gluons do not arise directly in the meson exchange model for the strong interaction under consideration, and the substance consists of mesons, nucleons, and their resonances. The meson exchange model successfully describes nucleon–nucleon interactions at energies $\lesssim 350\,\text{MeV}$, the properties of $^2$H and diluted nuclear systems, and operates on nuclear and meson fields $\psi$ and $\varphi$. In this case, mesons have a mass $\lesssim 1\,\text{GeV}/\text{c}^2$, which does not permit describing short-range interactions for $\lesssim 0.2\,\text{fm}$. Meson–nucleon interaction is described by the corresponding term in the Lagrangian, which depends on the properties of meson field symmetry under rotation and reflection. As regards symmetry, considered are the following types of mesons: pseudoscalar (the field $\varphi^{\text{ps}}$, the mass $m_{\text{ps}}$), scalar (the field $\varphi^{(\text{s})}$, the mass $m_{\text{s}}$) and vectorial (the field $\varphi_\mu^{(\text{v})}$, $\mu = 0, \ldots, 3$, the mass $m_{\text{v}}$).

Experiment gives [414]: $m_\pi c^2 = 138\,\text{MeV}$, $m_\eta c^2 = 548\,\text{MeV}$, $m_\rho c^2 = 769\,\text{MeV}$, $m_\omega c^2 = 783\,\text{MeV}$, and $m_\delta c^2 = 983\,\text{MeV}$. The terms in the corresponding Hamiltonian describe the prescribed types of interactions and field symmetries, and contain pure adjustable parameters for describing experimental data [414].

The contribution from single-particle exchange by $\pi$- and $\omega$-mesons reflects two important properties of nucleon–nucleon interaction. The $\pi$ exchange results in long-range ($\frac{\hbar}{(m_\pi c)} \approx 1.4\,\text{fm}$) forces, while the $\omega$ exchange is responsible for the short-range ($\frac{\hbar}{(m_\omega c)} \approx 0.25\,\text{fm}$) repulsion and for the reasonable magnitude of the spin-orbit term.

The two-pion exchange is responsible for the medium-range ($\frac{\hbar}{(2m_\pi c)} \approx 0.7\,\text{fm}$) attraction, in which excited resonance states also make their contribution. Summing the corresponding Feynman diagrams gives the scattering matrices, the calculations which may be compared with $^2$HH collision experiment data.

To perform multiparticle calculations requires interaction pseudopotentials, which are constructed proceeding from the field-theory meson-exchange methods described above. This is usually done in the framework of a single-boson exchange (SBE) model, when the multipion exchange is described by the scalar $\sigma$ meson ($m_\sigma c^2 \approx 550\,\text{MeV}$) exchange with the corresponding $\sigma$-nucleonic coupling constant.

The interaction term depends not only on the distance between the particles, but also on their neighbors and their wave functions.

The single-component exchange model permits describing with a good accuracy $\approx 4300$ results of $pp$ and $np$ collisional experiments at energies $\lesssim 350\,\text{MeV}$. Detailed data on the potentials may be found in Refs. [956, 1046].

Data on nucleon–hyperon and hyperon–hyperon interactions are quite scarce. There are only a few points on nucleon–hyperon interactions and no points on hyperon–hyperon interactions. Available are data on hypernuclei with $A = 3$ and $A = 4$. Under these conditions, considerations about the types of interaction symmetries are invoked for constructing the corresponding potentials (for more details, see Ref. [414]).

The energy of nuclear matter is calculated in the following formulation [414]: a system of $A_b$ nucleons ($N$ neutrons and $Z$ protons) is considered in the limit $V \to \infty$, $Z/A_b = \text{constant}$. In homogeneous nuclear matter $n_n = \frac{N}{V}$, $n_p = \frac{Z}{V}$, and $n_b = \frac{A_b}{V} = n_n + n_p$. The Coulomb energy and proton–neutron mass difference are neglected.

To calculate the EOS of nuclear matter, use is made of the method of Green function, which is basically a version of the perturbation method, or variational methods, which involve minimizing the Hamiltonian in the space of multiparticle wave functions $\{\psi_{\text{var}}\}$ [11].

It is well known that the main properties of atomic nuclei may be described by two, on the face of it, different models: the shell model and the liquid-drop model. In the former model the multiparticle problem reduces to the problem of the motion of a single particle in some average field of other particles. In the liquid-drop model the nucleons interact with each other so strongly that their motion is completely collective, so that the nucleus resembles a droplet of strongly coupled nucleons. An analysis of the features of these models from the standpoint of field theory is given in Ref. [489], where the interaction of nucleons is effected by scalar and vector mesons. In this case, the scalar meson determines long- and medium-range attraction, which explains the formation of nuclei out of nucleons and the magnitude of potential minimum, while the vector meson defines repulsion and thereby the nuclear binding energy. Analogous models [224, 489] were developed by introducing similar meson fields and including relativistic effects.

An average field model with the introduction of electroneutral field of $\sigma$- and $\omega$-mesons, where $m_\sigma c^2 \approx 600\,\text{MeV}$, was proposed in Ref. [489]. The contribution of $\omega$ mesons to the interparticle potential is negative in character and increases with density. As a result, the potential has a minimum, which determines the nuclear binding energy.

The relativistic generalization of the model of Ref. [489] was set forth in Ref. [224], where the average field of the system is calculated in the Hartree approximation. The relativistic average field model has many attractive features. It is simple from the computational viewpoint even for multicomponent baryon matter consisting of nucleons and hyperons. Its Lorentzian invariance ensures that the perturbation propagation velocities are always lower than the velocity of light.

However, this model also has disadvantages. The meson approach implies that the average internucleon distance $r_{\text{NN}}$ is much shorter than the virtual meson path $r_\phi = \frac{\hbar}{m_\phi c}$, which is fulfilled for $n_{\text{b}} \gg 100 n_0$. There where the model corresponds to experiment, $r_{\text{NN}} \gg r_\phi$. There are other drawbacks as well (see Ref. [414]).

According to density functional method (DFM) [563], the energy of a multiparticle system is a functional of the single-particle density. The ground state energy is determined by minimizing this functional of the single-particle density. The model of the DFM is significantly simplified in the case of a homogeneous spin-nonpolarized nuclear matter, where everything depends on $n_p$ and $n_n$. The parametric form of the dependence $\varepsilon_{\text{N}}(n_n, n_p)$ is selected so as to satisfy experimental data or is inferred from more exact models [165, 166]. Several functionals for the EOS may be found in Refs. [414, 842].

Table 11.2   Masses, electric charges, strangenesses, and average measured lifetime of a baryon octet. The baryon number, spin, and parameter ratio of all these baryons are 1, 1/2 and +1 respectively [414].

| Baryon | $mc^2$, MeV | $Q, e$ | $S$ | $\tau$, s |
|---|---|---|---|---|
| p | 938.27 | 1 | 0 | $> 10^{32}$ |
| n | 939.56 | 0 | 0 | 886 |
| $\Lambda^0$ | 1115.7 | 0 | $-1$ | $2.6 \cdot 10^{-10}$ |
| $\Sigma^+$ | 1189.4 | 1 | 1 | $0.80 \cdot 10^{-10}$ |
| $\Sigma^0$ | 1192.6 | 0 | $-1$ | $7.4 \cdot 10^{-20}$ |
| $\Sigma^-$ | 1197.4 | $-1$ | $-1$ | $1.5 \cdot 10^{-10}$ |
| $\Xi^0$ | 1314.8 | 0 | $-2$ | $2.9 \cdot 10^{-10}$ |
| $\Xi^-$ | 1321.3 | $-1$ | $-2$ | $1.6 \cdot 10^{-10}$ |

With increasing density, hyperons begin to replace high-energy neutrons [414]. Introduced in lieu of the nucleon energy density $\varepsilon_{\mathrm{N}}(n_n, n_p)$ is a more general baryon functional $\varepsilon_{\mathrm{B}}(n_{\mathrm{B}})$, where $n_{\mathrm{B}}$ is the set of densities of baryon particles $B$:

$$\sum n_{\mathrm{B}} = n_{\mathrm{b}}.$$

For $n_{\mathrm{b}} \lesssim 10 n_0$ it would suffice to consider six light baryons: the nuclear charge doublet N, the charge singlet $\Lambda$, the charge triplet $\Sigma$, and the charge doublet $\Xi$. The parameters of these baryons are collected in Table 11.2. The baryon field possesses well-defined properties in an isospin space, nucleons and hyperon $\Xi$ are represented by an isospin field, $\Lambda$ by an isoscalar field, and $\Sigma$ by an isovector field.

To calculate the EOS [414] we consider a scheme consisting of nucleons, electrons, and muons. The nucleons are a strongly interacting Fermi liquid; the electrons and muons are a quasi-ideal Fermi gas:

$$\varepsilon(n_n, n_p, n_e, n_\mu) = \varepsilon_{\mathrm{N}}(n_n, n_p) + \varepsilon_e(n_e) + \varepsilon_\mu(n_\mu), \tag{11.16}$$

The equilibrium conditions

$$\mu_n = \mu_p + \mu_e, \quad \mu_\mu = \mu_e, \tag{11.17}$$

are reflective of the weak interaction processes

$$n \to p + e + \overline{\nu}_e, \quad p + e \to n + \nu_e, \tag{11.18}$$

$$n \to p + \mu + \overline{\nu}_\mu, \quad p + \mu \to n + \nu_\mu, \tag{11.19}$$

The substance is usually transparent to neutrinos, and they make no contribution to the thermodynamics.

For ultrarelativistic electrons

$$\mu_e = \hbar c p_{\mathrm{F}} \approx 122.1 \left( \frac{n_e}{0.05 n_0} \right)^{1/3} \mathrm{MeV}. \tag{11.20}$$

For medium-relativistic muons

$$\mu_\mu = m_\mu c^2 \sqrt{1 + \left( \frac{\hbar p_{\mathrm{F}}}{m_\mu c} \right)^2}. \tag{11.21}$$

Muons are present when $\mu_e > m_\mu c^2 = 105.65\,\mathrm{MeV}$.
The pressure

$$P = n_{\mathrm{b}}^2 \left( \frac{d\left( \frac{\varepsilon}{n_{\mathrm{b}}} \right)}{d n_{\mathrm{b}}} \right). \tag{11.22}$$

Figure 11.15 shows the EOS of the nuclear matter [414] calculated using five models. The BBB1 and BBB2 models are the BBG models calculated with different nucleon-nucleon potentials and triple interaction models. The APR and APR* models are based on the variational method with the same

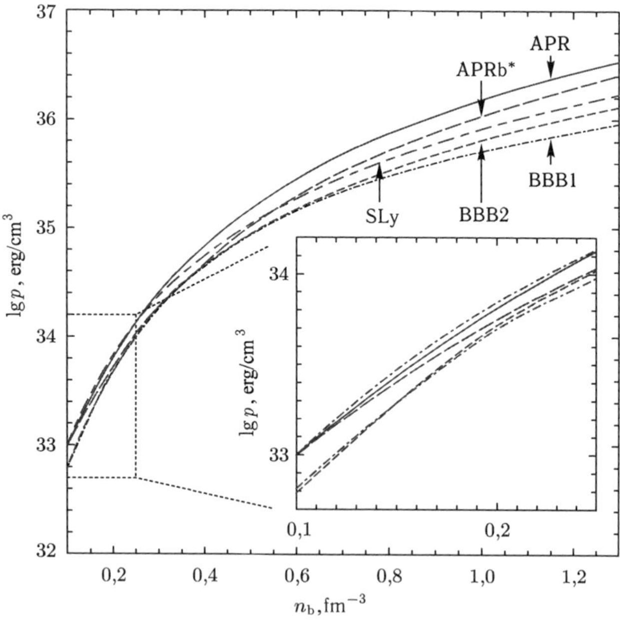


Fig. 11.15  Pressure in relation to the baryon number density for several EOS (Table 11.3) of *npeμ* matter in $\beta$ equilibrium. Reproduced from Ref. [414] with kind permission of the authors.

Table 11.3   Selected EOS of the cores of neutron stars [414].

| EOS | Model | Reference |
|---|---|---|
| BPAL12 | $npe\mu$ energy density functional | [120] |
| BGN1H1 | $np\Lambda\Xi e\mu$ energy density functional | [72] |
| FPS | $npe\mu$ energy density functional | [768] |
| BGN2H1 | $np\Lambda\Xi e\mu$ energy density functional | [72] |
| BGN1 | $npe\mu$ energy density functional | [72] |
| BBB2 | $npe\mu$ Brueckner theory, NN Paris potential + UVII NNN Urbana potential | [73] |
| BBB1 | $npe\mu$ Brueckner theory, A14 NN Argonne potential + UVII NNN Urbana potential | [73] |
| SLy | $npe\mu$ energy density functional | [220] |
| APR | $npe\mu$ variational theory, A18 NN Argonne potential + UVII NNN Urbana potential | [11] |
| APRb* | $npe\mu$ variational theory, A18 NN Argonne potential with stress correction + UVII NNN Urbana potential | [11] |
| BGN2 | $npe\mu$ effective nucleon energy functional | [72] |

nucleon–nucleon interaction potentials but with a new triple interaction potential improved using experimental data. The UIX* repulsive part of triple nucleon interactions is stronger than in UIX forces of the APR EOS. However, this has a weak effect on the EOS. Finally, the SLy EOS is based on the DFM and is close to the APR model. As we have already seen, the fraction of protons has a strong (cubic) dependence on the symmetry energy.

An assumption that the substance has the same composition for $\rho \gtrsim 2\rho_0$ as under lower compressions form the basis for the so-called "minimal" model [414], which underlies several models (Table 11.3 [414]). The results of calculations by the APR and APRb* EOS are compared in Fig. 11.15 [414]. One can see that the inclusion of relativistic calculations and improvement of the triple nucleon interaction potential has only a weak effect on the EOS even for $n_b \simeq 1.2\,\text{fm}^{-3}$. The APR and APR* models are "softer" than the models with older versions of the potential.

Figure 11.16 gives the density dependence of the adiabatic exponent:

$$\gamma = \frac{n_b}{p}\frac{dp}{dn_b} = \frac{p+\varepsilon}{p}\frac{dp}{d\varepsilon}, \qquad (11.23)$$

The oscillations of $\gamma(n_b)$ are reflective of the physical processes in compressed $npe\mu$ substance described above.

With increasing compression, for $\rho \gtrsim 2\rho_0$, as noted above, in the system there sets in the production of hyperons. Their condensation is

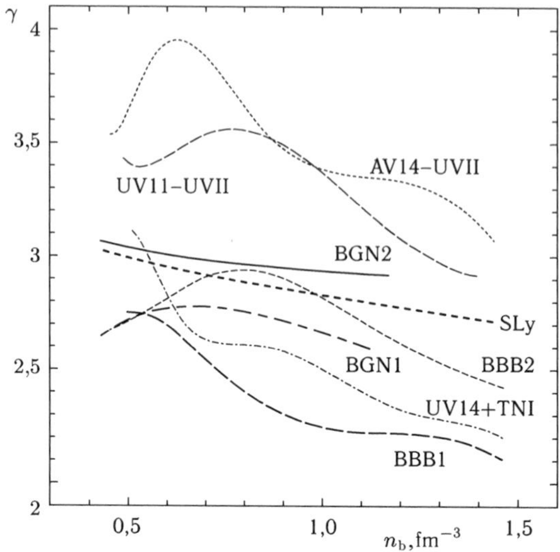



Fig. 11.16  Adiabatic exponent of $npe\mu$ matter in the inner core of a neutron star in relation to the baryon density for several EOS. All the equations, excepting three EOS, were borrowed from Table 11.3 and the remaining ones from [1045]. Reproduced from Ref. [414] with kind permission of the authors.

calculated by conventional thermodynamic methods [414] and is depicted in Figs 11.17 and 11.18. The appearance of hyperons in the system is easily seen in the composition curves (Figs 11.17 and 11.18). For instance, $n_\Lambda \simeq n_n$ for $n_b = 8n_0$. According to calculations [414], with increase in compression there initially emerge $\Sigma^-$ hyperons. The hyperonization of substance results in its deleptonization: the negatively charged particles e and $\mu$ are replaced by $\Sigma^-$ and other negative hyperons, and eventually there persist only $\Sigma^+$ hyperons. The so-called "baryon soup" emerges for $n_b \cong 1\,\mathrm{fm}^{-3}$.

The emergence of baryons "softens" the EOS, because high-energy neutrons are replaced by massive low-energy baryons (which produce a lower pressure). Figure 11.19 serves to illustrate this "softening" of the EOS. The change of adiabatic exponent $\gamma$ is shown in Fig. 11.20.

We note that the EOS depends heavily on the model for $n_b \geq 5n_0$. In particular, little is known about three-particle interactions including hyperons.

General constraints on the EOS are required in the domain $\rho > \rho_0$, where our knowledge about the behavior of nuclear substance are especially limited. The main requirements of general nature on the EOS are causality

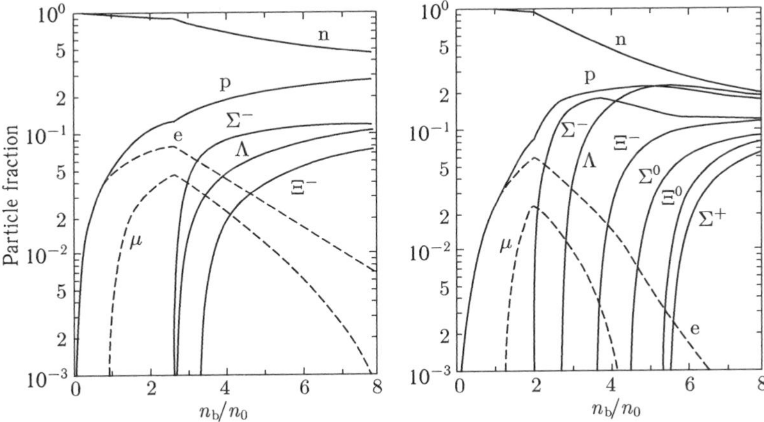

Fig. 11.17  Particle fractions $x_j = n_j/n_b$ as functions of the baryon number density $n_b$ (in units of $n_0 = 0.16\,\text{fm}^{-3}$) are calculated by a formula from Ref. [418] for two relativistic baryon interaction models. At the left: effective chiral model [418]. At the right: relativistic average field model TM1 [958]. Reproduced from Ref. [414] with kind permission of the authors.

and Lorentz invariance. They may be formulated in different ways. For instance, by introducing the requirements of

$$\text{ultrabaricity,} \quad p > \varepsilon \tag{11.24}$$

$$\text{and causality}(v_s > c), \quad \frac{dp}{d\varepsilon} > 1. \tag{11.25}$$

The circumstances of the violation $v_s > c$ for several EOS are marked by points in Fig. 11.21.

The problem of interrelation between the constraint $v_s < c$, Lorentzian invariance, and causality was analyzed in Refs. [106, 107, 156, 759, 869]. Two questions are raised:

(a) does Lorentzian invariance impose constraints on the equations of state?
(b) is the condition $\frac{dp}{d\varepsilon} > 1$ sufficient for the fulfillment of causality condition?

How is the EOS under discussion related to the principle of causality? The condition $v_s > c$ may be indicative of the violation of causality, because a signal propagates faster than light in this case. However, $v_s > c$ may result from the interference of several special components of sound waves, when high-frequency components are amplified and low-frequency ones attenuate.

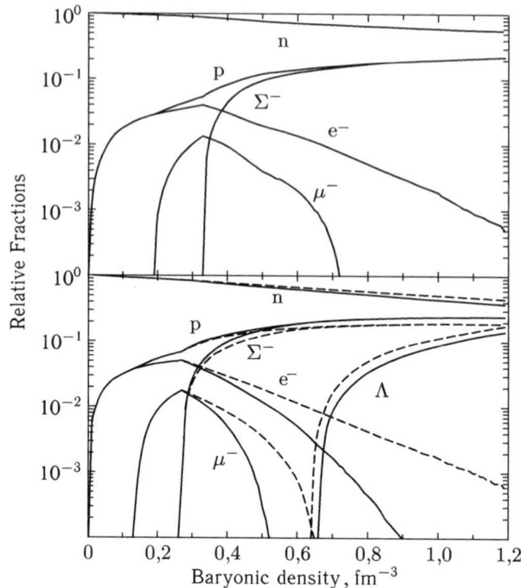

Fig. 11.18  Particle fractions $x_j = n_j/n_b$ as functions of $n_b$ calculated by a formula from Ref. [1016] in the BHF approximation for two baryon interaction models. The upper panel: Nijmegen model $E$ [841], only $\Sigma^-$ hyperon is present in the cores of neutron stars. The lower panel: the APR model for the nucleon sector (Table 11.3, [11]) and Nijmegen model E [841] for the NH and HH interactions. Unlike the upper panel, $\Lambda$ is present in the dense matter. Solid lines in both panels: all baryon–baryon (NN, NH, HH) interactions are included. Dashed lines in the lower panel: HH interactions are (artificially) disengaged. Reproduced from Ref. [414] with kind permission of the authors.

To summarize, it is valid to say that Lorentzian invariance and causality do not rule out the possibility that $v_s > c$. However, the corresponding disconfirmative example pertains to ultrahigh densities, which are higher than the characteristic quark deconfinement densities.

In summary of this section, we emphasize the following. As discussed above, our experimental data about the properties of nuclear matter are quite scarce when we are dealing with temperatures and densities exceeding the ground state with $\rho_0 \simeq 0.15\,\mathrm{fm}^{-3}$ [954]. In this domain the hadronic matter may exhibit a great variety of properties and structures, quite often of a highly exotic nature. Possible for $T < 20\,\mathrm{MeV}$ and $\rho < \rho_0$ is nuclear evaporation [954] and for high densities $\rho > 3\rho_0$ the emergence of hadronic nuclear substance (pion condensate and isomers).

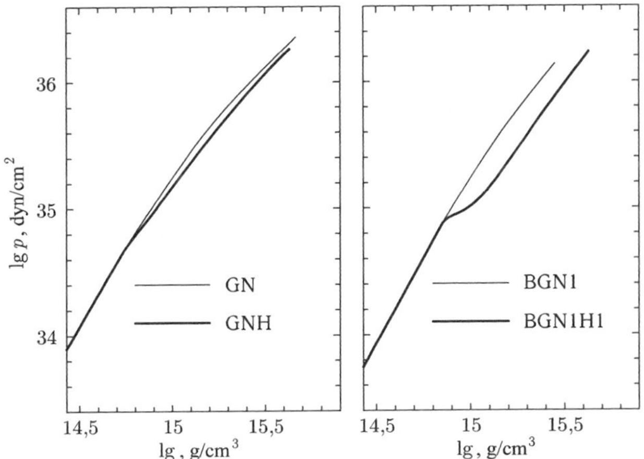

Fig. 11.19 "Softening" of the EOS by the presence of hyperons. Each panel shows the EOS with (bold line) and without hyperons (thin line). At the left: EOS model of [349]. At the right: BGN1 and BGN1H1 EOS from Ref. [72] (see Table 11.3). Reproduced from Ref. [414] with kind permission of the authors.

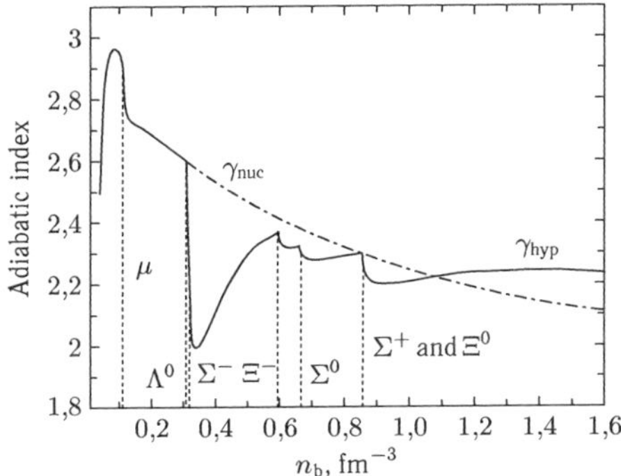

Fig. 11.20 Adiabatic exponent $\gamma$ as a function of $n_b$ in the core of a neutron star. The calculations were performed for a model EOS [349]. Solid line ($\gamma_{hyp}$): hyperonic substance (vertical dotted lines indicate the thresholds of the emergence of muons and hyperons); the dash-dotted line ($\gamma_{nuc}$) corresponds to the case when the appearance of hyperons is artificially forbidden. Reproduced from Ref. [414] with kind permission of the authors.

The pionization of nuclear matter is predicted to take place for $T > 50\,\mathrm{MeV}$. The phase transition caused by the deconfinement of quarks and transition of nuclear matter to the state of quark–gluon plasma takes place for $\rho > 5 - 10\rho_0$ and $T \approx 150 - 250\,\mathrm{MeV}$. These exotic states of matter supposedly emerged in the first milliseconds after the Big Bang and emerge

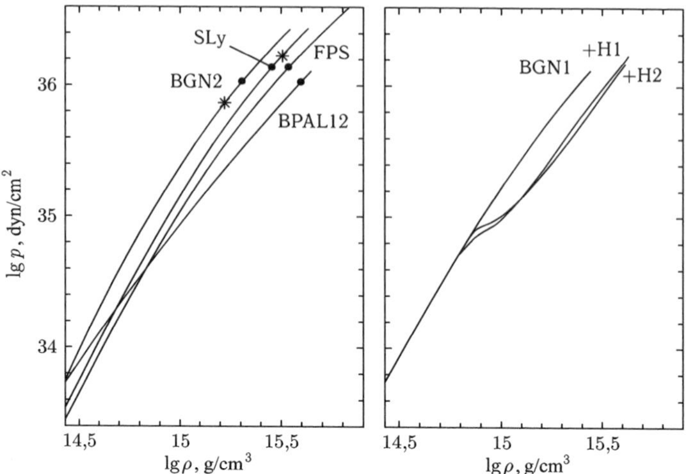


Fig. 11.21   Model EOS of the core of a neutron star (see Table 11.3). Black points correspond to the highest densities in stable neutron stars, while asterisks indicate the densities above which the EOS are superluminal ($v_s > c$). Reproduced from Ref. [414] with kind permission of the authors.

in the course of supernova explosions [294, 295] and neutron star production [414] as well as in the relativistic collisions of heavy nuclei [954].

Of special interest are the works on the prediction of heavy compressions in nuclear collisions (see Chapter 12) performed by the "Frankfurt" school of physics headed by W. Greiner and H. Stöcker [954]. Not only was a beautiful hydrodynamic theory of heavy nuclear collisions constructed in these elegant works, but they also were concerned with the development of a method for obtaining EOS proceeding from observable collisional data (fragments, pions, etc.).

Chapter 12

# Quark–Gluon Plasma
# and Strange Matter

The Equation of State (EOS) of degenerate hadronic matter becomes
substantially simpler in the limit of high densities. Quarks are no longer
bound in hadrons and constitute a weakly interacting Fermi gas [189].

## 12.1 Quark deconfinement and quark–gluon plasma

The Gell-Mann and Zweig quark model of hadrons started developing
rapidly after the mid-1960s [414]. Discovery of scaling in deep-inelastic
electron–nucleon reactions in the late 1960s has shown that at short range or
at very large momentum transfers the nucleon constituents (valence quarks)
behave as weakly interacting point-like particles. However, the interaction
between quarks must be very strong at long distances or at small momentum transfers, so that quarks would stay inside hadrons and could not exist
in a free state. Politzer, Gross, and Wilczek [379, 1017], awarded the 2004
Nobel Prize, have shown that only a renormalizable quark field theory, in
which the interaction forces increase at long range and decrease at short
range, can provide a sufficiently good description of the properties discovered by Yang and Mills [1056]. For classification of hadrons it was assumed
that a quark should be a spin-1/2 fermion, have a fractional electric charge,
and be in one of the color states (a color is a new quantum charge similar
to the electric charge). The interaction between the quarks is mediated by
gluons (gluons serve as a glue which keeps the quarks together). The gluons are massless bosons whose spin is equal to 1 like the spin of a photon,
but unlike photons gluons are carriers of the color charge. Therefore they
interact pointwise with each other. Such theory is called nonabelian gauge

theory [414]. The theory of quarks and gluons — Quantum Chromodynamics (QCD) — is the conventional theory of strong interactions.

After the mid-1970s it became clear that the increase in temperature and density should induce a qualitative change of the properties of hadronic matter. Such compressed system can be described in terms of pions, nucleons, and other hadrons. In a very dense matter these extended and complex particles will overlap, while the quarks and gluons will be able to move freely in the space. It may even happen that a phase transition to color deconfinement occurs when the temperature is several hundreds of MeV or the baryon density is approximately 10 times higher than the normal nuclear density (see Fig. 1.13). The phase transition from the hadronic gas to the quark–gluon plasma (see Fig. 1.12) needs a very high energy density. Such transition could occur in a very early Universe during the few first microseconds after The Big Bang. It can occur in neutron starts and in the relativistic collisions of heavy nuclei at very high energies in particle accelerators [294].

The quark-gluon plasma (QGP) is generated in relativistic collisions of heavy ions and arises from quark deconfinement — at energies $\geq 200$ MeV [410, 720, 723]. The experimental scheme is as follows: In the collision of two nuclei [843] their kinetic energy of motion transforms into the internal energy of nucleons that, according to the QCD [720] predictions, gives rise to the production of the so-called "color glass condensate" and, subsequently, as the matter thermalizes, to the formation of a new state of matter, quark–gluon plasma, or "quark soup" [410, 720] (Fig. 1.12). It is assumed in this case that the collision time is long enough for the matter to thermalize, so that the kinetic energy can transform into the internal energy of the produced plasma (this is a subject of special consideration).

The QGP formed in these collisions consists of quarks, antiquarks, and gluons [350, 419, 746, 867]. Such a plasma is sometimes called as the "oldest" form of matter because it already existed a few seconds after the Big Bang. As a result of its expansion and cooling hadrons were formed. The QGP has a maximal density, approximately $9$–$10\rho_0$ ($\rho_0 = 2.8 \cdot 10^{14}$ g/cm$^3$ is nuclear density) and may arise at the center of neutron stars and black holes or in the collapse of ordinary stars. To search for the QGP, large-scale experimental programs have been recently launched to study ultrarelativistic ion collisions at the accelerators HERA and RHIC in Brookhaven, at GSI in Darmstadt, and SPS and LHC at CERN (see Fig. 1.13). The programs are aimed at the investigation of collisions of heavy nuclei with energies of the order of 100 GeV per nucleus and higher in the center-of-mass

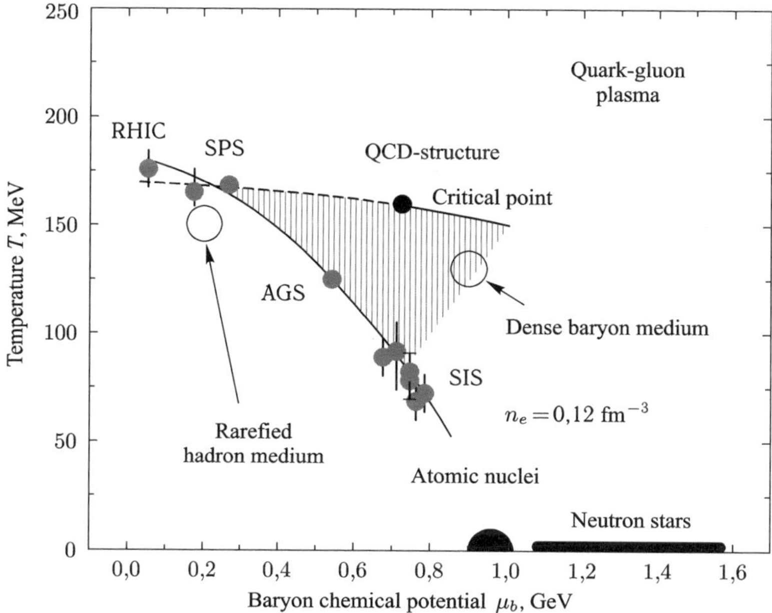



Fig. 12.1  Phase diagram of nuclear matter.

system or with energies of $20\,\mathrm{TeV}$ per nucleus in the laboratory frame. The conditions attainable with the existing accelerators are presented on the phase diagrams of nuclear matter (Figs. 12.1 and 1.13). The low-temperature and low-baryon-density area is occupied by hadrons (nuclei and mesons). The limiting case of high densities (5–10 times the nuclear density of large nuclei — approximately 0.17 particles per cubic femtometer, $\rho_0 \approx 2.8 \cdot 10^{14}\,\mathrm{g/cm^3}$) and high temperatures ($T > 200\,\mathrm{MeV} \approx 10^{12}\,\mathrm{K}$) corresponds to quarks and gluons which are not bound in hadrons under these conditions and form "quark-gluon plasma". The transition between these states may be smeared or abrupt, similar to the phase transition of the first kind with a critical point (Fig. 12.1). To describe the behavior of compressed baryon matter in the corresponding area of the phase diagram, the QCD approach is used which in this case is in itself a subject of experimental verification.

QGP provides an essential experimental study of matter transformation right after the birth of our Universe [294, 295]. During the first microseconds [746, 867] after the Big Bang the temperature decreased as $T(\mathrm{MeV}) \sim 1/\sqrt{t}$. Here $t$ is the time measured in seconds, so a QGP

Table 12.1    Characteristics of quarks.

| Notation | Name of the quark, its flavor | Charge, $\|e\|$ | Mass, MeV |
|---|---|---|---|
| $u$ | *up*–sup* | 2/3 | 5 |
| $d$ | *down*–down* | −1/3 | 10 |
| $s$ | *strange* – strange<br>(strangeness, $S = -1$) | −1/3 | 150 |
| $c$ | *charm* – charm<br>(charm, $C = +1$) | 2/3 | 1300 |
| $b$ | *beauty* – beauty<br>(beauty, $B = +1$) | −1/3 | 4200 |
| $t$ | *top* – top **,<br>*truth* – truth | 2/3 | 175000 |

*The name of the flavor is absent and the corresponding quantum number is not used.
**There is no conventional name for the flavor.

whose temperature was hundreds of MeV could exist during the first 5–10 microseconds after the Big Bang. The baryon density then was not too high. As the Universe expanded, the plasma was cooling inducing "hadronization" of the matter and subsequent formation of pions. If the phase transition of the first kind took place, "bubbles" of hadrons — neutrons, protons, and pions — could be formed in the plasma.

The QGP [497,511,910,911] is a superdense and superhot form of nuclear matter with unbound quarks and gluons, which at lower energies are bound in hadrons (Fig. 1.12).

Table 12.1 presents the characteristics of quarks (possessing a fractional charge) which make up hadrons. For example,

$$\text{Proton} = u + u + d,$$

$$\text{Neutron} = u + d + d,$$

$$\pi^+ = u + \overline{d}, etc.$$

## 12.2    Observed manifestations of the QGP

The existence of QGP follows from the property of the asymptotic freedom of QCD [85, 172, 189, 331] which yields a value of $1$–$10\,\text{GeV}/\text{fm}^3$ for the energy density of the corresponding transition, which is close to the energy density inside a proton and exceeds the nuclear energy density. Detailed numerical calculations give the critical conditions for the emergence of QGP: $T_\mathrm{c} \approx 150$–$200\,\text{MeV} \approx (1.8$–$2.4) \cdot 10^{12}\,\text{K}$ (Fig. 12.2).

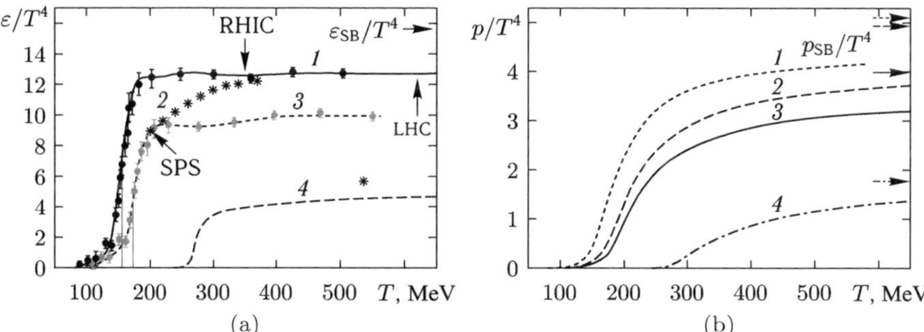


Fig. 12.2  QCD calculations of the temperature dependence of the energy density (*a*) and pressure (*b*) [410]. The most realistic case corresponds to $2+1$ flavors.

The appearance of this plasma manifests itself as an increase in the number of the degrees of freedom from 3 inherent to hadrons to 8 for gluons, times 2 helicity degrees of freedom, plus 2–3 degrees of freedom due to the quark light flavors which in turn possess 2 spins and 3 colors. Therefore, according to the quantum electrodynamics (QED), quarks possess 24–26 degrees of freedom, and in the QGP at $T \approx (1\text{–}3)T_{\rm c}$ 40–50 degrees of freedom are excited compared to 3 in a low-temperature ($T < T_{\rm c}$) pion gas. Since the energy density, pressure, and entropy are approximately proportional to the number of the excited degrees of freedom of the system, a sharp variation of these thermodynamic parameters in a narrow temperature range near $T_{\rm c}$ accounts for the large energy difference between ordinary nuclear matter and QGP (Fig. 12.2).

In view of the aforesaid, the EOS of the ideal ($g \to 0$) QGP has the form [409]

$$p_{\rm SB}^{\rm QCD}(T) = \left( \underbrace{2_s \times 8_c}_{\text{gluons}} + \frac{7}{8} \times \underbrace{2_s \times 3_c \times 2_{q\bar{q}} \times n_f}_{\text{quarks}} \right) \frac{\pi^2 T^4}{90} - \underbrace{B}_{\text{vacuum}} .$$

For hadronic matter [409]:

$$p^{\rm H}(T) = (\underbrace{3_{\rm ISO}}_{\text{pions}} + \underbrace{O(e^{-M/T})}_{\rho,\,\omega,\dots}) \frac{\pi^2 T^4}{90} .$$

A comparison of these expressions defines the critical temperature $T_{\rm c}$ of the QGP transition.

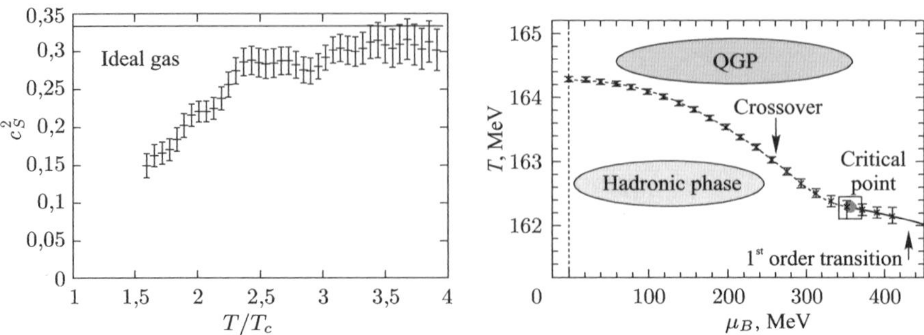

Fig. 12.3 Peculiarities of the QGP EOS [410]. Left: Temperature dependence of the speed of sound. Right: phase boundary and the critical point according to Refs. [198,279].

Like our customary "electromagnetic" plasma (EMP), the QGP may be ideal at $T \gg T_c$ and nonideal at $T \approx (1\text{–}3)T_c$. The proper nonideality parameter — the ratio of the interparticle interaction energy to the kinetic energy — in this case has the form $\Gamma = 2Cg^2/4\pi aT = 1.5\text{–}5$, where $C$ is the Casimir invariant ($C = 4/3$ for quarks and $C = 3$ for gluons) and $a \approx 0.5\,\text{fm}$ is the interparticle distance $a \sim 1/T$, $T = 200\,\text{eV}$, $g \approx 2$ is the strong interaction constant. The factor 2 in the numerator takes into account the magnetic interaction which in the relativistic case is of the same order of magnitude as the Coulomb interaction.

At present it is hard to say unambiguously if the transition to QGP is a true thermodynamic phase transition with an energy density jump or a sharp but continuous transition [410]. It is possible (Fig. 12.3 [410]) that for low baryon densities $\mu_B$ it might be a continuous transition and for high $\mu_B$ a phase transition of the first kind (Fig. 12.4). In any case, the theory [410] predicts a low value of the sound velocity in the transition region (Figs. 12.3 and 12.2) which is reflected in the observable hydrodynamic anomalies in the relativistic collisions of heavy nuclei. The specified peculiarities of the adiabatic compressibility of the QGP give an evidence for a "softer" QGP EOS at $T \approx T_c$ and a "stiffer" one at high temperatures and as well as at $T \leq T_c$. In the limit $T \ll T_c$, the hadronic matter EOS becomes "softer", although the uncertainty in this case is quite high and the description of this matter by QED technique encounters serious problems.

A picture of relativistic collisions of heavy nuclei is shown in Fig. 12.5. Under the conditions of the RHIC experiment, the longitudinal Lorenz compression of the size of colliding nuclei is of the order of 100. The

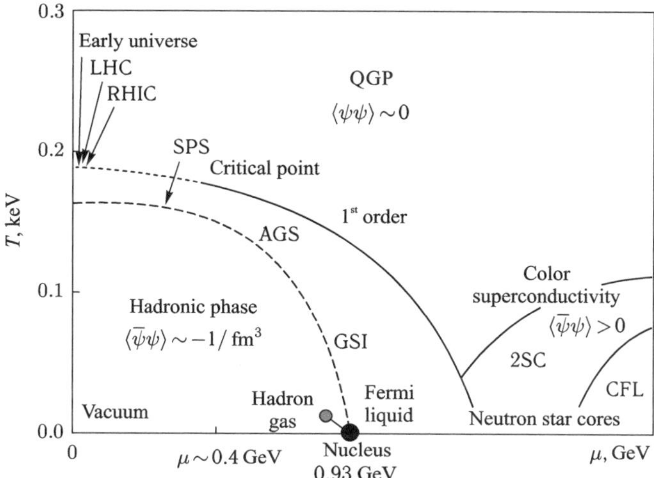



Fig. 12.4   Phase diagram of QGP [84, 409].

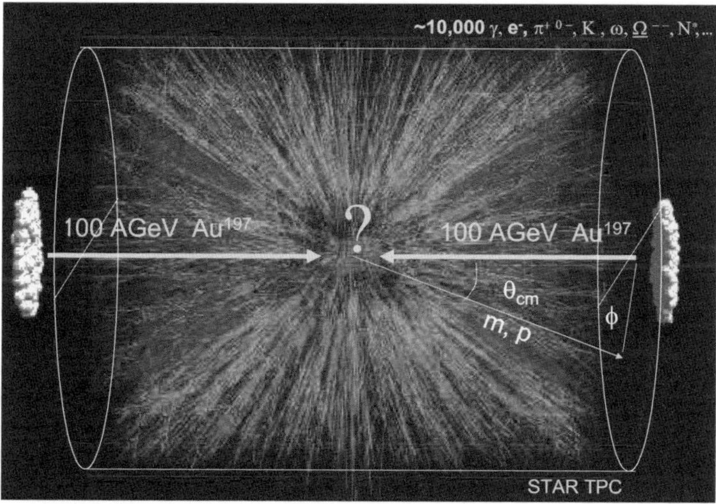



Fig. 12.5   Collision of relativistic hadrons — "burning of Vacuum" [601].

characteristic volume of the U+U collision range is approximately $3000\,\mathrm{fm}^3$, it contains about $10,000$ quarks and gluons, while the characteristic collision time is $\tau_0 \approx 0.2\text{–}2\,\mathrm{fm}/c \approx (5\text{–}50) \cdot 10^{-25}\,\mathrm{s}$. Therefore some of the high-energy processes presumably take place in the expanding matter when the nuclear bunches have already passed through each other.

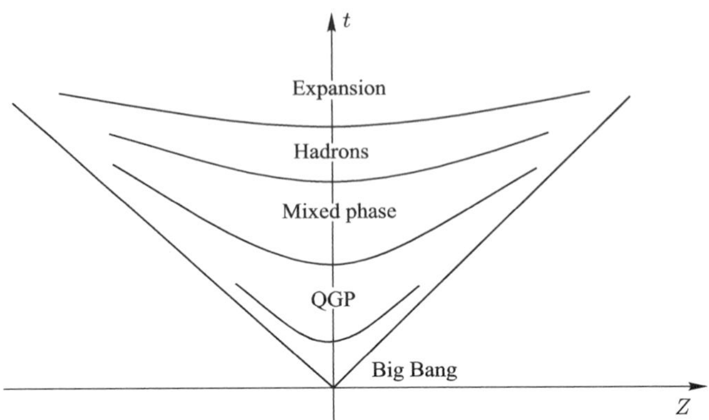


Fig. 12.6   Space–time evolution of matter after a collision [720].

The authors of Ref. [410] called attention to the fact that production of fast particles in the expanding plasma after nuclear collisions is similar to the production of new forms of matter after the Big Bang [294, 295]. However, the difference is that the expansion in nuclear collisions is one-dimensional rather than three-dimensional as in cosmology. A space–time evolution of matter after a relativistic collision is presented in Fig. 12.6.

During the collision, as the nuclear matter is expanding and cooling down, the produced quarks and gluons are thermalized (time $\tau_{\mathrm{eq}} \leq 1\,\mathrm{fm}/s \approx 3 \cdot 10^{-24}$ s) and may reach the local thermodynamic equilibrium during the plasma lifetime $\tau_0 \approx (1-2)R/c \approx 10\,\mathrm{fm}/c$. In this case, the medium will be set in the hydrodynamic motion; detection of it may provide experimental information on the properties of hadronic or quark–gluon matter and on the limits of their back and forth transition which according to QED should occur at an energy density about $1\,\mathrm{GeV}/\mathrm{fm}^3$.

Figure 12.7 [410] illustrates the characteristic energy densities in nuclear collisions as a function of time. An analysis of the collision and expansion dynamics shows [410] that the transition from a relatively slow one-dimensional expansion to a faster three-dimensional expansion occurs in a characteristic time of the order of $0.3\,\mathrm{fm}/c$. The upper part of the dark band in Fig. 12.7 corresponds to the assumption that the system is in thermodynamic equilibrium and that it is an ideal massless gas, while the lower part corresponds to nonequilibrium "frozen" conditions. By the time moment $3\,\mathrm{fm}/c$ plasma becomes a mixture of quarks, gluons, and hadrons, and at

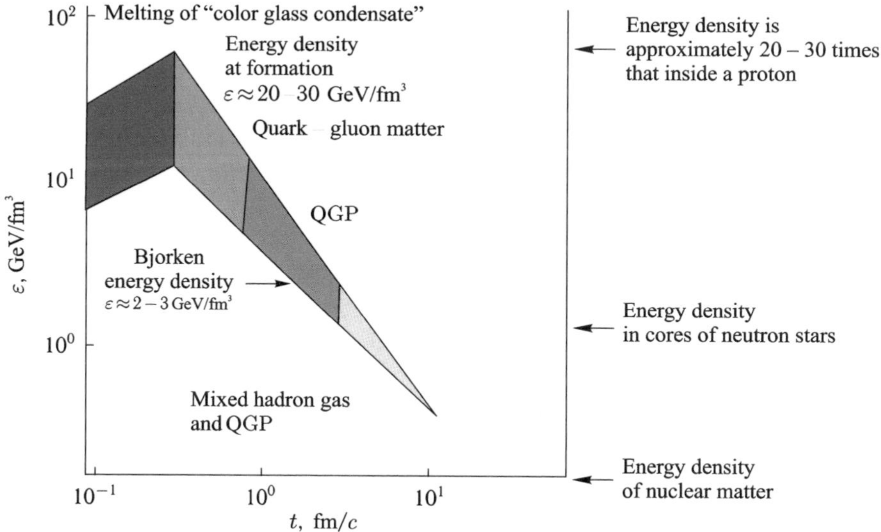


Fig. 12.7 Characteristic energy densities in nuclear collisions as a function of the interaction time [410].

$10\,\mathrm{fm}/c$ quarks and gluons recombine in hadrons. The lower bound of the attainable energy densities is realized at $t \approx 1\,\mathrm{fm}/c$, and the upper bound (the massless gas) is realized at $0.3\,\mathrm{fm}/c$. The general estimate of the energy densities is of the form [410]:

$$2\text{–}3\ \mathrm{GeV/fm}^3 \leq E \leq 20\text{–}30\ \mathrm{GeV/fm}^3.$$

For comparison, the energy density in neutron starts [414] is approximately $1\,\mathrm{GeV/fm}^3$.

The generation of a new form of matter — QGP — should be accompanied by new physical phenomena which should manifest themselves in experiments.

First, the appearance of new degrees of freedom in plasma must be reflected in the collision and expansion relativistic hydrodynamics which in local thermodynamic equilibrium is in turn described by the equations of motion of a viscous fluid. This formalism is simplified for a nonviscous fluid (the Euler equation), while the experimental manifestation of the collective (viscous) effects may be an evidence for plasma effects.

The second manifestation of the QGP is the difference between the experiment and the parameters of the hydrodynamic phenomena calculated from the given EOS [294, 295].

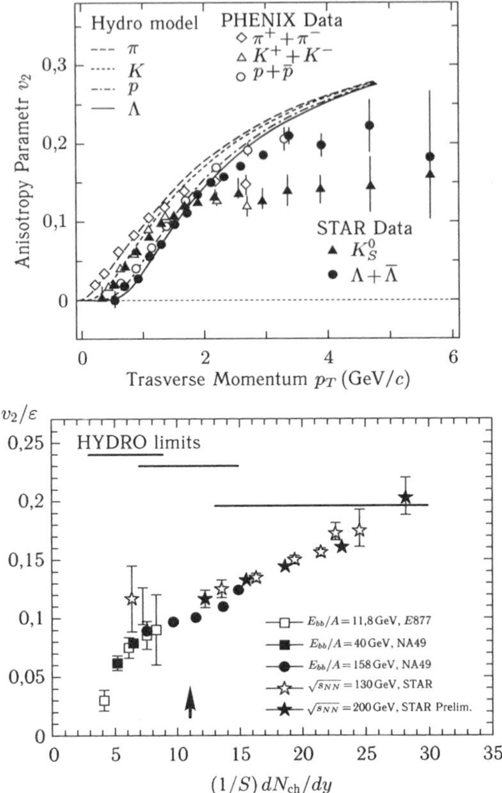


Fig. 12.8 Experimental manifestation of QGP [410]. Top: STAR [6, 7, 937] and FENIX [8] experimental results. Bottom: data from Ref. [25].

The results of such a comparison for the azimuthal flow components for $\pi$, K, p, $\Lambda$, and Au + Au in the 200 AGeV Au–Au collisions are shown in Fig. 12.8 [6–8, 25, 937]. One can see that up to energies of the order of $1\,\mathrm{GeV/fm^3}$ the calculation results agree well with the measurements, but at higher energies it is not the case. This disagreement is attributed to the appearance of QGP. Taking into account the decrease of the sound velocity in the vicinity of $T \approx T_\mathrm{c}$, which is caused by the appearance of this plasma, and the corresponding "softening" of the EOS would improve the agreement between the results of calculations and experiments.

The emergence of QGP may show up not only in the peculiarities of the EOS but also in the behavior of the viscousity in the hydrodynamic motion.

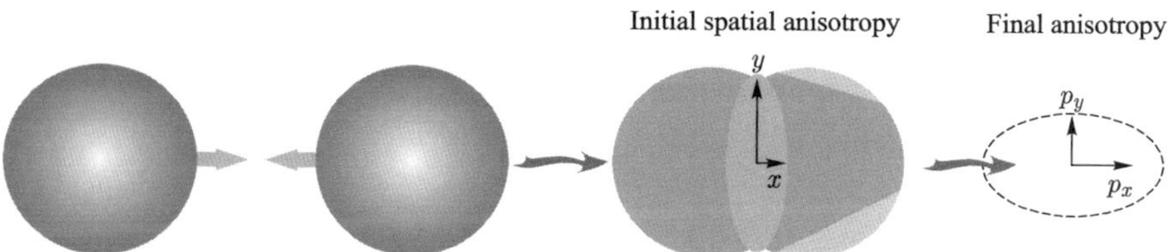


Fig. 12.9 Relativistic collision of nuclei: generation of elliptical flows. The high-energy region has elliptical shape, so that the spatial anisotropy generates the anisotropy of momenta of the expanding medium [758].

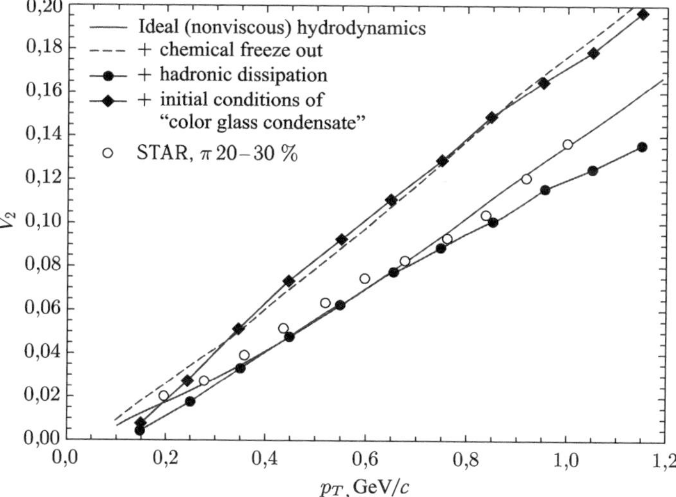


Fig. 12.10 Comparison of measured (circles) and calculated expansion velocities in nuclear collisions [84].

It is precisely these effects which are supposed to be responsible for the lower (compared to calculations) values of the elliptic expansion velocities measured at the SPS accelerator (Fig. 12.9) (this discrepancy is unavoidable in the framework of the three-dimensional nonstationary (3 + 1D) hydrodynamics) and for the reduction of this discrepancy by the increase of the collision impact parameter [410]. This is due to the lower efficiency of the pressure transfer to the hydrodynamic flow by hadrons as against plasma. In any case, the ideal nonviscous flow calculations by the Euler equations [84] are in better agreement with the experimental results (Fig. 12.10) than the calculations using the equations with viscous dissipation.

Interestingly, at moderate energies the RHIC experiments yield anomalies in QGP dissipation and give effective viscousity values which are up to a factor of 10 lower than those following from the models of Debye weakly nonideal plasma. This is assumed [410] to be due to the plasma nonideality effects.

The quenching of the jets produced in the relativistic nuclear collisions also contains information about the properties of shock-compressed matter [411, 412, 1031] and the emergence of QGP. By the order of magnitude this quenching is determined by the radiative losses of gluons, while the contribution given by elastic losses is rather low.

The results of such "tomography" for RHENIX experiments are presented on the left-hand panel of Fig. 12.11 [410, 1018]. They show that to account for the observed jet quenching the initial reduced density of gluons should be $dN^g/dy \approx 1000 \pm 200$. These values are in reasonable agreement with other independent measurements [410]:

(1) With initial entropy values determined from plasma expansion after a collision.

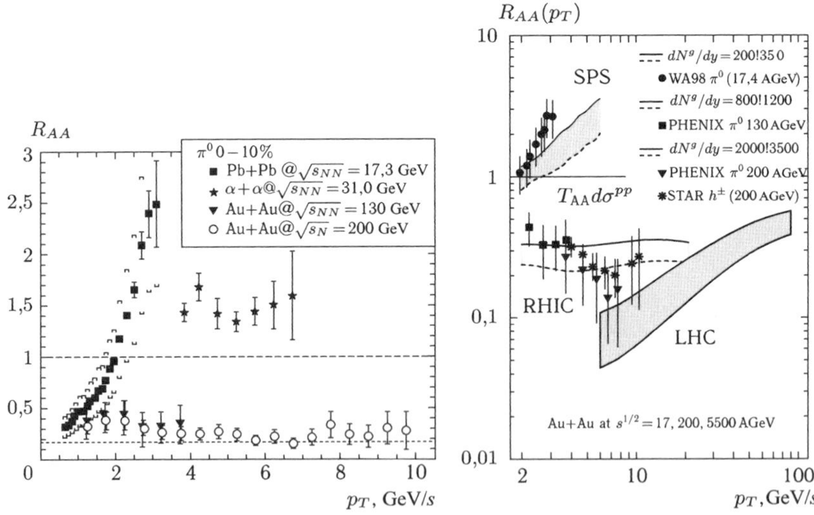


Fig. 12.11    Left: $\pi^0$ quenching data detected at PHENIX, RHIC in comparison with earlier data from ISR and SPS. Right: Factor $R_{AA} = dN_{AA}/T_{aa}(b)d\sigma_{pp}$ measured at SPS, RHIC, and LHC in comparison with the results of radiative energy loss models [410, 1018].

(2) With plasma initial parameters following from hydrodynamic calculations of "elliptical" flows.

(3) With density variation rates calculated within the framework of QED.

These four data sets enable the calculation of the initial energy density in relativistic collisions:

$$E_0 = E\left(\frac{1}{\rho_0}\right) \approx \frac{\rho_0^2}{\pi R^2} \cdot \frac{dN^{\mathrm{g}}}{dy} \approx 20 \text{ GeV/fm}^3 \approx 100\, E_a,$$

with the characteristic gluon momentum of about $\approx 1.0$–$1.4\,$GeV, which in turn determines the formation time $\hbar/p_0 \approx 0.2\,$fm$/c$ of the primary nonequilibrium QGP. Under these conditions, the local thermodynamic equilibrium required for the application of hydrodynamics is reached at

$$\tau_{\mathrm{eq}} \approx (1\text{–}3)B/P_0 < 0.6 \text{ fm}/c.$$

By this time the temperature will be

$$T(\tau_{\mathrm{eq}}) \approx \left(\frac{\varepsilon_0}{(1\text{–}3)\cdot 12}\right)^{1/4} \approx 2T_{\mathrm{c}}.$$

According to one of the models [410], for $P_0 \approx 2$–$2.2\,$GeV the number of minijets should be of the order of 1000.

More detailed information on plasma properties can be obtained from the experimental detection of correlated double jets in nuclear collisions [294, 295]. The correspondence of the characteristics of these jets obtained under different conditions (central and peripheral collisions, protons and gold nuclei) is considered [410] to be a substantial argument for the applicability of the QCD methods and for QGP production.

Thus, the jet quenching effects observed in nuclear collisions allow one to determine the energy density of nuclear matter and to draw conclusions about strong collective interaction (nonideality) of such plasma analyzing the energy losses of the jets in their motion through the QGP.

Today, the observational manifestations of QGP [409], which are being actively analyzed, include

— "Barometric" effects: from the parameters of the collective flows (elliptic, longitudinal, radial),

— "Thermodynamic" effects: photons, lepton pairs, vector mesons,

— "Critical" phenomena: hadron density fluctuations,

— "Tomography": short jets, heavy quark flows,

— "Exotics": multiquark jets, femtometer-dimensional "fullerens".

QGP formation signals include heat emission of photons and lepton pairs, $J/\psi$ jeneration, strangeness production, relative multiplicity of various mesons and baryons.

Among the interesting hydrodynamic phenomena, special mention should be made of the elegant and beautiful idea by Stöcker [955] on formation of conic Mach shock waves (Fig. 1.21) whose properties may provide information concerning the characteristics of compressed nuclear matter.

The search for QGP and the investigation of its physical properties are currently the subject of active theoretical and experimental studies in many laboratories throughout the world. These works received a new impetus after the Large Hadron Collider became operational.

## 12.3    Analogies between the QGP and EMP

The QGP and the ordinary plasma, which constitutes (98–99) % of the visible part of the Universe and is referred to as the EMP, have many differences and many similarities [720]. In particular, the QGP is different from EMP in that QGP contains antiparticles the existence of which cannot be neglected. Therefore, the total number of leptons rather than the number of particles is conserved. The particle density is no longer an adequate characteristic. It is replaced, instead, by the density of baryons and "strange" particles.

In EMP the large difference in the masses of electrons and ions leads to the difference in their dynamics and kinetics and determines, in particular, the difference of electron and ion temperatures in relaxation processes. In QGP there are also heavy (charm, top, and bottom) particles whose number is, however, lower than the number of light quarks and gluons and the lifetime is shorter. Therefore the contribution of heavy quarks to the QGP dynamics is small. The QGP is governed by QCD, while the EMP by QED. The latter theory is Abelian, unlike chromodynamics. In QCD gluons not only carry the color charge being responsible for quark–quark and quark–antiquark interaction but, also, interact with each other. Gluons, in contrast to photons, contribute to the color charge density and to the color current.

The most important common feature of QGP and EMP is the collective character of interparticle interaction [720]. The range of the effective electromagnetic interaction, in spite of the screening, is usually much larger than the interparticle distance. Therefore there are many particles in the Debye sphere and their motion is highly correlated. QED gives the solution which corresponds to the Debye one: $\Phi(r) = (q/r) \cdot e^{-rm_\mathrm{D}}$. Here, the Debye mass (playing the role of the Debye radius in the atomic system of units)

$m_{\rm D}^2 = g^2 T^2/3$ is of the order of $(gT)^2$, where $g$ is the QED constant. Since the particle density is $\sim T^3$ in this theory, the number of particles in the Debye sphere is $\sim 1/g^3$ in the weak compression limit $(1/g \gg 1)$. Interestingly, the pseudopotential of interparticle interaction for equal charges in QGP may become attractive in some cases [720]. The long-range interparticle interactions characteristic of EMP and QGP result in that an important role is played by collective effects like screening, plasma oscillations, instability, etc.

In laboratory experiments with EMG external electromagnetic or gravitational fields are applied, while for QGP it is hard to imagine fields of required intensity and in relativistic collisions only self-induced fields are important.

The description of the dynamics for EMP is also very different from that for QGP [720]. For EMP, under the corresponding time restraints, a wide use is made $((m_{\rm i} \gg m_{\rm e}))$ of the two-fluid (electron-ion) model with different electron and ion temperatures. The local electroneutrality condition hinders considerable charge separation which leads to equations of magnetic hydrodynamics where the pressure gradients and magnetic field drive the plasma motion.

For QGP there is no such a magnetohydrodynamic analog because each quark or gluon can carry different color charges. Therefore, when the local thermodynamic equilibrium is reached, different color components of the plasma will have equal temperatures and velocities. Besides, the quark-gluon system becomes color neutral even before the local thermodynamic equilibrium is reached and for a QGP we deal with the hydrodynamics of neutral fluid where chromodynamic fields are absent. Naturally, in the absence of local thermodynamic equilibrium, more complicated kinetic equations are applied incorporating various forms of the "collisional" term [410].

To close the equations of motion, which express the mass, momentum, and energy conservation laws, the EOS must be used. In the simplest case of an ideal ultrarelativistic gas of massless particles it is assumed that $E_{(x)} = 3p(X)$.

Many hydrodynamical and kinetic plasma instabilities typical of EMP may presumably show up in QGP [720], but it is extremely difficult to observe them there. However, the development of these perturbations is assumed to be the reason of the small $((\leq 1\,{\rm fm}/c)$ value of the measured QGP thermalization time and of the jet quenching effect in relativistic nuclear collisions.

The experimentally observed [720] fast thermalization of matter, the parameters of elliptic flows, the spectrum of outgoing particles, jet quenching, and low viscousity are attributed [720] to nonideality of the QGP close to the deconfinement threshold. The estimate of the nonideality parameter $\Gamma \approx 1.5$–$5.0$ given above may increase by an order of magnitude if the higher-order terms are taken into account in the interaction potentials [720]. This, in turn, may induce a "plasma" phase transition similar to that observed in the nonrelativistic strongly nonideal plasma [287, 293, 301, 302, 305].

In Ref. [720] the plasma nonideality effects found for the compressed EMP are used in the analysis of the behavior of the viscousity, collision cross sections, and stopping power of QGP. In QGP the ratio of the Landau length $((\Gamma_c \sim q^2/E)\,)$ to the Debye radius varies from 1 to 5. This increases the collision cross section by a factor of 2–9, which decreases the free path $\lambda$ and, consequently, the viscousity $((\eta \sim \lambda))$ also by an order of magnitude. This is consistent with the measured parameters of elliptic flows and particle spectra in nuclear collisions as well as with detected growth of the collisional losses.

Interesting analogies [720] arise between the strongly nonideal QGP and strongly nonideal "dust" plasma [287, 302, 305]. In both cases, we presumably deal with a Newtonian liquid in which the shear viscousity depends on the velocity of motion. Besides, the QGP possesses features of a nanoliquid [720]. In particular, the initial QGP size immediately after a collision is approximately 10 fm, i.e., 20 interparticle distances, which distinguishes this system from a continuous medium. This is also characteristic of a nonideal "dust" plasma.

The authors of Ref. [84] analyzed the analogies between a QGP and a strongly nonideal EMP consisting of ions in electrostatic traps cooled by laser and evaporation methods to ultralow temperatures. In both cases we deal with strongly interacting systems with a limited volume and a large number ($\approx 10^4$–$10^7$) of degrees of freedom. Figure 12.12 presents the expansion of a cryogenic ion EMP cloud [751] which shows pronounced features of an "elliptic" flow typical of the superdense nuclear matter expansion in relativistic collision experiments (Fig. 12.9). Kindred energy characteristics of neutron matter and supercold fermionic atoms (Fig. 12.13) [344] demonstrate an amazing similarity at temperatures differing by approximately 20 orders of magnitude.

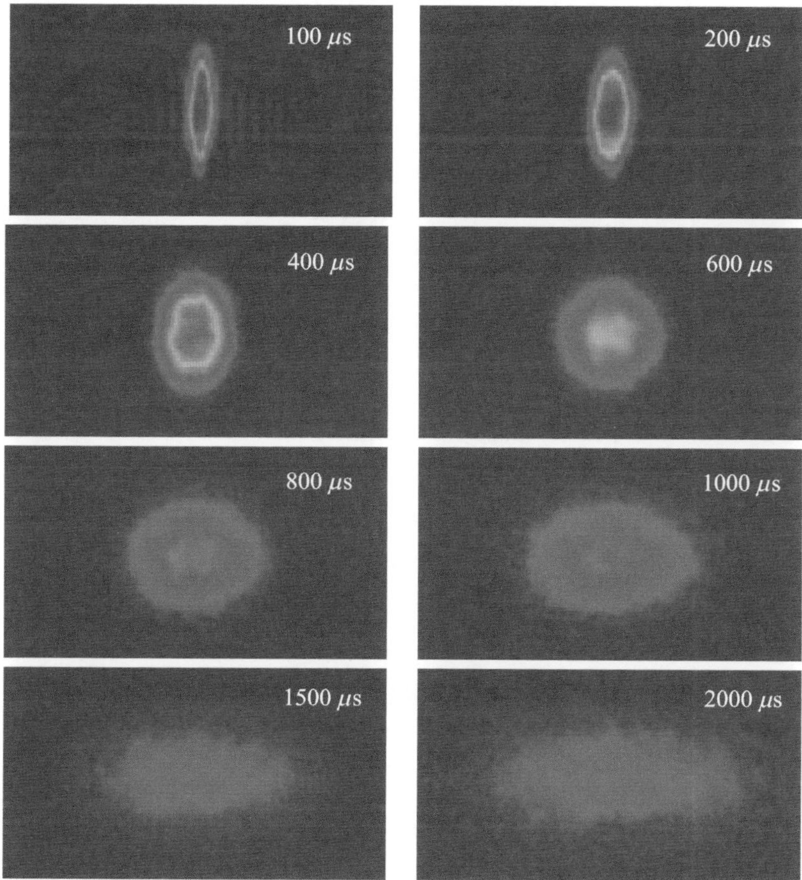


Fig. 12.12   Asymmetric ("elliptical") dynamics of expansion of the nonideal cryogenic lithium $^6$Li plasma after switching off the electrostatic trap [751]. An analogy with an "elliptical" flow in nuclear collisions.

These similarities may be helpful in understanding of a number of important properties of the QGP, especially those which are hard to obtain directly from relativistic ion collision experiments, i.e.,:

— Kinetic energy distribution in compressed medium,
— Collective motions and their measures,
— Screening,
— Collisions and expansion,

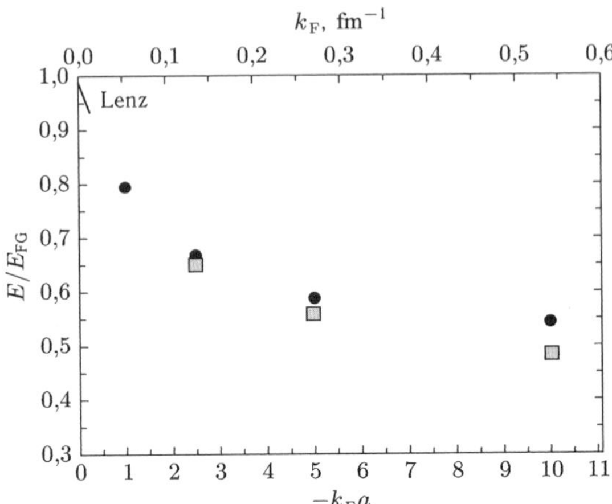


Fig. 12.13 Reduced energy of neutron matter (circles) and ultracold atoms (squares) [344].

— Flows, hydrodynamics,
— Thermalization,
— Correlation,
— Interaction with fast particles, stopping power,
— Viscousity, dissipation.

This list of analogies can be easily continued.

To conclude this section, we present a scheme of matter transformation [294, 295] at high energy densities (Fig. 12.14) which in a sense is an extension of the concept on matter simplification as one moves towards extremely high pressures and temperatures.

## 12.4　EOS of the QGP

To build up an EOS one needs to know the quark masses and the quark-gluon coupling parameters [414].

Masses of light quarks $u$, $d$, and $s$ are presented in [1057]:

$$m_u^{(c)}c^2 = (1.5\text{--}3.0)\ \text{MeV}, \quad m_d^{(c)}c^2 = (3\text{--}7)\ \text{MeV}, \quad m_s^{(c)}c^2 = (70\text{--}120)\ \text{MeV}.$$

Since the chemical potential of the $u$, and $d$ quarks in quark matter is much larger than $m_u^{(c)}c^2$ and $m_d^{(x)}c^2$, these quarks may be considered as ultrarelativistic and massless.

Breaking of the flavor symmetry assumes that the electroneutral quark plasma contains electrons. To write EOS one must find a thermodynamic

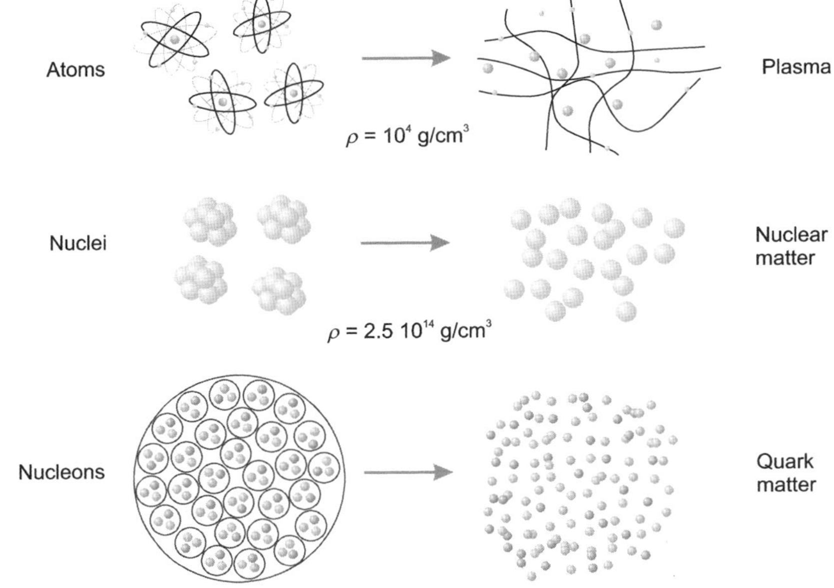


Fig. 12.14   Matter transformation at high energy densities.

equilibrium in the four-component plasma with respect to four thermodynamic variables $n_j$, where $j = u$, $d$, $s$, and $e$. It is assumed that the plasma is in equilibrium with respect to the weak interaction processes [414].

$$d \to u + e + \bar{\nu}_e, \quad u + e \to d + \nu_e, \tag{12.1}$$

$$u + d \to s + u, \quad s + u \to u + d. \tag{12.2}$$

The medium is considered to be transparent to the neutrino:

$$\mu_d = \mu_u + \mu_e, \quad \mu_d = \mu_s. \tag{12.3}$$

The electrons are considered to be free and ultrarelativistic:

$$\Omega_e = \frac{-\mu_e^4}{[12\pi(\hbar c)^3]}. \tag{12.4}$$

For massless noninteracting quarks $u$ and $d$ color degrees of freedom give factor 3:

$$\Omega_i^{(0)} = -\frac{\mu_i^4}{[4\pi(\hbar c)^3]}, \quad \text{where} \quad i = u, d. \tag{12.5}$$

In the weak coupling limit, the contribution given by the $\Omega_q$ and $\Omega$ quarks can be calculated within the framework of QCD by writing a coupling constant expansion ($\alpha_s = \frac{g_c^2}{4\pi}$, where $g_c$ is the quark–gluon coupling constant).

In the lowest order for nonrelativistic $u$ and $d$ quarks

$$\frac{\Omega_i}{\Omega_i^{(0)}} = 1 - \frac{2\alpha_s}{\pi} \quad \text{where} \quad i = u, d. \tag{12.6}$$

The contribution given by the massive $s$ quarks is [251]

$$\begin{aligned}
\frac{\Omega_s}{\Omega_s^{(0)}} = {} & \sqrt{1 - y_s^2}\left(1 - \frac{5}{2}y_s^2\right) + \frac{3}{2}y_s^4 L \\
& - \frac{2\alpha_s}{\pi}\left[3\left(\sqrt{1 - y_s^2} - y_s^2 L\right)^2 - 2(1 - y_s^2)^2 + 3y_s^4(\ln y_s)^2\right] \\
& - \frac{\alpha_s}{\pi}\ln\left(\frac{\rho_R}{\mu_s}\right)\left[\sqrt{1 - y_s^2} - y_s^2 L\right],
\end{aligned} \tag{12.7}$$

where $y_s = m_s c^2/\mu_s$ and $L = \ln\{(1 + \sqrt{1 - y_s^2})/y_s\}$.

The electroneutrality condition gives:

$$\frac{1}{3}(n_u + n_d + n_s) = n_b, \quad n_e + \frac{1}{3}(n_d + n_s - 2n_u) = 0. \tag{12.8}$$

After solving the system of equations (12.3 and 12.8) one can calculate the EOS employing

$$p(n_b) = -\Omega_e - \Omega_q, \tag{12.9}$$

$$\varepsilon(n_b) = \Omega + \sum_i n_i \mu_i = \Omega_e + \Omega_q + \sum_i n_i \mu_i, \tag{12.10}$$

where $i = u,\, d,\, s,$ and $e$.

The concentration of electrons in nuclear matter $n_e/n_b$ tends to zero when $m_s = 0$, but grows rapidly when $m_s$ increases. The approximate formula for $n_e/n_b$ is given in Ref. [415] for $\alpha_s \lesssim 0.2$:

$$\frac{n_e}{n_b} \simeq \frac{0.002}{1 - 2\varepsilon}\left\{1 - \frac{4}{3}\varepsilon - 4\varepsilon\ln\left(\frac{2\mu_u}{\rho_R}\right)\right\}^2\left(\frac{n_0}{n_b}\right)^2\left(\frac{m_s c^2}{200\,\text{MeV}}\right)^6, \tag{12.11}$$

where $\varepsilon = \alpha_s/\pi$. In particular, this relationship is valid for noninteracting quarks ($\alpha_s \to 0$). At $n_b \gg n_0$ and $m_s c^2 \approx 200\,\text{MeV}$ the fraction of electrons is small ($n_e/n_b \ll 10^{-3}$) and does not affect the EOS.

Making use of the perturbation theory and renormalization procedure the authors of Refs. [77, 331] obtained a one-parameter EOS

$$\Omega_q^{(2)} = -p(\mu) = -\frac{3\mu^4}{4\pi^2(\hbar c)^3}(1 - 2\varepsilon - 3\varepsilon^2 \ln \varepsilon - 7.46\varepsilon^2), \qquad (12.12)$$

where $\varepsilon = \alpha_s(\mu)/\pi$.

It is shown in Ref. [414] that the phase transition of the baryon matter to the quark matter is the transition of the first kind. According to Ref. [77], it takes place at $\mu_b \approx (1\text{–}2)\,\text{GeV}$ and $\rho \approx 10^{15}\,\text{g/cm}^3$. In this case, first the quark matter drops are formed [714] and the density jump may be so large that it may lead to the destabilization of a star whose core is composed of such a condensate [414].

A more delicate statement is that the qualitative phenomena like confinement and deconfinement cannot be described in the framework of the perturbation theory. The expansion parameters at the transition boundary cannot be considered small.

This disadvantage is absent in the lattice model [179] which describes the quark behavior by the lattice constant $B$. In this case the quarks are confined by the lattice size and cannot move beyond the lattice. This model provides the mass values of mesons and baryons and their magnetic moments and describes the thermodynamics of the system. In the lattice model

$$\varepsilon_q = B + \varepsilon_q^{\text{kin}} + \varepsilon_q^{\text{int}}, \qquad (12.13)$$

where $\varepsilon_q^{\text{int}}$ is calculated within the perturbation theory approach. The pressure has the form

$$p_q = -B + p_q^{\text{kin}} + p_q^{\text{int}}. \qquad (12.14)$$

The obtained expressions contain the coupling constant $\alpha_s$ and the strange quark mass. In this model the B–Q-transition is a transition of the first kind with a double jump of the baryon and mass density.

It is a controversial subject now [414] whether the QGP exists in neutron stars which in this case would become "quark" stars.

These astrophysical questions as well as high energy physics problems stimulate the development of sophisticated QGP models which are reviewed in Ref. [414].

The bulk of the QGP phenomenological models in the range $\rho \approx 10^{15}\text{–}10^{16}\,\text{g/cm}^3$ is a reflection of the fact that the respective energies are not sufficiently high for the use of the QCD methods and we are far away from the applicability of the asymptotic freedom principle (valid at $\mu_f \gg 1\,\text{GeV}$).

Since the attraction between quarks is weak, the QGP may be supercon-
ductive [79, 640, 823]. The corresponding pairs form the boson condensate.
Since quarks carry color, this phenomenon is called as color superconductivity [414]. The corresponding gap is $\Delta \approx 100\,\mathrm{GeV}$ [640]. However, the
effect of this superconductivity is very small [414]: it does not exceed several percent [824].

The description of the phase transition in QGP is hindered by many
reasons. Both phases are described by essentially different models. One of
them is applied well beyond its validity region. The quark matter is assumed
to be a Fermi gas, while in the transition area strong correlations should
be expected. There are also other problems in the description of the phase
transitions in the QGP. They are being actively investigated now.

## 12.5   Neutron crystallization and strange matter

Crystallization of neutron matter [414] is induced by the strong short-range
repulsion of neutrons approximated by the potential $U_{ij} = \infty$ at $r_{ij} \leq r_{\mathrm{core}}$.
This happens at $n_\mathrm{b} \sim \left( \frac{4\pi r_{\mathrm{core}}}{3} \right)^{-1}$ when neutrons are localized in the lattice state. This effect was estimated for the first time using the principle
of corresponding states on the base of the $^3$He data [47]. More advanced
models gave $\rho \approx (5\text{--}30)10^{14}\,\mathrm{g/cm^3}$ [155]. Later this estimate was substantially refined and it was revealed that the crystallization region moves to
the range $\rho \gtrsim 3\rho_0$, the density jump is of 20% and the phase transition of
the first kind "softens" the EOS. The review of mechanical properties of
neutron crystals is presented in Refs. [414, 761].

Strange matter arises at ultrahigh pressures when baryons lose their
individuality and transform into quasifree $u$- and $d$- quarks. At so high
densities matter is strongly degenerate and the temperature effects are of
little importance. Let us denote as $p_D$ the quark deconfinement pressure
in nuclear matter. At $p > p_D$ the baryon matter is unstable with respect
to the transition to the $ud$-plasma. However, the transition pressure may
be reduced when about half of all $d$-quarks transform into $s$-quarks. This
transformation proceeds via the weak interaction, while the baryon transformation into quarks is a strong interaction process. When the $ud$- and
$uds$-quark matter is extrapolated to the normal conditions $p \to 0$, the
equilibrium plasma consists of nuclei and does not contain free quarks.
Actually, the ground state of ordinary matter is a $^{56}$Fe crystal with $^{56}$Fe
$E_0 \equiv \mu_\mathrm{b}(p = 0) = 930.4\,\mathrm{MeV}$.

The strange matter hypothesis corresponds to the situation when the *uds*-matter stays stable even at $p = 0$. This means that heavy droplets of *uds*- matter whose baryon number is $A > A_{\min} \gg 10$ have lower energy per baryon than $^{56}$Fe. In their turn, ordinary heavy nuclei become metastable with respect to such strange matter, their lifetime being about $10^{100}$ years [414]. However, it is shown experimentally that at small $p$ the *ud*-matter transforms into nuclei via a strong interaction channel during time $\tau \approx 10^{-22}$ s.

Despite the extravagance of the strange matter theory, one should bear in mind that this matter cannot arise under terrestrial conditions. It can be formed in a supercompressed state: in neutron star cores, in supernovae, or at the early stages of the Universe.

The strange matter concept was developed in Refs. [114, 176, 204, 610], where quark nuclei, quark droplets, etc. were considered. The author of Ref. [1047] suggested a cosmological scenario in which quark "nuggets" with equal numbers of *u*-, *d*-, and *f*-quarks are produced. They arise in the hadronization epoch in the early Universe $10^{-5}$ s after the Big Bang. The lattice model calculations showed that a fraction of the *uds*-matter could have stayed in a stable state at $p = 0$. Similar results were obtained later in Ref. [251]. Nowadays, the quest of the strange matter signatures in cosmic rays and in relativistic heavy ion collisions is underway. The review of the respective papers is presented in Refs. [350, 644, 645, 1033, 1034]. Note that this simple model assumes $0.962 < B_{60} < 1.53$.

There are much more sophisticated models suggested for the description of the strange matter [215, 251, 414, 1032, 1057]. Each of them allows in one form or another the existence of QGP under normal conditions.

A characteristic feature of the strange matter EOS is that the pressure is negative at $\rho < \rho_s$. In the vicinity of $\rho \sim \rho_c$, a linear EOS $p \sim (\rho - \rho_s)$ is valid which also follows from the lattice model [1075] and other strange matter models.

The lattice model gives

$$p = ac^2(\rho - \rho_s), \tag{12.15}$$

where $\rho_s$ is slightly different from density $\rho$ at $p = 0$. Such form of EOS is precise for massless or noninteracting quarks at $a = 1/3$. When $m_s c^2 \lesssim 300\,\text{MeV}$ and $\alpha_c \leq 0.6$ and $\rho$ have the values typical of strange stars, EOS (12.15) is valid within the accuracy of a few percent. The parameters of the linear EOS model following from other models are presented in Table 12.2 [414].

Table 12.2  Parameters $a$ and $\rho_s$ of the EOS linear model (12.15). SQM1 and SQM2 are from Ref. [1075]. SS1 and SS2 are obtained in Ref. [215].

| Model EOS | $a$ | $\rho_s$, $10^{14}$ g/cm$^3$ |
|---|---|---|
| SQM1 | 0.301 | 4.50 |
| SQM2 | 0.324 | 3.06 |
| SS1 | 0.463 | 11.54 |
| SS2 | 0.455 | 13.32 |

The calculation of the strange matter adiabatic exponent $\gamma = \frac{n_{\mathrm{b}}}{p}\left(\frac{dp}{dn_{\mathrm{b}}}\right)$ demonstrates a sharp difference from $\gamma$ for neutron matter, especially at low $\rho$. In the range of the strange matter existence $\gamma$ appreciably exceeds $4/3$ given by the asymptotic freedom theory [414].

It is assumed [414] that the ground state of massless quarks in the asymptotic region of the weak interaction corresponds to the color-flavor superconductivity. According to Ref. [824], the same is also valid for massive $s$-quarks with $m_s \lesssim 200\,\mathrm{MeV}/c^2$ and the gap value $\Delta \gtrsim 100\,\mathrm{MeV}$. In this case we deal with flavor-symmetric quark matter which is characterized by $n_u = n_d = n_s = n_b/3$ and has no electron component.

It is shown [414] that superconductivity increases the stability of the strange matter. However, although in toto the superconductivity effect on EOS is small at large $p$, in the $p \approx 0$ range, it is essential for the stability of the strange matter phase.

There are also other hypothetical predictions of self-coupled matter which exists at $p \approx 0$ [414].

One of the examples was suggested in Refs. [428, 686, 687]. It is related to the superdense pion condensate which may lead to the production of anomalous nuclei with large mass numbers $A$ whose densities significantly exceed $\rho_0$.

The idea of the anomalous states of nuclear matter suggested in Refs. [609, 610] is based on the model of strongly interacting nuclear matter in which at high $\rho \gtrsim 3 \div 5\rho_0$ nucleons become massless. This zero mass result follows from the nonlinear model of scalar mesons (chiral sigma model).

Another mechanism inducing the appearance of the second minimum on the $E(\rho)$ curve is associated with spin–isospin collective excitations carrying the pion quantum numbers ("pion" condensation). As a result, the curve presenting the baryon energy dependence on the density shows the second minimum appearing at $n_b > n_0$, where $n_0$ corresponds to the normal state. The respective binding energy turns out to be very large — hundreds of MeV/nuclear nucleon. The criticism of this model can be found in Refs. [716, 769].

The supersymmetric version of the Standard Model of elementary particles allows the existence [594, 643] of the baryonic Q-spheres. They are macroscopic self-coupled superdense domains filled with a scalar field condensate, which possesses electric and baryon charge; these Q-spheres are treated as a hypothetical component of the dark matter. If so, it is assumed that the energy of the Q-sphere matter is lower than that of $^{56}$Fe. Should the Q-matter be real, we would deal with a transformation of ordinary neutron stars into Q-stars in $10^7$–$10^{11}$ years [594, 643].

The EOS of the Q-matter was suggested in Ref. [69] where the transition density is obtained to be $\rho(p = 0) \equiv \rho_s = 10^{14}$ g/cm$^3$ and the lower density limit on the Q-matter existence is by a factor of 3–5 smaller than the nuclear density.

The hypotheses on anomalous and Q-matter formed the basis of the works which analyze possible "strange" stars built up of such exotic matter forms [414, 415]. As distinct from neutron stars, in this case the matter density on the surface of strange stars is extremely high ($\rho \approx 10^{15}$ g/cm$^3$ ). The density profile is quite flat — the density inside the nucleus is only five times that on the surface (while for neutron stars they differ by 14 orders of magnitude) and at the star boundary strong (up to $10^{18}$ V/cm) electric fields arise.

It should be said in the conclusion that we can clearly see that moving up to higher values on the matter density scale we have in fact reached the limit of our present understanding of high energy physics and are far ahead of the region of experimentally attainable parameters. This makes our considerations less and less reliable but at the same time they become more and more intriguing opening the way to various heuristic hypotheses and the most exotic approaches.

# Chapter 13

# Semi-Empiric Nuclear Models

In this chapter, we will follow Ref. [667]. First of all, note that the very concept of "nuclear matter" as an infinite homogeneous system consisting of nucleons is sort of conventional. In practice, we deal with the nuclei which have up to $N \approx 300$ nucleons and the concept of nuclear matter is an approximation of their interior domains. A prolonged nuclear matter presumably arises in neutron and quark stars and at the early stages of the Big Bang.

A general form of the nuclear matter diagram is presented in Fig. 1.13. Some parts of this diagram are attainable for neutron, quark–gluon, and strange stars as well as in experiments on relativistic collisions of heavy nuclei with the accelerators RHIC, LHC, and FAIR GSI SIS300.

In the early investigations nuclear matter was assumed to be incompressible. However, the data on the giant monopole resonance (on spherically-symmetric oscillations) provided the first information on nuclear compressibility. These first experiments gave the value $c_s \approx 0.2c$ for sound velocity in nuclear matter and the application of the simplest model of nonideal matter, Fermi gas model, confirmed these data.

Experiments on the collisions of heavy nuclei will make it possible to achieve deep compression of nuclear matter with formation of nuclear shock waves [955] at energies of several GeV per nucleon [294, 295].

The "evergreen" problems encountered in this case (the thermodynamic equilibrium in the shock-compressed and heated plasma as well as the applicability of the hydrodynamic description when the number of colliding particles is $\approx 500$) need a continuing nontrivial analysis. The first problem arises due to the fact that at energies 100–200 MeV per nucleon the nucleon free path in a nucleus is of the order of 2–4 fm and it increases with the energy decrease according to the Pauli principle. This fact requires a careful study of the specific experimental conditions.

For nuclei, under normal conditions, it is presumed that the density value is $\rho_0 \approx 0.15\,\mathrm{fm}^{-3}$ and the binding energy per nucleon is $B_0 \approx -16\,\mathrm{MeV}$. When $\rho < \rho_0$, a phase transition to free nucleons takes place and the energy dependence becomes $E \sim \rho^{2/3}$ as it follows from the Fermi gas theory. The giant monopole resonance data give $K \approx 200\text{--}400\,\mathrm{MeV}$ [105] for the compressibility. The results provided by the supernova observations, by the properties of neutron stars, and Big Bang models may give information on the nuclear matter equation of state (EOS), but it needs some additional considerations.

The results given by the strong interaction theory and quantum chromodynamics (QCD) provide information which is useful for the construction of the EOS of nuclear matter.

To describe the QGP (Chapter 12) let us assume that the gluon component is a boson gas whose degeneracy factor is 2 (for the spin) multiplied by 8 (for the color), so that the energy density is

$$\varepsilon_{\mathrm{g}} = \frac{\pi^2 T^4}{30}.$$

The degeneracy factor for quarks and antiquarks is 2 (spin) $\times$ 3 (color) $\times$ 2 (flavor). If so,

$$\varepsilon_{\mathrm{q}} + \varepsilon_{\bar{\mathrm{q}}} = \frac{7\pi^2 T^4}{120} + \frac{\mu^2 T^2}{4} + \frac{\mu^4}{8\pi^2}.$$

As we have seen in Chapter 12, the QGP is produced at the characteristic energies $2\text{--}3\,\mathrm{GeV/fm}^3$, while the typical energy of cold nuclei is $150\,\mathrm{MeV/fm}^3$.

## 13.1 "Cold" components

Following the semi-empirical approach to the EOS construction (see Chapter 9), let us decompose the thermodynamic functions in potential (or "cold" $T = 0\,\mathrm{K}$) and thermal components [667]:

$$W(\rho, s) = W_0(\rho) + W_{\mathrm{th}}(\rho, s),$$

where $W_0$ is the contribution given by noninteracting fermions and bosons which includes the rest energy of nucleons at the equilibrium density $W_{\mathrm{th}} \approx 938\,\mathrm{MeV}$.

The pressure has the form:

$$p(\rho, s) = \rho^2 \frac{\partial W_0}{\partial \rho} + \rho^2 \frac{\partial W_{\text{th}}}{\partial \rho}\bigg|_s = p_0(\rho) + p_{\text{th}}(\rho, s).$$

Function $W_0(\rho)$ should be chosen in such a way as to reproduce the known properties of the atomic nucleus. In equilibrium $W_0(\rho_0) = -B_0$ and $\dfrac{\partial W_0}{\partial \rho}\bigg|_{\rho_0} = 0$ the compressibility at the equilibrium point $\rho = \rho_0$ is $K = 9\dfrac{\partial^2 W_0}{\partial \rho^2}\bigg|_{\rho_0}$.

In principle, $W_0(\rho)$ must possess quadratic asymptotics when $\rho \to 0$ and the system must transform into a gas of noninteracting nucleons. However, during the expansion process the nuclear matter becomes unstable at $\rho \sim (1/3 - 1/2)\rho_0$ and decays into nucleons and smaller nuclei. Therefore for the description of this expansion domain more detailed methods are employed.

Another restraint on the EOS is caused by the causality principle. The velocity of sound is given by the expression $c_{\text{s}}^2 = \dfrac{\partial W_{(\rho,\text{s})}}{\partial e}\bigg|_{\text{s}}$.

In case of zero enthropy it takes the form: $c_{\text{s}}^2 = \dfrac{dW_0(\rho)}{de}$, where $e$ is the energy density with account of the rest mass.

The relativistic condition $c_s < c$ ($c$ is the velocity of light) results in that $W_0(\rho)$ cannot increase with $\rho$ faster than proportionally to $\rho$.

This leads to the quadratic EOS [667] ($x = \rho/\rho_0$):

$$W_0(x) = B_0 + \frac{K}{18}(x - 1)^2, \quad p_0(x) = \frac{K}{9}\rho_0 x^2(x - 1),$$

$$c_{\text{s}}^2 = \frac{2x(3x - 2)}{\frac{18(mc^2 + B_0)}{K + 3x^2 - 4x + 1}}.$$

This EOS violates the causality principle when $x \gtrsim 5.2x \gtrsim 5.2$, when $K = 210\,\text{MeV}$, $B_0 = -16\,\text{MeV}$ and $mc^2 = 938\,\text{MeV}$.

A linearized EOS has the form:

$$W_0(x) = B_0 + \frac{K}{18x}(x - 1)^2, \quad p_0(x) = \frac{K}{9}\rho_0(x - 1),$$

$$c_{\text{s}}^2 = \frac{x}{\frac{9(mc^2 + B_0)}{K + x - 1}}.$$

It does not violate the causality condition if the parameters are chosen correctly.

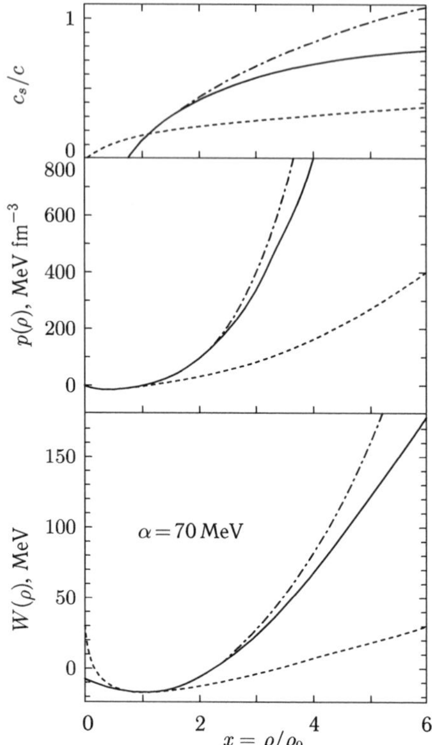



Fig. 13.1   Comparison of the energy per nucleon $W$, the pressure $p$, and the sound velocity $c_s$ for the hyperbolic (solid line), quadratic (dashed-dotted line), and linear (dashed line) EOS at $K = 210\,\mathrm{MeV}$ [667].

A hyperbolic EOS can be written as

$$W_0(x) = B_0 + \sqrt{\alpha^2(x-1)^2 + \beta^2} - \beta,$$

where $\alpha$ is the asymptotic slope and $\beta$ is related to the compressibility as $\beta = 9\alpha^2/K$.

The comparison of various EOS of nuclear matter is shown in Fig. 13.1.

The ideal gas model serves as the simplest approximation for the description of the thermal part of EOS:

$$p_\mathrm{th} = \frac{2}{3}\rho E_\mathrm{th}, \tag{13.1}$$

which is applied to the calculation of nonviscous hydrodynamics of nuclear collisions and hadron mixtures. The above equation is valid not only for a

Fermi gas in the low-$T$ limit, but in the relativistic case it is also applicable for an ideal gas at all values of $T$.

However, when the temperature is close to 50 MeV, pion production should be taken into account in the EOS. At the first stage one must calculate the Fermi integrals for nucleons and their excited states (resonances, etc.) and Bose integrals for pions and heavy muons (Chapter 11). In this case the computations are usually so complicated that in hydrodynamic calculations the corresponding EOS parameters are introduced in a tabular form.

## 13.2 Thermal excitations

To go beyond approximation (13.1) we introduce a different thermal contribution to the semi-empiric EOS suggested in Ref. [954]. In this model, the thermal contribution is a sum of the interaction energy and the kinetic energy of particles. The latter can be described by the Fermi–Dirac and Bose–Einstein relativistic distribution:

$$
W = U + \sum_{i=1}^{\sigma_{\mathrm{b}}} \left( \frac{\rho_i^0 m_i c^2}{\rho} + \frac{4\pi g_i}{\rho(2\pi hc)^3} \int_{m_i c^2}^{\infty} \frac{\varepsilon^2 \sqrt{\varepsilon^2 - m_i^2 c^4}}{\exp\left[\frac{\varepsilon}{T}\right] - 1} d\varepsilon \right)
$$

$$
+ \sum_{i=\sigma_{\mathrm{b}}+1}^{\sigma} \frac{4\pi g_i}{\rho(2\pi hc)^3} \int_{m_i c^2}^{\infty} \frac{\varepsilon^2 \sqrt{\varepsilon^2 - m_i^2 c^4}}{\exp\left[\frac{(\varepsilon+U-\mu)}{T}\right] + 1} d\varepsilon, \tag{13.2}
$$

here, the first sum is associated with the degrees of freedom of Bose pions and heavy mesons. The summation in the second term is done over all excited states of nucleons. The $\Delta(1232)$ resonance is the most important at energies of the order of several GeV/nucleon. It is assumed in this case that all the resonances have the same interaction energy $U$ which in its turn depends on the total baryon density $\rho$. Due to the chemical thermal equilibrium, the chemical potentials of nucleons are equal to each other. $\rho_i^0$ is the contribution given by the ground state of bosons in the boson density phase.

The baryon density and the chemical potential are related as

$$
\rho = \sum_{i=\sigma_{\mathrm{b}}+1}^{\sigma} \frac{4\pi g_i}{(2\pi hc)^3} \int_{m_i c^2}^{\infty} \frac{\varepsilon \sqrt{\varepsilon^2 - m_i^2 c^4}}{\exp\left[\frac{(\varepsilon+U-\mu)}{T}\right] + 1} d\varepsilon. \tag{13.3}
$$

The number of mesons has the form:

$$N_i = \frac{g_i}{\left\{ \exp\left(\frac{m_i c^2}{T}\right) - 1 \right\}} + \frac{4\pi g_i V}{(2\pi \hbar c)^3} \int\limits_{m_i c^2}^{\infty} \frac{\varepsilon \sqrt{\varepsilon^2 - m_i^2 c^4}}{\exp\left[\frac{\varepsilon}{T}\right] - 1} d\varepsilon. \qquad (13.4)$$

According to (13.3) the relationship of $U$ with the compression energy $E_{\rm c}$ at $T \to 0$ has the form:

$$(\mu - U)^2 = m^2 c^4 + \left(\frac{\rho C}{g}\right)^{2/3}, \qquad (13.5)$$

where $C = 6\pi^2 (\hbar c^3)$.

In the same limit for (13.2) we have

$$W(T = 0) = 0.75X + U + \frac{m^2 c^4}{8}$$
$$\times \left\{ \frac{3X}{X_1^2} - \frac{3m^2 c^4}{X_1^3} \ln\left[\frac{(X + X_1)}{mc^2}\right] \right\}, \qquad (13.6)$$

where $g = 4$, $mc^2 = 939\,\text{MeV}$, $X = \sqrt{m^2 c^4 + X_1^2}$, $X_1 = (\rho C/g)^{1/3}$.

It is assumed that at $T = 0$ only one level is populated which is valid at small densities. Expansion (13.5, 13.6) in the low-density limit gives

$$\mu - U = mc^2 + \frac{1}{2}\frac{X_1^2}{mc^2} + \cdots . \qquad (13.7)$$

$$W = U + mc^2 + 0.3 \left(\frac{\rho}{\rho_0}\right)^{2/3} \left(\frac{h^2}{m}\right) \left(\frac{6\pi^2 \rho_0}{g}\right)^{2/3} + \cdots . \qquad (13.8)$$

The difference between the precise formula (13.6) and approximation (13.8) is about $1\,\text{MeV}$ at $\rho/\rho_0 = 3$. The final expression relating $U$ and $E_{\rm C}$ is of the form:

$$U(\rho) = E_{\rm C} + W_0 - 0.75X - \frac{m^2 c^4}{8}$$
$$\times \left\{ \frac{3X}{X_1^2} - \frac{3m^2 c^4}{X_1^3} \ln\left[\frac{(X + X_1)}{mc^2}\right] \right\}, \qquad (13.9)$$

We obtain the flowing expression for pressure:

$$p = -T \sum_{i=1}^{\sigma_{\mathrm{b}}} \frac{4\pi g_i}{\rho(2\pi hc)^3} \int_{m_i c^2}^{\infty} \varepsilon \sqrt{\varepsilon^2 - m_i^2 c^4} \ln\left(1 - \exp\left[-\frac{\epsilon}{T}\right]\right) d\epsilon$$

$$+ T \sum_{i=\sigma_{\mathrm{b}}+1}^{\sigma} \frac{4\pi g_i}{\rho(2\pi hc)^3} \int_{m_i c^2}^{\infty} \varepsilon \sqrt{\varepsilon^2 - m_i^2 c^4}$$

$$\times \ln\left(1 + \exp\left[\frac{(\mu - U - \epsilon)}{T}\right]\right) d\epsilon + \rho^2 \frac{\partial U}{\partial \rho}. \tag{13.10}$$

The entropy per baryon can be written as

$$\frac{S}{N_{\mathrm{B}}} = \frac{p}{(\rho T)} - \frac{\rho}{T} \frac{\partial U}{\partial \rho} - \frac{1}{N_{\mathrm{B}}} \sum_{i=1}^{\sigma_{\mathrm{b}}} g_i$$

$$\times \ln\left(1 - \exp\left[-\frac{m_i c^2}{T}\right]\right) + \frac{(W - \mu)}{T}. \tag{13.11}$$

These simplified EOS may be employed [954] for the determination of the nuclear matter temperature from the experimentally measured pion and kaon yields produced in the relativistic collisions of heavy nuclei. This result in its turn may be used as the experimental information for the verification of the nuclear matter EOS models [954]. For this purpose, one usually [954] makes use of the experimentally measured yields of pions generated in compressed and heated matter in nuclear collisions. The corresponding dependence is presented in Figures 13.2–13.4 [954]. One can see that the pion yield shows a sharp growth up to $T \approx 100\,\mathrm{MeV}$ which then slows down due to the transformation of nuclear matter into hadron plasma. It can be even better seen in Fig. 13.3 which presents pion distribution over different channels. At small $T \approx 50\,\mathrm{MeV}$ and below this value the bulk of the pions are in the Bose-condensate state and possess a zero momentum. At large $T$ pions are generated via nuclear resonance states. Figure 13.4 illustrates the nuclear matter temperature calculated using the above-mentioned EOS and the measured pion yields.

Simplifications of the EOS are related to the form of the thermal energy term [954]. Going beyond the simplest approximation (13.1) is associated

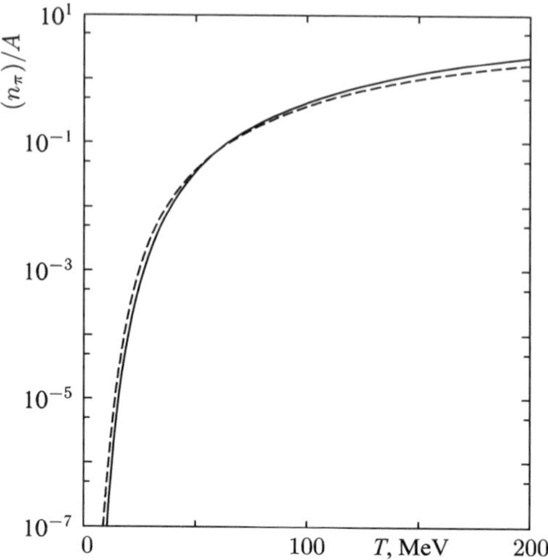

Fig. 13.2   Temperature dependence of the pion yields for the baryon density exceeding the density of normal nuclear matter by a factor of two (full curve) and by a factor of four (dashed curve). The curves describe the properties of the hot and dense fraction of infinite nuclear matter. The figure is adopted from Ref. [954].

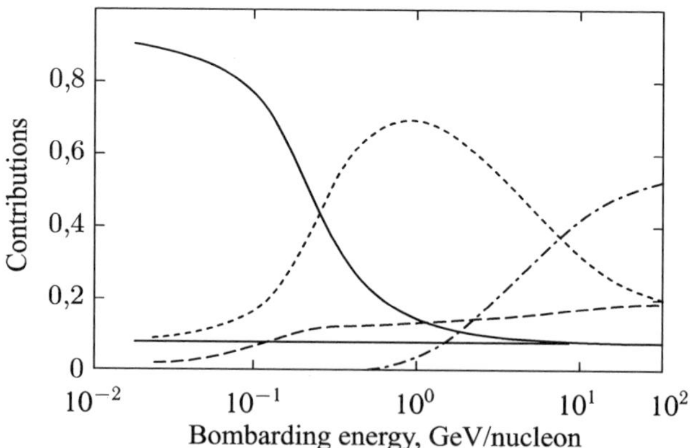

Fig. 13.3   Pion distributions for the reaction $C + C$. Dotted curve: Delta resonance; dashes: free pion gas, full curve: Bose-condensate, dashed-dotted curve: strong resonances. The figure is adopted from Ref. [954].

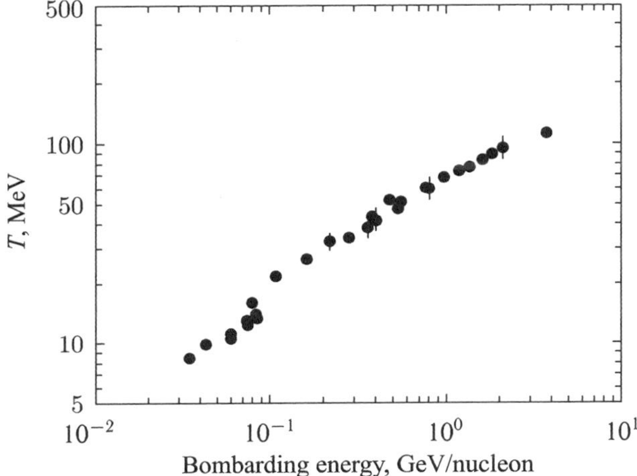


Fig. 13.4 "Freeze-out" (yield) temperature for pions. At $E_{\text{lab}} < 400\,\text{MeV}/\text{nucleon}$ pions escape at $\rho_0 < \rho < 3\rho_0$. At higher energies pions escape at values of $(2\text{–}5)\,\rho_0$. The figure is adopted from Ref. [954].

with taking into account the Fermi statistics [954] at $T < T_{\text{F}}$:

$$E_{\text{T}}(\rho, T) = \frac{\beta}{2}\rho^{-2/3}T^2 = \frac{S^2}{2\beta}\rho^{2/3} = E_{\text{T}}(\rho, S), \qquad (13.12)$$

where

$$T = \left.\frac{\partial W}{\partial S}\right|_\rho = \frac{S}{\beta}\rho^{2/3}, \quad \beta = \left(\frac{g\pi}{6}\right)^{2/3}\frac{mc^2}{\hbar c}.$$

As we have seen, at higher $T$ resonances become essential. The mass being $m_i c^2$, they can be described with $E_{\text{T}_i} = (3/2)T$.

The pion energy [954] equals $E_\pi = 1.85\rho_0\,(T/m_\pi)^{9/2}\,m_\pi$.

The compression energy $p_c = \dfrac{K_0}{18\rho_0}(\rho^2 - \rho_0^2)$, is summed with the thermal energy (13.1).

The dependence (13.1) can be generalized also to the case of resonances: $p_T = \alpha(\rho, E_T)\rho E_T$, where $\alpha(\rho, E_T)$ takes into account the resonance contribution.

A peculiarity of the pionic diagnostics of EOS is that because pions have large inelastic scattering cross section, they undergo several cycles of absorption on nuclear resonances before they leave the reaction domain. Therefore their sensitivity to nuclear compression processes decreases.

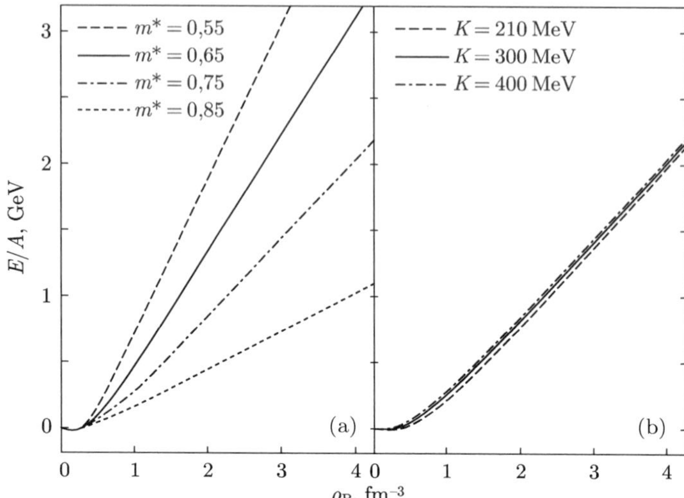


Fig. 13.5   (a) EOS calculated in the mean field model for fixed noncompressibility value $K = 210\,\mathrm{MeV}$ and different effective mass $m^*$ values at the equilibrium density $K(\rho_0) = 300\,\mathrm{MeV}$: (b) The same for fixed $m^* = 0.55$ and different noncompressibility $K$ values, $m^*(\rho_0) = 0.75\,m$.

As distinct from pions, $K^+$-mesons have a considerable free path in compressed nuclear matter. Moreover, at energies typical for the accelerators BEVALAC and SIS $K^+$-mesons are mainly produced in the high compression domain. Owing to these features the $K^+$-mesons become an attractive tool for the EOS diagnostics.

Figure 13.5 presents the EOS calculation results in the mean field model (see Chapter 11) for $K = 210\,\mathrm{MeV}$ and different effective mass $m^*$ values. Interestingly, in this case the potential and thermal parts of the EOS are not separated.

## 13.3   Hydrodynamics of nuclear collisions

Semi-empiric EOS are widely used for the description of the collisions of heavy nuclei at relativistic velocities. Today it is the main experimental tool for the generation of record high concentrations of energy, pressure, and temperature of matter under terrestrial laboratory conditions [294].

Relativistic collisions of heavy nuclei are described in the framework of the hydrodynamic model [294, 667, 955]. The medium is assumed to be continuous, in local thermodynamic equilibrium and described by a EOS.

Adopting notation from Ref. [698], we have in the center-of-mass system

$$x^\mu = \{x^0, x^1, x^2, x^3\} = \{ct, \mathbf{x}\} = \{ct, x^k\},$$

where the Greek indices take the values from 0 through 3, and the latin indices vary from 1 through 3. The four-dimensional gradient can be written as $\partial_\mu = \left\{\dfrac{1}{c}\dfrac{\partial}{\partial t}, \nabla\right\}$.

The equation of motion of the relativistic hydrodynamics follows from the conservation of the four-dimensional current $\partial_\mu j^\mu = 0$ and energy–momentum tensor $\partial_\mu T^{\mu\nu} = 0$.

$j^\mu$ is determined in terms of the density in the rest frame: $j^\mu = nu^\mu/c$ where $u^\mu$ is the velocity: $\mathbf{v}$: $u^0 = \gamma$, $u^i = \gamma v_i/c$; $u_\mu u^\mu = 1$.

The relativistic factor $\gamma$ is normalized as $\gamma = 1/\sqrt{1 - v^2/c^2}$. For a perfect fluid [599] $T^{\mu\nu} = (e + p)u^\mu u^\nu + pq^{\mu\nu}$, where $e$ is the energy density in the rest frame, $p$ is the pressure.

Let us rewrite these equations in the conventional form for the laboratory frame.

The energy density has the form: $E = T^{00} = \gamma^2(e + p) - p$.

The momentum is $M_k = cT^{0k} = \gamma^2(e + p)v_k$.

In the laboratory system $\rho = j^0/c = \gamma n$ and the spatial part is $j^k = \rho u^k = \gamma \rho v^k/c$.

Adopting this notation, we can write the equations of motion in a traditional form:

$$\partial_t \rho + \nabla \cdot (\rho \mathbf{v}) = 0, \quad \partial_t \mathbf{M} + \nabla \cdot (\mathbf{v}\mathbf{M}) = -\nabla p,$$
$$\partial_t E + \nabla \cdot (\mathbf{v}E) = -\nabla \cdot (\mathbf{v}p), \quad p = p(n, e).$$

The form of these equations is the same as that of the nonrelativistic equation, but the EOS is written in the local, rest system of coordinates. The calculation procedure for this case can be found in EOS Ref. [367].

The typical calculation results for relativistic C + Au collisions at $84\,\mathrm{MeV/nucleon}$ are presented in Figs. 13.6 and 1.21. There can be well seen the compression domain, propagation of nuclear shock waves, and shock-wave Mach cone [955], as well as a nuclear matter jet produced at an angle to the collision axis.

It is stated by now that the hydrodynamic model provides reasonable results at energies up to $2.1\,\mathrm{GeV}$ per nucleon.

The comparison of the hydrodynamic calculations with the experimental data may give information about the nuclear matter EOS and its parameters [667].

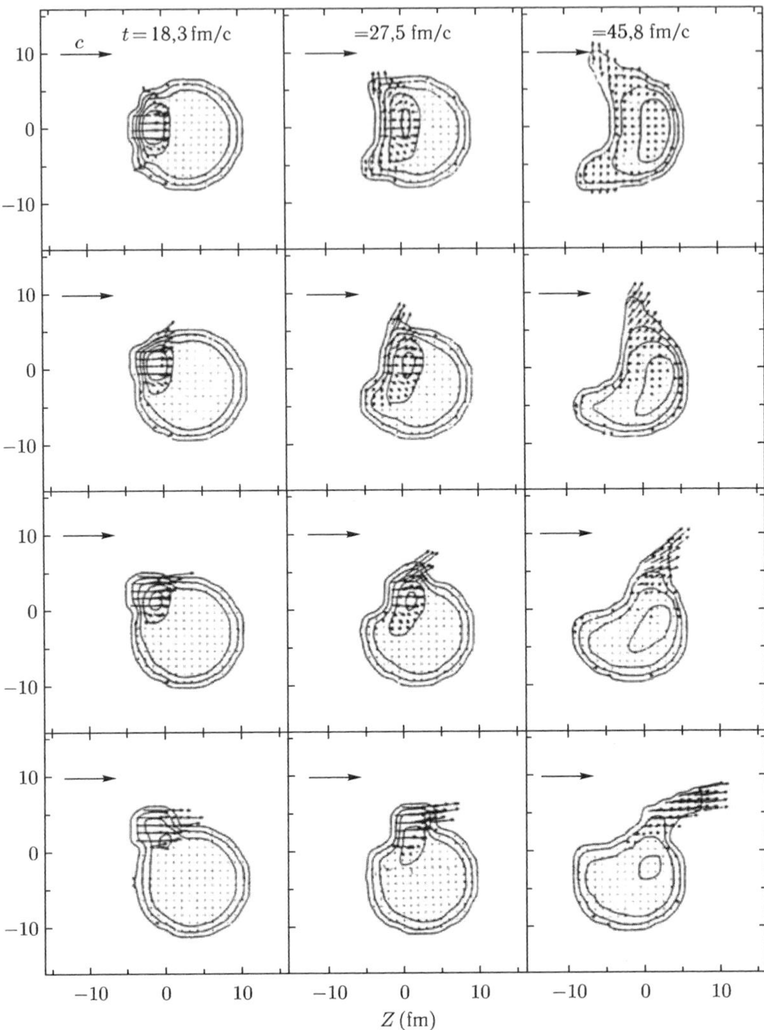



Fig. 13.6   Density isolines. Hydrodynamic simulation of the C + Au collision at the energy 84 MeV per nucleon. Time increases from left to right and from top to bottom [955].

For instance, in Refs. [427, 952, 953] information on EOS was obtained making use of the pion yield data [876]. This made it possible to estimate $W_0$ as a function of the shock energy employing the one-dimensional model of shock compression. It was found in Ref. [416] that the compression energy turns out to be almost the same for linear and quadratic EOS, though the densities estimated within the accuracy 30% were appreciably different. The following simplifications were used:

(1) The one-dimensional compressibility model is applicable and the pion yield is determined by the parameters of the shock-compressed zone.
(2) A smooth dependence of the intrinsic energy $E$ on the shock energy $E_c(E)$ is used, although the experiment gives discrete and rare values.
(3) The thermal contribution is described by the Fermi gas formulae $p_\mathrm{T} = (2/3)\rho E_\mathrm{th}$.

Using $W_0(E)$ and the experimental curve one must determine $\rho(E)$ and then build $W_0(\rho)$.

For a one-dimensional problem

$$e(\rho, s) = \rho(m + W_0(\rho) + E_\mathrm{th}(\rho, s)),$$

where $\rho$ is density and $s$ is enthropy. Here, the nucleon mass $m$ is replaced by the nucleon mass minus the ground state energy. The enthalpy density has the form

$$w(\rho, s) = e(\rho, s) + p(\rho, s),$$

where the pressure is

$$p(\rho, s) = \rho^2 \frac{\partial}{\partial \rho} \left( W_0(\rho) + E_\mathrm{th}(\rho, s) \right) \Bigg|_s.$$

The Hugoniot equation [1081] has the form:

$$\frac{w_1^2}{\rho_1^2} - \frac{w_2^2}{\rho_2^2} + (p_2 - p_1) \left( \frac{w_1}{\rho_1^2} + \frac{w_2}{\rho_2^2} \right) = 0. \tag{13.13}$$

For the nonperturbed nuclei we have $\rho_1 = \rho_0$, $p_1 = 0$ and $e_1 = w_1 = \rho m$. For the compressed matter domain $\rho_2 = x\rho_1$, $e_2 = \rho_2 E$.

Taking into account the energy formula given by the nonrelativistic Fermi gas, $p_2$ can be written as

$$p_2 = \rho^2 \frac{dW_0(\rho)}{d\rho} + \frac{2}{3}\rho(E - W_0(\rho)).$$

Using (13.13) we arrive at the equation for $W_0(\rho)$:

$$\frac{dW_0(\rho)}{dx} = \frac{2}{3} \frac{W_0 - E}{x} + \frac{E(E + 2m)}{x[m(x - 1) - E]}, \tag{13.14}$$

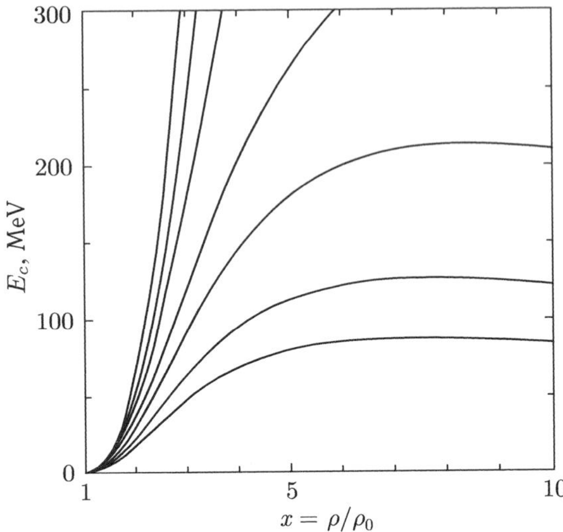

Fig. 13.7   EOS reproducing the experimental data in the simplified model. The curves can be obtained by the integration of equation (13.14) with quadratic behavior of $W_0(x)$ at density values below the experimental data region [416].

The obtained results were approximated [416] as

$$E(W) = 2.26W - 8.61 \text{ MeV}.$$

The calculation results are shown in Fig. 13.7. The curves were approximated to the ground state by the polynomial

$$W_0(x) = \frac{K}{18}(x - 1)^2 + A(x - 1)^3.$$

The additional restraints must be imposed on the found EOS. If the curve $W_0(\rho)$ is too soft, it will describe experimental data only at low energies. For the "stiff" part of the curves the causality condition $c_s^2 = dW_0(\rho)/de < c^2$ must be satisfied.

# Bibliography


[1] Abbamonte, P., Blumberg, G., Rusydi, A., Gozar, A., Evans, P. G., Siegrist, T., Venema, L., Eisaki, H., Isaacs, E. D. and Sawatzky, G. A. (2004). Crystallization of charge holes in the spin ladder of $Sr_{14}Cu_{24}O_{41}$, *Nature* **431**, pp. 1078–1081.

[2] Abe, R. (1959). Giant cluster expansion theory and its application to high-temperature plasma, *Progr. Theor. Phys.* **22**, pp. 213–226.

[3] Abrikosov, A. A. (1960). Some properties of strongly compressed matter, *Zh. Eksp. Teor. Fiz.* **39**, pp. 1797–1805.

[4] Abrikosov, A. A. (1962). Contribution to the theory of highly compressed matter. II, *J. Exp. Theor. Phys.* **14**, 2, p. 408.

[5] Abrikosov, A. A., Gorkov, L. P. and Dzyaloshinskii, I. E. (1977). *Methods of Quantum Field Theory in Statistical Physics* (Dover Publications, New York).

[6] Adams, J., Adler, C., Aggarwal, M. M., Ahammed, Z., Amonett, J., Anderson, B. D., Anderson, M., Arkhipkin, D., Averichev, G. S., Badyal, S. K., Balewski, J., Barannikova, O., Barnby, L. S., Baudot, J., Bekele, S., Belaga, V. V., Bellwied, R., Berger, J., Bezverkhny, B. I., Bhardwaj, S., Bhaskar, P., Bhati, A. K., Bichsel, H., Billmeier, A., Bland, L. C., Blyth, C. O. and Bonner, B. E. (2004a). Azimuthal anisotropy at the relativistic heavy ion collider: The first and fourth harmonics, *Phys. Rev. Lett.* **92**, 6, p. 062301.

[7] Adams, J., Adler, C., Aggarwal, M. M., Ahammed, Z., Amonett, J., Anderson, B. D., Anderson, M., Arkhipkin, D., Averichev, G. S., Badyal, S. K., Balewski, J., Barannikova, O., Barnby, L. S., Baudot, J., Bekele, S., Belaga, V. V., Bellwied, R., Berger, J., Bezverkhny, B. I., Bhardwaj, S., Bhaskar, P., Bhati, A. K., Bichsel, H., Billmeier, A., Bland, L. C., Blyth, C. O. and Bonner, B. E. (2004b). Particle-type dependence of azimuthal anisotropy and nuclear modification of particle production in $Au + Au$ collisions at $sNN = 200GeV$, *Phys. Rev. Lett.* **92**, 5, p. 052302.

[8] Adler, S. S., Afanasiev, S., Aidala, C., Ajitanand, N. N., Akiba, Y., Alexander, J., Amirikas, R., Aphecetche, L., Aronson, S. H., Averbeck, R., Awes, T. C., Azmoun, R., Babintsev, V., Baldisseri, A., Barish, K. N.,

Barnes, P. D., Bassalleck, B., Bathe, S., Batsouli, S., Baublis, V., Bazilevsky, A., Belikov, S., Berdnikov, Y., Bhagavatula, S., Boissevain, J. G., Borel, H. and Borenstein, S. (2003). Elliptic Flow of Identified Hadrons in $Au + Au$ Collisions at $sNN = 200GeV$, *Phys. Rev. Lett.* **91**, 18, p. 182301.

[9] Ailawadi, N. K. (1980). Equilibrium theories of simple liquids, *Phys. Rep.* **57**, 4, pp. 241–306.

[10] Akhiezer, A. I. and Berestetskii, V. B. (1965). *Quantum Electrodynamics* (Interscience Publishers, New York).

[11] Akmal, A., Pandharipande, V. R. and Ravenhall, D. G. (1987). Equation of state of nucleon matter and neutron star structure, *Phys. Rev. C* **58**, pp. 1804–1828.

[12] Alastuey, A., Cornu, F. and Perez, A. (1994). Virial expansions for quantum plasmas: Diagrammatic resummations, *Phys. Rev. E* **49**, pp. 1077–1093.

[13] Alastuey, A., Cornu, F. M. C. and Perez, A. (1995). Virial expansions for quantum plasmas: Maxwell–Boltzmann statistics, *Phys. Rev. E* **51**, pp. 1725–1744.

[14] Alastuey, A. and Perez, A. (1992). Virial expansion of the equation of state of a quantum plasma, *EPL (Europhys. Lett.)* **20**, 1, p. 19.

[15] Alastuey, A. and Perez, A. (1996). Virial expansions for quantum plasmas: Fermi-Bose statistics, *Phys. Rev. E* **53**, pp. 5714–5728.

[16] Albers, R. C. and Gubernatis, J. E. (1986). Low-order anharmonic contributions to the internal energy of the one-component plasma, *Phys. Rev. B* **33**, pp. 5180–5185.

[17] Alder, B. J. and Wainwright, T. E. (1957). *J. Chem. Phys.* **27**, p. 1208.

[18] Alekseev, V., Velikhov, E. and Lopantseva, G. B. (1967). *MHD generators* **1**, p. 17.

[19] Alekseev, V. A. (1968). *Teplofiz. Vys. Temp.* **6**, p. 961.

[20] Alekseev, V. A. and Vedenov, A. A. (1971). Electric conductivity of dense cesium vapor, *Phys. Usp.* **13**, 6, pp. 830–831. URL http://ufn.ru/en/articles/1971/6/l/.

[21] Alekseev, V. A., Velikhov, E. P. and Lopantseva, G. B. (1966). *Electricity from MHD*, p. 617.

[22] Alekseev, Y. L., Ratnikov, B. P. and Rybakov, A. P. (1971). *Prikl. Mekh. Tekh. Fiz.* **2**, p. 101.

[23] Alemasov, V., Dregalin, A. F., P., T. A. and Khudyakov, V. A. (1971). *Thermodynamic and thermophysical properties of combustion products* (VINITI, Moscow).

[24] Allen, M. and Tildesley, D. (1989). *Computer simulation of liquids* (Oxford University Press, New York, Oxford).

[25] Alt, C., Anticic, T., Baatar, B., Barna, D., Bartke, J., Behler, M., Betev, L., Białkowska, H., Billmeier, A., Blume, C., Boimska, B., Borghini, N., Botje, M., Bracinik, J., Bramm, R., Brun, R., Bunčić, P., Cerny, V., Chvala, O., Cooper, G. E., Cramer, J. G., Csató, P., Dinh, P. M., Dinkelaker, P., Eckardt, V., Filip, P. and Fodor, Z. (2003). Directed

and elliptic flow of charged pions and protons in $Pb + Pb$ collisions at $4A$ and $158AGeV$, *Phys. Rev. C* **68**, 3, p. 034903.

[26] Al'tshuler, L., Krupnikov, K. K., Fortov, V. E. and Funtikov, A. I. (2004a). The beginning of megabar pressure physics, *Vestnik RAS* **74**, 11, p. 1011.

[27] Al'tshuler, L. V. (1965). Use of shock waves in high-pressure physics, *Phys. Usp.* **8**, 1, pp. 52–91.

[28] Al'tshuler, L. V. and Bakanova, A. A. (1969). Electronic structure and compressibility of metals at high pressures, *Phys. Usp.* **11**, 5, pp. 678–689.

[29] Al'tshuler, L. V., Bakanova, A. A., Bushman, A. V., Dudoladov, I. P. and Zubarev, V. N. (1977a). Evaporation of shock-compressed lead in release waves, *J. Exp. Theor. Phys.* **46**, 5, p. 980.

[30] Al'tshuler, L. V., Bakanova, A. A. and Dudoladov, I. P. (1968). Effect of electron structure on the compressibility of metals at high pressure, *J. Exp. Theor. Phys.* **26**, 6, p. 1115.

[31] Al'tshuler, L. V., Bakanova, A. A., Dudoladov, I. P., Dynin, E. A., Trunin, R. F. and Chekin, B. S. (1981a). *Prikl. Mekh. Tekh. Fiz.*, **2**, p. 3.

[32] Al'tshuler, L. V., Bakanova, A. A. and Trunin, R. F. (1962). Shock Adiabats and zero isotherms of seven metals at high pressures, *J. Exp. Theor. Phys.* **15**, 1, p. 65.

[33] Al'tshuler, L. V., Brusnikin, S. E. and Marchenko, A. I. (1989). Determination of the gruneisen coefficient of a strongly nonideal plasma, *High Temp.* **27**, p. 492.

[34] Al'tshuler, L. V., Bushman, A. V., Zhernokletov, M. V., Zubarev, V. N., Leont'ev, A. A. and Fortov, V. E. (1980). Unloading isentropes and the equation of state of metals at high energy densities, *J. Exp. Theor. Phys.* **51**, 2, p. 373.

[35] Al'tshuler, L. V., Kalitkin, N. N., Kuz'mina, L. V. and Chekin, B. S. (1977b). Shock adiabats for ultrahigh pressures, *J. Exp. Theor. Phys.* **45**, 1, p. 167.

[36] Al'tshuler, L. V., Kormer, S. B., Bakanova, A. A. and Trunin, R. F. (1960). *ZhETF* **38**, p. 790.

[37] Al'tshuler, L. V., Krupnikov, K. K. and Brazhnik, M. I. (1958). *ZhETF* **34**, p. 886.

[38] Al'tshuler, L. V., Krupnikov, K. K., Fortov, V. E. and Funtikov, A. I. (2004b). Origins of megabar pressure physics, *Herald Russ. Acad. Sci.* **74**, 6, p. 613.

[39] Al'tshuler, L. V., Trunin, R. F., Krupnikov, K. K. and Panov, N. V. (1996). Explosive laboratory devices for shock wave compression studies, *Phys. Usp.* **39**, 5, pp. 539–544.

[40] Al'tshuler, L. V., Trunin, R. F., Urlin, V. D., Fortov, V. E. and Funtikov, A. I. (1999). Development of dynamic high-pressure techniques in Russia, *Phys. Usp.* **42**, 3, 261.

[41] Al'tshuler, L. V., Voropinov, A. I., Gandel'man, G. M. *et al.* (1981b). *Fiz. Met. Metalloved.* **51**, p. 76.

[42] Al'tshuller, L. V., Moiseev, B. N., Popov, L. V., Simakov, G. V. and Trunin, R. F. (1968). Relative compressibility of iron and lead at pressures of 31 to 34 Mbar, *J. Exp. Theor. Phys.* **27**, 3, p. 420.

[43] Ambartsumyan, V. A. and Saakyan, G. S. (1960). *Astron. Zh.* **37**, p. 193.

[44] Ambartsumyan, V. A. and Saakyan, G. S. (1963). *Vopr. Kosmogon.* **9**, p. 91.

[45] Andersen, H. C., Chandler, D. and Weeks, J. D. (1976). *Roles of Repulsive and Attractive Forces in Liquids: The Equilibrium Theory of Classical Fluids* (John Wiley & Sons, Inc.), pp. 105–156.

[46] Andersen, O. K. (1975). Linear methods in band theory, *Phys. Rev. B* **12**, 8, pp. 3060–3083.

[47] Anderson, P. W. and Palmer, R. G. (1971). Solidification pressure of nuclear and neutron star matter, *Nature Phys. Sci.* **231**, pp. 145–145.

[48] Andrei, E. Y., Deville, G., Glattli, D. C., Williams, F. I. B., Paris, E. and Etienne, B. (1988). Observation of a magnetically induced wigner solid, *Phys. Rev. Lett.* **60**, 26, pp. 2765–2768.

[49] Anisichkin, V. F. (1978). *Prikl. Mekh. Tekh. Fiz.*, **3**, p. 117.

[50] Anisimov, S. I., Prokhorov, A. M. and Fortov, V. E. (1984). Application of high-power lasers to study matter at ultrahigh pressures, *Sov. Phys. — Usp.* **27**, 3, pp. 181–205.

[51] Antia, H. M. (1993). Rational function approximations for Fermi–Dirac integrals, *Astrophys. J. Suppl. Ser.* **84**, pp. 101–108.

[52] Apfelbaum, E. M., Vorob'ev, V. S. and Martynov, G. A. (2004). Virial expansion providing of the linearity for a unit compressibility factor, *J. Phys. Chem. A* **108**, 47, pp. 10381–10385.

[53] Arp, O., Block, D., Piel, A. and Melzer, A. (2004). Dust coulomb balls: Three-dimensional plasma crystals, *Phys. Rev. Lett.* **93**, 16, p. 165004.

[54] Asaumi, K., Mori, T. and Kondo, Y. (1982). Effect of very high pressure on the optical absorption edge in solid Xe and its implication for Metallization, *Phys. Rev. Lett.* **49**, 11, pp. 837–840.

[55] Ashcroft, N. W. and Lekner, J. (1966). Structure and resistivity of liquid metals, *Phys. Rev.* **145**, 1, pp. 83–90.

[56] Ashcroft, N. W. and Stroud, D. (1978). *Theory of the Thermodynamics of Simple Liquid Metals*, (Academic Press), pp. 1–81.

[57] Ashkroft, N. W. (2002). Condensed Matter at Higher Densities, in G. L. Chiarotti, R. J. Hemley, M. Bernasconi and L. Ulivi (eds.), *High Pressure Phenomena, Proceedings of the International School of Physics "Enrico Fermi" Course CXLVII* (IOS Press, Amsterdam), p. 151.

[58] Ashkroft, N. W. and Mermin, N. (1975). Vol. 1, 2, *Solid State Physics*. (Wiley & Sons).

[59] Atzeni, S. and Meyer-ter-Vehn, J. (2004). *The Physics of Inertial Fusion* (Oxford University Press, Oxford).

[60] Auluck, F. C. and Jain, A. (1970). Thomas–Fermi equation of state with exchange, correlation, and quantum corrections, *Phys. Rev. B* **1**, 2, pp. 931–932.

[61] Averin, A. B., Dremov, V. V., Samarin, S. I. and Sapozhnikov, A. T. (1996). Equation of state and phase diagram of carbon, *AIP Conference Proceedings* **370**, 1, pp. 65–68.

[62] Avrorin, E. N., Simonenko, V. A. and Shibarshov, L. I. (2006). Physics research during nuclear explosions, *Phys. Usp.* **49**, 4, p. 432.

[63] Avrorin, E. N., Vodolaga, B. K., Simonenko, V. A. and Fortov, V. E. (1993). Intense shock waves and extreme states of matter, *Phys. Usp.* **36**, 5, pp. 337–364.

[64] Avrorin, E. N., Vodolaga, B. K., Volkov, L. P., Vladimirov, A. S., Simonenko, V. A. and Chernovolyuk, B. T. (1980). Shock compressibility of lead, quartzite, aluminum, and water at a pressure of 100 Mbar, *JETP Lett.* **31**, 12, p. 685.

[65] Avrorin, E. N., Vodolaga, B. K., Voloshin, N. P., Kuropatenko, V. F., Kovalenko, G. V., Simonenko, V. A. and Chernovolyuk, B. T. (1986). Experimental confirmation of shell effects on the shock adiabats of aluminum and lead, *JETP Lett.* **43**, p. 308.

[66] Ayukov, S. V., Baturin, A., Gryaznov, V. K., Iosilevskii, I. L., Starostin, A. N. and Fortov, V. E. (2004). Analysis of the presence of small admixtures of heavy elements in the solar plasma by using the SAHA-S equation of state, *JETP Lett.* **80**, 3, p. 141.

[67] Baade, W. and Zwicky, F. (1934). Cosmic rays from super-novae, *Proc. Natl. Acad. Sci. USA* **20**, 5, pp. 259–263.

[68] Bahcall, J. N. and Wolf, R. A. (1965). Neutron stars. II. Neutrino-cooling and observability, *Phys. Rev* **140**, pp. B1452–B1466.

[69] Bahcall, S., Lynn, B. W. and Selipsky, S. B. (1990). New models for neutron stars, *Astrophys. J.* **362**, pp. 251–255.

[70] Bailey, A. E., Poon, W. C. K., Christianson, R. J., Schofield, A. B., Gasser, U., Prasad, V., Manley, S., Segre, P. N., Cipelletti, L., Meyer, W. V., Doherty, M. P., Sankaran, S., Jankovsky, A. L., Shiley, W. L., Bowen, J. P., Eggers, J. C., Kurta, C., Lorik, T., Pusey, P. N. and Weitz, D. A. (2007). Spinodal decomposition in a model colloid-polymer mixture in microgravity, *Phys. Rev. Lett.* **99**, p. 205701.

[71] Bakanova, A. A., Dudoladov, I. P. and Sutulov, Y. N. (1974). *Prikl. Mekh. Tekh. Fiz.*, **2**, p. 117.

[72] Balberg, S. and Gal, A. (1997). An effective equation of state for dense matter with strangeness, *Nucl. Phys. A* **625**, pp. 435–472.

[73] Baldo, M., Bombaci, I. and Burgio, G. F. (1997). Microscopic nuclear equation of state with three-body forces and neutron-star structure, *Astron. Astrophys.* **328**, pp. 274–282.

[74] Baldo, M., Fiasconaro, A., Song, H. Q., Giansiracusa, G. and Lombardo, U. (2001). High density symmetric nuclear matter in Bethe–Brueckner–Goldstone approach, *Phys. Rev. C* **65**, p. 017303.

[75] Balescu, R. (1975). *Equilibrium and Nonequilibrium Statistical Mechanics* (Wiley, N.Y.).

[76] Balescu, R. (1978). Vol. 1, *Equilibrium and Nonequilibrium Statistical Mechanics* (John Wiley & Sons).

[77] Baluni, V. (1978). Non-Abelian gauge theories of Fermi systems: Quantum-chromodynamic theory of highly condensed matter, *Phys. Rev. D* **17**, 8, pp. 2092–2121.

[78] Barker, J. A. and Henderson, D. (1976). What is "liquid"? Understanding the states of matter, *Rev. Mod. Phys.* **48**, 4, pp. 587–671.

[79] Barrois, B. (1977). Superconducting quark matter, *Nucl. Phys. B* **129**, pp. 390–396.

[80] Basharin, A. Y. (2002). Investigation of non-congruent evaporation of uranium dioxide, in *Theses of the Conference, TSB-2002* (Kazan'), p. 172.

[81] Basko, M. M. (1985). Metallic equation of state in the mean ion approximation, *High Temp.* **23**, 3, p. 388.

[82] Baumung, K., Bluhm, H. J., Goel, B., Hoppe, P., Karow, H. U., Rusch, D., Fortov, V. E., Kanel, G. I., Razorenov, S. V., Utkin, A. V. and Vorobjev, O. Y. (1996). Shock-wave physics experiments with high-power proton beams, *Laser Part. Beams* **14**, pp. 181–209.

[83] Baus, M. and Hansen, J.-P. (1980). Statistical mechanics of simple coulomb systems, *Phys. Rep.* **59**, 1, pp. 1–94.

[84] Baym, G. (2007). Matter under extreme conditions.

[85] Baym, G. and Chin, S. A. (1976). Can a neutron star be a giant MIT bag? *Phys. Lett. B* **62**, 2, pp. 241–244.

[86] Baym, G., Pethick, C., Pines, D. and Ruderman, M. (1969). Spin up in neutron stars: The future of the Vela pulsar, *Nature* **224**, pp. 872–874.

[87] Baym, G., Pethick, C. and Sutherland, P. (1971). The ground state of matter at high densities: Equation of state and stellar models, *Astrophys. J.* **170**, pp. 299–317.

[88] Bazanov, O. V., Bespalov, V. E., Zharkov, A. P., Rumyantsev, B. V., Fedotova, T. B., Fortov, V. E. and Misonochnikov, A. L. (1985). Irregular reflection of conically converging shock-waves in plexiglas and copper, *High Temp.* **23**, 5, p. 781.

[89] Bazavov, A., Bhattacharya, T., Cheng, M., Christ, N. H., DeTar, C., Ejiri, S., Gottlieb, S., Gupta, R., Heller, U. M., Huebner, K., Jung, C., Karsch, F., Laermann, E., Levkova, L., Miao, C., Mawhinney, R. D., Petreczky, P., Schmidt, C., Soltz, R. A., Soeldner, W., Sugar, R., Toussaint, D. and Vranas, P. (2009). Equation of state and QCD transition at finite temperature, *Phys. Rev. D* **80**, p. 014504.

[90] Belonoshko, A. B., Rosengren, A., Burakovsky, L., Preston, D. L. and Johansson, B. (2009). Melting of fe and $fe_{0.9375}si_{0.0625}$ at earth's core pressures studied using *ab initio* molecular dynamics, *Phys. Rev. B* **79**, 22, p. 220102.

[91] Belov, S., Boriskov, G., Bykov, A., Il'kaev, R., Luk'yanov, N., Matveev, A., Mikhailova, O., Selemir, V., Simakov, G., Trunin, R., Trusov, I., Urlin, V., Fortov, V. and Shuikin, A. (2002). Shock compression of solid deuterium, *JETP Lett.* **76**, 7, pp. 433–435.

[92] Belov, S. I., Boriskov, G. V. *et al.* (2003). *Substances, materials and constructions under intense dynamic impacts* (VNIIEF, Sarov).

[93] Bendt, P. and Zunger, A. (1982). New approach for solving the density-functional self-consistent-field problem, *Phys. Rev. B* **26**, 6, pp. 3114–3137.

[94] Bernal, J. D. (1960). Packing of Spheres: Co-ordination of randomly packed spheres, *Nature* **188**, pp. 910–911.

[95] Bernal, J. D. (1964). *Proc. R. Soc. London* **A280**, p. 299.

[96] Berne, B. J., Ciccotti, G. and Coker, D. F. (eds.) (1998.). *Quantum Dynamics of Condensed Phase Simulation* (World Scientific, Singapore).

[97] Bespalov, V. E., Gryaznov, V. K., Dremin, A. N. and Fortov, V. E. (1975). Dynamic compression of non-ideal argon plasma, *J. Exp. Theor. Phys.* **42**, 6, p. 1046.

[98] Beth, E. and Uhlenbeck, G. E. (1937). The quantum theory of the non-ideal gas. II: Behaviour at low temperatures, *Physica, The Hague* **4**, pp. 915–924.

[99] Bethe, H. A. (1953). What holds the nucleus together, *Sci. Am.* **189**, pp. 58–63.

[100] Bethe, H. A. (1973). *Theory of nuclear matter* (Nauka, Moscow).

[101] Beule, D., Ebeling, W., Förster, A., Juranek, H., Nagel, S., Redmer, R. and Röpke, G. (1999). Equation of state for hydrogen below 10000 K: From the fluid to the plasma, *Phys. Rev. B* **59**, 22, pp. 14177–14181.

[102] Bezkrovniy, V., Filinov, V. S., Kremp, D., Bonitz, M., Schlanges, M., Kraeft, W. D., Levashov, P. R. and Fortov, V. E. (2004a). Monte Carlo results for the hydrogen Hugoniot, *Phys. Rev. E* **70**, 5, p. 057401.

[103] Bezkrovniy, V., Schlanges, M., Kremp, D. and Kraeft, W. D. (2004b). Reaction ensemble Monte Carlo technique and hypernetted chain approximation study of dense hydrogen, *Phys. Rev. E* **69**, 6, p. 061204.

[104] Binder, K. (ed.) (1979). *Monte Carlo Methods in Statistical Physics* (Springer-Verlag, Berlin).

[105] Blaizot, J. P. (1980). Nuclear compressibilities, *Phys. Rep.* **64**, 4, pp. 171–248.

[106] Bludman, S. A. and Ruderman, M. A. (1968). Possibility of the speed of sound exceeding the speed of light in ultradense matter, *Phys. Rev.* **170**, pp. 1176–1184.

[107] Bludman, S. A. and Ruderman, M. A. (1970). Noncausality and instability in ultradense matter, *Phys. Rev. D* **1**, pp. 3243–3246.

[108] Boates, B. and Bonev, S. A. (2009). First-order liquid–liquid phase transition in compressed nitrogen, *Phys. Rev. Lett.* **102**, p. 015701.

[109] Bober, M., Breitung, W., Karow, H. U. and Schretzmann, K. (1976). On the interpretation of vapor pressure measurements on oxide fuel at very-high temperatures for fast reactor safety analysis, *J. Nucl. Materials* **60**, 1, pp. 20–30.

[110] Bober, M. and Singer, J. (1987). *Nuclear Science & Engineering* **97**, p. 344.

[111] Bobrovskii, S. V., Gogolev, V. M., Menzhulin, M. G. and Shilova, R. P. (1978). *Prikl. Mekh. Tekh. Fiz.* , 5, p. 130.

[112] Bobrovskii, S. V., Gogolev, V. M., Zamyshljaev, V. V. and Lozhkina, V. P. (1979). *Prikl. Mekh. Tekh. Fiz.* , 6, p. 112.

[113] Bocquet, M., Bonazzola, S., Gourgoulhon, E. and Novak, J. (1995). Rotating neutron star models with a magnetic field. *Astron. Astrophys.* **301**, p. 757.

[114] Bodmer, A. R. (1971). Collapsed Nuclei, *Phys. Rev. D* **4**, 6, pp. 1601–1606.

[115] Boehler, R. and Ramakrishnan, J. (1980). Experimental results on the pressure dependence of the Grüneisen parameter, *J. Geophys. Res. Ser. B.* **85**, pp. 6996–7002.

[116] Boehler, R. and Ross, M. (1997). Melting curve of aluminum in a diamond cell to 0.8 Mbar: Implications for iron, *Earth Planet. Sci. Lett.* **153**, 3–4, pp. 223–227.

[117] Bogolyubov, N. N. (1947). To the superfluidity theory, *Izv. Akad. Nauk. SSSR* **11**, 1, p. 77.

[118] Boguslavskii, Y. A., Voronov, F. F., Il'ina, M. A. and Stal'gorova, O. V. (1979). Ba I–Ba II phase boundary and the maximum on the melting curve of Bal at high pressures, *J. Exp. Theor. Phys.* **50**, 3, p. 476.

[119] Boissiere, C. and Fiorese, G. (1977). *Rev. Phys. Appl.* **12**, p. 857.

[120] Bombaci, I. (1995). An equation of state for asymmetric nuclear matter and the structure of neutron stars, in I. Bombaci, A. Bonaccorso, A. Fabrocini *et al.* (eds.), *Perspectives on Theoretical Nuclear Physics* (Edizioni ETS, Pisa), pp. 223–237.

[121] Bonche, P. and Vautherin, D. (1981). A mean field calculation of the equation of state of supernova matter, *Nucl. Phys. A* **372**, pp. 496–526.

[122] Bonev, S. A., Militzer, B. and Galli, G. (2004a). *Ab initio* simulations of dense liquid deuterium: Comparison with gas-gun shock-wave experiments, *Phys. Rev. B* **69**, 1, p. 014101.

[123] Bonev, S. A., Schwegler, E., Ogitsu, T. and Galli, G. (2004b). A quantum fluid of metallic hydrogen suggested by first-principles calculations, *Nature* **431**, pp. 669–672.

[124] Bonitz, M., Filinov, V. S., Fortov., V. E., Levashov, P. R. and Fehske, H. (2006). *J. Phys. A: Math. Gen.* **39**, p. 4717.

[125] Bonitz, M., Mulenko, I. A. *et al.* (2001). Phase transition in superdense hydrogen and deuterium plasma, *Phys. Rep.* **27**, p. 1025.

[126] Bonitz, M., Semkat, D., Filinov, A., Golubnychyi, V., Kremp, D., Gericke, D. O., Murillo, M. S., Filinov, V., Fortov, V., Hoyer, W. and S. W. Koch, J. (2003). *J. Phys. A: Math. Gen.* **36**, pp. 5921–5930.

[127] Boriskov, G. V., Bykov, A. I., Il'kaev, R. I., Selemir, V. D., Simakov, G. V., Trunin, R. F., Urlin, V. D., Fortov, V. E. and Shuikin, A. I. (2003). Shock-wave compression of liquid deuterium under pressure of 120 GPa, *Dokl. Akad. Nauk* **392**, 6, pp. 755–757.

[128] Boriskov, G. V., Bykov, A. I., Il'kaev, R. I., Selemir, V. D., Simakov, G. V., Trunin, R. F., Urlin, V. D., Shuikin, A. N. and Nellis, W. J. (2005). Shock compression of liquid deuterium up to 109 GPa, *Phys. Rev. B* **71**, p. 9, 092104.

[129] Born, M. (1940). On the stability of crystal lattices. I. *Proc. Cambridge Phil. Soc.* **36**, pp. 160–172.

[130] Born, M. and Huang, K. (1954). *Dynamical Theory of Crystal Lattices* (Clarendon Press, Oxford).

[131] Bouchet, J., Bottin, F., Jomard, G. and Zérah, G. (2009). Melting curve of aluminum up to 300 GPa obtained through *ab initio* molecular dynamics simulations, *Phys. Rev. B* **80**, 9, p. 094102.

[132] Boys, S. F. (1950). *Proc. R. Soc. London Ser. A.*

[133] Breitung, W. and Reil, K. O. (1985). Report KfK 3939, *Tech. Rep.*, KfK, Karlsruhe.

[134] Breitung, W. and Reil, K. O. (1989). Vapor pressure measurements on liquid uranium oxide and (U,Pu) mixed oxide, *Nucl. Sci. Eng.* **101**, 1, pp. 26–40.

[135] Breitung, W. and Reil, K. O. (1990). The density and compressibility of liquid (U,Pu)-mixed oxide, *Nucl. Sci. Eng.* **105**, 3, pp. 205–217.

[136] Brillouin, L. (1934). *Quantum Statistics* (NKTP Scientific-Technical Publishers of Ukraine, Kharkov, Kiev).

[137] Brinkley, S. (1947). *J. Chem. Phys.* **15**, p. 107.

[138] Brovman, E. G. and Kagan, Y. M. (1974). Phonons in nontransition metals, *Phys. Usp.* **17**, 2, pp. 125–152.

[139] Brown, G. E., Gelman, B. A. and Rho, M. (2006). Matter formed at the bnl relativistic heavy ion collider, *Phys. Rev. Lett.* **96**, p. 132301.

[140] Brush, S. G., Sahlin, H. L. and Teller, E. (1966). Monte Carlo study of a one-component plasma. I, *J. Chem. Phys.* **45**, 6, pp. 2102–2118.

[141] Buchler, J.-R. and Barkat, Z. (1971). Properties of low-density neutron-star matter, *Phys. Rev. Lett.* **27**, pp. 48–51.

[142] Bulanov, S. V. (2006). New epoch in the charged particle acceleration by relativistically intense laser radiation, *Plasma Phys. Control. Fusion* **48**, 12B, pp. B29–B37.

[143] Bundy, F. P. (1989). Pressure-temperature phase diagram of elemental carbon, *Physica A: Statistical Mechanics and its Applications* **156**, 1, pp. 169–178.

[144] Burke, D. L., Field, R. C., Horton-Smith, G., Spencer, J. E., Walz, D., Berridge, S. C., Bugg, W. M., Shmakov, K., Weidemann, A. W., Bula, C., McDonald, K. T., Prebys, E. J., Bamber, C., Boege, S. J., Koffas, T., Kotseroglou, T., Melissinos, A. C., Meyerhofer, D. D., Reis, D. A. and Ragg, W. (1997). Positron production in multiphoton light-by-light scattering, *Phys. Rev. Lett.* **79**, 9, pp. 1626–1629.

[145] Burrows, A., Hubbard, W. B., Lunine, J. I. and Liebert, J. (2001). The theory of brown dwarfs and extrasolar giant planets, *Rev. Mod. Phys.* **73**, 3, pp. 719–765.

[146] Bushman, A., Kanel', G. I., Ni, A. L. and Fortov, V. E. (1989). *Thermal Physics and Dynamics of Intense Pulsed Effects* (Institute of Chemical Physics of the USSR Academy of Science, Chernogolovka).

[147] Bushman, A. V. and Fortov, V. E. (1983). Model equations of state, *Phys. Usp.* **26**, 6, pp. 465–496.

[148] Bushman, A. V., Fortov, V. E. and Sharipdzhanov, I. I. (1977). *Teplofiz. Vys. Temp.* **15**, 5, p. 1095.

[149] Bushman, A. V., Lomakin, B. N., Sechenov, V. A., Fortov, V. E., Shchekotov, O. E. and Sharipdzhanov, I. I. (1975). Thermodynamics of nonideal cesium plasma, *J. Exp. Theor. Phys.* **42**, 5, pp. 828–831.

[150] Bushman, A. V., Lomonosov, I. V. and Fortov, V. E. (1992). *Equations of State of Metals at High Energy Densities* (Institute of Chemical Physics of RAS, Chernogolovka).

[151] Bushman, A. V., Ni, A. L. and Fortov, V. E. (1981). *Equation of State in Extreme Conditions*, In G. V. Gadiyak (ed.), (Inst. of Theor. and Appl. Mechan. SB of the USSR Acad. of Sci.), p. 3.

[152] Calderola, P. and Knopfel, H. (eds.) (1971). *Physics of High Energy Density* (Academic, New York).

[153] Calderola, P. and Knopfel, H. (eds.) (1974). *High energy density physics* (Mir, Moscow).

[154] Cameron, A. G. (1959). Neutron Star Models. *Ap. J.* **130**, p. 884.

[155] Canuto, V. (1975). Equation of state at ultrahigh densities. Part 2, *Annu. Rev. Astron. Astrophys.* **13**, pp. 335–380.

[156] Caporaso, G. and Brecher, K. (1979). Must ultrabaric matter be superluminal? *Phys. Rev. D* **20**, pp. 1823–1831.

[157] Car, R. and Parrinello, M. (1985). Unified approach for molecular dynamics and density-functional theory, *Phys. Rev. Lett.* **55**, 22, pp. 2471–2474.

[158] Carnahan, N. F. and Starling, K. E. (1969). *J. Chem. Phys.* **51**, p. 635.

[159] Carr, W. J. (1961). Energy, specific heat, and magnetic properties of the low-density electron gas, *Phys. Rev.* **122**, 5, pp. 1437–1446.

[160] Carter, B. and Quintana, H. (1972). Foundations of general relativistic high-pressure elasticity theory, *Proc. Roy. Soc. London Ser. A* **331**, pp. 57–83.

[161] Carter, W. J., Fritz, J. N., Marsh, S. P. and McQueen, R. G. (1975). Hugoniot equation of state of the lanthanides, *J. Phys. Chem. Solids.* **36**, 7–8, pp. 741–752.

[162] Ceperley, D. (1978). Ground state of the fermion one-component plasma: A Monte Carlo study in two and three dimensions, *Phys. Rev. B* **18**, 7, pp. 3126–3138.

[163] Ceperley, D. (1996). *The Monte Carlo and Molecular Dynamics of Condensed Matter Systems* (Bologna, SIF).

[164] Ceperley, D. M. and Alder, B. J. (1980). Ground state of the electron gas by a stochastic method, *Phys. Rev. Lett.* **45**, 7, pp. 566–569.

[165] Chabanat, E., Bonche, P., Haensel, P., Meyer, J. and Schaeffer, R. (1997a). A Skyrme parameterization from subnuclear to neutron star densities, *Nucl. Phys. A* **627**, pp. 710–746.

[166] Chabanat, E., Bonche, P., Haensel, P., Meyer, J. and Schaeffer, R. (1997b). A Skyrme parameterization from subnuclear to neutron star densities. Part II. Nuclei far from stabilities, *Nucl. Phys. A* **635**, pp. 231–256.

[167] Chabrier, G. (1993). Quantum effects in dense Coulombic matter — Application to the cooling of white dwarfs, *Astrophys. J.* **414**, pp. 695–700.

[168] Chabrier, G. (1999). Review on white dwarf cooling theory, in S. E. Solheim and E. G. Meistas (eds.), *11th European Workshop on White Dwarfs*, ASP Conf. Ser., Vol. 169, pp. 369–377.

[169] Chabrier, G. and Potekhin, A. Y. (1998). Equation of state of fully ionized electron–ion plasmas, *Phys. Rev. E* **58**, pp. 4941–4949.

[170] Chabrier, G., Saumon, D., Hubbard, W. and Lunine, J. (1992a). The molecular–metallic transition of hydrogen and the structure of Jupiter and Saturn, *Astrophys. J.* **391**, pp. 817–826.

[171] Chabrier, G., Saumon, D., Hubbard, W. and van Horn, H. (1992b). The role of the molecular–metallic transition of hydrogen in the evolution of Jupiter, Saturn and brown dwarfs, *Astrophys. J.* **391**, pp. 827–831.

[172] Chapline, G. and Nauenberg, M. (1977). Asymptotic freedom and the baryon-quark phase transition, *Phys. Rev. D* **16**, 2, pp. 450–456.

[173] Chekin, B. S. (1978). *Prikl. Mekh. Tekh. Fiz.* **2**, p. 89.

[174] Chen, F. F. (1984). *Introduction to Plasma Phys. Control. Fusion*, Vol. 1, 2nd edn. (Springer, New York).

[175] Chernin, A. D. (2001). Cosmic vacuum, *Phys. Usp.* **44**, 11, 1099.

[176] Chin, S. A. and Kerman, A. K. (1979). Possible long-lived hyperstrange multiquark droplets, *Phys. Rev. Lett.* **43**, 18, pp. 1292–1295.

[177] Cho, S. and Zahed, I. (2009a). Classical strongly coupled quark-gluon plasma. III. The free energy, *Phys. Rev. C* **79**, p. 044911.

[178] Cho, S. and Zahed, I. (2009b). Classical strongly coupled quark-gluon plasma. IV. Thermodynamics, *Phys. Rev. C* **80**, p. 014906.

[179] Chodos, A., Jaffe, R. L., Johnson, K., Thorn, C. B. and Weisskopf, V. F. (1974). New extended model of hadrons, *Phys. Rev. D* **9**, pp. 3471–3495.

[180] Christensen-Dalsgaard, J., Däppen, W., Dziembowski, W. A. and Guzik, J. A. (2000). An Introduction to Helioseismology, in C. Ibanoğlu (ed.), *Variable Stars as Important Astrophysical Tools, Proc. NATO Advanced Study Institute, Cesme, Turkey*, August 30–September 11, 1998 (Kluwer Academic, Dordrecht), pp. 59–167.

[181] Chu, J. H. and I. L. (1994). Direct observation of coulomb crystals and liquids in strongly coupled rf dusty plasmas, *Phys. Rev. Lett.* **72**, 25, pp. 4009–4012.

[182] Clark, E. L., Krushelnick, K., Davies, J. R., Zepf, M., Tatarakis, M., Beg, F. N., Machacek, A., Norreys, P. A., Santala, M. I., Watts, I. and Dangor, A. E. (2000). Measurements of energetic proton transport through magnetized plasma from intense laser interactions with solids, *Phys. Rev. Lett.* **84**, 4, pp. 670–673.

[183] Clendenen, G. L. and Drickamer, H. G. (1964). Effect of pressure on the volume and lattice parameters of magnesium, *Phys. Rev.* **135**, 6A, pp. A1643–A1645.

[184] Cmeron, A. G. W. (1969). *Neutron Stars, Lectures, Preprint*.

[185] Cohen, E. G. D. and Murphy, T. J. (1969). New results in the theory of the classical electron gas, *Phys. Fluids* **12**, pp. 1403–1411.

[186] Cohen, M. L., Heine, V. and Phillips, J. C. (1982). The quantum mechanics of materials, *Sci. American* **246**, 6, pp. 66–78.

[187] Cohen, R., Lodenquai, J. and Ruderman, M. (1970). Atoms in superstrong magnetic fields, *Phys. Rev. Lett.* **25**, 7, pp. 467–469.

[188] Collins, G. W., Da Silva, L. B., Celliers, P., Gold, D. M., Foord, M. E., Wallace, R. J., Ng, A., Weber, S. V., Budil, K. S. and Cauble, R. (1998). Measurements of the equation of state of deuterium at the fluid insulator–metal transition, *Science* **281**, 5380, pp. 1178–1181.

[189] Collins, J. C. and Perry, M. J. (1975). Superdense Matter: Neutrons or asymptotically free quarks? *Phys. Rev. Lett.* **34**, 21, pp. 1353–1356.

[190] Collins, L., Kwon, I., Kress, J., Troullier, N. and Lynch, D. (1995). Quantum molecular dynamics simulations of hot, dense hydrogen, *Phys. Rev. E* **52**, 6, pp. 6202–6219.

[191] Constantinescu, D. H. and Rehák, P. (1973). Ground state of atoms and molecules in a superstrong magnetic field, *Phys. Rev. D* **8**, 6, pp. 1693–1706.

[192] Cook, M. A., Keyes, R. T. and Udy, L. L. (1959). Propagation characteristics of detonation-generated plasmas, *J. Appl. Phys.* **30**, 12, p. 1881.

[193] Correa, A. A., Bonev, S. A. and Galli, G. (2006). Carbon under extreme conditions: Phase boundaries and electronic properties from first-principles theory, *Proc. Natl. Acad. Sci. USA* **103**, 5, pp. 1204–1208.

[194] Cowan, R. D. and Ashkin, J. (1957). Extension of the Thomas–Fermi–Dirac statistical theory of the atom to finite temperatures, *Phys. Rev.* **105**, 1, pp. 144–157.

[195] Cowley, E. R. and Barker, J. A. (1980). A self-consistent cell model for the melting transition in hard sphere systems, *J. Chem. Phys.* **73**, 7, pp. 3452–3455.

[196] Csernai, L. P. and Kapusta, J. I. (1986). Entropy and cluster production in nuclear collisions, *Phys. Rep.* **131**, 4, pp. 223–318.

[197] Csikor, F. (2004). *J. High Energy Phys.* **046**, p. 0405.

[198] Csikor, F., Egri, G. I., Fodor, Z., Katz, S. D., Szabo, K. K. and Toth, A. I. (2004). The QCD Equation of State at Finite $T\mu$ on the Lattice, *Prog. Theor. Phys. Supp.*, 153, pp. 93–105.

[199] Cummins, H. Z. and Pike, E. R. (eds.) (1974). *Photon Correlation and Light Beating Spectroscopy* (Plenum, New York).

[200] Da Silva, L. B., Celliers, P., Collins, G. W., Budil, K. S., Holmes, N. C., Barbee Jr., T. W., Hammel, B. A., Kilkenny, J. D., Wallace, R. J., Ross, M., Cauble, R., Ng, A. and Chiu, G. (1997). Absolute equation of state measurements on shocked liquid deuterium up to 200 GPa (2 Mbar), *Phys. Rev. Lett.* **78**, 3, pp. 483–486.

[201] Datchi, F., Loubeyre, P. and LeToullec, R. (2000). Extended and accurate determination of the melting curves of argon, helium, ice ($H_2O$), and hydrogen ($H_2$), *Phys. Rev. B* **61**, pp. 6535–6546.

[202] de Wette, F. W. (1964). Note on the electron lattice, *Phys. Rev.* **135**, 2A, pp. A287–A294.

[203] Debye, P. and Hückel, E. (1923). On the theory of electrolytes. I. Freezing point depression and related phenomena, *Physikalische Zeitschrift* **24**, p. 185.

[204] DeGrand, T., Jaffe, R. L., Johnson, K. and Kiskis, J. (1975). Masses and other parameters of the light hadrons, *Phys. Rev. D* **12**, 7, pp. 2060–2076.

[205] Delannoy, M. and Perrin, G. (1980). On the determination of the volume dependence of Grüneisen parameters in cubic and non-cubic solids, *J. Phys. Chem. Solids* **41**, 1, pp. 11–16.

[206] Demarest Jr., H. H. (1974). Lattice model calculation of hugoniot curves and the Grüneisen parameter at high pressure for the alkali halides, *J. Phys. Chem. Solids* **35**, 10, pp. 1393–1404.

[207] Demyanko, Y. G., Konyukhov, G. V., Koroteev, A. S., Kuzmin, E. P. and Pavel'ev, A. A. (2001). *Nuclear Rocket Engines* (Norma-Info, Moscow).

[208] Desjarlais, M. P. (2003). Density-functional calculations of the liquid deuterium hugoniot, reshock, and reverberation timing, *Phys. Rev. B* **68**, 6, p. 064204.

[209] DeWitt, H. and Slattery, W. (1999). Screening enhancement of thermonuclear reactions in high density stars, *Contrib. Plasma Phys.* **39**, pp. 97–100.

[210] DeWitt, H., Slattery, W., Baiko, D. and Yakovlev, D. (2001). *Contrib. Plasma Phys.* **41**, pp. 251–254.

[211] DeWitt, H., Slattery, W. and Chabrier, G. (1996). Numerical simulation of strongly coupled binary ionic plasmas, *Physica B* **228**, pp. 21–26.

[212] DeWitt, H. E. (1961). Bound states in nonideal plasmas: Formulation of the partition function and application to the solar interior, *J. Nucl. Energy C* **2**, p. 27.

[213] DeWitt, H. E. (1962). Evaluation of the quantum-mechanical ring sum with Boltzmann statistics, *J. Math. Phys.* **3**, pp. 1216–1228.

[214] DeWitt, H. E. (1976). Asymptotic form of the classical one-component plasma fluid equation of state, *Phys. Rev. A* **14**, 3, pp. 1290–1293.

[215] Dey, M., Bombaci, I., Dey, J., S., R. and Samanta, B. C. (1998). Trange stars with realistic quark vector interaction and phenomenological density-dependent scalar potential, *Phys. Lett. B* **438**, 123–128.

[216] Dick, R. D. and Kerley, G. I. (1980). Shock compression data for liquids. II. Condensed hydrogen and deuterium, *J. Chem. Phys.* **73**, 10, pp. 5264–5271.

[217] Dmitriev, N. A. and Podval'nyi, V. G. (1981). *Fiz. Met. Metalloved.*

[218] Dolgov, O. V. and Maksimov, E. G. (1981). Local field effects and violation of the Kramers–Kronig permittivity relations, *Usp. Fiz. Nauk.* **135**, 11, pp. 441–477.

[219] Douchin, F. and Haensel, P. (2000). Inner edge of neutron-star crust with sly effective nucleonnucleon interactions, *Phys. Lett. B* **485**, pp. 107–114.

[220] Douchin, F. and Haensel, P. (2001). A unified equation of state of dense matter and neutron star structure, *Astron. Astrophys.* **380**, pp. 151–167.

[221] Drake, R. P. (2006). *High-Energy-Density Physics* (Springer, Berlin, Heidelberg).

[222] Dreizler, R. M. and Gross, E. K. U. (1990). *Density Functional Theory: An Approach to the Quantum Many-body Problem* (Springer-Verlag, Berlin).

[223] Dubin, D. H. E. and O'Neil, T. M. (1999). Trapped nonneutral plasmas, liquids and crystals (the thermal equilibrium states), *Rev. Mod. Phys.* **71**, p. 87.

[224] Duerr, H. P. (1956). Relativistic effects in nuclear forces, *Phys. Rev.* **103**, pp. 469–480.

[225] Dugdale, J. S. and Franck, J. P. (1964). The thermodynamic properties of solid and fluid helium-3 and helium-4 above 3 degrees K at High Densities, *Phil. Trans. R. Soc. Lond. A* **257**, 1076, pp. 1–29.

[226] Dusling, K. and Zahed, I. (2010). Parton energy loss in a classical strongly coupled QGP, *Nucl. Phys. A* **833**, 1–4, pp. 172–183.

[227] Duthie, J. C. and Pettifor, D. G. (1977). Correlation between $d$-band occupancy and crystal structure in the rare earths, *Phys. Rev. Lett.* **38**, 10, pp. 564–567.

[228] Ebeling, W. (1967). Statistische thermodynamik der gebundenen Zustande in Plasmen, *Annalen der Physik* **474**, 1–2, pp. 104–112.

[229] Ebeling, W., Foerster, A., Fortov, V., Gryaznov, V. and Polishchuk, A. (2007). *Thermophysical Properties of Hot Dense Plasma* (SRC "Regular and chaotic dynamics", Moscow, Izhevsk).

[230] Ebeling, W., Foerster, A., Fortov, V., Gryaznov, V. and Polishchuk, A. (1991). *Thermophysical Properties of Hot Dense Plasmas* (Teubner, Stuttgart-Leipzig).

[231] Ebeling, W., Förster, A., Richert, W. and Hess, H. (1988). Thermodynamic properties and plasma phase transition of xenon at high pressure and high temperature, *Physica A* **150**, pp. 159–171.

[232] Ebeling, W., Hoffmann, H. J. and Kelbg, G. (1967). Quantenstatistik des Hochtemperatur-Plasmas in thermodynamischen Gleichgewicht, *Beitr. Plasmaphys.* **7**, pp. 233–248.

[233] Ebeling, W., Kelbg, G. and Rohde, K. (1968). Binäre Slatersummen und Verteilungsfunktionen für quantenstatistische Systeme mit Coulomb-Wechselwirkung i. *Ann. der Phys.* **21**, 5–6, pp. 235–243.

[234] Ebeling, W., Kraeft, W. D. and Kremp, D. (1976). *Theory of Bound States and Ionization Equilibrium in Plasmas and Solids* (Akademie-Verlag, Berlin).

[235] Ebeling, W. and Norman, G. (2003). Coulombic phase transitions in dense plasmas, *J. Stat. Phys.* **110**, 3–6, pp. 861–877.

[236] Ebeling, W., Redmer, R., Reinholz, H. and Röpke, G. (2008). Thermodynamics and phase transitions in dense hydrogen — the role of bound state energy shifts, *Contrib. Plasma Phys.* **48**, p. 670.

[237] Ebeling, W. and Richert, W. (1965). Plasma phase transition in hydrogen, *Phys. Lett. A* **108**, 2, p. 80.

[238] Ebeling, W. and Richert, W. (1985a). Plasma phase transition in hydrogen, *Phys. Lett. A* **108**, pp. 80–82.

[239] Ebeling, W. and Richert, W. (1985b). Thermodynamic properties of liquid hydrogen metal, *Phys. Stat. Sol.* (B) **128**, 2, pp. 467–474. URL http://dx.doi.org/10.1002/pssb.2221280211.

[240] Ebeling, W. and Sandig, R. (1973). *Ann. Physik (Leipzig)* **483**, p. 289.

[241] Ebelting, W. and Förster, A. (1990). Thermodynamics and ionization kinetics for high pressure plasmas, *High Press. Res.* **4**, 1–6, pp. 484–486.

[242] Ecker, G. and Kröll, W. (1963). Lowering of the ionization energy for a plasma in thermodynamic equilibrium, *Phys. Fluids.* **6**, 1, p. 62.

[243] Enstrom, J. E. (1971). Solution of the Thomas–Fermi model with quantum corrections, *Phys. Rev. A* **4**, 2, pp. 460–464.

[244] Eremets, M. I., Gavriliuk, A. G., Trojan, I. A., Dzivenko, D. A. and Boehler, R. (2004). Single-bonded cubic form of nitrogen, *Nat. Mater.* **3**, pp. 558–563.

[245] Eremets, M. I. and Trojan, I. A. (2009). Evidence of maximum in the melting curve of hydrogen at megabar pressures, *JETP Lett.* **89**, 4, pp. 174–179.

[246] Errandonea, D., Meng, Y., Hausserman, D. and Uchida, T. (2003). *J. Phys. Condens. Mat.* **15**, p. 1277.

[247] Errandonea, D., Schwager, B., Boehler, R. and Ross, M. (2002). Phase behavior of krypton and xenon to 50 GPa, *Phys. Rev. B* **65**, p. 214110.

[248] Eyring, H. and Jhon, M. S. (1969). Significant liquid structures, in *Significant Liquid Structures* (J. Wiley, N.Y.), p. 16.

[249] Faber, T. E. (1977). *Fluid Dynamics for Physicists* (Cambridge University Press, Cambridge).

[250] Faddeev, L. D. (1961). Scattering theory for a three-particle system, *Sov. Phys. JETP* **12**, p. 1014.

[251] Farhi, E. and Jaffe, R. L. (1984). Strange matter, *Phys. Rev. D* **30**, 11, pp. 2379–2390.

[252] Farouki, R. T. and Hamaguchi, S. (1993). Thermal energy of the crystalline one-component plasma from dynamical simulations, *Phys. Rev. E* **47**, pp. 4330–4336.

[253] Feinberg, E. L. (1935a). *Zh. Eksp. Teor. Fiz.* **5**, p. 519.

[254] Feinberg, E. L. (1935b). *Zh. Eksp. Teor. Fiz.* **5**, p. 926.

[255] Fermi, E. (1924). Über die Wahrscheinlichkeit der Quantenzustände, *Z. Phys.* **26**, p. 54.

[256] Fermi, E. (1927). *Rend. Accad. Nazl. Lincei* **6**, p. 602.

[257] Feynman, R. P. (1939). Forces in molecules, *Phys. Rev.* **56**, 4, pp. 340–343.

[258] Feynman, R. P. and Hibbs, A. R. (1965). *Quantum Mechanics and Path Integrals* (McGraw-Hill, New York, Moscow).

[259] Feynman, R. P., Metropolis, N. and Teller, E. (1949). Equations of state of elements based on the generalized fermi-thomas theory, *Phys. Rev.* **75**, 10, pp. 1561–1573.

[260] Filinov, A. V., Bonitz, M. and Lozovik, Y. E. (2001a). Wigner crystallization in mesoscopic 2D electron systems, *Phys. Rev. Lett.* **86**, 17, pp. 3851–3854.

[261] Filinov, S., Ivanov, Y. B., Bonitz, M., Levashov, P. R. and Fortov, V. E. (2010). arXiv:1006.3390v1.

[262] Filinov, V. (2001). Cluster expansion for ideal fermi systems in the "fixed-node approximation," *J. Phys. A* **34**, 8, p. 1665.

[263] Filinov, V. S., Bonitz, M., Ivanov, Y. B., Skokov, V. V., Levashov, P. R. and Fortov, V. E. (2011). Quantum color dynamic simulations of the strongly coupled quark–gluon plasma, *Contrib. Plasma Phys.* 4, pp. 322–327.

[264] Filinov, V. S., Bonitz, M., Levashov, P., Fortov, V. E., Ebeling, W., Schlanges, M. and Koch, S. W. (2003a). Plasma phase transition in dense hydrogen and electron–hole plasmas, *J. Phys. A* **36**, 22, pp. 6069–6076.

[265] Filinov, V. S., Fehske, H., Bonitz, M., Fortov, V. E. and Levashov, P. (2007). Correlation effects in partially ionized mass asymmetric electron-hole plasmas, *Phys. Rev. E* **75**, 3, p. 036401.

[266] Filinov, V. S., Fortov, V. E., Bonitz, M. and Levashov, P. R. (2001b). *Pis'ma v ZhETF* **74**, 7, pp. 422–425.

[267] Filinov, V. S., Fortov, V. E., Bonitz, M. and Levashov, P. R. (2001c). Phase transition in strongly degenerate hydrogen plasma, *JETP Lett.* **74**, 7, p. 384.

[268] Filinov, V. S., Fortov, V. E., Bonitz, M. and Levashov, P. R. (2001d). Phase transition in strongly degenerate hydrogen plasma, *JETP Lett.* **74**, 7, p. 384.

[269] Filinov, V. S., Levashov, P. R., Bonitz, M. and Fortov, V. E. (2004). in *AIP Conference Proceedings*, Vol. 731 (Melville, New York), pp. 239–247.

[270] Filinov, V. S., Levashov, P. R., Bonitz, M. and Fortov, V. E. (2005a). Calculation of the shock Hugoniot of deuterium atpressures above 1 Mbar by the path-integral Monte Carlo method, *Plasma Phys. Rep.* **31**, 8, pp. 700–704.

[271] Filinov, V. S., Levashov, P. R., Bonitz, M. and Fortov, V. E. (2005b). Thermodynamics of hydrogen and hydrogen–helium plasmas: Path Integral Monte Carlo calculations and chemical picture, *Contrib. Plasma Phys.* **45**, 3–4, pp. 258–265.

[272] Filinov, V. S., M., B., Levashov, P. R., Fortov, V. E., Ebeling, W., Schlanges, M. and Koch, S. W. (2003b). *J. Phys. A: Math. Gen.* **36**, pp. 6069–6079.

[273] Fisch, N. J., Gladush, M. G., Petrushevich, Y. V., Quarati, P. and Starostin, A. N. (2012). Enhancement of fusion rates due to quantum effects in the particles momentum distribution in nonideal plasma media, *Eur. Phys. J. D* **66**, 6, p. 154.

[274] Fischer, E. A. (1987). Reports KfK 4084, *Tech. Rep.* KfK, Karlsruhe.

[275] Fischer, E. A. (1989). *Journal of Nuclear Science and Engineering* **101**, p. 97.

[276] Fischer, E. A. (1992). Reports KfK 4889, *Tech. Rep.* KfK, Karlsruhe.

[277] Fisher, I. Z. (1961). *Statisticheskaya Teoriya Zhidkostei (Statistical Theory of Liquids)* (Nauka, Moscow).

[278] Fock, V. A. (1935). Zur Theorie des Wasserstoffatoms, *Z. Physik* **98**, 3–4, pp. 145–154.

[279] Fodor, Z. and Katz, S. D. (2004). Critical point of QCD at finite T and $\mu$, lattice results for physical quark masses, *Journal of High Energy Physics* **2004**, 04, p. 050.

[280] Fodor, Z. and Katz, S. D. (2009). The phase diagram of quantum chromodynamics, *arXiv:0908.3341*.

[281] Fontaine, G., Graboske, H. C. J. and van Horn, H. M. (1977). Equation of state for stellar partial ionization zones, *Astrophys. J. Suppl. Ser.* **35**, pp. 293–358.

[282] Forsëter, A., Ebeling, W. and Richert, W. (1988). *High Pressure Geosciences and Material Synthesis* In H. Vollstadt (ed.), (Akademie-Verlag, Berlin), p. 84.

[283] Forster, A., Kahlbaum, T. and Ebeling, W. (1991a). Thermodynamic properties and the plasma phase transition of dense helium plasma, *High Pressure Res.* **7**, 1–6, pp. 375–377.

[284] Forster, A., Kahlbaum, T. and Rickert, A. (1991b). Mean ionization state and composition of a dense high-Z plasma, *Zeitschrift fur physik D-atoms molecules and clusters* **21**, pp. S171–S173.

[285] Fortney, J. and Hubbard, W. (2003). *Icarus* **164**, 1, pp. 228–243.

[286] Fortov, V., Bespalov, V. *et al.* (1990). Experimental Study of Optical Properties of Strongly Coupled Plasmas, in S. Ishimaru (ed.), *Strongly Coupled Plasma Physics* (Elsevier), pp. 571–578.

[287] Fortov, V., Iakubov, I. and Khrapak, A. (2006). *Physics of Strongly Coupled Plasma* (Oxford University Press, Oxford).

[288] Fortov, V., Lebedev, M., Dyabilin, K., Vorobiev, O., Smirnov, V. and Grabovskij, E. (1996a). *Shock Compression of Condensed Matter-1995*, in S. C. Schmidt and W. C. Tao (eds.), AIP Conf. Proc., p. 1255.

[289] Fortov, V. E. (1972). *Prikl. Mekh. Tekh. Fiz.*, **8**, p. 156.

[290] Fortov, V. E. (1982). Dynamic methods in plasma physics, *Phys. Usp.* **25**, 11, pp. 781–809.

[291] Fortov, V. E. (ed.) (2000). *Encyclopedia of Low-Temperature Plasma* (Nauka, Moscow).

[292] Fortov, V. E. (2005). *Intense Shock Waves and Extreme States of Matter* (Bukos, Moscow).

[293] Fortov, V. E. (2007). Intense shock waves and extreme states of matter, *Phys. Usp.* **50**, 4, 333.

[294] Fortov, V. E. (2009a). *Extreme States of Matter* (Fizmatlit, Moscow), The Frontiers Collection (Springer, Berlin-Heidelberg).

[295] Fortov, V. E. (2009b). Extreme states of matter on earth and in space, *Phys. Usp.* **52**, 6, pp. 615–647.

[296] Fortov, V. E., Al'tshuler, L. V., Trunin, R. F. and Funtikov, A. I. (2004a). *High-Pressure Shock Compression of Solids VII: Shock Waves and Extreme States of Matter* (Springer, New York).

[297] Fortov, V. E., Dremin, A. N. and Leont'ev, A. A. (1975). *Teplofiz. Vys. Temp.* **13**, p. 1072.

[298] Fortov, V. E., Gavrikov, A. V., Petrov, O. F., Shakhova, I. A. and Vorob'ev, V. S. (2007a). Investigation of the interaction potential and thermodynamic functions of dusty plasma by measured correlation functions, **14**, 4, p. 040705.

[299] Fortov, V. E., Gryaznov, V. K., Mintsev, V. B., Ternovoi, V. Y., Iosilevski, I. L., Zhernokletov, M. V. and Mochalov, M. A. (2001a). Thermophysical properties of shock compressed argon and xenon, *Contrib. Plasma Phys.* **41**, 2–3, pp. 215–218.

[300] Fortov, V. E., Hoffmann, D. H. H. and Sharkov, B. Y. (2008a). Intense ion beams for generating extreme states of matter, *Phys. Usp.* **51**, 2, p. 109.

[301] Fortov, V. E., Ilkaev, R. I., Arinin, V. A., Burtzev, V. V., Golubev, V. A., Iosilevskiy, I. L., Khrustalev, V. V., Mikhailov, A. L., Mochalov, M. A., Ternovoi, V. Y. and Zhernokletov, M. V. (2007b). Phase transition in a strongly nonideal deuterium plasma generated by quasi-isentropical compression at megabar pressures, *Phys. Rev. Lett.* **99**, p. 185001.

[302] Fortov, V. E., Ivlev, A. V., Khrapac, S. A., Khrapac, A. G. and Morfill, G. E. (2005). Complex (Dusty) plasmas: Current status, Open issues, perspectives, *Phys. Report* **421**, pp. 1–103.

[303] Fortov, V. E., Khishchenko, K. V., Levashov, P. R. and Lomonosov, I. V. (1998). Wide-range multi-phase equations of state for metals, *Nucl. Instr. Meth. Phys. Res. A* **415**, 3, pp. 604–608.

[304] Fortov, V. E., Khrapak, A. G., Khrapak, S. A., Molotkov, V. I. and Petrov, O. F. (2004b). Dusty plasmas, *Phys. Usp.* **47**, 5, 447.

[305] Fortov, V. E., Khrapak, A. G. and Yakubov, I. T. (2004c). *Physics of Nonideal Plasma* (Fizmatlit, Moscow).

[306] Fortov, V. E. and Krasnikov, Y. G. (1970). *Zh. Eksp. Teor. Fiz.* **59**, 1, p. 1645.

[307] Fortov, V. E., Leont'ev, A. A., Dremin, A. N. and Gryaznov, V. K. (1976). Shock-wave production of a non-ideal plasma, *J. Exp. Theor. Phys.* **44**, p. 116.

[308] Fortov, V. E., Leont'ev, A. A., Dremin, A. N. and Pershin, S. V. (1974a). Isentropic expansion of shock-compressed lead, *JETP Lett.* **20**, 1, p. 13.

[309] Fortov, V. E., Lomakin, B. N. and Krasnikov, Y. G. (1971). *Teplofiz. Vys. Temp.* **9**, 5, pp. 869–878.

[310] Fortov, V. E. and Lomonosov, I. V. (2010). Shock waves and equations of state of matter, *Sock Waves* **20**, 1, pp. 53–71.

[311] Fortov, V. E., Mintsev, V. B., Ternovoi, V. Y., Gryaznov, V. K., Iosilevskii, I. L., Mochalov, M. A. and Zhernokletov, M. V. (2004d). Conductivity of nonideal plasma, *High Temp. Mater. Processes* **8**, 3, pp. 447–459.

[312] Fortov, V. E., Molodets, A. M., Postnov, V. I., Shakhrai, D. V., Kagan, K. L., Maksimov, E. G., Ivanov, A. V. and Magnitskaya, M. V. (2004e). Electrophysical properties of calcium at high pressures and temperatures, *JETP Lett.* **79**, 7, p. 346.

[313] Fortov, V. E. and Morfill, G. E. (eds.) (2009). *Complex and Dusty Plasmas: From Laboratory to Space* (CRC Press).

[314] Fortov, V. E., Musyankov, S. I., Yakushev, V. V. and Dremin, A. N. (1974b). *Teplofiz. Vys. Temp* **12**, p. 957.

[315] Fortov, V. E., Nefedov, A. P., Petrov, O. F., Samarian, A. A. and Chernyschev, A. V. (1996b). Emission properties and structural ordering of strongly coupled dust particles in a thermal plasma, *Phys. Lett. A* **219**, 1–2, pp. 89–94.

[316] Fortov, V. E., Nefedov, A. P., Torchinskii, V. M., Molotkov, V. I., Khrapak, A. G., Petrov, O. F. and Volykhin, K. F. (1996c). Crystallization of a dusty plasma in the positive column of a glow discharge, *JETP Lett.* **64**, 2, pp. 92–98.

[317] Fortov, V. E., Petrov, O. F. and Vaulina, O. S. (2008b). Dusty-plasma liquid in the statistical theory of the liquid state, *Phys. Rev. Lett.* **101**, p. 195003.

[318] Fortov, V. E., Ternovoi, V. Y., Kvitov, S. V., Mintsev, V. B., Nikolaev, D. N., Pyalling, A. A. and Filimonov, A. S. (1999a). Electrical conductivity of nonideal hydrogen plasma at megabar dynamic pressures, *JETP Lett.* **69**, 12, pp. 926–931.

[319] Fortov, V. E., Ternovoi, V. Y., Zhernokletov, M. V., Mochalov, M. A., Mikhailov, A. L., Filimonov, A. S., Pyalling, A. A., Mintsev, V. B., Gryaznov, V. K. and Iosilevskii, I. L. (2003). Pressure-produced ionization of nonideal plasma in a megabar range of dynamic pressures, *J. Exp. Theor. Phys.* **97**, 2, pp. 259–278.

[320] Fortov, V. E. and Yakubov, I. T. (1984). *Nonideal Plasma Physics* (Institute of Chemistry Physics of RAS, Chernogolovka).

[321] Fortov, V. E. and Yakubov, I. T. (1994). *Nonideal Plasma* (Energoatomizdat, Moscow).

[322] Fortov, V. E., Yakushev, V. V., Kagan, K. L., Lomonosov, I. V., Maksimov, E. G., Magnitskaya, M. V., Postnov, V. I. and Yakusheva, T. I. (2002). Lithium at high dynamic pressure, *J. Phys. Condens. Mat.* **14**, 44, p. 10809.

[323] Fortov, V. E., Yakushev, V. V., Kagan, K. L., Lomonosov, I. V., Postnov, V. I. and Yakusheva, T. I. (1999b). Anomalous electric conductivity of lithium under quasi-isentropic compression to 60 GPa (0.6 Mbar). Transition into a molecular phase? *JETP Lett.* **70**, 9, pp. 628–632.

[324] Fortov, V. E., Yakushev, V. V., Kagan, K. L., Lomonosov, I. V., Postnov, V. I., Yakusheva, T. I. and Kuryanchik, A. N. (2001b). Anomalous resistivity of lithium at high dynamic pressure, *JETP Lett.* **74**, 8, p. 418.

[325] Fradkin, E. S. (1959). *Zh. Eksp. Teor. Fiz.* **36**, p. 1533.

[326] Fradkin, E. S. (1965). Metod funktsii Grina v teorii kvantovannyh polei i v kvantovoi statistike (diss. d.f.-m.n.), in *Kvantovaya teoriya polya i gidrodinamika, Trudy FIAN*, Vol. 29 (Nauka, Moscow), pp. 5–138.

[327] Franck, E. U. and Hensel, F. (1966). Metallic conductance of supercritical mercury gas at high pressures, *Phys. Rev.* **147**, pp. 109–110.

[328] Franck, S. (1980). *Ann. Physik (Leipzig)* **492**, p. 349.

[329] Francl, M. M., Pietro, W. J., Hehre, W. J., Binkley, J. S., Gordon, M. S., DeFrees, D. J. and Pople, J. A. (1982). Self-consistent molecular orbital methods. xxiii. a polarization-type basis set for second-row elements, *J. Chem. Phys.* **77**, 7, pp. 3654–3665.

[330] Frank-Kamenetskii, D. A. (1959). *Fizicheskie Protsessy Vnutri Zvezd (Physical Processes inside Stars)* (Fizmatlit, Moscow).

[331] Freedman, B. A. and McLerran, L. D. (1977). Fermions and gauge vector mesons at finite temperature and density. III. The ground-state energy of a relativistic quark gas, *Phys. Rev. D* **16**, 4, pp. 1169–1185.

[332] Frenkel, Y. I. (1975). *Kineticheskaya Teoriya Zhidkostei* (*Kinetic Theory of Liquids*) (Nauka, Leningrad).

[333] Gadiyak, G. V. and Chigisheva, T. M. (1975). In *Numerical Methods of Continuous Medium Mechanics* (Novosibirsk).

[334] Gadiyak, G. V., Kirzhnits, D. A. and Lozovik, Y. E. (1975). Collective levels in heavy atoms, *JETP Lett.* **21**, 2, p. 61.

[335] Gadiyak, G. V. and Lozovik, Y. E. (1975). In *Numerical Methods of Continuous Medium Mechanics* (Novosibirsk).

[336] Galam, S. and Hansen, J. P. (1976). Statistical mechanics of dense ionized matter. VI. Electron screening corrections to the thermodynamic properties of the one-component plasma, *Phys. Rev. A* **14**, pp. 816–832.

[337] Galitskii, V. M. (1958). The energy spectrum of a non-ideal Fermi gas, *Sov. Phys. JETP* **7**, p. 104.

[338] Gandel'man, G. M., Ermachenko, V. M. and Zel'dovich, Y. B. (1963). *Zh. Eksp. Teor. Fiz.* **44**, p. 386.

[339] Gantmakher, V. F. (2005). *Electrons in Disordered Media* (*lecture course*) (Nauka, Moscow).

[340] Gathers, G. R., Shaner, J. W., Hixson, R. S. and Young, D. A. (1979). *High Temp. High Press.*

[341] Gathers, G. R., Shaner, J. W. and Young, D. A. (1974). Experimental, very high-temperature, liquid-uranium equation of state, *Phys. Rev. Lett.* **33**, 2, pp. 70–72.

[342] Gehrels, T. (ed.) (1978). *Jupiter. Collection of papers* (Mir publishers).

[343] Gelman, B. A., Shuryak, E. V. and Zahed, I. (2006). Classical strongly coupled quark–gluon plasma. II. Screening and equation of state, *Phys. Rev. C* **74**, 4, p. 044909.

[344] Gezerlis, A. and Carlson, J. (2008). Strongly paired fermions: Cold atoms and neutron matter, *Phys. Rev. C* **77**, 3, p. 032801.

[345] Ginzburg, V. L. (2001). *The Physics of a Lifetime: Reflections on the Problems and Personalities of 20th Century Physics* (Springer, Berlin, Heidelberg).

[346] Ginzburg, V. L. (2004). On superconductivity and superfluidity (what I have and have not managed to do), as well as on the "physical minimum" at the beginning of the XXI Century (December 8, 2003), *Phys. Usp.* **47**, 11, p. 1155.

[347] Gitterman, M. and Steinberg, V. (1975). First-order phase transition in metallic vapors, *Phys. Rev. Lett.* **35**, 23, pp. 1588–1591.

[348] Glembotskii, N. and Petkyavichus, Y. (1973). *Lit. fiz. collection* **13**, p. 51.

[349] Glendenning, N. K. (1985). Neutron stars are giant hypernuclei? *Astrophys. J.* **293**, pp. 470–493.

[350] Glendenning, N. K. (2000). *Compact Stars: Nuclear Physics, Particle Physics, and General Relativity*, 2nd edn. (Springer, New York).

[351] Glötzel, D. (1978). *J. Phys. Ser. F* **8**, p. L163.

[352] Glötzel, D. and McMahan, A. K. (1979). Relativistic effects, phonons, and the isostructural transition in cesium, *Phys. Rev. B* **20**, 8, pp. 3210–3216.

[353] Glotzel, D. T. (1981). *Physics of Solids under High Pressure* in J. Schilling and R. Shelton (eds.), (North Holland, Amsterdam), p. 263.

[354] Glukhodedov, V. D., Kirshanov, S. I., Lebedeva, T. S. and Mochalov, M. A. (1999). Properties of shock-compressed liquid krypton at pressures of up to 90 GPa, *J. Exp. Theor. Phys.* **89**, p. 292.

[355] Glushak, B. L., Zharkov, A. P., Zhernokletov, M. V. *et al.* (1989). abc, *Sov. Phys. JETP* **69**, p. 739.

[356] Godwal, B. K., Sikka, S. K. and Chidambaram, R. (1979). Electronic grüneisen parameter in the shock hugoniot equation of state of aluminum, *Phys. Rev. B* **20**, 6, pp. 2362–2365.

[357] Godwall, B. K. (1980). Electronic thermal grüneisen parameters for elements with high atomic numbers, *J. Phys. F: Met. Phys.* **10**, pp. 377–382.

[358] Golden, S. (1957a). Statistical theory of many-electron systems. Discrete bases of representation, *Phys. Rev.* **107**, 5, pp. 1283–1290.

[359] Golden, S. (1957b). Statistical theory of many-electron systems. General considerations pertaining to the Thomas–Fermi theory, *Phys. Rev.* **105**, 2, pp. 604–615.

[360] Gombash, P. (1953). Statistical theory of atomic nuclei, *Usp. Fiz. Nauk* **49**, 3, pp. 385–448.

[361] Gombash, P. (1956). *Statistical Theory of the Atom and its Application* (IL, Moscow).

[362] Goncharov, A. F., Crowhurst, J. C., Struzhkin, V. V. and Hemley, R. J. (2008). Triple point on the melting curve and polymorphism of nitrogen at high pressure, *Phys. Rev. Lett.* **101**, p. 095502.

[363] Goncharov, A. F., Gregoryanz, E., Mao, H.-K., Liu, Z. and Hemley, R. J. (2000). Optical evidence for a nonmolecular phase of nitrogen above 150 GPa, *Phys. Rev. Lett.* **85**, pp. 1262–1265.

[364] Gorobchenko, V. D. and Maksimov, E. G. (1980). The dielectric constant of an interacting electron gas, *Phys. Usp.* **23**, 1, pp. 35–58.

[365] Graboske, H. C., Harwood, D. J. and Rogers, F. J. (1969). Thermodynamic properties of nonideal gases. I. Free-energy minimization method, *Phys. Rev.* **186**, 1, pp. 210–225.

[366] Gradshtein, I. S. and Ryzhik, I. M. (1971). *Tables of Integrals, Sums, Series, and Products*, 5th edn. (Nauka, Moscow).

[367] Graebner, G. (1985). Ph.D. thesis, University of Frankfurt.

[368] Green, D. W. and Leibowitz, L. (1982). Vapor pressures and vapor compositions in equilibrium with hypostoichiometric uranium dioxide at high temperatures, *J. Nucl. Materials* **105**, 2–3, pp. 184–195.

[369] Gregoryanz, E., Degtyareva, O., Somayazulu, M., Hemley, R. J. and Mao, H.-K. (2005). Melting of dense sodium, *Phys. Rev. Lett.* **94**, p. 185502.

[370] Gregoryanz, E., Goncharov, A. F., Hemley, R. J. and Mao, H.-K. (2001). High-pressure amorphous nitrogen, *Phys. Rev. B* **64**, p. 052103.

[371] Gregoryanz, E., Goncharov, A. F., Matsuishi, K., Mao, H.-K. and Hemley, R. J. (2003). Raman spectroscopy of hot dense hydrogen, *Phys. Rev. Lett.* **90**, p. 175701.

[372] Grigor'ev, F., Kormer, S. B., Mikhailova, . L., Mochalov, M. A. and Urlin, V. D. (1985). Shock compression and brightness temperature of a shock wave front in argon. Electron screening of radiation, *J. Exp. Theor. Phys.* **61**, 4, p. 751.

[373] Grigor'ev, F. V., Kormer, S. B., Mikhailova, O. L., Tolochhko, A. P. and Urlin, V. D. (1978). Equation of state of molecular hydrogen. Phase transition into the metallic state, *J. Exp. Theor. Phys.* **48**, 5, p. 847.

[374] Grigor'ev, F. V., Kormer, S. B., Mikhailova, O. L., Tolochko, A. P. and Urlin, V. D. (1972). Experimental determination of the compressibility of hydrogen at densities $0.5 - 2$ g/cm$^3$ metallization of hydrogen, *JETP Lett.* **16**, p. 201.

[375] Grigor'ev, F. V., Kormer, S. B., Mikhailova, O. L., Tolochko, A. P. and Urlin, V. D. (1975). Equation of state of the molecular phase of hydrogen in the solid and liquid states at high pressure, *J. Exp. Theor. Phys.* **42**, 2, p. 378.

[376] Grimes, C. C. and Adams, G. (1979). Evidence for a liquid-to-crystal phase transition in a classical, two-dimensional sheet of electrons, *Phys. Rev. Lett.* **42**, 12, pp. 795–798.

[377] Grishechkin, S. K., Gruzdev, S. K., Gryaznov, V. K., Zhernokletov, M. V., Il'kaev, R. I., Iosilevskii, I. L., Kashintseva, G. N., Kirshanov, S. I., Manachkin, S. F., Mintsev, V. B., Mikhailov, A. L., Mezhevov, A. B., Mochalov, M. A., Fortov, V. E., Khrustalev, V. V., Shuikin, A. N. and Yukhimchuk, A. A. (2004). Experimental measurements of the compressibility, temperature, and light absorption in dense shock-compressed gaseous deuterium, *JETP Lett.* **80**, 6, pp. 398–404.

[378] Grjaznov, V. K., Iosilevskij, I. L. and Fortov, V. E. (1973). *Prikl. Mekh. Tekh. Fiz.* **3**, pp. 70–76.

[379] Gross, D. D., Politzer, D. and Wilczek, F. (2005). Nobel lectures on physics, *Usp. Fiz. Nauk* **175**, 12, p. 1305.

[380] Grosshans, W. A., Vohra, Y. K. and Holzapfel, W. B. (1982). Evidence for a soft phonon mode and a new structure in rare-earth metals under pressure, *Phys. Rev. Lett.* **49**, 21, pp. 1572–1575.

[381] Grover, R. (1971). Liquid metal equation of state based on scaling, *J. Chem. Phys* **55**, 7, pp. 3435–3441.

[382] Grover, R. (1977). High temperature equation of state for simple metals, in A. Cezairliyan (ed.), *Proceedings of the Seventh Simposium on Thermophysical Properties* (ASME, New-York).

[383] Grover, R. (1979). *High Pressure Science and Technology*, Vol. 1 (Plenum Press, New-York).

[384] Grover, R. and Alder, B. J. (1974). Absence of first order electronic transitions in liquid metals, *J. Phys. Chem. Solids* **35**, 6, pp. 753–757.

[385] Grumbach, M. P. and Martin, R. M. (1996). Phase diagram of carbon at high pressures and temperatures, *Phys. Rev. B* **54**, pp. 15730–15741.

[386] Gryaznov, V., Iosilevskiy, I., Fortov, V., Zhernokletov, M. and Mochalov, M. (2006a). *Equation of state and Hugoniots of shock compressed gases on the basis of chemical model* (Darmstadt).

[387] Gryaznov, V. K., Ayukov, S. V., Baturin, V. A., Iosilevskiy, I. L., Starostin, A. N. and Fortov, V. E. (2006b). Solar plasma: Calculation of thermodynamic functions and equation of state, *J. Phys. A* **39**, 17, p. 4459.

[388] Gryaznov, V. K. and Fortov, V. E. (1987). *Teplofiz. Vys. Temp.* **25**, p. 1208.

[389] Gryaznov, V. K., Fortov, V. E., Zhernokletov, M. V., Simakov, G. V., Trunin, R. F., Trusov, L. I. and Iosilevski, I. L. (1998a). Shock compression and thermodynamics of highly nonideal metallic plasma, *J. Exp. Theor. Phys.* **87**, 4, pp. 678–690.

[390] Gryaznov, V. K., Iosilevski, I. L. and Fortov, V. E. (1996). Calculation of Porous Metal Hugoniots, in W. D. Kraeft and M. Schlanges (eds.), *Proc. of the International Conference on Physics of Strongly Coupled Plasmas* (World Scientific, Singapore), pp. 277–356.

[391] Gryaznov, V. K. and Iosilevskii, I. L. (1976). Some problems of thermodynamic calculation of multicomponent nonideal plasma, in *Thermophysical Properties of Low-Temperature Plasma* (Nauka, Moscow), pp. 25–30.

[392] Gryaznov, V. K., Iosilevskii, I. L. and Fortov, V. E. (1982a). *Pisma ZhTF* **8**, p. 1378.

[393] Gryaznov, V. K., Iosilevskii, I. L. and Fortov, V. E. (1982b). Thermodynamics of a highly compressed plasma in the megabar range, *Sov. Tech. Phys. Lett.* **8**, 22, p. 592.

[394] Gryaznov, V. K., Iosilevskii, I. L. and Fortov, V. E. (1995). Calculation of thermodynamic properties of shock-compressed plasma of metals, in *Fizika nizkotemperaturnoi plazmy* (Petrozavodsk), p. 105.

[395] Gryaznov, V. K., Iosilevskii, I. L. and Fortov, V. E. (1999a). *Contrib. Plasma Phys.* **39**, pp. 89–92.

[396] Gryaznov, V. K., Iosilevskii, I. L. and Fortov, V. E. (2000). Thermodynamics of shock-compressed plasma in the representation of the chemical model, in V. E. Fortov, L. V. Al'tshuler, R. F. Trunin and A. I. Funtikov (eds.), *Shock Waves and Extreme States of Matter* (Nauka, Moscow), p. 342.

[397] Gryaznov, V. K., Iosilevskii, I. L. and Fortov, V. E. (2004a). Thermodynamic Properties of Shock-Compressed Plasma, in V. E. Fortov (ed.), *Encyclopedia of Low-Temperature Plasma. V. III-I, "Thermodynamic Properties of Low-Temperature Plasma"* (Fizmatlit, Moscow), p. 111.

[398] Gryaznov, V. K., Iosilevskii, I. L., Krasnikov, Y. G., Kuznetsova, N. I., Kucherenko, V. I., Lappo, G. B., Lomakin, B. N., Pavlov, G. A., Son, E. E. and Fortov, V. E. (1980a). *Thermophysical Properties of Working Media of the Gas-Phase Nuclear Reactor* (Atomizdat, Moscow).

[399] Gryaznov, V. K., Iosilevskii, I. L., Semenov, A. M., Yakub, E. S., Fortov, V. E., Hyland, G. J. and Ronchi, C. (1999b). Equation of state of uranium dioxide, *Izvestiya RAS, Ser. Fiz.* **63**, 11, p. 2258.

[400] Gryaznov, V. K. and Iosilevskiy, I. L. (1973). *Chisl. Met. Mekh. Splosh. Sredy* **4**, 5, p. 166.

[401] Gryaznov, V. K., Iosilevskiy, I. L. and Fortov, V. E. (2004b). Thermodynamic properties of shock compressed plasmas based on a chemical picture, in V. E. Fortov, L. V. Al'tshuler, R. F. Trunin and A. I. Funtikov (eds.), *High-Pressure Shock Compression of Solids VII Shock Waves and Extreme States of Matter* (Springer-Verlag), pp. 437–489.

[402] Gryaznov, V. K., Ivanova, A. N., Gutsev, G. L., Levin, A. A. and Krestinin, A. V. (1989). *Zh. Struk. Khim.* **30**, p. 132.

[403] Gryaznov, V. K., Nikolaev, D. N., Ternovoi, V. Y., Fortov, V. E., Filimonov, A. S. and Keeler, N. (1998b). Generation of a nonideal plasma by shock compression of a highly porous $SiO_2$ aerogel, *Chem. Phys. Rep.* **17**, 1–2, pp. 239–245.

[404] Gryaznov, V. K., Zhernokletov, M. V., Zubarev, V. N., Iosilevskii, I. L. and Tortov, V. E. (1980b). Thermodynamic properties of a nonideal argon or xenon plasma, *J. Exp. Theor. Phys.* **51**, 2, p. 288.

[405] Gudkova, T. V. and Zharkov, V. N. (1999). Models of Jupiter and Saturn after Galileo mission, *Planet. Space Sci.* **47**, 10–11, pp. 1201–1210.

[406] Gupta, U. and Rajagopal, A. K. (1980). Inhomogeneous electron gas at nonzero temperatures: Exchange effects, *Phys. Rev. A* **21**, 6, pp. 2064–2068.

[407] Gust, W. H. and Royce, E. B. (1973). New electronic interactions in rare-earth metals at high pressure, *Phys. Rev. B* **8**, 8, pp. 3595–3609.

[408] G.V. Gadiyak, Y. E. L., D. A. Kirzhnits (1975). Collective excitations of a heavy atom, *J. Exp. Theor. Phys.* **42**, 1, p. 62.

[409] Gyulassy, M. (2008). Quark gluon plasmas: Femto cosmology with A+A @ LHC.

[410] Gyulassy, M. and McLerran, L. (2005). New forms of QCD matter discovered at RHIC, *Nucl. Phys. A* **750**, 1, pp. 30–63.

[411] Gyulassy, M. and Plumer, M. (1991). Jet quenching as a probe of dense matter, *Nucl. Phys. A* **527**, pp. 641–644.

[412] Gyulassy, M., Plumer, M., Thoma, M. and Wang, X. N. (1992). High $p_t$ probes of nuclear collisions, *Nucl. Phys. A* **538**, pp. 37–49.

[413] Haensel, P. and Pichon, B. (1994). Experimental nuclear masses and the ground state of cold dense matter, *Astron. Astrophys.* **283**, pp. 313–318.

[414] Haensel, P., Potekhin, A. Y. and Yakovlev, D. G. (2007). *Neutron Stars 1: Equation of State and Structure, Astrophysics and Space Science Library*, Vol. 326 (Springer, N.Y.).

[415] Haensel, P., Zdunik, J. L. and Schaeffer, R. (1986). Strange quark stars, *Astron. Astrophys.* **160**, pp. 121–128.

[416] Hahn, D. and Stöcker, H. (1986). Temperatures in heavy-ion collisions from pion multiplicities, *Nucl. Phys. A* **452**, 4, pp. 723–737.

[417] Hamaguchi, S., Farouki, R. T. and Dubin, D. H. E. (1997). Triple point of yukawa systems, *Phys. Rev. E* **56**, pp. 4671–4682.

[418] Hanauske, M., Zschiesche, D., Pal, S., Schramm, S., Stöcker, H. and Greiner, W. (2000). Neutron star properties in a chiral SU(3) model, *Astrophys. J.* **537**, pp. 958–963.

[419] Hands, S. (2001). The phase diagram of QCD, *Contemp. Phys.* **42**, 4, pp. 209–225.

[420] Hansen, B. M. S. (2004). The astrophysics of cool white dwarfs, *Phys. Rep.* **399**, pp. 1–70.

[421] Hansen, J. P. (1973). Statistical mechanics of dense ionized matter. I. Equilibrium properties of the classical one-component plasma, *Phys. Rev. A* **8**, 6, pp. 3096–3109.

[422] Hansen, J. P. and McDonald, I. R. (1976). *Theory of Simple Liquids* (Academic Press, N.Y.).

[423] Hansen, J. P. and McDonald, I. R. (1981). Microscopic simulation of a strongly coupled hydrogen plasma, *Phys. Rev. A* **23**, 4, pp. 2041–2059.

[424] Hänström, A. and Lazor, P. (2000). *J. Alloys Compd.* **305**, p. 209.

[425] Hariharan, P. C. and Pople, J. A. (1973). *Theor. Chim. Ada* **28**, p. 213.

[426] Haronska, P., Kremp, D., Schlanges, M. and Wise, Z. W.-P. (1987). *Univ. Rostock, N-Reihe* **36**, p. 98.

[427] Harris, J. W., Bock, R., Brockmann, R., Sandoval, A., Stock, R., Stroebele, H., Odyniec, G., Pugh, H. G., Schroeder, L. S., Renfordt, R. E., Schall, D., Bangert, D., Rauch, W. and Wolf, K. L. (1985). Pion production as a probe of the nuclear matter equation of state, *Phys. Lett. B* **153**, 6, pp. 377–381.

[428] Hartle, J. B., Sawyer, R. F. and Scalapino, D. J. (1975). Pion condensed matter at high densities: Equation of state and stellar models, *Astrophys. J.* **199**, pp. 471–481.

[429] Hartree, D. (1960). *Calculations of Atomic Structures* (Izd. Inostr. Lit., Moscow).

[430] Hartree, D. and Hartree, W. (1935). *Proc. Roy. Soc.* **149**, p. 210.

[431] Hawke, P. S., Burgess, T. J., Duerre, D. E., Huebel, J. G., Keeler, R. N., Klapper, H. and Wallace, W. C. (1978). Observation of electrical conductivity of isentropically compressed hydrogen at megabar pressures, *Phys. Rev. Lett.* **41**, 14, pp. 994–997.

[432] Hayashi, Y. and Tachibana, K. (1994). *Jpn. J. Appl. Phys.* **33**, pp. L804–L806.

[433] Heine, V., Cohen, M. L. and Weaire, D. (1973). In *Solid State Physics*, Vol. 24, (Academic Press, New York).

[434] Hemley, R. J. and Ashcroft, N. W. (1998). The revealing role of pressure in the condensed matter sciences, *Phys. Today* **51**, 8, pp. 26–32.

[435] Hemley, R. J. and Mao, H. K. (2002). Overview of Static High Pressure Science, in R. J. Hemley, G. L. Chiarotti, M. Bernasconi and L. Ulivi (eds.), *High Pressure Phenomena, Proceedings of the International School of Physics "Enrico Fermi" Course CXLVII* (IOS Press, Amsterdam), p. 3.

[436] Hirschfelder, J. O., Curtiss, C. F. and Bird, R. B. (1961). *Molecular theory of liquids and gases* (IIL, Moscow).

[437] Hoffmann, D. H. H., Fortov, V. E., Lomonosov, I. V., Mintsev, V., Tahir, N. A., Varentsov, D. and Wieser, J. (2002). Unique capabilities of an intense heavy ion beam as a tool for equation-of-state studies, *Phys. Plasmas* **9**, 9, pp. 3651–3654.

[438] Hofmann, M., Bleicher, M., Scherer, S., Neise, L., Stocker, H. and Greiner, W. (2000). Statistical mechanics of colored objects, *Phys. Lett. B* **478**, 1–3, pp. 161–171.

[439] Hogan, W. J. (ed.) (1995). *Energy from Inertial Fusion* (IAEA, Vienna, Austria).

[440] Hohenberg, P. and Kohn, W. (1964). Inhomogeneous electron gas, *Phys. Rev.* **136**, 3B, pp. B864–B871.

[441] Holian, K. S. (1986). A new equation of state for aluminum, *J. Appl. Phys.* **59**, pp. 149–157.

[442] Holmes, N. C., Ross, M. and Nellis, W. J. (1995). Temperature measurements and dissociation of shock-compressed liquid deuterium and hydrogen, *Phys. Rev. B* **52**, 22, pp. 15835–15845.

[443] Holst, B., Nettlemann, N. and Redmer, R. (2007). *Contrib. Plasma Phys.* **47**, p. 368.

[444] Holst, B., Redmer, R. and Desjarlais, M. P. (2008). Thermophysical properties of warm dense hydrogen using quantum molecular dynamics simulations, *Phys. Rev. B* **77**, 18, p. 184201.

[445] Hoover, W. G. and Ross, M. (1971). *Contemp. Phys.*

[446] Hoover, W. G., Stell, G., Goldmark, E. and Degani, G. D. (1975). Generalized van der waals equation of state, *J. Chem. Phys.* **63**, 12, pp. 5434–5438.

[447] Hostler, L. and Pratt, R. H. (1963). Coulomb green's function in closed form, *Phys. Rev. Lett.* **10**, 11, pp. 469–470.

[448] Hubbard, W. B. (1990). In *Strongly Coupled Plasma Physics* (Elsevier), p. 21.

[449] Hubbard, W. B. and DeWitt, H. E. (1985). Statistical mechanics of light elements at high pressure. VII — A perturbative free energy for arbitrary mixtures of H and He, *Astrophys. J.* pp. 388–393.

[450] Hubbard, W. B., Podolak, M. and Stevenson, D. J. (1995). The interior of neptune, in D. P. Cruikshank (ed.), *Neptune and Triton* (Univ. of Arizona Press, Tucson, AZ), pp. 109–138.

[451] Hubbard, W. B. and Slallery, W. L. (1971). *Astrophys. J.* **168**, p. 131.

[452] Huff, V. N., Gordon, S. and Morell, V. Z. (1951). *Tech. Rep.* 1037, NASA Report.

[453] Hurwicz, L. V., V., W. I., Khachkuruzov, G. A., Yungman, V. S., Bergman, G. A., Baibuz, V. F., Iorish, V. S., Yurkov, G. N., Gorbov, S. I., Kuratova, L. F., Rtishcheva, N. P., Przheval'skii, I. N., Zitserman, V. Y., Leonidov, V. Y., Ezhov, Y. S., Tomberg, S., Nazarenko, I. I., Rogatskii, A. L., Dorofeeva, O. V. and M., D. (1978). *Thermodynamic properties of individual substances*, Vol. 1 (Nauka, Moscow).

[454] I. N. Makarenko, S. M. S., A. M. Nikolaenko (1978). The thermodynamics of the melting of the alkali metals, *J. Exp. Theor. Phys.* **47**, 6, p. 1132.

[455] Ichimaru, S. (1982). Strongly coupled plasmas: High-density classical plasmas and degenerate electron liquids, *Rev. Mod. Phys.* **54**, pp. 1017–1059.

[456] Ichimaru, S. (ed.) (1990). *Strongly Coupled Plasma Physics* (North-Holland Publ. Co., Amsterdam).

[457] Ichimaru, S., Iyetomi, H. and Ogata, S. (1988). *Astrophys. J.* p. L17.

[458] Ichimaru, S., Iyetomi, H. and Tanaka, S. (1987). Statistical physics of dense plasmas: Thermodynamics, transport coefficients and dynamic correlations, *Phys. Rep.* **149**, 2–3, pp. 91–205.

[459] Iglesias, C. A., DeWitt, H. E., Lebowitz, J. L., MacGowan, D. and Hubbard, W. B. (1985). Low-frequency electric microfield distributions in plasmas, *Phys. Rev. A* **31**, pp. 1698–1702.

[460] Imshennik, V. S. (1995a). Experimental possibilities for studying the neutronization of matter in stars, *Phys. At. Nucl.* **58**, 5, pp. 823–831.

[461] Imshennik, V. S. (1995b). Explosion mechanism in supernovae collapse, *Space Sci. Rev.* **74**, 3–4, pp. 325–334.

[462] Imshennik, V. S. and Nadezhdin, D. K. (1965). *Astron. Zh.* **42**, p. 1159.

[463] Imshennik, V. S. and Nadyozhin, D. K. (1988). Supernova-1987A, and the emergence of the blast wave at the surface of the compact presupernova, *Sov. Astron. Lett.* **14**, 6, pp. 449–452.

[464] Inglis, D. R. (1939). Ionic depression of series limits in one-electron spectra, *Astrophys. J.* **90**, p. 439.

[465] Iosilevski, I. and Chigvintsev, A. (2000). *Journal de Physique* **4**, 10, p. 451.

[466] Iosilevski, I. L., Hyland, G. J., Ronchi, C. and Yakub, E. S. (1999). *T. Am. Nucl. Soc.* **81**, p. 122.

[467] Iosilevskii, I. L. (1979). *Thermodynamic Properties of Uranium Hexafluoride Heating Products* (NTO NIITP N9165, GONTI-8, Moscow).

[468] Iosilevskii, I. L. (1980). *Teplofiz. Vys. Temp.* **18**, 3, pp. 447–452.

[469] Iosilevskii, I. L. (1981). *The Equation of State in Extreme Conditions* (ITAM SB USSR Acad. of Sci., Novosibirsk).

[470] Iosilevskii, I. L. (2004). Nonideality Effects in Thermodynamics of Low-Temperature Plasma, in V. E. Fortov (ed.), *Encyclopedia of low-temperature plasma. V. III-I, "Thermodynamic Properties of Low-Temperature Plasma"* (Fizmatlit, Moscow), p. 349.

[471] Iosilevskii, I. L. and Gryaznov, V. K. (1981). *Teplofiz. Vys. Temp.* **19**, p. 1121.

[472] Iosilevskii, I. L., Gryaznov, V. K., Semenov, A. M., Yakub, E. S., Fortov, V. E., Ronchi, C. and Hyland, G. J. (2002). Non-congruent phase equilibrium in high-temperature uranium dioxide heating products, in *Appendix to the Journal "Butlerov Communications. (Khim. i komp. modelir.)"*, **10**, pp. 49–54.

[473] Iosilevskii, I. L., K., G. V. *et al.* (2003). *Vopr. Atomn. Nauki i Tekhn* **61**, 1, p. 3.

[474] Iosilevskii, I. L. and Kuznetsova, N. I. (1975). *Raket. i kosm. tekhn*, 25–26, p. 341.

[475] Iosilevskii, I. L. and Starostin, A. N. (2000). Thermodynamic Properties of Low-Temperature Plasma, in V. E. Fortov (ed.), *Encyclopedia of Low-Temperature Plasma. Introductory volume I* (Fizmatlit, Moscow), p. 327.

[476] Iosilevskiy, I., Gryaznov, V., Yakub, E., Ronchi, C. and Fortov, V. (2003). *Contrib. Plasma Phys.* **43**, 5–6, p. 316.

[477] Istomin, Y. N. (2008). Electron–positron plasma generation in the magnetospheres of neutron stars, *Phys. Usp.* **51**, 8, 844.

[478] Itano, W. M., Bollinger, J. J., Tan, J. N., Jelenkovic, B., Huang, X.-P. and Wineland, D. J. (1998). Bragg diffraction from crystallized ion plasmas, *Science* **279**, 5351, pp. 686–689.

[479] Ivanov, Y. B., Skokov, V. V. and Toneev, V. D. (2005). Equation of state of deconfined matter within a dynamical quasiparticle description, *Phys. Rev. D* **71**, 1, p. 014005.

[480] Ivlev, V. M. (1977). *Izv. AN SSSR, Ser. Energetics and transport* **6**, 6, p. 24.

[481] Jakob, B. (2006). *Die Beschreibung von dichtem Wasserstoff mit der Methode der Wellenpaket-Molekulardynamik*, Ph.D. thesis, Erlangen University.

[482] Jakob, B., Reinhard, P.-G., Toepffer, C. and Zwicknagel, G. (2007). Wave packet simulation of dense hydrogen, *Phys. Rev. E* **76**, 3, p. 036406.

[483] Jan, J. P. and Skriver, H. L. (1981). *J. Phys. Ser. F.*

[484] Jancovici, B. (1962). On the relativistic degenerate electron gas, *Nuovo Cimento* **25**, pp. 428–455.

[485] Jayaraman, A., Klement, W. and Kennedy, G. C. (1963). Phase diagrams of calcium and strontium at high pressures, *Phys. Rev.* **132**, pp. 1620–1624.

[486] Jeffries, C. D. and Keldysh, L. V. (eds.) (1988). *Electron–Hole Drops in Semiconductors* (Nauka, M.).

[487] Jin, W., Reno, J. P. *et al.* (1993). *Strongly Coupled Plasma Physics* (University Rochester Press).

[488] Johansson, B. (1978). Energy difference between trivalent and tetravalent metallic cerium and other f-elements: Stability study of various lanthanide- and actinide-halides and -oxides, *J. Phys. Chem. Solids* **39**, 5, pp. 467–484.

[489] Johnson, M. H. and Teller, E. (1955). Classical field theory of nuclear forces, *Phys. Rev.* **98**, pp. 783–787.

[490] Jones, A. H., Isbell, W. M. and Maiden, C. J. (1966). Measurement of the very-high-pressure properties of materials using a light-gas gun, *J. Appl. Phys.* **37**, 9, pp. 3493–3499.

[491] Jones, M. D. and Ceperley, D. M. (1996). Crystallization of the one-component plasma at finite temperature, *Phys. Rev. Lett.* **76**, pp. 4572–4575.

[492] Kadanoff, L. and Baym, G. (1962). *Quantum Statistical Mechanics* (W. A. Benjamin, Inc., New York).

[493] Kadomtsev, B. B. (1970). Heavy atom in an ultrastrong magnetic field, *J. Exp. Theor. Phys.* **31**, 5, p. 945.

[494] Kadomtsev, B. B. and Kudryavtsev, V. S. (1971). Atoms in a superstrong magnetic field, *JETP Lett.* **13**, 1, p. 42.

[495] Kahlbaum, T. and Forster, A. (1990). Thermodynamic properties of non-ideal plasmas with multiple ionization and coulomb and hard-core interactions, *Laser Part. Beams* **8**, 4, pp. 753–762.

[496] Kahlbaum, T. and Forster, A. (1992). Generalized thermodynamic functions for electrons in a mixture of hard spheres: Application to partially ionized nonideal plasmas, *Fluid Phase Equilibr.* **76**, pp. 71–86.

[497] Kalashnikov, O. K. and Klimov, V. V. (1979). Phase transition in the quark-gluon plasma, *Phys. Lett. B* **88**, 3–4, pp. 328–330.

[498] Kalitkin, N. N. (1960). *Zh. Eksp. Teor. Fiz.* **38**, p. 1534.

[499] Kalitkin, N. N. (1968). *Preprint of IAM of the USSR Academy of Sciences* (IAM of the USSR Academy of Sciences, Moscow).

[500] Kalitkin, N. N. (1975). *Doctor's thesis* (IAM of the USSR Academy of Sciences).

[501] Kalitkin, N. N. (1989). *Mat. Mod.* **1**, p. 64.

[502] Kalitkin, N. N. and Govoruhina, I. A. (1965). *Fiz. Tverd. Tela* **7**, pp. 355–362.

[503] Kalitkin, N. N. and Kuz'mina, L. V. (1961). *Preprint of IAM of the USSR Academy of Sciences N55* (IAM of the USSR Academy of Sciences, Moscow).

[504] Kalitkin, N. N. and Kuz'mina, L. V. (1971). *Fiz. Tverd. Tela* **12**, p. 2314.

[505] Kalitkin, N. N. and Kuz'mina, L. V. (1975). *Tables of Thermodynamic Functions of Matter with High Energy Density*, No. 35 in Preprint of the IAM of the USSR Academy of Sciences (IAM of the USSR Academy of Sciences, Moscow).

[506] Kalitkin, N. N., Kuz'mina, L. V. and Sharipdzhanov, I. I. (1976). *Preprint of IAM of the USSR Academy of Sciences N43* (IAM of the USSR Academy of Sciences, Moscow).

[507] Kalman, G. (ed.) (1978). *Strongly Coupled Plasmas, NATO — Advanced Study Institutes Series. Ser. B*, Vol. 36 (Plenum Press, N.Y.).

[508] Kanel, G. I., Fortov, V. E. and Razorenov, S. V. (2007). Shock waves in condensed-state physics, *Phys. Usp.* **50**, 8, pp. 771–791.

[509] Kanel, G. I., Razorenov, S. V. and Fortov, V. E. (2004). *Shock-wave phenomena and properties of condensed matter* (Springer, New York).

[510] Kaplan, D. B. and Nelson, A. E. (1986). Strange goings in dense nucleonic matter, *Phys. Lett. B* **175**, pp. 57–63.

[511] Kapusta, J. I. (1979). Quantum chromodynamics at high temperature, *Nucl. Phys. B* **88**, 3–4, pp. 461–498.

[512] Karpenko, A. S., Mapkevich, A. M. and Ryabinin, Y. N. (1952). *Zh. Eksp. Teor. Fiz.* **23**, p. 468.

[513] Karsch, F. and Kitazawa, M. (2009). Quark propagator at finite temperature and finite momentum in quenched lattice qcd, *Phys. Rev. D* **80**, p. 056001.

[514] Keeler, R. K., Van Thiel, M. and Alder, B. J. (1965). *Physica* **31**, p. 1437.

[515] Kelbg, G. (1963). *Ann. Physik* **12**, p. 219.

[516] Keldysh, L. V. (1965). Diagram technique for nonequilibrium processes, *Sov. Phys. JETP* **20**, pp. 1018–1022.

[517] Kerley, G. (1993). Multiphase equation of state for iron. Report SAND93-0027.

[518] Kerley, G. I. (1972a). In *A Theoretical Equation of State for Deuterium. NTIS Document No. LA-47766* (National Technical Information Service, Springfield, VA).

[519] Kerley, G. I. (1972b). *Phys. Earth Planet Inter.* **6**, p. 78.

[520] Kerley, G. I. (1980a). Perturbation theory and the thermodynamic properties of fluids. I. General theory, *J. Chem. Phys.* **469–477**, 73, p. 1, doi: 10.

[521] Kerley, G. I. (1980b). Perturbation theory and the thermodynamic properties of fluids. II. The CRIS model, *J. Chem. Phys.* **478–486**, 73, p. 1.

[522] Kerley, G. I. (1980c). Perturbation theory and the thermodynamic properties of fluids. III. Inverse-power and 6-12 potentials, *J. Chem. Phys.* **487–494**, 73, p. 1.

[523] Kerley, G. I. (1991). *User's manual for Panda II: a computer code for calculating equations of state.* Report SAND88-2291 (Sandia Natl. Labs., Albuquerque).

[524] Kerley, G. I. and Abdallah, J. (1980). Theoretical equations of state for molecular fluids: Nitrogen, oxygen, and carbon monoxide, *J. Chem. Phys.* **73**, 10, pp. 5337–5350.

[525] Khazanov, E. A. and Sergeev, A. M. (2008). Petawatt laser based on optical parametric amplifiers: Their state and prospects, *Phys. Usp.* **51**, 9, 969.

[526] Khishchenko, K. V. (1997). Temperature and heat capacity of polymethyl methacrylate behind the front of strong shock waves, *High Temp.* **35**, 6, p. 991.

[527] Khishchenko, K. V. (2008). Equation of state and phase diagram of tin at high pressures, *J. Phys. Conf. Ser.* **121**, p. 022025.

[528] Khishchenko, K. V. and Fortov, V. E. (2002a). To the problem of the equation of state of aluminum in the negative pressure range, in V. E. Fortov *et al.* (eds.), *Physics of extreme states of matter* (Institute of Problems of Chemical Physics RAS, Chernogolovka), p. 68.

[529] Khishchenko, K. V. and Fortov, V. E. (2002b). To the question of the equation of state of aluminum in the negative pressure range, in V. Fortov, V. P. Efremov, V. K. K, V. G. Sultanov, A. I. Temrokov, G. I. Kanel' and V. B. Mintsev (eds.), Institute of Problems of Chemical Physics RAS, Chernogolovka, pp. 68–70.

[530] Khishchenko, K. V., Lomonosov, I. V. and Fortov, V. E. (1998). Equations of state for polymethylmethacrylate and polytetrafluoroethylene in a wide range of densities and temperatures, *High Temp.–High Press.* **30**, 3, pp. 373–378.

[531] Khishchenko, K. V., Lomonosov, I. V., Fortov, V. E. and Shlenskii, O. F. (1996). Thermodynamic properties of plastics in a broad range of densities and temperatures, *Dokl. Akad. Nauk* **349**, 3, pp. 322–325.

[532] Khishtsenko, K. V. (2004). Equation of state of magnesium in the high pressure range, *Pisma ZhTF* **30**, 19, pp. 65–71.

[533] Khomskii, D. I. (1979). The problem of intermediate valency, *Phys. Usp.* **22**, 11, pp. 879–903.

[534] Kietzmann, A., Holst, B., Redmer, R., Desjarlais, M. P. and Mattsson, T. R. (2007). Quantum molecular dynamics simulations for the nonmetal-to-metal transition in fluid helium, *Phys. Rev. Lett.* **98**, 19, p. 190602.

[535] Kifonidis, K., Plewa, T., Janka, H.-T. and Müller, E. (2000). Nucleosynthesis and clump formation in a core-collapse supernova, *Astrophys. J. Lett.* **531**, pp. L123–L126.

[536] Kikoin, I. K., Senchenkov, A. P., Gel'man, E. V., Korsunskii, M. M. and Naurzakov, S. P. (1966). Electrical conductivity and density of a metal vapor, *J. Exp. Theor. Phys.* **22**, 1, p. 89.

[537] Killian, T. (2004). Plasmas put in order, *Nature* **429**, pp. 815–817.

[538] Kirzhnits, D. A. (1957). Quantum corrections to the Thomas–Fermi equation, *Zh. Exp. Teor. Fiz.* **32**, 1, pp. 115–123.

[539] Kirzhnits, D. A. (1958). Applicability limits of the quasi-classical equation of state of matter, *Zh. Eksp. Teor. Fiz.* **35**, 6, pp. 1545–1557.

[540] Kirzhnits, D. A. (1960). *Zh. Eksp. Teor. Fiz.* **38**, 503.

[541] Kirzhnits, D. A. (1961). *Trudy FIAN SSSR* **16**, p. 3.

[542] Kirzhnits, D. A. (1963a). *Field Methods in the Many-Body Theory* (Gosatomizdat, Moscow).

[543] Kirzhnits, D. A. (1963b). Supplement to the Theory of Strongly Compressed Matter, in *Field Methods of Many-Particle Theory*, Chapter 6 (Gosatomizdat, Moscow).

[544] Kirzhnits, D. A. (1967). *Field-Theoretical Methods in Many-Body Systems* (Pergamon Press, Oxford).

[545] Kirzhnits, D. A. (1972). Extremal states of matter (ultrahigh pressures and temperatures), *Sov. Phys. Usp.* **14**, 4, pp. 512–523.

[546] Kirzhnits, D. A. (2006). Lectures on Physics (Nauka, Moscow).

[547] Kirzhnits, D. A., Lozovik, Y. E. and Shpatakovskaya, G. V. (1975). Statistical model of matter, *Sov. Phys. Usp.* **18**, 9, pp. 649–672.

[548] Kirzhnits, D. A. and Shpatakovskaya, G. V. (1972). Atomic structure oscillation effects, *J. Exp. Theor. Phys.* **35**, 6, p. 1088.

[549] Kirzhnits, D. A. and Shpatakovskaya, G. V. (1974). Oscillations of the elastic parameters of compressed matter, *J. Exp. Theor. Phys.* **39**, 5, p. 899.

[550] Kitamura, H. and Ichimaru, S. (1998). Metal-insulator transitions in dense hydrogen: Equations of state, phase diagrams and interpretation of shock-compression experiments, *J. Phys. Soc. Jpn.* **67**, pp. 950–963.

[551] Knaup, M., Reinhard, P. and Topffer, C. (1999). *Contrib. Plasma Phys.* **39**, p. 57.

[552] Knaup, M., Reinhard, P. G., Toepferr, C. and Zwicknagel, G. (2003). *J. Phys. A. Math. Gen.* **36**, p. 6165.

[553] Knudson, M. D., Assay, J. R. and Deeney, C. (2005). *J. Appl. Phys.* **97**, p. 073514.

[554] Knudson, M. D. and Desjarlais, M. P. (2009). Shock compression of quartz to 1.6 TPa: Redefining a pressure standard, *Phys. Rev. Lett.* **103**, 22, p. 225501.

[555] Knudson, M. D., Desjarlais, M. P. and Dolan, D. H. (2008). Shock-wave exploration of the high-pressure phases of carbon, *Science* **322**, 5909, pp. 1822–1825.

[556] Knudson, M. D., Hanson, D. L., Bailey, J. E., Hall, C. A. and Asay, J. R. (2003). Use of a wave reverberation technique to infer the density compression of shocked liquid deuterium to 75 GPa, *Phys. Rev. Lett.* **90**, 3, p. 035505.

[557] Knudson, M. D., Hanson, D. L., Bailey, J. E., Hall, C. A., Asay, J. R. and Anderson, W. W. (2001). Equation of state measurements in liquid deuterium to 70 GPa, *Phys. Rev. Lett.* **87**, 22, p. 225501.

[558] Knudson, M. D., Hanson, D. L., Bailey, J. E., Hall, C. A., Asay, J. R. and Deeney, C. (2004). Principal Hugoniot, reverberating wave, and mechanical reshock measurements of liquid deuterium to 400 GPa using plate impact techniques, *Phys. Rev. B* **69**, 14, p. 144209.

[559] Koch, V., Majumder, A. and Randrup, J. (2005). Baryon-strangeness correlations: A diagnostic of strongly interacting matter, *Phys. Rev. Lett.* **95**, 18, p. 182301.

[560] Koester, D. (2007). White dwarfs: Recent developments, *Astron. Astrophys. Rev.* **11**, 1, pp. 33–66.

[561] Kohanoff, J. and Hansen, J. P. (1996). Statistical properties of the dense hydrogen plasma: An *ab initio* molecular dynamics investigation, *Phys. Rev. E* **54**, pp. 768–781.

[562] Kohn, W. (1999). Nobel lecture: Electronic structure of matter — wave functions and density functionals, *Rev. Mod. Phys.* **71**, 5, pp. 1253–1266.

[563] Kohn, W. and Sham, L. J. (1965). Self-consistent equations including exchange and correlation effects, *Phys. Rev.* **140**, 4A, pp. A1133–A1138.

[564] Kolehmainen, K. and Baym, G. (1982). Pion condensation at finite temperature (ii). Simple models including thermal excitations of the pion field, *Nucl. Phys. A* **382**, pp. 528–541.

[565] Kolomeitsev, E. E. and Voskresensky, D. N. (2003). Negative kaons in dense baryonic matter, *Phys. Rev. C* **68**, 1, p. 015803.

[566] Kompaneets, A. S. (1953a). *Zh. Eksp. Teor. Fiz.* **25**, p. 540.

[567] Kompaneets, A. S. (1953b). *Zh. Eksp. Teor. Fiz.* **26**, p. 153.

[568] Kompaneets, A. S. and Pavlovski, E. S. (1956). *Zh. Eksp. Teor. Fiz.* **31**, p. 427.

[569] Konyukhov, A. V., Likhachev, A. P., Oparin, A. M., Anisimov, S. I. and Fortov, V. E. (2004). Numerical modeling of shock-wave instability in thermodynamically nonideal media, *J. Exp. Theor. Phys.* **98**, 4, pp. 811–819.

[570] Kopyshev, V. P. (1968). The second virial coefficient of a plasma, *Zh. Éksp. Teor. Fiz.* **55**, pp. 1304–1310.

[571] Kopyshev, V. P. (1977). *Chisl. Met. Mekh. Splosh. Sredy* **8**, p. 54.

[572] Kopyshev, V. P. and Khrustalev, V. V. (1980). *Prikl. Mekh. Tekh. Fiz.* **1**, p. 122.

[573] Kopyshev, V. P., Medvedev, A. B. and Khrustalev, V. V. (2004). Thermodynamics of Low-Temperature Plasma. Two Aspects, in V. E. Fortov (ed.), *Encyclopedia of Low-Temperature Plasma. V. III-I, "Thermodynamic properties of low-temperature plasma"* (Fizmatlit, Moscow), p. 59.

[574] Kopyshev, V. P. and Urlin, V. D. (2000). *Shock Waves and Extreme States of Matter*, in V. E. Fortov, L. V. Al'tshuler, R. F. Trunin and A. I. Funtikov (eds.) (Nauka, Moscow), pp. 297–315.

[575] Kormer, S. B. (1968). Optical study of the characteristics of shock-compressed condensed dielectrics, *Phys. Usp.* **11**, 2, pp. 229–254.

[576] Kormer, S. B., Funtikov, A. I., Urlin, V. D. and Kolesnikova, A. N. (1962). Dynamic compression of porous metals and the equation of state with variable specific heat at high temperatures, *J. Exp. Theor. Phys.* **15**, 3, p. 477.

[577] Kormer, S. B., Sinitsyn, M. V., Kirillov, G. A. and Urlin, V. D. (1965). Experimental determination of temperature in shock-compressed NaCl and KCl and of their melting curves at pressures up to 700 kbar, *J. Exp. Theor. Phys.* **21**, 4, p. 689.

[578] Kormer, S. B., Urlin, V. D. and Popova, L. T. (1961). Interpolation equation of state and its application to the description of experimental data on shock compression of metals, *Fiz. Tverd. Tela* **3**, pp. 2131–2140.

[579] Kouveliotou, C., Duncan, R. C. and Thompson, C. (2003). Intensely magnetic neutron stars alter the quantum physics of their surroundings, *Sci. Am.* **288**, 2, p. 35.

[580] Kovalenko, G. V. and Sapozhnikov, A. T. (1979). *Vopr. atom. nauki i tekhniki. Ser. Metod. i progr. reshen. zadach mat. fiz.* **3**, p. 93.

[581] Kraeft, W.-D., Ebeling, W., Kremp, D. and Ropke, G. (1988). *Ann. Physik (Leipzig)* **500**, p. 429.

[582] Kraeft, W.-D., Kremp, D., Ebeling, W. and Röpke, G. (1986). *Quantum Statistics of Charged Particle Systems* (Plenum Press, New York).

[583] Krasnikov, Y. G. (1967). On the thermodynamics of a dense plasma, *Zh. Éksp. Teor. Fiz.* **53**, pp. 2223–2232.

[584] Krasnikov, Y. G. (2004). Basic Principles of Plasma Thermodynamics, in V. E. Fortov (ed.), *Encyclopedia of Low-Temperature Plasma. V. III-I, "Thermodynamic Properties of Low-Temperature Plasma"* (Fizmatlit, Moscow), p. 1.

[585] Krasnikov, Y. G. and Kucherenko, V. I. (1978). *Teplofiz. Vys. Temp* **16**, p. 43.

[586] Kresse, G. and Joubert, D. (1999). From ultrasoft pseudopotentials to the projector augmented-wave method, *Phys. Rev. B* **59**, 3, pp. 1758–1775.

[587] Krishnan, R., Frisch, M. J. and Pople, J. A. (1980). Contribution of triple substitutions to the electron correlation energy in fourth order perturbation theory, *J. Chem. Phys.* **72**, 7, pp. 4244–4245.

[588] Krishnan, R. and Pople, J. A. (1978). *Int. X Quantum Chem.* **14**, p. 91.

[589] Kruer, W. L. (1988). *The Physics of Laser Plasma Interactions* (Addison-Wesley, Reading, MA).

[590] Kudrin, L. P. and Tarasov, Y. A. (1963). Energy-level shifts and the equation of state of a plasma, *J. Exp. Theor. Phys.* **16**, 4, p. 1062.

[591] Kulish, M., Gryaznov, V. *et al.* (1997). Experimental Study of Al Line in Dense Xenon Plasma, in *XXIII Int. Conf. on Phenomena in Ionized Gases*, Vol. I (Toulouse), pp. 212–213.

[592] Kunihiro, T., Takatsuka, T. and Tamagaki, R. (1993). Neutral pion condensation in hot and dense nuclear matter, *Prog. Theor. Phys. Suppl.* **112**, pp. 197–219.

[593] Kuropatenko, V. F. and Minaeva, I. S. (1982). *Chisl. Met. Mekh. Splosh. Sredy* **13**, 6, pp. 69–76.

[594] Kusenko, A., Shaposhnikov, M., Tinyakov, P. G. and Tkachev, I. I. (1998). Star wreck, *Phys. Lett. B* **423**, pp. 104–108.

[595] Labzowsky, L. N. and Lozovik, Y. E. (1972). Spin rearrangement of atoms in strong magnetic fields, *Phys. Lett. A* **40**, 4, pp. 281–282.

[596] Lai, D. and Shapiro, E. E. (1991). Cold equation of state in strong magnetic field — Effects of inverse beta-decay, *Astrophys. J.* **383**, p. 745.

[597] Landau, L. D. (1937). *Zh. Eksp. Teor. Fiz.* **7**, p. 627.

[598] Landau, L. D. and Lifshitz, E. M. (1980). *Statistical Physics* (Pergamon Press, Oxford).

[599] Landau, L. D. and Lifshitz, E. M. (1987). *Fluid Mechanics* (*Course of Theoretical Physics*), vol. 6 (Butterworth-Heinemann, Oxford).

[600] Landau, L. D., Lifshitz, E. M. and Pitaevskii, L. P. (1984). *Electrodynamics of Continuous Media* (*Course of Theoretical Physics*), Vol. 8 (Butterworth-Heinemann, Oxford).

[601] Langanke, L. (2007). A FAIR Chance for Nuclear Astrophysics, [Kick-off event and symposium on the physics at FAIR].

[602] Langer, W. D. and Rosen, L. C. (1970). Hyperonic equation of state, *Astr. Space Sci.* **6**, pp. 217–227.

[603] Larkin, A. I. (1960). Thermodynamic functions of a low temperature plasma, *Sov. Phys. JETP* **11**, p. 1363.

[604] Latter, R. (1955). Temperature behavior of the Thomas–Fermi statistical model for atoms, *Phys. Rev.* **99**, 6, pp. 1854–1870.

[605] Latter, R. (1956). Thomas–Fermi model of compressed atoms, *J. Chem. Phys* **24**, 2, pp. 280–292.

[606] Lattimer, J. M., Pethick, C. J., Ravenhall, D. G. and Lamb, D. Q. (1985). Physical properties of hot, dense matter: The general case, *Nucl. Phys. A* **432**, pp. 646–742.

[607] Lebowitz, J. L. and Lieb, E. H. (1969). Existence of thermodynamics for real matter with coulomb forces, *Phys. Rev. Lett.* **22**, 13, pp. 631–634.

[608] Lee, C. M. and Thorsos, E. I. (1978). Properties of matter at high pressures and temperatures, *Phys. Rev. A* **17**, 6, pp. 2073–2076.

[609] Lee, T. D. (1975). Abnormal nuclear states and vacuum excitation, *Rev. Mod. Phys.* **47**, 2, pp. 267–275.

[610] Lee, T. D. and Wick, G. C. (1974). Vacuum stability and vacuum excitation in a spin-0 field theory, *Phys. Rev. D* **9**, pp. 2291–2316.

[611] Leont'ev, A. A. and Fortov, V. E. (1974). *Prikl. Mekh. Tekh. Fiz.*, **3**, p. 162.

[612] Levashov, P. R., Fortov, V. E., Khishchenko, K. V. and Lomonosov, I. V. (2002). Analysis of isobaric expansion data based on soft-sphere equation of state for liquid metals, in M. D. Furnish, N. N. Thadhani and Y. Horie (eds.), *Shock Compression of Condensed Matter, 2001* (AIP, Melville, NY), pp. 71–74.

[613] Levy, M. (1982). Electron densities in search of hamiltonians, *Phys. Rev. A* **26**, 3, pp. 1200–1208.

[614] Liao, J. and Shuryak, E. V. (2006). What do lattice baryonic susceptibilities tell us about quarks, diquarks, and baryons at $t > t_c$? *Phys. Rev. D* **73**, 1, p. 014509.

[615] Liberman, D. (1979a). *Report LA–UR–77–15950* (LANL, Los Alamos).

[616] Liberman, D. A. (1979b). Self-consistent field model for condensed matter, *Phys. Rev. B* **20**, 12, pp. 4981–4989.

[617] Lieb, E. (1983). Density functionals for coulomb systems, *Int. J. Quant. Chem.* **24**, 243.

[618] Lieb, E. H. and Simon, B. (1973). Thomas–Fermi Theory revisited, *Phys. Rev. Lett.* **31**, 11, pp. 681–683.

[619] Lifshitz, E. M. (1960). *Zh. Eksp. Teor. Fiz.* **38**, p. 1569.

[620] Lifshitz, E. M. and Pitaevskii, L. P. (1981). *Physical Kinetics*. Pergamon Press, Oxford.

[621] Likal'ter, A. A. (1969). Interaction of atoms with electrons and ions in a plasma, *J. Exp. Theor. Phys.* **29**, 1, p. 133.

[622] Likal'ter, A. A. (1992). Gaseous metals, *Physics-Uspekhi* **35**, 7, pp. 591–605.

[623] Likal'ter, A. A. (2000). Critical points of condensation in coulomb systems, *Phys. Usp.* **43**, 8, pp. 777–797.

[624] Likal'ter, A. A. (2004). Metal and semiconductor plasma in the neighborhood of the critical point of condensation, in V. E. Fortov (ed.), *Encyclopedia of Low-Temperature Plasma. V. III-I, "Thermodynamic Properties of Low-Temperature Plasma"* (Fizmatlit, Moscow), p. 140.

[625] Lindl, J. D. (1998). *Inertial Confinement Fusion* (Springer, New York).

[626] Lipaev, A. M., Molotkov, V. I., Nefedov, A. P., Petrov, O. F., Torchinskii, V. M., Fortov, V. E., Khrapak, A. G. and Khrapak, S. A. (1997). Ordered structures in a nonideal dusty glow-discharge plasma, *J. Exp. Theor. Phys.* **85**, 6, pp. 1110–1118.

[627] Lipp, M. J., Klepeis, J. P., Baer, B. J., Cynn, H., Evans, W. J., Iota, V. and Yoo, C.-S. (2007). Transformation of molecular nitrogen to nonmolecular phases at megabar pressures by direct laser heating, *Phys. Rev. B* **76**, p. 014113.

[628] Litim, D. F. and Manuel, C. (1999a). Effective transport equations for non-Abelian plasmas, *Nucl. Phys. B* **562**, 1–2, pp. 237–274.

[629] Litim, D. F. and Manuel, C. (1999b). Mean field dynamics in non-abelian plasmas from classical transport theory, *Phys. Rev. Lett.* **82**, pp. 4981–4984.

[630] Litim, D. F. and Manuel, C. (2000). Fluctuations from dissipation in a hot non-Abelian plasma, *Phys. Rev. D* **61**, p. 125004.

[631] Litim, D. F. and Manuel, C. (2002). Semi-classical transport theory for non-Abelian plasmas, *Phys. Rep.* **364**, 6, pp. 451–539.

[632] Lomakin, B. N. and Fortov, V. E. (1973). Equation of State of a nonideal cesium plasma, *J. Exp. Theor. Phys.* **36**, 1, p. 48.

[633] Lomakin, B. N. and Fortov, V. E. (1975). Equation of state of nonideal cesium plasma, *Zh. Eksp. Teor. Fiz.* **63**, 7, p. 1972.

[634] Lomonosov, I. V. (2007). Multi-phase equation of state for aluminum, *Laser Part. Beams* **25**, 4, pp. 567–584.

[635] Lorenz, C. P., Ravenhall, D. G. and Pethick, C. J. (1993). Neutron star crusts, *Phys. Rev. Lett.* **70**, pp. 379–382.

[636] Loubeyre, P., Le Toullec, R. and Pinceaux, J. P. (1987). Binary phase diagrams of $h_2$–he mixtures at high temperature and high pressure, *Phys. Rev. B* **36**, 7, pp. 3723–3730.

[637] Loubeyre, P., Occelli, F. and Le Toulec, R. (2002). Optical studies of solid hydrogen to 320 GPa and evidence for black hydrogen, *Nature* **416**, 6881, pp. 613–617.

[638] Ludwig, P., Kosse, S. and Bonitz, M. (2005). Structure of spherical three-dimensional Coulomb crystals, *Phys. Rev. E* **71**, 4, p. 046403.

[639] Lyutikov, M. (2006). Magnetar giant flares and afterglows as relativistic magnetized explosions, *Month. Notices R. Astron. Soc.* **367**, 4, pp. 1594–1602.

[640] M., A., Rajagopal, K. and Wilczek, F. (1998). Qcd at finite baryon density: Nucleon droplets and color superconductivity, *Phys. Lett. B* **422**, pp. 247–256.

[641] Macgillavry, C. H. and Rieck, G. D. (eds.) (1962). *International Tables for X-ray Crystallography*, Vol. 3 (Kynoch Press, Birmingham, England).

[642] Mackinnon, A. J., Borghesi, M., Hatchett, S., Key, M. H., Patel, P. K., Campbell, H., Schiavi, A., Snavely, R., Wilks, S. C. and Willi, O. (2001). Effect of plasma scale length on multi-MeV proton production by intense laser pulses, *Phys. Rev. Lett.* **86**, 9, pp. 1769–1772.

[643] Madsen, J. (1998). Detecting supersymmetric Q-balls with neutron stars, *Phys. Lett. B* **435**, pp. 125–130.

[644] Madsen, J. (1999). Physics and Astrophysics of Strange Quark Matter, in J. Cleymans (ed.), *Hadrons in Dense Matter and Hadrosynthesis* (Springer, Berlin), pp. 162–203.

[645] Madsen, J. and Haensel, P. (eds.) (1991). Strange quark matter in physics and astrophysics, *Nucl. Phys. (Proc. Suppl.) B*, **24**.

[646] Magro, W. R., Ceperley, D. M., Pierleoni, C. and Bernu, B. (1996). Molecular Dissociation in Hot, Dense Hydrogen, *Phys. Rev. Lett.* **76**, pp. 1240–1243.

[647] Maksimchuk, A., Gu, S., Flippo, K., Umstadter, D. and Bychenkov, V. Y. (2000). Forward ion acceleration in thin films driven by a high-intensity laser, *Phys. Rev. Lett.* **84**, 18, pp. 4108–4111.

[648] Maksimov, E. G., Magnitskaya, M. V. and Fortov, V. E. (2005). Non-simple behavior of simple metals at high pressure, *Phys. Usp.* **48**, 8, 761.

[649] Maksimov, E. G. and Shilov, Y. I. (1999). Hydrogen at high pressure, *Physics-Uspekhi* **42**, 11, pp. 1121–1138.

[650] Mansoori, C. F., Carnahan, V., Starling, K. E. and Leland, T. W. J. (1971). *J. Chem. Phys.* **54**, p. 1523.

[651] Mansoori, G. A. and Canfield, F. B. (1969). Variational approach to the equilibrium thermodynamic properties of simple liquids. i, *J. Chem. Phys.* **51**, 11, pp. 4958–4967.

[652] Mao, H. K., Hazen, R. M., Bell, P. M. and Willig, J. (1981). *J. Appl. Phys.* **52**, p. 4572.

[653] Mao, H. K. and Hemley, R. J. (1989). Optical studies of hydrogen above 200-gigapascals — evidence for metallization by band overlap, *Science* **244**, 4911, pp. 1462–1465.

[654] March, N. and Murray, A. (1961). *Proc. Roy. Soc.* **A261**, p. 119.

[655] March, N. H. and Tosi, M. P. (1984). *Coulomb Liquids* (Academic Press, London).

[656] March, N. H. and Tosi, M. P. (2002). *Introduction to liquid state physics* (World Scientific, New Jersey, London, Singapore, Hong Kong).

[657] March, N. H., Young, W. H. and Sampanthar, S. (1967). *The Many-Body Problem in Quantum Mechanics* (Cambridge University Press, Cambridge).

[658] March, S. P. (ed.) (1979). *Los Alamos Shock Hugoniot Data* (University of California Press, Berkeley).

[659] Marcos, S., Barranco, M. and Buchler, J.-R. (1982). Low entropy adiabats for stellar collapse, *Nucl. Phys. A* **381**, pp. 507–518.

[660] Margenau, H. and Lewis, M. (1959). Structure of Spectral Lines from Plasmas, *Rev. Mod. Phys.* **31**, pp. 569–615.

[661] Marley, M. S. and Hubbard, W. B. (1988). Thermodynamics of dense molecular-hydrogen helium mixtures at high-pressure, *Icarus* **73**, 3, pp. 536–544.

[662] Marsh, S. P. (ed.) (1980). *LASL Shock Hugoniot Data* (Univesrsity of California Press, Berkeley-LA-London).

[663] Martin, R. M. (2004). *Electronic Structure: Basic Theory and Practical Methods* (University Press, Cambridge).

[664] Martynov, G. A. (1992). *Fundamental Theory of Liquid* (Adam Hilger, Dristol, U. K.).

[665] Martynov, G. A. (1999). The problem of phase transitions in statistical mechanics, *Phys. Usp.* **42**, 6, pp. 517–543.

[666] Martynov, G. A. and Sarkisov, G. N. (1981). *Dokl. Akad. Nauk SSSR* **260**, p. 1348.

[667] Maruhn, J. A. (1991). The Nuclear Eguation of State, in S. Eliezer and R. A. Ricci (eds.), *High Pressure Equation of State: Theory and Applications. Course CXIII*, Proc. of the International School of Phisics "Enrico Fermi" (North Holland), pp. 507–533.

[668] Matsubara, T. (1955). A new approach to quantum-statistical mechanics, *Progr. Theor. Phys.* **14**, pp. 351–378.

[669] Mattsson, T. R. and Desjarlais, M. P. (2006). Phase diagram and electrical conductivity of high energy-density water from density functional theory, *Phys. Rev. Lett.* **97**, 1, p. 017801.

[670] Matzubara, T. (1955). A new approach to quantum-statistical mechanics, *Progr. of Theoret. Phys.* **14**, 4, pp. 351–378.

[671] Mayer, J. and Goeppert-Mayer, M. (1940). *Statistical Mechanics* (New York).

[672] McCarthy, S. I. (1965). *Preprint UCRL-14364* (Livermore, California).

[673] McMahan, A. K. (1978). Calculation of the electronic transition in cesium, *Phys. Rev. B* **17**, 4, pp. 1521–1527.

[674] McMahan, A. K., Hord, B. L. and Ross, M. (1977). Experimental and theoretical study of metallic iodine, *Phys. Rev. B* **15**, 2, pp. 726–737.

[675] McMahan, A. K. and Ross, M. (1977). High-temperature electron-band calculations, *Phys. Rev. B* **15**, 2, pp. 718–725.

[676] McMahan, A. K. and Ross, M. (1979). *High Pressure Science and Technology*, (Plenum Press).

[677] McMahan, A. K., Skriver, H. L. and Johansson, B. (1981). The $s-d$ transition in compressed lanthanum, *Phys. Rev. B* **23**, 10, pp. 5016–5029.

[678] McQueen, R. G. and Marsh, S. P. (1960). *J. Appl. Phys.* **31**, p. 1253.

[679] McQueen, R. G., Marsh, S. P., Taylor, J. W., Fritz, J. N. and Carter, W. J. (1970). The Equation of State of Solids from Shock Wave Studies, in R. Kinslow (ed.), *High Velocity Impact Phenomena* (Academic Press, N.Y.), pp. 293–417.

[680] Medvedev, A. B. (1992). Model of the equation of state with allowance for evaporation, *Vopr. atom. nauki i tekhn, ser. Teor. and prikl. fiz.* **26**, 1, pp. 23–29.

[681] Mejson, E. and Sperling, T. (1972). *Virial Equations of State* (Mir, Moscow).

[682] Melzer, A., Trottenberg, T. and Piel, A. (1994). Experimental determination of the charge on dust particles forming coulomb lattices, *Phys. Lett. A* **191**, 3–4, pp. 301–308.

[683] Mermin, N. D. (1965). Thermal properties of the inhomogeneous electron gas, *Phys. Rev.* **137**, 5A, pp. A1441–A1443.

[684] Mestel, L. and Ruderman, M. A. (1967). The Energy Content of a White Dwarf and its Rate of Cooling, *Mon. Not. R. Astron. Soc.* **136**, p. 27.

[685] Meyer-ter Vehn, J. and Zittel, W. (1988). Electronic structure of matter at high compression: Isostructural transitions and approach of the fermi-gas limit, *Phys. Rev. B* **37**, 15, pp. 8674–8688.

[686] Migdal, A. B. (1971). Stability of vacuum and limiting fields, *Zh. Eksp. Teor. Fiz.* **61**, pp. 2209–2224.

[687] Migdal, A. B. (1972). Phase transitions in nuclear matter and non-pair nuclear forces, *Sov. Phys. JETP* **36**, pp. 1052–1055.

[688] Migdal, A. B. (1974). Meson condensation and anomalous nuclei, *Phys. Lett. B* **52**, pp. 172–174.

[689] Migdal, A. B., Saperstein, E. E., Troitsky, M. A. and Voskresensky, D. N. (1990). Pion degrees of freedom in nuclear matter, *Phys. Rep.* **192**, pp. 179–437.

[690] Mihalas, D., Däppen, W. and Hummer, D. G. (1988). The equation of state for stellar envelopes. II. Algorithm and selected results, *Astrophys. J.* **331**, pp. 815–825.

[691] Militzer, B. (2006). First principles calculations of shock compressed fluid helium, *Phys. Rev. Lett.* **97**, 17, p. 175501.

[692] Militzer, B. and Ceperley, D. M. (2000). Path integral Monte Carlo calculation of the Deuterium Hugoniot, *Phys. Rev. Lett.* **85**, 9, pp. 1890–1893.

[693] Militzer, B. and Ceperley, D. M. (2001). Path integral Monte Carlo simulation of the low-density hydrogen plasma, *Phys. Rev. E* **63**, 6, p. 066404.

[694] Militzer, B. and Pollock, E. L. (2005). Equilibrium contact probabilities in dense plasmas, *Phys. Rev. B* **71**, p. 134303.

[695] Mintsev, V. B. and Fortov, V. E. (1979). Electrical conductivity of xenon under supercritical conditions, *JETP Lett.* **30**, p. 375.

[696] Mintsev, V. B., Ternovoi, V. Y., Gryaznov, V. K., Pyalling, A. A., Fortov, V. E. and Iosilevskii, I. L. (2000). Electrical conductivity of shock compressed Xenon, in S. C. Shhmidt, D. P. Dandekar and J. W. Forbes (eds.), *Shock Compression of Condensed Matter-1999* (Woolbury, NY), pp. 987–990.

[697] Mirwald, P. W. (1979). in R. Kinslow (ed.), *High Pressure Science and Technology*, Vol. 1 (Plenum Press, N.Y.), p. 361.

[698] Misner, C. W., Thorne, K. S., Wheeler, J. A. and Chandrasekhar, S. (1974). Gravitation, **27**, 8, pp. 47–48.

[699] Mitchell, A. C. and Nellis, W. J. (1981). *J. Appl. Phys.* **52**, p. 3363.

[700] Mitchell, A. C. *et al.* (1991). *J. Appl. Phys.* **69**, p. 2891.

[701] Mochalov, M. A., Zhernokletov, M. V. *et al.* (2005). Study of Thermodynamical and Optical Properties of Deuterium under Shock and Adiabatic Compression, *In Theses Int. Conf. on Strongly Coupled Coulomb Systems, Moscow*, p. 35.

[702] Mochkovich, R. and Hansen, J. P. (1979). Fluid–solid coexistence curve of dense coulombic matter, *Phys. Lett. A* **73**, pp. 35–38.

[703] Møller, C. and Plesset, M. S. (1934). Note on an approximation treatment for many-electron systems, *Phys. Rev.* **46**, 7, pp. 618–622.

[704] Molodets, A. M. and Fortov, V. E. (2004). Phase transitions in uranium dioxide at high pressures and temperatures, *JETP Lett.* **80**, 3, p. 172.

[705] Molodets, A. M., Molodets, M. A. and Nabatov, S. S. (1998). Free energy and shock compression of diamond, *AIP Conference Proceedings* **429**, 1, pp. 91–94.

[706] Mon, K. K., Chester, G. V. and Ashcroft, N. W. (1980). Simulation studies of a model of high-density metallic hydrogen, *Phys. Rev. B* **21**, 6, pp. 2641–2646.

[707] Mon, K. K., Gann, R. and Stroud, D. (1981). Thermodynamics of liquid metals: The hard-sphere versus one-compoent-plasma reference systems, *Phys. Rev. A* **24**, 4, pp. 2145–2150.

[708] Montroll, E. W. and Ward, J. C. (1958). Quantum statistics of interacting particles; general theory and some remarks on properties of an electron gas, *Phys. Fluids* **1**, pp. 55–72.

[709] Moore, C. E., Minnaert, M. G. J. and Houtgast, J. (1966). *The Solar Spectrum 2935 Å to 8770 Å, NBS Monograph*, Vol. 61 (US Government Printing Office, Washington).

[710] More, R. M. (1979). Quantum-statistical model for high-density matter, *Phys. Rev. A* **19**, 3, pp. 1234–1246.

[711] More, R. M. and Skupsky, S. (1976). Nuclear-motion corrections to the thomas-fermi equation of state for high-density matter, *Phys. Rev. A* **14**, 1, pp. 474–479.

[712] More, R. M., Warren, K. H., Young, D. A. and Zimmerman, G. B. (1988). A new quotidian equation of state (QEOS) for hot dense matter, *Fluids* **31**, 10, pp. 3059–3078.

[713] Moriarty, J. A. and McMahan, A. K. (1982). High-pressure structural phase transitions in Na, Mg, and Al, *Phys. Rev. Lett.* **48**, pp. 809–812.

[714] Morley, P. D. and Kislinger, M. B. (1979). Relativistic many-body theory, quantum chromodynamics, and neutron stars/supernova, *Phys. Rep.* **51**, pp. 63–110.

[715] Mostovych, A. N., Chan, Y., Lehecha, T., Schmitt, A. and Sethian, J. D. (2000). Reflected shock experiments on the equation-of-state properties of liquid deuterium at 100–600 GPa (1–6 Mbar), *Phys. Rev. Lett.* **85**, 18, pp. 3870–3873.

[716] Moszkowski, S. A. and Källman, C. G. (1977). Abnormal neutron star matter at ultrahigh densities, *Nucl. Phys. A* **287**, pp. 495–500.

[717] Mott, N. F. (1949). Metallic conductance of supercritical mercury gas at high pressures, *Proc. Phys. Soc. Sec. A* **62**, p. 416.

[718] Mott, N. F. (1964). Electrons in transition metals, *Adv. Phys.* **13**, 51, p. 325.

[719] Mourou, G. A., Tajima, T. and Bulanov, S. V. (2006). Optics in the relativistic regime, *Rev. Mod. Phys.* **78**, 2, pp. 1804–1816.

[720] Mrowczynski, S. and Thoma, M. H. (2007). What do electromagnetic plasmas tell us about the quark–gluon plasma? *Annual Rev. Nucl. Part. Sci.* **57**, 1, pp. 61–94.

[721] Mueller, R. O., Rau, A. R. P. and Spruch, L. (1973). Simple atomic model and its associated wave function, *Phys. Rev. A* **8**, 3, pp. 1186–1194.

[722] Mulenko, I. A., Olejnikova, E. N., Khomkin, A. L., Filinov, V. S., Bonitz, M. and Fortov, V. E. (2001). Phase transition in dense low-temperature molecular gases, *Phys. Lett. A* **289**, 3, pp. 141–146.

[723] Müller, B. (1985). *The Physics of the Quark–Gluon Plasma* (Springer-Verlag).

[724] Murray, C. A. and Wenk, R. A. (1993). Observation of Order–Disorder Transitions and Particle Trajectories in a Model One-Component Plasma: Time Resolved Microscopy of Colloidal Spheres, in H. M. van Horn and S. Ichimaru (eds.), *Strongly Coupled Plasma Physics* (University of Rochester Press, Rochester, NY), p. 367.

[725] Nabatov, S. S., Dremin, A. M., Postnov, V. I. and Yakushev, V. V. (1979). Measurement of the electrical conductivity of sulfur under super-high dynamic pressures, *JETP Lett.* **29**, 7, p. 369.

[726] Nagara, H., Nagata, Y. and Nakamura, T. (1987). Melting of the wigner crystal at finite temperature, *Phys. Rev. A* **36**, pp. 1859–1873.

[727] National Research Council (2003). *Frontiers in High Energy Density Physics* (National Academies Press, Washington, DC).

[728] Naumann, R. J. (1971). Equation of state for porous metals under strong shock compression, *J. Appl. Phys.* **42**, pp. 4945–4954.

[729] Neal, T. (1976). Dynamic determinations of the Grüneisen coefficient in aluminum and aluminum alloys for densities up to 6 $Mg/m^3$, *Phys. Rev. B* **14**, 12, pp. 5172–5181.

[730] Neaton, J. B. and Ashcroft, N. W. (1999). Pairing in dense lithium, *Nature* **400**, p. 141.

[731] Nefedov, A. P., Petrov, O. F. and Fortov, V. E. (1997). Quasicrystalline structures in strongly coupled dusty plasma, *Phys. Usp.* **40**, 11, pp. 1163–1173.

[732] Negele, J. W. and Vautherin, D. (1973). Neutron star matter at subnuclear densities, *Nucl. Phys. A* **207**, pp. 298–320.

[733] Nellis, W. J. (2002). Shock Compression of Deuterium near 100 GPa Pressures, *Phys. Rev. Lett.* **89**, 16, p. 165502.

[734] Nellis, W. J. (2006). Dynamic compression of materials: Metallization of fluid hydrogen at high pressures, *Rep. Prog. Phys.* **69**, 5, pp. 1479–1580.

[735] Nellis, W. J., van Thiel, M. and Mitchell, A. C. (1982). Shock compression of liquid xenon to 130 GPa (1.3 Mbar), *Phys. Rev. Lett.* **48**, 12, pp. 816–818.

[736] Nellis, W. J. *et al.* (1983). *J. Chem. Phys.* **79**, p. 1480.

[737] Nigmatulin, R. I. and Bolotnova, R. K. (2008a). Wide-range equation of state for water and steam: Calculation results, *High Temp.* **46**, 3, p. 325.

[738] Nigmatulin, R. I. and Bolotnova, R. K. (2008b). Wide-range equation of state for water and steam: Method of construction, *High Temp.* **46**, 2, p. 182.

[739] Nikiforov, A. F., Novikov, V. G., Orlov, Y. N. and Uvarov, V. B. (1979a). *Preprint of IAM of the USSR Academy of Sciences*, 172 (IAM of the USSR Academy of Sciences, Moscow).

[740] Nikiforov, A. F., Novikov, V. G. and Uvarov, V. B. (1979b). A Modified Hartree–Fock–Slater model for a substance with given temperature and density, *Vopr. Atom. Nauki i Tekhn. Met. i progr. chisl. resh. zadach mat. fiz.*, **4**(6), pp. 16–26.

[741] Nikiforov, A. F., Novikov, V. G. and Uvarov, V. B. (2000). *Quantum-Statistical Models of High-Temperature Plasma and Methods of Calculation of Rosseland Lengths and Equations of State* (Fizmatlit, Moscow).

[742] Nikiforov, A. F., Orlov, N. Y. and Uvarov, V. B. (1982). *Preprint of IAM of the USSR Academy of Sciences N114* (IAM of the USSR Academy of Sciences, Moscow).

[743] Nikiforov, A. F. and Uvarov, V. B. (1974). *Preprint of IAM of the USSR Academy of Sciences N13* (IAM of the USSR Academy of Sciences, Moscow).

[744] Norman, G. E. and Starostin, A. N. (1968). Groundlessness of the classical description of nondegenerate dense plasma, *Teplofiz. Vys. Temp.* **6**, p. 410.

[745] Norman, H. E. and Starostin, A. N. (1970). Thermodynamics of a strongly nonideal plasma, *High Temp.* **8**, 2, p. 381.

[746] Novikov, I. D. (2001). "Big Bang" echo (cosmic microwave background observations), *Phys. Usp.* **44**, 8, p. 817.

[747] Novikov, V. G. (1985). *Preprint of IAM of the USSR Academy of Sciences N133* (IAM of the USSR Academy of Sciences, Moscow).

[748] Nozières, P. and Pines, D. (1958). Correlation energy of a free electron gas, *Phys. Rev.* **111**, 2, pp. 442–454.

[749] Ogasawara, R. and Sato, K. (1983). Nuclei in the neutrino-degenerate dense matter. II, *Prog. Theor. Phys.* **70**, pp. 1569–1582.

[750] Ogata, S. and Ichimaru, S. (1990). First-principles calculations of shear moduli for monte carlo simulated coulomb solids, *Phys. Rev. A* **42**, pp. 4867–4870.

[751] O'Hara, K. M., Hemmer, S. L., Gehm, M. E., Granade, S. R. and Thomas, J. E. (2002). Observation of a strongly interacting degenerate fermi gas of atoms, *Science* **298**, 5601, pp. 2179–2182.

[752] Ohse, R. W., Babelot, J.-F. *et al.* (1985). *Journal of Nuclear Materials* **130**, p. 165.

[753] Ohse, R. W. and Tippelskirch, H. (1977). *High Temp. High Press.* **9**, p. 376.

[754] Okun', L. B. (1990). *Leptons and Quarks.* (North-Holland, Amsterdam).

[755] Olijnik, H. (2004). *J. Phys. Condens. Mat.* **16**, p. 8791.

[756] Olijnyk, H. and Holzapfel, W. B. (1984). Phase transitions in alkaline earth metals under pressure, *Phys. Lett. A* **100**, 4, pp. 191–194.

[757] Olijnyk, H. and Holzapfel, W. B. (1985). High-pressure structural phase transition in mg, *Phys. Rev. B* **31**, 7, pp. 4682–4683.

[758] Ollitrault, J.-Y. (1992). Anisotropy as a signature of transverse collective flow, *Phys. Rev. D* **46**, 1, pp. 229–245.

[759] Olson, T. S. (2000). Maximally incompressible neutron star matter, *Phys. Rev. C* **63**, p. 015802.

[760] Ossipyan, Y. A., Avdonin, B. V. *et al.* (2005). Nonmonotonic variation of the electrical conductivity of $C_{60}$ fullerene crystals dynamically compressed to 300 kbar as evidence of anomalously strong reduction of the energy barrier of $C_{60}$ polymerization at high pressures, *JETP Lett.* **81**, 9, p. 471.

[761] Owen, B. J. (2005). Maximum elastic deformations of compact stars with exotic equations of state, *Phys. Rev. Lett.* **95**, 21, p. 211101.

[762] Oyamatsu, K. (1993). Nuclear shapes in the inner crust of a neutron star, *Nucl. Phys. A* **561**, pp. 431–452.

[763] Page, D. and Applegate, J. H. (1992). The cooling of neutron stars by the direct Urca process, *Astrophys. J.* **394**, pp. L17–L20.

[764] Palciauskas, V. V. (1976). Lindemann's criterion and the melting of solids at high pressures, *J. Phys. Chem. Solids* **37**, 6, pp. 571–576.

[765] Palciauskas, V. V. (1979). Lindemanns' criterion and the necessary condition for a melting curve maximum, *J. Phys. Chem. Solids* **40**, 10, pp. 787–790.

[766] Palmer, D. M., Barthelmy, S., Gehrels, N., Kippen, R. M., Cayton, T., Kouveliotou, C., Eichler, D., Wijers, R. A. M. J., Woods, P. M., Granot, J., Lyubarsky, Y. E., Ramirez-Ruiz, E., Barbier, L., Chester, M., Cummings, J., Fenimore, E. E., Finger, M. H., Gaensler, B. M., Hullinger, D., Krimm, H., Markwardt, C. B., Nousek, J. A., Parsons, A., Patel, S., Sakamoto, T., Sato, G., Suzuki, M. and Tueller, J. (2005). A giant gamma-ray flare from the magnetar SGR 1806-20, *Nature* **434**, 7037, pp. 1107–1109.

[767] Pandharipande, V. R., Pethick, C. J. and Thorsson, V. (1995). Kaon energies in dense matter, *Phys. Rev. Lett.* **75**, 25, pp. 4567–4570.

[768] Pandharipande, V. R. and Ravenhall, D. G. (1989). Hot nuclear matter, in M. Soyeur, H. Flocard, B. Tamain and M. Porneuf (eds.), *Nuclear Matter and Heavy Ion Collisions* (Reidel, Dordrecht), pp. 103–132.

[769] Pandharipande, V. R. and Smith, R. A. (1975). Nuclear matter calculations with mean scalar fields, *Phys. Lett. B* **59**, pp. 15–18.

[770] Pariser, R. and Parr, R. G. (1953a). A semi-empirical theory of the electronic spectra and electronic structure of complex unsaturated molecules. I. *J. Chem. Phys.* **21**, 3, pp. 466–471.

[771] Pariser, R. and Parr, R. G. (1953b). A semi-empirical theory of the electronic spectra and electronic structure of complex unsaturated molecules. II, *J. Chem. Phys.* **21**, 5, pp. 767–776.

[772] Parr, R. G. (1952). *Chem. Phys.* **20**, p. 1499.

[773] Parr, R. G. and Yang, W. (1989). *Density-Functional Theory of Atoms and Molecules* (Oxford University Press, New York).

[774] Pavlovski, A. I., Boriskov, G. V. *et al.* (1987). Isentropic Solid Hydrogen Compression by Ultrahigh Magnetic Field Pressure in Megabar Range, in C. M. Fowler *et al.* (eds.), *Megagauss Technology and Pulsed Power Applications* (Plenum, N.Y., London), pp. 255–262.

[775] Payne, M. C., Teter, M. P., Allan, D. C., Arias, T. A. and Joannopoulos, J. D. (1992). Iterative minimization techniques for *ab initio* total-energy calculations: molecular dynamics and conjugate gradients, *Rev. Mod. Phys.* **64**, 4, pp. 1045–1097.

[776] Peierls, R. E. (1988). *Quantum statistics* (Nauka, Moscow).

[777] Perdew, J. P. and Burke, K. (1996). Comparison shopping for a gradient-corrected density functional, *Int. J. Quant. Chem.* **57**, pp. 309–319.

[778] Perdew, J. P. and Kurth, S. (1998). Density Functionals: Theory and Applications, in D. Joubert (ed.), *Lecture Notes in Physics* (Springer, Berlin), p. 8.

[779] Perez, A., Mussack, K., Dappen, W. and Mao, D. (2009). The solar-interior equation of state with the path-integral formalism, *Astron. Astrophys.* **505**, 2, pp. 735–742.

[780] Perez-Albuerne, E. A., Clendenen, R. L., Lynch, R. W. and Drickamer, H. G. (1966). Effect of very high pressure on the structure of some hcp metals and alloys, *Phys. Rev.* **142**, 2, pp. 392–399.

[781] Perrot, F. (1979a). Gradient correction to the statistical electronic free energy at nonzero temperatures: Application to equation-of-state calculations, *Phys. Rev. A* **20**, 2, pp. 586–594.

[782] Perrot, F. (1979b). Zero-temperature equation of state of metals in the statistical model with density gradient correction, *Physica A* **98**, 3, pp. 555–565.

[783] Perrot, F. (1980). *Phys. Stat. Sol. (B)* **101**, p. 741.

[784] Perrot, F. and Dharma-wardana, C. (1994). Equation of State of Dense Hydrogen and the Plasma Phase Transition; A Microscopic Calculational Model for Complex Fluids, in G. Chabrier and E. Schatzman (eds.), *The Equation of State in Astrophysics* (Cambridge University Press, Cambridge), pp. 272–286.

[785] Peshier, A., Kämpfer, B., Pavlenko, O. P. and Soff, G. (1996). Massive quasiparticle model of the su(3) gluon plasma, *Phys. Rev. D* **54**, 3, pp. 2399–2402.

[786] Pethick, C. J. and Ravenhall, D. G. (1995). Matter at large neutron excess and the physics of neutronstar crusts, *Annu. Rev. Nucl. Sci.* **45**, pp. 429–484.

[787] Pethick, C. J., Ravenhall, D. G. and Lorenz, C. P. (1995). The inner boundary of a neutron-star crust, *Nucl. Phys. A* **584**, pp. 675–703.

[788] Petreczky, P., Karsch, F., Laermann, E., Stickan, S. and Wetzorke, I. (2002). *Nucl. Phys. Proc. Suppl.* **106**, p. 513.

[789] Pettifor, D. G. (1970). Theory of the crystal structures of transition metals, *J. Phys. C* **3**, 2, p. 367.

[790] Pfaffenzeller, O., Hohl, D. and Ballone, P. (1995). Miscibility of hydrogen and helium under astrophysical conditions, *Phys. Rev. Lett.* **74**, 13, pp. 2599–2602.

[791] Pickard, C. J. and Needs, R. J. (2010). Aluminium at terapascal pressures, *Nature Mat.* **9**, pp. 624–627.

[792] Pieranski, P. (1983). Colloidal crystals, *Contemp. Phys.* **24**, 1, pp. 25–73.

[793] Pierleoni, C., Ceperley, D. M., Bernu, B. and Magro, W. R. (1994). Equation of state of the hydrogen plasma by path integral Monte Carlo simulation, *Phys. Rev. Lett.* **73**, 16, pp. 2145–2149.

[794] Planck, M. (1924). Quantum statistics of Bohr atom model, *Ann. der Phys.* **75**, pp. 673–684.

[795] Pohl, T., Pattard, T. and Rost, J. M. (2004). Coulomb crystallization in expanding laser-cooled neutral plasmas, *Phys. Rev. Lett.* **92**, 15, p. 155003.

[796] Pokrant, M. A. (1977). Thermodynamic properties of the nonzero-temperature, quantummechanical, one-component plasma, *Phys. Rev. A* **16**, pp. 413–423.

[797] Pollock, E. L. and Hansen, J. P. (1973). Statistical mechanics of dense ionized matter. II. Equilibrium properties and melting transition of the crystallized one-component plasma, *Phys. Rev. A* **8**, 6, pp. 3110–3122.

[798] Pons, J. A., Miralles, J. A., Prakash, M. and Lattimer, J. M. (2001). Evolution of proto-neutron stars with kaon condensates, *Astrophys. J.* **553**, pp. 382–393.

[799] Pons, J. A., Reddy, S., Ellis, P. J., Prakash, M. and Lattimer, J. M. (2000). Kaon condensation in proto-neutron star matter, *Phys. Rev. C* **62**, 3, p. 035803.

[800] Pople, J. A. (1953). *Trans. Faraday Soc.* **49**, p. 1375.

[801] Pople, J. A. (2002). Quantum-physical models, *Usp. Fiz. Nauk* **172**, 3, p. 349.

[802] Pople, J. A. and Hehre, W. J. (1978). *Comput. Phys.* **11**, p. 161.

[803] Pople, J. A. and Nesbet, R. K. (1954). *Chem. Phys.* **22**, p. 571.

[804] Pople, J. A., Santry, D. P. and Segal, G. A. (1965). Approximate self-consistent molecular orbital theory. i. invariant procedures, *The Journal of Chemical Physics* **43**, 10, pp. S129–S135.

[805] Pople, J. A. and Segal, G. A. (1965). Approximate self-consistent molecular orbital theory. ii. calculations with complete neglect of differential overlap, *The Journal of Chemical Physics* **43**, 10, pp. S136–S151.

[806] Popov, S. B. and Prohomov, M. E. (2007). Stars: A Life After Death, in V. G. Surdin (ed.), *Astronomy: XXIst Century* (Vek 2, Fryazino) p. 183.

[807] Popov, S. B. and Prokhorov, M. E. (2002). *Astrophysics of Single Neutron Stars: Radio Quiet Neutron Stars and Magnetars* (P.K. Shternberg State Astronomical Institute MSU, Moscow).

[808] Potekhin, A. Y. (2010). The physics of neutron stars, *Phys. Usp.* **53**, 12, pp. 1235–1256.

[809] Potekhin, A. Y. and Chabrier, G. (2000). Equation of state of fully ionized electron-ion plasmas. II. Extension to relativistic densities and to the solid phase, *Phys. Rev. E* **62**, pp. 8554–8563.

[810] Potekhin, A. Y., Chabrier, G. and Yakovlev, D. G. (1997). Internal temperatures and cooling of neutron stars with accreted envelopes, *Astron. Astrophys.* **323**, pp. 415–428.

[811] Poulsen, U. K., Kollar, J. and Andersen, O. K. (1976). *J. Phys. Ser. F.*

[812] Prieto, F. E. (1974). A law of corresponding states for materials at shock pressures, *Journal of Physics and Chemistry of Solids* **35**, 2, pp. 279–286.

[813] Prieto, F. E. (1975). System-independent release adiabats from shocked states*, *Journal of Physics and Chemistry of Solids* **36**, 9, pp. 871–875.

[814] Prieto, F. E. and Renero, C. (1976). The equation of state of solids, *Journal of Physics and Chemistry of Solids* **37**, 2, pp. 151–160.

[815] Prieto, F. E. and Renero, C. (1982). Test of a high pressure equation of state, *J. Phys. Chem. Solids* **43**, 2, pp. 147–154.

[816] Pudliner, B. S., Pandharipande, V. R., Carlson, J. and Wiringa, R. B. (1995). Quantum Monte Carlo calculations of $A \leq 6$ nuclei, *Phys. Rev. Lett.* **74**, pp. 4396–4399.

[817] Pukhov, A. (2003). Strong field interaction of laser radiation, *Rep. Prog. Phys.* **66**, 1, pp. 47–101.

[818] Quigg, C. (2008). The coming revolutions in particle physics, *Sci. Am.* **298**, 2, p. 46.

[819] Radousky, H. B. and Ross, M. (1988). Shock temperature measurements in high density fluid xenon, *Phys. Lett. A* **129**, 1, pp. 43–46.

[820] Ragan, C. E. (1980). Ultrahigh-pressure shock-wave experiments, *Phys. Rev. A* **21**, 2, pp. 458–463.

[821] Ragan, C. E. (1982). Shock compression measurements at 1 to 7 TPa, *Phys. Rev. A* **25**, 6, pp. 3360–3375.

[822] Ragan, C. E., Silbert, M. G. and Diven, B. C. (1977). Shock compression of molybdenum to 2.0 TPa by means of a nuclear explosion, *J. Appl. Phys.* **48**, 7, p. 2860.

[823] Rajagopal, K. and Wilczek, F. (2000). The Condensed Matter Physics of QCD, in M. Shifman (ed.), *At the frontier of particle physics/Handbook of QCD* (World Scientific), pp. 2061–2151.

[824] Rajagopal, K. and Wilczek, F. (2001). Enforced electrical neutrality of the color-flavor locked phase, *Phys. Rev. Lett.* **86**, 16, pp. 3492–3495.

[825] Ramos, A., Schaffner-Bielich, J. and Wambach, J. (2001). Kaon Condensation in Neutron Stars, in D. Blaschke, N. K. Glendenning and A. Sedrakian (eds.), *Physics of Neutron Star Interiors, Lecture Notes in Phys*, Vol. 578, pp. 175–202.

[826] Rapoport, E. (1967). Model for melting-curve maxima at high pressure, *J. Chem. Phys.* **46**, 8, pp. 2891–2895.

[827] Rapoport, E. (1968). Melting-curve maxima at high pressure. II. Liquid cesium. Resistivity, Hall effect, and composition of molten tellurium, *J. Chem. Phys.* **48**, 4, pp. 1433–1437.

[828] Raty, J.-Y., Schwegler, E. and Bonev, S. A. (2007). Electronic and structural transitions in dense liquid sodium, *Nature* **449**, p. 448.

[829] Raveché, H. J. and Mountain, R. D. (1972). Three atom correlations in liquid neon, *J. Chem. Phys* **57**, 9, pp. 3987–3992.

[830] Raveché, H. J. and Mountain, R. D. (1972). Three atom correlations in liquid neon, *J. Chem. Phys.* **57**, 9, pp. 3987–3992.

[831] Raveché, H. J., Mountain, R. D. and Streett, W. B. (1972). Three atom correlations in the lennard-jones fluid, *J. Chem. Phys.* **57**, 11, pp. 4999–5006.

[832] Redmer, R., Holst, B., Juranek, H., Nettelmann, N. and Schwarz, V. (2006). Equation of state for dense hydrogen and helium: Application to astrophysics, *J. Phys. A: Math. Gen.* **39**, 17, p. 4479.

[833] Ree, F., Ross, M. and Young, D. (1983). *J. Chem. Phys.* **79**, p. 1487.

[834] Ree, F. H. (1971). *Physical Chemistry An Advanced Treatise: Liquid State*, Vol. 8A, Chap. 3 (Academic Press, N.Y.).

[835] Reinholz, H., Redmer, R. and Nagel, S. (1995). Thermodynamic and transport properties of dense hydrogen plasmas, *Phys. Rev. E* **52**, pp. 5368–5386.

[836] Renaudin, P., Blancard, C., Clérouin, J., Faussurier, G., Noiret, P. and Recoules, V. (2003). Aluminum equation-of-state data in the warm dense matter regime, *Phys. Rev. Lett.* **91**, 7, p. 075002.

[837] Renaudin, P., Blancard, C., Faussurier, G. and Noiret, P. (2002). Combined pressure and electrical-resistivity measurements of warm dense aluminum and titanium plasmas, *Phys. Rev. Lett.* **88**, 21, p. 215001.

[838] Renaudin, P., Recoules, V., Noiret, P. and Clérouin, J. (2006). Electronic structure and equation of state data of warm dense gold, *Phys. Rev. E* **73**, 5, p. 056403.

[839] Renkert, H., Hensel, F. and Franck, E. U. (1969). Metal-nonmetal transition in dense cesium vapour, *Phys. Lett. A* **30**, 9, pp. 494–495.

[840] Richardson, J. L. (1979). *Phys. Lett. B* **82**, 2, pp. 272–274.

[841] Rijken, T. A., Stoks, V. G. J. and Yamamoto, Y. (1999). Soft-core hyperon-nucleon potentials, *Phys. Rev. C* **59**, pp. 21–40.

[842] Rikovska Stone, J., Miller, J. C., Koncewicz, R., Stevenson, P. D. and Strayer, M. R. (2003). Nuclear matter and neutron-star properties calculated with the Skyrme interaction, *Phys. Rev. C* **68**, 3, p. 034324.

[843] Riordan, M. and Zajc, W. A. (2006). The first few microseconds, *Sci. Am.* **294**, 5, pp. 34A–41A.

[844] Robnik, M. and Kundt, W. (1983). Hydrogen at high-pressures and temperatures, *Astron. Astrophys.* **120**, 2, pp. 227–233.

[845] Rogers, F. J. (1971). Phase shifts of the static screened coulomb potential, *Phys. Rev. A* **4**, pp. 1145–1155.

[846] Rogers, F. J. (1974). Statistical mechanics of Coulomb gases of arbitrary charge, *Phys. Rev. A* **10**, 6, pp. 2441–2456.

[847] Rogers, F. J. (1981). Equation of state of dense, partially degenerate, reacting plasmas, *Phys. Rev. A* **24**, 3, pp. 1531–1543.

[848] Rogers, F. J., Swenson, F. J. and Iglesias, C. A. (1996). OPAL equation-of-state tables for astrophysical applications, *Astrophys. J.* **456**, pp. 902–908.

[849] Rogers, F. J., Wilson, B. G. and Iglesias, C. A. (1988). Parametric potential method for generating atomic data, *Phys. Rev. A* **38**, 10, pp. 5007–5020.

[850] Rogers, F. J. and Witt, H. E. D. (1973). Statistical mechanics of reacting Coulomb gases, *Phys. Rev. A* **8**, 2, pp. 1061–1076.

[851] Romain, J. P., Migault, A. and Jacquesson, J. (1976). Relation between the Grüneisen ratio and the pressure dependence of Poisson's ratio for metals, *J. Phys. Chem. Solids* **37**, 12, pp. 1159–1165.

[852] Romain, J. P., Migault, A. and Jacquesson, J. (1980). Melting curve and Grüneisen coefficient for aluminum, *J. Phys. Chem. Solids* **41**, 4, pp. 323–326.

[853] Ronchi, C., Iosilevskiy, I. and Yakub, E. (2004). *Equation of State of Uranium Dioxide* (Springer, Berlin).

[854] Roof, R. B., Haire, R. G., Schiferl, D., Schwalbe, L. A., Kmetko, E. A. and Smith, J. L. (1981). *Science* **207**, p. 1353.

[855] Roothaan, C. C. J. (1951). New developments in molecular orbital theory, *Rev. Mod. Phys.* **23**, 2, pp. 69–89.

[856] Ross, M. (1968). Shock compression of argon and xenon. IV. Conversion of xenon to a metal-like state, *Phys. Rev.* **171**, 3, pp. 777–784.

[857] Ross, M. (1969). Generalized lindemann melting law, *Phys. Rev.* **184**, 1, pp. 233–242.

[858] Ross, M. (1980a). Extension of liquid-metal theory to dense partially ionized plasmas, *Phys. Rev. B* **21**, 8, pp. 3140–3151.

[859] Ross, M. (1980b). The repulsive forces in dense argon, *J. Chem. Phys.* **73**, 9, pp. 4445–4450.

[860] Ross, M. (1998). Linear-mixing model for shock-compressed liquid deuterium, *Phys. Rev. B* **58**, pp. 669–677.

[861] Ross, M., DeWitt, H. E. and Hubbard, W. B. (1981). Monte Carlo and perturbation-theory calculations for liquid metals, *Phys. Rev. A* **24**, 2, pp. 1016–1020.

[862] Ross, M. and McMahan, A. K. (1980). Condensed xenon at high pressure, *Phys. Rev. B* **21**, 4, pp. 1658–1664.

[863] Ross, M., Ree, F. H. and Young, D. A. (1983). *J. Chem. Phys.* **79**, p. 1487.

[864] Ross, M. and Seale, D. (1974). Perturbation approximation to the screened coulomb gas, *Phys. Rev. A* **9**, 1, pp. 396–399.

[865] Rozsnyai, B. F. (1972). Relativistic Hartree–Fock–Slater calculations for arbitrary temperature and matter density, *Phys. Rev. A* **5**, 3, pp. 1137–1149.

[866] Rubakov, V. A. (2001). Large and infinite extra dimensions, *Phys. Usp.* **44**, 9, 871.

[867] Rubakov, V. A. (2005). Introduction to Cosmology, *Proc. Sci.* **RTN2005**.

[868] Ruderman, M. (1971). Matter in superstrong magnetic fields: The surface of a neutron star, *Phys. Rev. Lett.* **27**, 19, pp. 1306–1308.

[869] Ruderman, M. A. (1968). Causes of sound faster than light in classical models of ultradense matter, *Phys. Rev.* **172**, pp. 1286–1290.

[870] Russel, W. B., Saville, D. A. and Schowalter, W. R. (1989). *Colloidal Dispersions* (Cambridge University Press, Cambridge).

[871] Ryutov, D. D., Remington, B. A., Robey, H. F. and Drake, R. P. (2001). Magnetodynamic scaling: from astrophysics to the laboratory, *Phys. Plasmas* **8**, 5, pp. 1804–1816.

[872] Saha, M. N. (1920). LIII. Ionization in the solar chromosphere, *Philosophical Magazine Series 6* **40**, 238, pp. 472–488.

[873] Saha, M. N. (1921). On a physical theory of stellar spectra, *Proc. R. Soc. Lon. A* **99**, 697, pp. 135–153.

[874] Salpeter, E. E. (1961). Energy and pressure of a zero-temperature plasma, *Astrophys. J.* **134**, pp. 669–682.

[875] Salpeter, E. E. and van Horn, H. M. (1969). Nuclear reaction rates at high densities, *Ap. J.* **155**, p. 183.

[876] Sandoval, A., Stock, R., Stelzer, H. E., Renfordt, R. E., Harris, J. W., Brannigan, J. P., Geaga, J. V., Rosenberg, L. J., Schroeder, L. S. and Wolf, K. L. (1980). Energy dependence of multi-pion production in high-energy nucleus-nucleus collisions, *Phys. Rev. Lett.* **45**, pp. 874–877.

[877] Sapozhnikov, A. T. and Pershina, A. V. (1979a). *Vopr. atom. nauki i tekhn. Ser. Metod. i progr. resh. zadach mat. fiz.* **3**, p. 37.

[878] Sapozhnikov, A. T. and Pershina, A. V. (1979b). Semiempirical equation of state of metals in a broad range of densities and temperatures, *Vopr. atom. nauki i tekhn, ser. Metod. i progr. chisl. resh. zadach mat. fiz.* **3**, 4(6), pp. 47–56.

[879] Sarkar, S. K. and Sen, D. (1981). Unified study of a polyvalent simple metal-aluminium, *J. Phys. F Met. Phys.* **11**, 2, p. 377.

[880] Sarkisov, G. N. (1999). Approximate equations of the theory of liquids in the statistical thermodynamics of classical liquid systems, *Phys. Usp.* **42**, 6, pp. 545–561.

[881] Saumon, D. and Chabrier, G. (1989). Fluid hydrogen at high density: The plasma phase transition, *Phys. Rev. Lett.* **62**, pp. 2397–2400.

[882] Saumon, D. and Chabrier, G. (1991). Fluid hydrogen at high density: Pressure dissociation, *Phys. Rev. A* **44**, 8, pp. 5122–5141.

[883] Saumon, D. and Chabrier, G. (1992). Fluid hydrogen at high density: Pressure ionization, *Phys. Rev. A* **46**, 4, pp. 2084–2100.

[884] Saumon, D., Chabrier, G. and van Horn, H. M. (1995). *Astrophys. J. Suppl. Ser.* **99**, pp. 713–741.

[885] Sawyer, R. F. (1972). Condensed $\pi^-$ phase in neutron-star matter, *Phys. Rev. Lett.* **29**, pp. 382–385.

[886] Scalapino, D. J. (1972). $\pi^-$ condensate in dense nuclear matter, *Phys. Rev. Lett.* **29**, pp. 386–388.

[887] Scandolo, S. (2003). *Proc. Nat. Ac. Sci.* **100**, p. 3051.

[888] Schatz, T., Schramm, U. and Habs, D. (2001). Crystalline ion beams, *Nature* **412**, 6848, pp. 717–720.

[889] Schertlera, K., Greinera, C., Schaffner-Bielichc, J. and Thoma, M. H. (2001). Quark phases in neutron stars and a third family of compact stars as signature for phase transitions, *Nucl. Phys. A* **677**, 1–4, pp. 463–490.

[890] Schey, H. M. and Schwartz, J. L. (1965). Quantum Corrections in the Thomas-Fermi Model, *Phys. Rev.* **137**, 3A, pp. A709–A716.

[891] Schiavilla, R., Pandharipande, V. R. and Wiringa, R. B. (1986). Momentum distributions in $a = 3$ and 4 nuclei, *Nucl. Phys. A* **449**, pp. 219–242.

[892] Schlanges, M., Bonitz, M. and Tschttschjan, A. (1995). Plasma phase transition in fluid hydrogen-helium mixtures, *Contrib. Plasma Phys.* **35**, pp. 109–125.

[893] Schramm, U., Schatz, T., Bussmann, M. and Habs, D. (2003). Cooling and heating of crystalline ion beams, *J. Phys. B* **36**, 3, pp. 561–571.

[894] Schramm, U., Schatz, T. *et al.* (2001). Crystalline Ion Beams, *Nature* **412**, p. 717.

[895] Scott, G. D. (1960). Packing of spheres: Packing of equal spheres, *Nature* **188**, pp. 908–909.

[896] Segretain, L. (1996). *Astron. Astrophys.* **310**, pp. 485–488.

[897] Seldam, C. A. (1957). *Proc. Phys. Soc. A* **70**, p. 97.

[898] Sen, D. and Sarkar, S. K. (1980). Pseudopotential study of alkali metals: Unified approach, *Phys. Rev. B* **22**, pp. 1856–1865.

[899] SESAME (1992). *The Los Alamos National Laboratory Equation of State Database. LA-UR-92-3407* (LANL, Los Alamos).

[900] Shaner, J. W., Brown, J. M. and McQueen, R. G. (1984). *High Pressure in Science and Technology* (North Holland, Amsterdam).

[901] Shapiro, S. L. and Teukolsky, S. A. (1983). *Black Holes, White Dwarfs, and Neutron Stars: The Physics of Compact Objects* (Wiley, N.Y.).

[902] Shashkin, A. A. (2005). Metal-insulator transitions and the effects of electron–electron interactions in two-dimensional electron systems, *Phys. Usp.* **48**, 2, 129.

[903] Shaviv, G. and Kovetz, A. (1972). The thermodynamics of white dwarf matter. II, *Astron. Astrophys.* **16**, pp. 72–76.

[904] shion Wang, S. and Krumhansl, J. A. (1972). Superposition assumption. ii. high density fluid argon, *J. Chem. Phys.* **56**, 9, pp. 4287–4290.

[905] Shpatakovskaya, G. V. (1969). *Graduate work* (MSU).

[906] Shpatakovskaya, G. V. (1973). *Kratk. Soobshch. Fiz. (FIAN SSSR)*, **5**, p. 10.

[907] Shpatakovskaya, G. V. (1974). *Ph.D. thesis* (FIAN SSSR).

[908] Shpatakovskaya, G. V. (1975). *Preprint of IAM of the USSR Academy of Sciences N54* (IAM of the USSR Academy of Sciences, Moscow).

[909] Shuryak, E. (2009). *Prog. Part. Nucl. Phys.* **62**, p. 48.

[910] Shuryak, E. V. (1978). Quark–gluon plasma and hadronic production of leptons, photons and psions, *Phys. Lett. B* **78**, 1, pp. 150–153.

[911] Shuryak, E. V. (1980). Quantum chromodynamics and the theory of super-dense matter, *Phys. Rep.* **61**, 2, pp. 71–158.

[912] Shuryak, E. V. and Zahed, I. (2004). Rethinking the properties of the quark-gluon plasma at $t_c < t < 4t_c$, *Phys. Rev. C* **70**, 2, p. 021901.

[913] Singwi, K. S., Tosi, M. P., Land, R. H. and Sjölander, A. (1968). Electron Correlations at Metallic Densities, *Phys. Rev.* **176**, 2, pp. 589–599.

[914] Sin'ko, G. (1989). Description of Many-Particle Systems by the Density Functional Method, in N. Kalitkin (ed.), *Mathematical Simulation* (Nauka, M.), pp. 197–231.

[915] Sin'ko, G. V. (1979). *Chisl. met. mekh. splosh. sredy*, **10**, 1, p. 124.

[916] Sin'ko, G. V. (1981). *Chisl. met. mekh. splosh. sredy*, **12**, 1, p. 121.

[917] Sin'ko, G. V. (1983). Use of the self-consistent-field method to calculate the thermodynamic functions of electrons in simple materials, *High Temp.* **21**, 6, p. 783.

[918] Sin'ko, G. V. and Smirnov, N. A. (2005a). Effect of electronic topological transitions on the calculations of some Zn and Fe properites, *J. Phys. Condens. Mat.* **17**, pp. 559–569.

[919] Sin'ko, G. V. and Smirnov, N. A. (2005b). Relative stability and elastic properties of hcp, bcc, and fcc beryllium under pressure, *Phys. Rev. B* **71**, 21, p. 214108.

[920] Sin'ko, G. V. and Smirnov, N. A. (2009). *Ab initio* calculations for the elastic properties of magnesium under pressure, *Phys. Rev. B* **80**, 10, p. 104113.

[921] Sirnonenko, V. A., Voloshin, N. P., Vladirnirov, A. S., Nagibin, A. P., Nogin, V. N., Popov, V. A., Vasilenko, V. A. and Shoidin, Y. A. (1985). Absolute measurements of shock compressibility of aluminum at pressures $P \geq 1$ TPa, *J. Exp. Theor. Phys.* **61**, 4, p. 869.

[922] Skriver, H. L., Andersen, O. K. and Johansson, B. (1978). Calculated bulk properties of the actinide metals, *Phys. Rev. Lett.* **41**, 1, pp. 42–45.

[923] Skriver, H. L., Andersen, O. K. and Johansson, B. (1980). $5f$-electron delocalization in americium, *Phys. Rev. Lett.* **44**, 18, pp. 1230–1233.

[924] Skriver, H. L. and Jan, J. P. (1980). Electronic states in thorium under pressure, *Phys. Rev. B* **21**, 4, pp. 1489–1496.

[925] Skryshevskii, A. F. (1980). *Structural Analysis of Liquids and Amorphous Bodies* (Vysshaya Shkola, Moscow).

[926] Slater, J. C. (1937). Wave functions in a periodic potential, *Phys. Rev.* **51**, 10, pp. 846–851.

[927] Slater, J. C. (1951). A Simplification of the Hartree–Fock Method, *Phys. Rev.* **81**, 3, pp. 385–390.

[928] Slater, J. C. (1972). *Statistical Exchange-Correlation in the Self-Consistent Field*, (Academic Press), pp. 1–92.

[929] Slater, J. C. and Koster, G. F. (1954). Simplified LCAO method for the periodic potential problem, *Phys. Rev.* **94**, 6, pp. 1498–1524.

[930] Slater, J. C. and Krutter, H. M. (1935). The Thomas–Fermi method for metals, *Phys. Rev.* **47**, 7, pp. 559–568.

[931] Slattery, W. L., Doolen, G. D. and DeWitt, H. E. (1980). Improved equation of state for the classical one-component plasma, *Phys. Rev. A* **21**, 6, pp. 2087–2095.

[932] Slattery, W. L., Doolen, G. D. and DeWitt, H. E. (1982). $N$ dependence in the classical one-component plasma Monte Carlo calculations, *Phys. Rev. A* **26**, 4, pp. 2255–2258.

[933] Sobelman, I. I., Vainshtein, L. A. and Yukov, E. A. (1981). *Excitation of Atoms and Broadening of Spectral Lines* (Springer, Berlin).

[934] Soma, T. (1980). Equation of state and bulk modulus under pressure of alkali metals, *J. Phys. F* **10**, 7, p. 1401.

[935] Soma, T., Satoh, H. and Matsuo, H. (1981). Pressure-volume relations and elastic constants under pressure of Cu, Ag, and Au, *Physica Status Solidi (B)* **107**, 1, pp. K25–K29.

[936] Soma, T. and Satoh, T. (1980). The effective ion–ion potential and compressibility of alkali metals, *J. Phys. F* **10**, 6, p. 1081.

[937] Sorensen, P. R. (2003). *Kaon and Lambda Production at Intermediate $P_T$: Insights into the Hadronization of the Bulk Partonic Matter Created in Au+Au Collisions at RHIC. Ph.D. Thesis.*

[938] Springer, J. F., Pokrant, M. A. and F. A. Stevens, J. (1973). Integral equation solutions for the classical electron gas, *J. Chem. Phys.* **58**, 11, pp. 4863–4867.

[939] Stager, R. A. and Drickamer, H. G. (1964). Effect of pressure and temperature on the electrical resistance of eleven rare-earth metals, *Phys. Rev.* **133**, 3A, pp. A830–A835.

[940] Starostin, A. N. (1969). Correlation radius of the field fluctuations in plasma, in *9th Int. Conf. Phenom. Ionized Gases* (Bucharest), p. 366.

[941] Starostin, A. N. and Roerich, V. C. (2005). A converging equation of state of a weakly nonideal hydrogen plasma without mystery, *J. Exp. Theor. Phys.* **100**, 1, pp. 165–198.

[942] Starostin, A. N. and Roerich, V. C. (2006a). Bound states in nonideal plasmas: formulation of the partition function and application to the solar interior, *Plasma Sources Sci. Technol.* **15**, p. 410.

[943] Starostin, A. N. and Roerich, V. C. (2006b). Equation of state of weakly nonideal plasmas and electroneutrality condition, *J. Phys. A: Math. Gen.* **39**, 17, p. 4431.

[944] Starostin, A. N. and Roerich, V. C. (2010). Account of atomic and molecular contributions in the equation of-state for a weakly non-ideal hydrogen plasmas, *Contrib. Plasma Phys.* **50**, 1, pp. 88–92.

[945] Starostin, A. N., Roerich, V. C., Gryaznov, V. K., Fortov, V. E. and Iosilevskiy, I. L. (2009). The influence of electron degeneracy on the contribution of bound states to the non-ideal hydrogen plasma EOS, *J. Phys. A: Math. Theor.* **42**, 21, p. 214009.

[946] Starostin, A. N., Roerich, V. C. and More, R. M. (2003). How correct is the EOS of weakly nonideal hydrogen plasmas? *Contrib. Plasma Phys.* **43**, 5–6, pp. 369–372.

[947] Steffen, B. and Hosemann, R. (1976). Paracrystalline microdomains in monatomic liquids. II. Three-dimensional structure of microdomains in liquid lead, *Phys. Rev. B* **13**, 8, pp. 3232–3238.

[948] Stevenson, D. and Salpeter, E. (1977). *Astrophys. J.* **35**, p. 221.

[949] Stishov, S. M. (1969). Melting at high pressures, *Phys. Usp.* **11**, 6, pp. 816–830.

[950] Stishov, S. M. (1975). The thermodynamics of melting of simple substances, *Phys. Usp.* **18**, 5, pp. 625–643.

[951] Stishov, S. M., Makarenko, I. N. and Nikolaenko, A. M. (1980). *Zh. Eksp. Teor. Fiz.*

[952] Stock, R. (1986). Particle production in high energy nucleus–nucleus collisions, *Phys. Rep.* **135**, 5, pp. 259–315.

[953] Stock, R., Bock, R., Brockman, R., Harris, J. W., Sandoval, A., Stroebele, H., Wolf, K. L., Pugh, H. G., Schroeder, L. S., Maier, M., Renfordt, R. E., Dacal, A. and Ortiz, M. E. (1982). Compression effects in relativistic nucleus-nucleus collisions, *Phys. Rev. Lett.* **49**, pp. 1236–1239.

[954] Stöcker, H. and Greiner, W. (1986). High energy heavy ion collisions–probing the equation of state of highly excited hardronic matter, *Phys. Rep.* **137**, 5-6, pp. 277–392.

[955] Stocker, H., Hofmann, J., Maruhn, J. A. and Greiner, W. (1980). Shock waves in nuclear matter — proof by circumstantial evidence, *Prog. Part. Nucl. Phys.* **4**, pp. 133–195.

[956] Stoks, V. G. J., Klomp, R. A. M., Terheggen, C. P. F. and de Swart, J. J. (1994). Construction of high-quality NN potential models, *Phys. Rev. C* **49**, pp. 2950–2962.

[957] Stroud, D. and Ashcroft, N. W. (1976). Comment on the thermodynamics of a classical one-component plasma, *Phys. Rev. A* **13**, 4, pp. 1660–1663.

[958] Sugahara, Y. and Toki, H. (1994). Relativistic mean-field theory for unstable nuclei with non-linear $\sigma$ and $\omega$ terms, *Nucl. Phys. A* **579**, pp. 557–572.

[959] Sumiyoshi, K., Yamada, S., Suzuki, H. and Hillebrandt, W. (1998). The fate of a neutron star just below the minimum mass: does it explode? *Astron. Astrophys.* **334**, pp. 159–168.

[960] Syassen, K., Wortmann, G., Feldhaus, J., Frank, K. H. and Kaindl, G. (1982). Mean valence of Yb metal in the pressure range 0 to 340 kbar, *Phys. Rev. B* **26**, 8, pp. 4745–4748.

[961] Tabor, D. (1993). *Gases, Liquids and Solids* (University Press, Cambridge).

[962] Tahir, N. A., Deutsch, C., Fortov, V. E., Gryaznov, V., Hoffmann, D. H. H., Kulish, M., Lomonosov, I. V., Mintsev, V., Ni, P., Nikolaev, D., Piriz, A. R., Shilkin, N., Spiller, P., Shutov, A., Temporal, M., Ternovoi, V., Udrea, S. and Varentsov, D. (2005a). Proposal for the study of thermophysical properties of high-energy-density matter using current and future heavy-ion accelerator facilities at GSI Darmstadt, *Phys. Rev. Lett.* **95**, 3, p. 035001.

[963] Tahir, N. A., Deutsch, C., Fortov, V. E., Gryaznov, V., Hoffmann, D. H. H., Lomonosov, I. V., Piriz, A. R., Shutov, A., Spiller, P., Temporal, M., Udrea, S. and Varentsov, D. (2005b). Studies of strongly coupled plasmas using intense heavy ion beams at the future FAIR facility: The HEDgeHOB collaboration, *Contrib. Plasma Phys.* **45**, 3–4, pp. 229–235.

[964] Tanaka, S., Mitake, S. and Ichimaru, S. (1985). Parameterized equation of state for electron liquids in the Singwi-Tosi-Land-Sj.olander approximation, *Phys. Lett. A* **32**, pp. 1896–1899.

[965] Tanatar, B. and Ceperley, D. M. (1989). *Phys. Rev. Lett.*

[966] Temperley, H. N. V., Rowlinson, J. S. and Rushbrooke, G. S. (eds.) (1968). *Physics of Simple Liquids* (North-Holland, Amsterdam).

[967] Ternovoi, V. Y., Filimonov, A. S., Fortov, V. E., Kvitov, S. V., Nikolaev, D. N. and Pyalling, A. A. (1999). Thermodynamic properties and electrical conductivity of hydrogen under multiple shock compression to 150 GPa, *Physica B* **265**, 1–4, pp. 6–11.

[968] Ternovoi, V. Y., Filimonov, A. S., Pyalling, A. A., Mintsev, V. B. and Fortov, V. E. (2001). *Shock Compression of Condensed Matter*, in M. D. Furnish, N. N. Thardhani and Y. Horie (eds.), AIP Conference Proceedings (Melville, New York), pp. 107–110.

[969] Ternovoi, V. Y., Kvitov, S. V., Pyalling, A. A., Filimonov, A. S. and Fortov, V. E. (2004.). Experimental determination of the conditions for the transition of Jupiter's atmosphere to the conducting state, *JETP Lett.* **79**, 1, p. 6.

[970] Thoma, M. H. (2004). *IEEE Trans. Plasma Science* **32**, p. 738.

[971] Thomas, G. A., Rice, T. M. and Hensel, J. C. (1974). Liquid–gas phase diagram of an electron–hole fluid, *Phys. Rev. Lett.* **33**, 4, pp. 219–222.

[972] Thomas, H., Morfill, G. E., Demmel, V., Goree, J., Feuerbacher, B. and Möhlmann, D. (1994). Plasma Crystal: Coulomb Crystallization in a Dusty Plasma, *Phys. Rev. Lett.* **73**, 5, pp. 652–655.

[973] Thomas, L. H. (1927). The calculation of atomic fields, *Math. Proc. Cambridge*, **23**, pp. 542–548.

[974] Thomas, L. H. and Umeda, K. (1957). Atomic scattering factors calculated from the TFD atomic model, *J. Chem. Phys.* **26**, 2, pp. 293–303.

[975] Tietz, T. (1954). Approximate analytic solution of the Thomas–Fermi equation for atoms, *J. Chem. Phys.* **22**, 12, pp. 2094–2095.

[976] Timan, B. A. (1953). *Zh. Eksp. Teor. Fiz.* **25**, 6, p. 733.

[977] Togaya, M. (1996). Thermophysical Properties of Carbon at High Pressure, *Presented at the 3rd NIRIM ISAM'96, Tsukuba, Yapan.*

[978] Tonkov, E. Y. (1979). *Phase Diagrams of Elements at High Pressure* (Nauka, Moscow).

[979] Trainor, R. J., Shaner, J. W., Auerbach, J. M. and Holmes, N. C. (1979). Ultrahigh-pressure laser-driven shock-wave experiments in aluminum, *Phys. Rev. Lett.* **42**, 17, pp. 1154–1157.

[980] Trubitsyn, V. P. and Ulinich, F. P. (1962). *Dokl. Akad. Nauk SSSR* **142**, p. 578.

[981] Trubnikov, B. A. and Elesin, V. F. (1965). Quantum correlation functions in a Maxwellian plasma, *J. Exp. Theor. Phys.* **20**, 4, p. 866.

[982] Trunin, R. and Simakov, G. (1993). Shock compression of ultralow density nickel, *J. Exp. Theor. Phys.* **76**, 6, p. 1090.

[983] Trunin, R. F. (ed.) (1992). *Properties of Condensed Substances at High Pressures and Temperatures* (VNIIEF).

[984] Trunin, R. F., Podurets, M. A., Popov, L. V., Moiseev, B. N., Simakov, G. V. and Sevast'yanov, A. G. (1993). Determination of the shock compressibility of iron at pressures up to 10 TPa (100 Mbar), *J. Exp. Theor. Phys.* **103**, 2129–2141.

[985] Trunin, R. F., Podurets, M. A., Simakov, G. V., Popov, L. V. and Moiseev, B. N. (1972). An experimental verification of the Thomas–Fermi model for metals under high pressure, *J. Exp. Theor. Phys.* **35**, 3, p. 550.

[986] Trunin, R. F., Simakov, G. V. and Panov, N. V. (2001). Shock compression of porous aluminum and nickel at megabar pressures, *High Temp.* **39**, p. 401.

[987] Trunin, R. F., Simakov, G. V., Sutulov, Y. N., Medvedev, A. B., Rogozkin, B. D. and Fedorov, Y. E. (1989). Compressibility of porous metals in shock waves, *J. Exp. Theor. Phys.* **69**, 3, p. 580.

[988] Tsuruta, S. and Cameron, A. G. W. (1966). Some effects of nuclear forces on neutron-star models, *Canadian J. Phys.* **44**, pp. 1895–1922.

[989] Uhlenbeck, G. E. and Beth, E. (1936). The quantum theory of the non-ideal gas. I. Deviations from the classical theory, *Physica* **3**, pp. 729–745.

[990] Ukrainets, A. V. and Iosilevskii, I. L. (2005). Specific features of realization of hypothetic plasma phase transition in the depth of Saturn and Jupiter,

in *Physics of Extreme States of Matter* (Institute of Problems of Chemcial Physcis RAS, Chernogolovka).

[991] Unsöld, A. (1948). Zur Berechnung der Zustandsummen für Atome und Ionen in einem teilweise ionisierten Gas. Mit 2 Textabbildungen, *Z. Astrophys.* **24**, pp. 355–362.

[992] Urlin, V. D. (1966). Melting at ultra high pressures in a shock wave, *J. Exp. Theor. Phys.* **22**, p. 341.

[993] Urlin, V. D. and Ivanov, A. A. (1963). *Dokl. Akad. Nauk SSSR* **149**, p. 1303.

[994] Urlin, V. D., Mochalov, M. A. and Mikhailova, O. L. (1992). Liquid xenon study under shock and quasi-isentropic compression, *High Pressure Res.* **8**, pp. 595–605.

[995] Urtiew, P. A. and Grover, R. (1977). *J. Appl. Phys.* **48**, p. 1122.

[996] Utsumi, K. and Ichimaru, S. (1981). Dielectric formulation of strongly coupled electron liquids at metallic densities. IV. Static properties in the low-density domain and the Wigner crystallization, *Phys. Rev. B* **24**, 6, pp. 3220–3225.

[997] Vacca, J. R. (ed.) (2004). *The World's 20 Greatest Unsolved Problems* (Prentice Hall PTR, Englewood Cliffs, NJ).

[998] Vainshtein, L. A., Sobelman, I. I. and Yukov, E. A. (1979). *Excitation of Atoms and Spectral Line Broadening* (Nauka, Moscow).

[999] Vaks, V. G., Kravchuk, S. P. and Trefilov, A. V. (1977a). *Fiz. Tverd. Tela* **19**, p. 1271.

[1000] Vaks, V. G., Kravchuk, S. P. and Trefilov, A. V. (1977b). *Fiz. Tverd. Tela* **19**, p. 3396.

[1001] Vaks, V. G., Kravchuk, S. P. and Trefilov, A. V. (1979). *Fiz. Tverd. Tela* **21**, p. 3370.

[1002] Val'ko, V. V., Lomonosov, I. V., Ostrik, A. V., Fortov, V. E. and Khishchenko, K. V. (2010). Wide-Range Equations of State of Construction Materials, in *Physics of nuclear explosion. The effect of explosion* (Fizmatlit, Moscow), pp. 140–228.

[1003] van Horn, H. (1965). *Thesis* (Cornell University, CRSR 211).

[1004] van Horn, H. (1990). in *Strongly Coupled Plasma Physics* (Elsevier), p. 3.

[1005] van Horn, H. M. (1968). Crystallization of white dwarfs, *Ap. J.* **151**, p. 227.

[1006] van Horn, H. M. (1969). Crystallization of a classical, one-component coulomb plasma, *Phys. Lett. A* **28**, 10, pp. 706–707.

[1007] Van Thiel, M. (ed.) (1977). *Compendium of shock wave data, LLNL Report UCRL-50108*, Vol. 1–3 (LLNL).

[1008] Vanderbilt, D. (1990). Soft self-consistent pseudopotentials in a generalized eigenvalue formalism, *Phys. Rev. B* **41**, 11, pp. 7892–7895.

[1009] Vargaftik, N. B. (1972). *Spravochnik po teplofizicheskim svoistvam gazov i zhidkostei (Reference book on thermophysical properties of gases and liquids)* (Nauka).

[1010] Vashchenko, V. Y. and Zubarev, V. N. (1963). *Fiz. Tverd. Tela* **5**, p. 886.

[1011] Vaulina, O. S., Petrov, O. F., Fortov, V. E., Chernyshev, A. V., Gavrikov, A. V. and Shakhova, O. A. (2004). Three-particle correlations in nonideal dusty plasma, *Phys. Rev. Lett.* **93**, 3, p. 035004.

[1012] Vaulina, O. S., Petrov, O. F., Fortov, V. E., Khrapak, A. G. and Khrapak, S. A. (2009). *Dust Plasma (Experiment and Theory)* (Fizmatlit, Moscow).

[1013] Vaulina, O. S., Vladimirov, S. V., Petrov, O. F. and Fortov, V. E. (2002). Criteria of phase transitions in a complex plasma, *Phys. Rev. Lett.* **88**, p. 245002.

[1014] Vedenov, A. A. and Larkin, A. I. (1959). The State Equation of a Plasma, *Sov. Phys. JETP* **9**, p. 806.

[1015] Veeser, L. R., Solem, J. C. and Lieber, A. J. (1979). Impedance-match experiments using laser-driven shock waves, *Appl. Phys. Lett.* **35**, 10, p. 761.

[1016] Vidaña, I., Polls, A., Ramos, A., Engvik, L. and Hjorth-Jensen, M. (2000). Hyperon–hyperon interactions and properties of neutron star matter, *Phys. Rev. C* **62**, p. 035801.

[1017] Vil'chek, F. A. (2005). Asymptotical freedom: From paradoxes to paradigms, *Usp. Fiz. Nauk* **175**, 12, p. 1325.

[1018] Vitev, I. and Gyulassy, M. (2002). High-$P_T$ Tomography of d+Au and Au+Au at SPS, RHIC, and LHC, *Phys. Rev. Lett.* **89**, 25, p. 252301.

[1019] Vladimirov, A. S., Voloshin, N. P., Nogin, V. N., Petrovtsev, A. V. and Simonenko, V. A. (1984). Shock compressibility of aluminum at p > 1 Gbar, *JETP Lett.* **39**, 2, p. 82.

[1020] Vohra, Y. K., Grosshans, W. and Holzapfel, W. B. (1982). High-pressure phase transformation in scandium, *Phys. Rev. B* **25**, 9, pp. 6019–6021.

[1021] Vohra, Y. K. and Holzapfel, W. B. (1982). *Phys. Lett. Ser. A* **89**, p. 149.

[1022] Vohra, Y. K., Olijnik, H., Grosshans, W. and Holzapfel, W. B. (1981). Structural Phase Transitions in Yttrium under Pressure, *Phys. Rev. Lett.* **47**, 15, pp. 1065–1067.

[1023] Volkov, A. P., Voloshin, N. P., Vladimirov, A. S., Nogin, V. N. and Simonenko, V. A. (1980). Shock compressibility of aluminum at a pressure of 10 Mbar, *JETP Lett.* **31**, 11, p. 588.

[1024] von Barth, U. and Hedin, L. (1972). A local exchange-correlation potential for the spin polarized case. i, *J. Phys. C: Solid State* **5**, 13, p. 1629.

[1025] Vorobiev, V. S. (1995). Investigation of liquid-vapor equilibrium using an interpolation equation of state, *High Temp.* **33**, 4, p. 551.

[1026] Vorobiev, V. S. and Khomkin, A. L. (1971). Canonical transformation method in thermodynamics of partially ionized plasma, *Teor. Mat. Fiz.* **8**, pp. 109–118.

[1027] Voropinov, A. I., Gandel'man, G. M. and Podval'nyi, V. G. (1970). Electronic energy spectra and the equation of state of solids at high pressures and temperatures, *Phys. Usp.* **13**, 1, pp. 56–72.

[1028] Wachter, P., Bucher, B. and Malar, J. (2004). Possibility of a superfluid phase in a Bose condensed excitonic state, *Phys. Rev. B* **69**, 9, p. 094502.

[1029] Walsh, J. M., Rice, M. H., McQueen, R. G. and Yarger, F. L. (1957). Shock-wave compressions of twenty-seven metals. Equations of state of metals, *Phys. Rev.* **108**, 2, pp. 196–216.

[1030] Wang, X., Scandolo, S. and Car, R. (2005). Carbon phase diagram from *Ab Initio* molecular dynamics, *Phys. Rev. Lett.* **95**, p. 185701.

[1031] Wang, X.-N. and Gyulassy, M. (1992). Gluon shadowing and jet quenching in A+A collisions at $\sqrt{s} = 200$A GeV, *Phys. Rev. Lett.* **68**, 10, pp. 1480–1483.

[1032] Wang, Y., Chen, D. and Zhang, X. (2000). Calculated Equation of State of Al, Cu, Ta, Mo, and W to 1000 GPa, *Phys. Rev. Lett.* **84**, 15, pp. 3220–3223.

[1033] Weber, F. (1999). *Pulsars as Astrophysical Laboratories for Nuclear and Particle Physics* (IOP Publishing, Bristol).

[1034] Weber, F. (2005). Strange quark matter and compact stars, *Progress in Particle and Nuclear Physics* **54**, pp. 193–288.

[1035] Weber, F. and Weigel, M. K. (1989a). Baryon composition and macroscopic properties of neutron stars, *Nucl. Phys. A* **505**, pp. 779–822.

[1036] Weber, F. and Weigel, M. K. (1989b). Neutron star properties and the relativistic nuclear equation of state of many baryon matter, *Nucl. Phys. A* **493**, pp. 549–582.

[1037] Weir, S. T., Mitchell, A. C. and Nellis, W. J. (1996). Metallization of fluid molecular hydrogen at 140 GPa (1.4 Mbar), *Phys. Rev. Lett.* **76**, 11, pp. 1860–1863.

[1038] Wheeler, J. (1965). *Gravitatsiya i Otnositel'nost'* (Mir Publishers, Moscow).

[1039] Wigner, E. (1934). On the interaction of electrons in metals, *Phys. Rev.* **46**, 11, pp. 1002–1011.

[1040] Wigner, E. P. (1938). *Trans. Farad. Soc.* **34**, p. 678.

[1041] Wildhack, W. A. (1940). The proton–deuteron transformation as a source of energy in dense stars, *Phys. Rev.* **57**, 2, pp. 81–86.

[1042] Wineland, D. J., Bergquist, J. C., Itano, W. M., Bollinger, J. J. and Manney, C. H. (1987). Atomic–ion Coulomb clusters in an ion trap, *Phys. Rev. Lett.* **59**, 26, pp. 2935–2938.

[1043] Winisdoerffer, C. and Chabrier, G. (2005). Free-energy model for fluid helium at high density, *Phys. Rev. E* **71**, p. 026402.

[1044] Winzenick, N. and Holzaphel, W. B. (1996). *High Pressure Science and Technology* (World Scientific, Singapore).

[1045] Wiringa, R. B., Fiks, V. and Fabrocini, A. (1988). Equation of state for dense nucleon matter, *Phys. Rev. C* **38**, pp. 1010–1037.

[1046] Wiringa, R. B., Stoks, V. G. J. and R., S. (1995). Accurate nucleon-nucleon potential with charge-independence breaking, *Phys. Rev. C* **51**, pp. 38–51.

[1047] Witten, E. (1984). Cosmic separation of phases, *Phys. Rev. D* **30**, pp. 272–285.

[1048] Wong, S. K. (1970). *Nuovo Cimento A* **65**, p. 689.

[1049] Xu, H. and Hansen, J.-P. (1998). Density-functional theory of pair correlations in metallic hydrogen, *Phys. Rev. E* **57**, 1, pp. 211–223.

[1050] Y., P. A. and G., C. (2010). Thermodynamic functions of dense plasmas: Analytic approximations for astrophysical applications, *Contrib. Plasma Phys.* **50**, p. 82.

[1051] Y., P. A. and G., C. (2013). Equation of state for magnetized Coulomb plasmas, *Astron. Astrophys.* **550**, p. A43.

[1052] Yakovlev, D. G., Levenfish, K. P. and Shibanov, Y. A. (1999). Cooling of neutron stars and superfluidity in their cores, *Phys. Usp.* **42**, 8, 737.

[1053] Yakovlev, D. G. and Shalybkov, D. A. (1989). Degenerate cores of white dwarfs and envelopes of neutron stars: Thermodynamics and plasma screening in thermonuclear reactions, *Sov. Sci. Rev. Ser. E: Astrophys. Space Phys.* **7**, pp. 311–386.

[1054] Yakub, E. (1999). Diatomic fluids at high pressures and temperatures: A non-empirical approach, *Physica B* **265**, 1-4, pp. 31–38.

[1055] Yakub, E. S. (1990). Equation of state of shock-compressed liquid hydrogen, *High Temp.* **28**, p. 490.

[1056] Yang, C. N. and Mills, R. L. (1954). Conservation of isotopic spin and isotopic gauge invariance, *Phys. Rev.* **96**, pp. 191–195.

[1057] Yao, W.-M. *et al.* (2006). Review of particle physics, *J. of Phys. G* **33**, p. 1.

[1058] Yoo, C. S., Holmes, N. C., Ross, M., Webb, D. J. and Pike, C. (1993). Shock temperatures and melting of iron at earth core conditions, *Phys. Rev. Lett.* **70**, 25, pp. 3931–3934.

[1059] Young, D. and Corey, E. M. A. (1995). *A New Global Equation of State Model for Hot Dense Matter* (Lawrence Livermore Lab., Livermore).

[1060] Young, D. A. (1974). UCRL-51575, *Tech. Rep.*, University of California, LLNL.

[1061] Young, D. A. (1977). UCRL-52352, *Tech. Rep.*, LLNL, University of California.

[1062] Young, D. A. and Alder, B. J. (1971). Critical Point of Metals from the van der Waals Model, *Phys. Rev. A* **3**, 1, pp. 364–371.

[1063] Young, D. A. and Alder, B. J. (1979a). Studies in molecular dynamics. XVII. Phase diagrams for "step" potentials in two and three dimensions, *J. Chem. Phys.* **70**, 1, pp. 473–481.

[1064] Young, D. A. and Alder, B. J. (1979b). Studies in molecular dynamics. XVIII. The square-well phase diagram, *J. Chem. Phys.* **73**, 5, pp. 2430–2434.

[1065] Young, D. A., McMahan, A. K. and Ross, M. (1981). Equation of state and melting curve of helium to very high pressure, *Phys. Rev. B* **24**, 9, pp. 5119–5127.

[1066] Young, D. A. and Ross, M. (1981). Theoretical calculation of thermodynamic properties and melting curves for hydrogen and deuterium, *J. Chem. Phys.* **74**, 12, pp. 6950–6955.

[1067] Young, D. A., Wolford, J. K., Rogers, F. J. and Holian, K. S. (1985). Theory of the aluminum shock equation of state to $10^4$ Mbar, *Phys. Lett. A* **108**, 3, pp. 157–160.

[1068] Young, D. A., Zha, C.-S., Boehler, R., Yen, J., Nicol, M., Zinn, A. S., Schiferl, D., Kinkead, S., Hanson, R. C. and Pinnick, D. A. (1987). Diatomic melting curves to very high pressure, *Phys. Rev. B* **35**, pp. 5353–5356.

[1069] Yu. Kagan, A. K., Pushkarev, V. V. (1977). Equation of state of the metallic phase of hydrogen, *J. Exp. Theor. Phys.* **46**, 3, p. 511.

[1070] Zababakhin, E. I. and Zababakhin, I. E. (1988). *Unlimited Cumulation Phenomenon* (Nauka, Moscow).

[1071] Zamalin, V. M., Norman, G. E. and Filinov, V. S. (1977). *Monte Carlo Method in Statistical Thermodynamics* (Nauka, M.).

[1072] Zaporozhets, Y. B., B., M. V., Fortov, V. E. and Batovskii, O. M. (1984). *Pisma ZhTF* **10**, 21, p. 133.

[1073] Zasov, A. V. and Postnov, K. A. (2006). *General Astrophysics* (Vek 2, Fryazino).

[1074] Zasov, A. V. and Surdin, V. G. (2007). A Variety of Galaxies, in V. G. Surdin (ed.), *Astronomy: XXIst Century* (Vek 2, Fryazino). p. 329.

[1075] Zdunik, J. L., Haensel, P., Gondek-Rosinska, D. and Gourgoulhon, E. (2000). Innermost stable circular orbits around strange stars and kHz QPOs in low-mass X-ray binaries, *Astron. Astrophys.* **356**, pp. 612–618.

[1076] Zel'dovich, Y. B. (1957a). *Zh. Eksp. Teor. Fiz.* **33**, p. 991.

[1077] Zel'dovich, Y. B. (1957b). *Zh. Eksp. Teor. Fiz.* **32**, p. 1577.

[1078] Zel'dovich, Y. B. and Landau, L. D. (1944). *Zh. Eksp. Teor. Fiz.* **14**, p. 32.

[1079] Zel'dovich, Y. B. and Novikov, I. D. (1967). *Relativistic Astrophysics* (Nauka, Moscow).

[1080] Zel'dovich, Y. B. and Novikov, I. D. (1971). *The Theory of Gravity and Star Evolution* (Nauka, Moscow).

[1081] Zel'dovich, Y. B. and Raizer, Y. P. (1966). *Physics of Shock Waves and High-Temperature Hydrodynamic Phenomena* (Dover, Mineola, NY).

[1082] Zelener, B. V., Norman, G. E. and Filinov, V. S. (1972). *Teplofiz. Vys. Temp.* **10**, 6, pp. 1160–1170.

[1083] Zelener, B. V., Norman, G. E. and Filinov, V. S. (1973a). *Teplofiz. Vys. Temp.* **11**, p. 992.

[1084] Zelener, B. V., Norman, G. E. and Filinov, V. S. (1973b). *Teplofiz. Vys. Temp.* **13**, p. 913.

[1085] Zelener, B. V., Norman, G. E. and Filinov, V. S. (1974). *Teplofiz. Vys. Temp.* **12**, p. 20.

[1086] Zelener, B. V., Norman, G. E. and Filinov, V. S. (1975). *Teplofiz. Vys. Temp.* **13**, p. 712.

[1087] Zelener, B. V., Norman, G. E. and Filinov, V. S. (1981). *Pseudopotential in Statistical Thermodynamics* (Nauka, Moscow).

[1088] Zemtsov, Y. K., Sechin, A. Y. and Starostin, A. N. (1996). Resonance emission transfer in dense dispersing media, *J. Exp. Theor. Phys.* **83**, p. 909.

[1089] Zha, C.-S. and Boehler, R. (1985). Melting of sodium and potassium in a diamond anvil cell, *Phys. Rev. B* **31**, pp. 3199–3201.

[1090] Zharkov, V. N. and Kalinin, V. A. (1968a). *Equations of State of Solids at High Pressures and Temperatures* (Nauka, Moscow).

[1091] Zharkov, V. N. and Kalinin, V. A. (1968b). *Equations of State of Solids at High Pressures and Temperatures* (Nauka, Moscow).

[1092] Zhdanov, V. A. and Polyakov, V. V. (1976). *Prikl. Mekh. Tekh. Fiz.*, **5**, p. 123.

[1093] Zhdanov, V. A. and Polyakov, V. V. (1979). *Prikl. Mekh. Tekh. Fiz.*, **4**, p. 102.

[1094] Zhernokletov, M. V. (ed.) (2003). *The Methods of Investigation of the Properties of Materials Under Intense Dynamical Loadings* (Russ. Fed. Nuc. Center — All-Russia Sci.-Res. Inst. of Exper. Phys., Sarov).

[1095] Zhernokletov, M. V., Zubarev, V. N., Trunin, R. F. and Fortov, V. E. (1996). *Experimental Data on Shock Compressibility and Adiabatic Expansion of Condensed Substances at High Energy Densities* (Institute of Problems of Chemical Physics RAS, Chernogolovka).

[1096] Ziman, J. (1973). *Calculation of Bloch Functions* (Mir, Moscow).

[1097] Zink, J. W. (1968). Shell structure and the Thomas–Fermi equation of state, *Phys. Rev.* **176**, 1, pp. 279–284.

[1098] Zubarev, V. N., Podurets, M. A., Popov, L. V., Simakov, G. V. and Trunin, R. F. (1978). *Detonation* (ICP USSR Academy of Science, Chernogolovka, Moscow region).

[1099] Zuliëcke, L. (1973). *Grundlagen und allgemeine Methoden, Quantenchemie — Ein Lehrgang*, Vol. 1 (Deutscher Verlag der Wissenschaften, Berlin).

[1100] Zuliëcke, L. (1985). *Atombau, Chemische Bindung und molekulare Wechselwirkungen, Quantenchemie — Ein Lehrgang*, Vol. 2 (Deutscher Verlag der Wissenschaften, Berlin).